SOCIAL INSECTS AND THE ENVIRONMENT

Proceedings of the 11th International Congress of IUSSI, 1990
(International Union for the Study of Social Insects)

Editors
G.K. VEERESH
B. MALLIK
C.A. VIRAKTAMATH

OXFORD & IBH PUBLISHING CO. PVT. LTD.
New Delhi **Bombay** **Calcutta**

ISBN 81-204-0532-3

Published in India by Mohan Primlani for Oxford & IBH Publishing Co. Pvt. Ltd., 66 Janpath, New Delhi 110001, processed and printed from camera ready text by Rekha Printers Pvt. Ltd., A-102/1 Okhla Industrial Area Phase-II, New Delhi - 110020.

1-EO-8

PREFACE

In the past two decades we have witnessed a surge of research on social insects covering varied facets of biology, more than on any other group of arthropods. Like *Drosophila* in genetics, social insects have been extensively researched to develop and support various theories on the principles of sociality, strategies of reproduction, foraging, and life history, with parallel advancement in phylogenetics and management of economically important species. This is reflected in the 373 papers discussed under the thirty symposia themes at the 11th International Congress of IUSSI held at Bangalore. These proceedings provide an excellent overview of the present trends in research on social insects.

Several distinguished scientists have helped in organising this Congress. Special thanks are due to following symposia organisers and those who have presented plenary lectures (in alphabetical order): A.N. Andersen (Australia), T.N. Ananthakrishnan (India), D.E. Bignell (U.K.), J. Billen (Belgium), J.J. Boomsma (The Netherlands), M. Breed (U.S.A.), A. Buschinger (F.R.G.), P. Calabi (U.S.A.), N.F. Carlin (U.S.A.), K. Chandrashekhar (India), D.Cherix (Switzerland), J.L. Clement (France), F.C. Dyer (U.S.A.), G.W. Elmes (U.K.), V.L.I. Fonseca (Brazil), W.A. Foster (U.S.A.), R. Gadagkar (India), K.N. Ganeshaiah (India), M.H. Hansell (U.K.), K. Hartfelder (U.S.A.), A. Hefetz (Israel), J.H. Hunt (U.S.A.), Y. Ito (Japan), K. Jaffe (Venezuela), P. Jaisson (France), S. Jayaraj (India), R.L. Jeanne (U.S.A.), D.P. Jouvenaz (U.S.A.), L. Keller (Switzerland), N. Koeniger (F.R.G.), M. Lepage (France), R.H. Leuthold (Switzerland), L.W. Macior (U.S.A.), R. Menzel (F.R.G.), C.A. Nalepa (U.S.A.), M.G. Nielsen (Denmark), C. Noirot (France), G.J. Peakin (U.K.), A. Raman (India), R.H. Rembold (F.R.G.), S.W. Rissing (U.S.A.), S. Sakagami (Japan), P. Schmid-Hempel (Switzerland), M.S. Swaminathan (India), J. Tengo (Sweden), J.F.A. Traniello (U.S.A.), W.R. Tschinkel (U.S.A.), S. Turillazzi (Italy), R.K. Vander Meer (U.S.A.), T. Veena (India), H.H.W. Velthuis (The Netherlands), M.J. West-Eberhard (Costa-Rica), D. Wittman (F.R.G.) and P. Wright (U.K.).

We also gratefully acknowledge the support of Bhaba Atomic Research Centre, Council of Scientific and Industrial Research, Department of Science and Technology, Indian Council of Agricultural Research, Indian Council of Medical Research, Indian National Science Academy, India Convention Promotion Bureau, University Grants Commission, and University of Agricultural Sciences (Bangalore).

At various stages of preparation of this volume, Mr. A.R.V. Kumar, Mr. T.N. Raviprasad, Miss H.N. Rama and Mr. A.N. Shylesha have assisted us, our grateful thanks to them.

July 15, 1990

G.K. VEERESH
B. MALLIK
C.A. VIRAKTAMATH

CONTENTS

Symposium 2

Evolution of Sociality-Lessons from Primitively Eusocial Wasps

Symposium 3

Ant-Plant Associations

Symposium 4

Ecology and Evolution of Honey Bee Behaviour

Symposium 5

Evolution and Speciation in Social Parasites

Symposium 6

Caste Differentiation in Social Insects

Symposium 7

The Role of Learning and Memory in the Orientation of Social Insects

Symposium 8

Social Insects in Ecosystems

Symposium 9

Evolution and Significance of Polygyny in Social Insects

Symposium 10

Pest Ants: Present and Future

Symposium 11

Evolution of Eusociality in Arthropods Other Than the Hymenoptera and Isoptera

Symposium 12

Phylogeny and Evolution of the Formicidae

Symposium 13

Social Polymorphism : How and Why?

Symposium 14

Reproductive Fitness and Eusocial Organisation

Symposium 15

Behavioural Ontogeny of Individuals and Colonies

Symposium 16

Chemical Signature in Social Insects

Symposium 17

Pollination Ecology of Social Insects

Symposium 18

Parasite-Host Relationship of *Varroa jacobsoni* and Other Asian Honey Bee Mites

Symposium 19

The Benefits of Social Insects in Agriculture

Symposium 20

Evolution of Social Behaviour in Carpenter Bees

Symposium 21

Behavioural Genetics of Eusocial Hymenoptera

Symposium 22

Pheromonal Aspects of Nest and Nestmate Recognition in Primitively Social Bees

Symposium 23

Models and Theoretical Approaches to the Study of Social Insects

Symposium 24

Foraging Strategies of Social Insects

Symposium 25

Biogeography, Ecophysiology and Social Organization of Stingless Bees

Symposium 26

Harmful Effects of Social Insects

Symposium 27

Biological Control of Pestiferous Social Insects

Symposium 28

The Role of Nourishment in the Ontogeny and Evolution of Insect Societies

Symposium 29

Nests and Nest Building in Social Insects

Symposium 30

Ant Community Structure

PLENARY LECTURES

NUTRITIONAL FACTORS CONTROLLING CASTE FORMATION IN *APIS MELLIFERA*

H. Rembold

Insect Biochemistry Unit, Max-Planck-Institute for Biochemistry, Martinsried, Fed. Rep. Germany

A fully developed honey bee colony consists of 40 - 50 thousand worker bees and several hundred drones. All are natives of one sexually reproductive female, the queen. During the winter season, the colony is reduced to a population of less than 10 thousand workers. Their only function is to guarantee the queen's survival till the colony initiates its next growth cycle. What is the secret of such a super-organism which for its whole survival is able to rely on the presence of a single queen which by its enormous reproductive power guarantees the colony's survival?

It is general knowledge that after the colony has lost its queen, nurse bees within a very short time period start constructing reproductive queen cells around a few young worker larvae. There is obviously a mechanism existing which allows them to direct the morphogenetic program of the young female honey bee larva into queen or worker development. What is the secret behind such a dual genetic program which makes the production of two types of female castes possible? The single reproductive queen and the many thousands of sterile workers, both of them being perfectly adapted to their function in the colony: the queen for egg production and the workers for brood and food care. Is there a queen determining substance contained in the Royal Jelly? Is it the food at all or is it just the difference in brood care by the nurse bees? Whereas a queen cell is abundantly visited and supplied with food, the worker larva receives scarce food and its nutritional quality is rapidly decreasing after the larva has reached its third instar. This loss of nutritional quality is due to addition of pollen and honey to the glandular secretion of the nurse bees. Does such a food affect caste formation?

Growth and development of each organism is based on genetic programs which act as blueprints for construction of the adult's final shape. Regulation of insect development works in a hierarchic manner. In close analogy to the hypothalamo-hypophysial system of mammals, neuropeptides from the central nervous system control synthesis of steroid (moulting) and sesquiterpenoid (juvenile) hormones in peripheric glands. Peptide factors from the neurosecretory cells in the insect brain are released through a neurohemal organ, the corpus cardiacum, into the hemolymph. Here they activate hormone biosynthesis - for the moulting hormone or ecdysone in the prothoracic gland and for the juvenile hormone in the corpus allatum. Consequently, each morphogenetic step is regulated by these hormones. Larval - larval moults are controlled by high juvenile and low moulting hormone, larval - pupal moults by lower juvenile and higher moulting hormone, and pupal - adult moults exclusively by the moulting hormone titer.

In view of the formation of female castes in the honey bee, attention falls on the question of how such a network of hormonal control can be influenced by nutritional factors. During the first three instars of its larval growth period, the female honey bee larva is open to develop either into a reproductive queen or into a sterile worker. It is long apicultural practice to use young worker larvae for growing perfect queens by implanting them into cells of queenless colonies. The young larvae are accepted for queen production by the colony only before they have reached their third instar. This clearly proves the plasticity of the caste determining developmental program in *Apis mellifera*. How can this program be switched on by nutritional factors?

The honey bee makes use of a rather restricted amount of nutrients. Pollen is the only source of protein, vitamins and inorganic ions, and nectar is the primary energy source by its carbohydrate contents; water collected from puddles also is another source of inorganic salts. All these nutrients are needed for maintenance and development of the honey bee colony. This pool of pollen, nectar and water is the starting material from which the nurse bees finally produce a larval food of constant quality. It is mainly a mixture of their hypopharyngeal and mandibular gland secretions to which honey is added. The growing larva has a closed intestine during its whole feeding period. Only at the end of its larval development, when spinning of the cocoon starts, the fifth instar larva defecates. From a biochemical point of view, the growing larva is some sort of a living test tube. All the nutrients which it takes up must be digestible and all the metabolic waste which cannot be released as carbon dioxide, water, or to some extent through the epidermis, must be produced in lowest amounts. As such, the nurse bee has the function of a living filter which protects the brood from fluctuations in food quality, mainly of the pollen collected from different sources, and from changes in temperature and humidity. Is it possible to simulate the situation in a honey bee colony by growing the newly hatched larva in an incubator on a diet of constant composition?

What nutritional or other factors are involved in the process of caste determination? Newly hatched worker larvae were transplanted and reared on a diet which in its total composition was carefully controlled and which was fed to the experimental larvae during their whole development in the thermostat. Consequently, all such parameters like amount or quality of food, cell size, temperature, are constant. Queen larval food, the so-called Royal Jelly, has in its chemical composition a fairly constant quality. However, in the *in vitro* growth test dramatic differences in the rate of caste determination come out with the jellies collected from different sources. We have estimated the gross composition of the low-molecular components. By modification of its overall composition it was found that minute imbalances in food quality already decide on queen or worker formation. It is not the amount of food which is taken up by the rapidly growing larva and it is not a special determinator contained in the queen larval food which finally controls caste formation in the honey bee colony. It is the through the nurse bees controlled nutrional homeostasis which finally is the prerequisite for queen bee establishment. However, although all analytical data demonstrate the controlled production of Royal Jelly by the nurse bees, commercial Royal Jelly differs in its queen bee determining activity. Even the by the nurse bees produced larval food therefore can be critical in its queen determining capacity.

Individual modification of different food components like sugar, amino acids, nucleotides shows, that minute deviations from the nutritional optimum immediately result in establishment of either intercastes or normal workers. Like under the colony conditions, queen and worker castes differ in the duration of pupal and consequently of adult development by almost 25 percent. There are also caste-specific biochemical and histological differences during larval development already. One typical physiological effect of caste differentiation is the rate of respiration. The queen larvae are typically higher in total cytochrome c during their whole larval development which, by extrapolation, seems to build up from the beginning already. Such differences reflect differences in hormonal control pools. The titers of juvenile hormone indeed become diverse during the first larval instar already. Whereas the worker larva maintains a fairly constant titer, it reaches a maximum in the queen larva during its third instar. A second maximum builds up in the last, fifth instar of both the castes. Again, the juvenile hormone titer is much higher in the queen than in the worker larva. These caste-specific hormone titers are finally the response on the different nutrition of the young honey bee larvae.

REFERENCES

Rembold, H. (1987). Caste differentiation of the honey bee - fourteen years of biochemical research at Martinsried. In: Chemistry and Biology of Social Insects (J. Eder & H. Rembold, eds.), J. Peperny, Publ., Munich, pp. 3-13.

Rembold, H. (1987). Caste specific modulation of juvenile hormone titers in *Apis mellifera.* Insect Biochem. 17, 1003-1006.

Rembold, H. (1985). Sequence of caste differentiation steps in *Apis mellifera.* In: Caste Differentiation in Social Insects (J.A.L. Watson et al., eds.) Pergamon Press Oxford, pp. 347-359.

SOCIOMETRY AND SOCIOGENESIS OF FIRE ANT COLONIES : THE SIZE, SHAPE AND DEVELOPMENT OF AN ANT SOCIETY

Walter R. Tschinkel
Department of Biological Science
Florida State University
Tallahassee, FL 32306 USA

Sociometry is the inventory of the quantifiable attributes of social insect colonies. Sociometric data are as much a part of the species description of social insects as morphological descriptions are of non-social insects. When sociometric analysis is carried out on a series of colonies, from founding to full size, the patterns of sociogenesis appear. Sociogenesis is the social analog of embryogenesis. During embryogenesis, the rules and interactions of growth and development transform the single-celled zygote into an organism. Patterns of embryogenesis illuminate the evolutionary history of structures, and structural differences among organisms result from differences in their embryogenesis. During growth, shape changes when dimensions grow at different rates. By analogy, the rules and interactions of sociogenesis guide the growth of colonies from the single, inseminated queen to colony maturity. The patterns of sociogenesis are likely to illuminate their evolution, and differences among societies result from differences in their sociogenesis. Insect societies can also be seen to have "shape", and this "shape" changes when "social dimensions" (sociometric attributes) grow at different rates. The analogy between embryogenesis and sociogenesis suggests that: (1) Rules may limit the combinations of characters which can evolve, i.e. characters may be linked; (2) The "shape" of a society may be size specific, and should be specified in all studies; (3) Study of sociogenesis and sociometry can contribute in important ways to the study of the evolution of differences among species.

The fire ant, Solenopsis invicta, is used for a paradigm study of seasonal sociometry and sociogenesis. Colonies of the full range of sizes were sampled and analyzed 8 times during the year. Colony growth is logistic with a strong, superimposed seasonal fluctuation. Nest volume and territory size grow in parallel. The changes occurring in worker polymorphism illustrate the contributions such studies can make to understanding the development and evolution of social traits. As in many ants, fire ant colonies are founded by inseminated queens whose first workers are the minims (nanitics), the smallest of the life cycle. As colonies grow, workers increase gradually in mean size, as follows. Beginning at about 1 mo., some of the larvae are reprogrammed to pupate at larger size. This discrete developmental event separates the workers into two distinct, normal subpopulations (the minors and majors) with very different means and variances. Both minors and majors increase in mean size and variance during the first 6 mo. of colony growth, and remain constant thereafter. Majors are more size-variable than expected from simply extending the growth of minors to major size, suggesting that some additional mechanism increases their size variability. Larval attractiveness to workers increases disproportionately with larval size, resulting in disproportionate feeding of initially larger larvae and increasing major variability. The proportion of larvae reprogrammed to majors increases as long as colonies grow causing both minor and major subpopulations to grow logistically. Colonies are about 10% majors (by weight) when these first appear, and increase to 70% in full-sized colonies. A possible mechanism for this shift may involve declining reproductive efficiency--- the ratio of workers per larva increases with colony size, leading to improved larval care, which in turn leads to an increased proportion of larvae reprogrammed to majors. In experiments, higher worker per larva ratios result in larger pupae. The outcome of these mechanisms is that mean pupal weight is tightly linked to colony size over 5 orders of magnitude, from founding queen to mature colonies.

This sociometric analysis suggests that the processes upon which natural selection acted to create the observed worker polymorphism are: (1) Adjustment of the threshold size for reprogramming of minors to majors; (2) Adjustment of the mean threshold size for pupation of majors; (3) The disproportionate attractiveness of larger larvae. All other attributes of worker polymorphism are the statistical outcome of this probabilistic system. The proportion of majors changes with reprogramming threshold, their mean size with pupation threshold and their size-variability with larval attractiveness.

To the extent that worker size is linked with division of labor, these sociogenetic shifts must influence the type and amount of labor available to the colony, and ultimately, must affect the niche occupied by the fire ant. Worker longevity also increases with worker size, suggesting that the mean lifespan of workers increases with colony size, and that parallel shifts in age structure of the worker population take place. This must have important effects on the division of labor by age, but these are as yet unknown.

The annual cycle of production rate seems mostly determined by temperature and varies 100 to 200 fold. Temperature has a threshold-like effect upon brood production, which ceases below 24 deg.C, rises to an optimum at 31 deg. C and drops at higher temperatures. In addition to seasonal effects on rate, there is a spring period of sexual production with low worker production, and a fall period of worker production during which few sexuals are produced. This is because any resources invested in sexuals must be taken from investment in workers. Thus, during sexual production, mature colony size falls from its winter maximum (ca. 220,000) to its summer minimum (ca. 120,000), and regrows during the fall. The degree of this size fluctuation is proportional to colony size, and is zero for colonies not producing sexuals. It follows that the majority of workers are produced in the fall and die in the spring. Because major workers live longer than minors, this results in a spring enrichment of the proportion of majors and a peak in mean worker weight.

Sociogenesis is also characterized by changing patterns of production and allocation. As noted above, the production efficiency declines as colony size increases, and this decline is reflected in all stages of brood rearing from egg to pupa and may be the ultimate factor limiting colony size. To estimate allocation patterns, production must be cumulated over an annual cycle because of distinct periods of sexual and worker production. The cumulation was carried out for size classes determined in January and tracked through the seasonal changes in colony size. Annual production increases with colony size class, but less rapidly than colony size, so that production rate declines. Such a decline is observed even in groups of founding queens and may represent very basic processes. Allocation (in joules produced) to sexual production rises from 0% in very small colonies to about 35% in colonies with January sizes between 50 and 100 thousand workers. There it remains as colonies grow to their maximum size. Thus, large colonies invest the same fraction of their resources in sexuals, but produce more by virtue of their larger colony size.

Based on total energy costs, 30% of sexual investment is in males, a value not significantly different from the theoretical 25%. Other bases for this calculation differ significantly from 25%. On the basis of numbers of individuals, sex ratio is 50%, a value perhaps forced by the single mating of females. The size and energetic difference in males and females (80 j vs. 280 j) may be the solution to these two forced values.

Queens use stored sperm at the rate of 3 per adult worker. A founding queen exhausts her supply of 7 million sperm in about 5 to 7 yr, having produced about 2½ million workers. Thereafter, she cannot produce workers to maintain the colony anymore and is probably replaced by an inseminated, as yet non-reproductive queen already present in the nest. Acceptance of a newly mated queen by a mature colony can thus be seen to be another route to success for newly-mated queens. Colonies are potentially immortal, though maternity of the workers changes with each queen.

When the variables of this study are thought of as "social dimensions", their relative values can be analysed as "social shape". A plot of the ratio of two social dimensions vs. colony size can be used to visualize how such social shapes change during colony growth. A slope of zero on such plots indicates isometry (no change of shape), a positive slope means a positive allometry and a negative slope a negative one. To select 3 examples, the ratio of major to minor workers shows a positive allometry, the ratio of larvae to workers a negative one, and the mound volume a nearly isometric one. Such changes of shape can also be tracked through the seasons and bring the distinct periods of sexual and worker production into stark relief. Such methods borrowed from morphometry are useful for visualizing the complex changes which insect societies undergo during the annual cycle and during colony growth.

SOCIAL BIOLOGY OF *ROPALIDIA* : INVESTIGATIONS INTO THE ORIGINS OF EUSOCIALITY

RAGHAVENDRA GADAGKAR
Centre for Ecological Sciences and Centre for Theoretical Studies, Indian Institute of Science, Bangalore-560 012, INDIA.

The evolution of sterile worker castes found in most social insects presents an obvious challenge to Darwin's theory of natural selection. Highly eusocial insects such as ants and honeybees have morphologically differentiated worker and reproductive castes and may be studied with the aim of understanding how eusociality is maintained by natural selection or why highly eusocial species do not revert to the solitary state. In primitively eusocial species such as many kinds of bees and wasps reproductive and worker castes are morphologically identical and social roles are left flexible to be decided by social interactions amongst the adults. Such species may therefore be studied with the hope of understanding the forces that promote the origin of eusociality. Independent founding species of the tropical wasp genus *Ropalidia* provide exceptionally good model systems for such investigations. Female wasps eclosing on nests of *R.marginata* and *R.cyathiformis* have the option of leaving their natal nests to found their own single foundress nests or remaining on their natal nests to assume the role of a worker The question then is, why such a large number of females remain on their natal nests after eclosion?

At the time we began our investigations, three classes of theories namely Kin selection (or Haplodiploidy hypothesis) (1), Parental manipulation (2) or Subfertility hypothesis (3) and Mutualism (4) were widely known to provide potential answers to this question. The measurement of sex-investment ratios and of the productivities of single - and multiple-foundress nests were the two methods that were then being used in attempts to distinguish between these potential theories. The few attempts that had been made in this direction had not been very successful (5). Besides, these methods seemed inappropriate for species such as *R.marginata* and *R.cyathiformis* because neither sex investment ratios nor differential productivities can be measured accurately in these species. This is because all females cannot be unambiguously classified into workers and reproductives; any female has a certain probability of becoming either a worker or a queen. For this reason, it became necessary to develop alternate methods to distinguish between the competing theories. The method I have used is primarily based on quantitative ethology. Constructing time activity budgets for individually identified members of several colonies and subjecting them to multivariate analysis revealed a behavioural caste differentiation into Sitters,

Fighters and Foragers (6). Although the foragers seem to have little or no chance of future reproduction, both Sitters and Fighters appear to keep their option for direct reproduction open while contributing towards the welfare of the colony. Social organisation based on such behavioural caste differentiation rather than one based on a rigid dominance hierarchy led by a despotic queen, suppressing all her nestmates into worker roles (7), prepared the stage for our enthusiasm for mutualistic models for the origin of eusociality. The complex behaviour of the wasps including the ability of some individuals to leave their natal nests along with a few workers and establish themselves as queens of new colonies and the ability of wasps within a colony to behave as two coordinated groups (8) removed any doubt about the ability of the wasps to adopt complex strategies that may be required for social evolution through mutualism.

At the same time we have accumulated evidence against the haplodiploidy hypothesis. We have established that queens of *R.marginata* mate multiply, use sperm simultaneously from different males and thus break down the genetic asymmetries created by haplodiploidy (9). Frequent queen replacements further reduce worker-brood genetic relatedness. Indeed, we have evidence that workers often rear complex mixtures of full-siblings, half-siblings, nieces, nephews and cousins (10). This may not be a serious problem for kin selection if workers can discriminate different levels of genetic relatedness within the colony and give preferential aid to close relatives (11). We therefore set up experiments to test this possibility. The results of these experiments suggest that labels and templates used in kin recognition are acquired by the wasps from such common sources outside their bodies, as, their nest or nestmates. This makes it unlikely that genetic heterogeneity within a colony will be recognised and used in dispensing altruism (12). In addition, these experiments showed that factors other than genetic relatedness modulate tolerance and acceptance of foreign conspecifics. Even genetically unrelated individuals sometimes cooperate in the founding of new colonies (13). These findings, while suggesting an insignificant role for haplodiploidy further strengthened our faith in mutualistic theories.

The ideas of parental manipulation and subfertility have been poorly tested. We therefore set up experiments to test the hypothesis that all eclosing females are potentially capable of becoming egg-layers. It turns out that about 50% of eclosing females do not initiate nests and lay eggs even if rescued from any inhibition by conspecifics. The queen and other adults in a colony appear to influence the future caste of, the brood by channelling them either into a developmental pathway leading to adults programmed to be good eaters and good egg-layers or into an alternate pathway leading to poor eaters and poor egg-layers (14). However, the extent of such pre-imaginal caste bias is obviously insufficient to explain the fact that most individuals become workers and only a few become queens.

The clear choice before us was therefore mutualism. But, mutualistic models have not been taken seriously because they are expected to fall short of giving rise to sterile castes. The

"Gambling" hypothesis is one way out of this difficulty. The advantages of group living may be so great that even if the roles of egg-layer and sterile worker are decided by chance, an average member of a group may do better than a solitary individual. Thus the hypothetical "Gambling" allele which programmes its bearers to take the risk of being part of a group may spread in the population at the cost of the wild type allele which programmes its bearer to be risk-averse and remain solitary. If fertility or sterility are thus the result of phenotypic plasticity, we may see a fraction of sterile individuals in every generation although there is no allele for sterility (15)! The perennial colony cycle of *R.marginata* with its frequent queen replacements is undoubtedly responsible for providing wasps, that stay back on their natal nests, a finite chance of direct reproduction. The answer to our original question then would be that most individuals stay back on their natal nests because of some hope of direct reproduction. Those that realize such hope must compensate in some measure for the remaining that die as workers (10).

These investigations and a survey of the literature on other kinds of social insects suggests a route to eusociality that has three stages : the "Gambling" stage where group living is brought about due to the advantages of mutualistic interactions followed by the "Manipulation" stage where parental manipulation and subfertility begin to act and finally the "Recognition" stage where genetic asymmetries created by haplodiploidy may be recognised and the species may be locked into the highly eusocial state. The time has come, however, when we should be looking beyond the three theories that we began with. Towards this goal, our recent modelling efforts have shown that additional factors such as "Assured fitness returns", "Delayed reproduction" and "Variation in age at reproductive maturity" may all act to select for worker behaviour in such species as *R.marginata* (unpublished).

1. Hamilton, W.D. 1964. J.theor.Biol., 7:1-52.
2. Alexander, R.D. 1974. Ann.Rev.Ecol.Syst., 5:325-383.
3. West-Eberhard, M.J.W. 1975. Q.Rev.Biol., 50:1-33.
4. Lin, N. and C.D.Michener. 1972. Q.Rev.Biol., 47:131-159.
5. Gadagkar, R. 1985a. Proc.Indian Acad.Sci.(Anim.Sci.), 94:309-324.
6. Gadagkar, R. and N.V.Joshi. 1983. Anim.Behav., 31:26-31.
7. Chandrashekara, K. and R.Gadagkar. *this volume*.
8. Gadagkar, R. and N.V.Joshi. 1985. Curr.Sci., 54:57-62.
9. Muralidharan, K., Shaila, M.S. and R.Gadagkar. 1986. J.Genet., 65:153-158.
10. Gadagkar, R., Chandrashekara, K., Chandran, S. and S.Bhagavan. *this volume*.
11. Gadagkar, R. 1985b. Proc.Indian Acad.Sci.(Anim.Sci.), 94:587-621.
12. Venkataraman, A.B., Swarnalatha, V.B., Nair, P. and R.Gadagkar. 1988. Behav.Ecol.Sociobiol., 23:271-279.
13. Venkataraman, A.B. and R.Gadagkar. *this volume*
14. Gadagkar, R., Vinutha, C., Shanubhogue, A. and A.P.Gore. 1988. Proc.R.Soc.Lond.B., 233:175-189.
15. Gadagkar, R. In: Social Biology of Wasps, K.G. Ross and R.W. Matthews, Eds., Cornell University Press, New York, *in press*.

THE RELATION BETWEEN BEHAVIORAL-DEVELOPMENTAL SWITCHES AND MAJOR STEPS IN SOCIAL INSECT EVOLUTION

M. J. West-Eberhard

Smithsonian Tropical Research Institute
(Correspondence: Escuela de Biologia, Universidad de Costa Rica, Ciudad Universitaria, Costa Rica, Centroamerica)

Studies of social insects have traditionally been compartmentalized into subfields. One of the strongest and most complex deals with the physiology of caste determination; another, similarly filled with its own obsessions and specialized jargon, deals with evolutionary theories of insect social behavior. The first studies the mechanisms, or "proximate causation" of behavior development; the second studies the principles of natural selection and the "ultimate" (evolutionary) causation of behavior and morphology. These two subfields have traditionally communicated very little with each other, for it has not been clear how developmental mechanisms are related to the process of natural selection, if at all.

Here I will show how regulatory architecture focusses natural selection, and in a sense directs the process of evolution. I will then give evidence that behavioral and developmental plasticity depending on that architecture has contributed to the origin of four major steps in the evolution of the social insects.

DEVELOPMENTAL/BEHAVIORAL SWITCHES AND SELECTION

Alternative phenotypes characterize the social biology of the wasps, ants, bees and termites. Societies of all of these groups contain contrasting alternative forms and behaviors, including workers vs. queens as well as a multitude of other behavioral alternatives, such as dominance vs. subordinance, group nesting vs. solitary nesting, and, among workers, different alternative tasks (nurse behavior, foraging, corpse removal, nest defense, etc.).

Alternative phenotypes, whether morphological or behavioral, have a common basic genetic-regulatory structure [1]: each individual carries the genes for the expression of each alternative, and a genetically specified regulatory (or switch) mechanism governing which set is expressed. The switch mechanism frequently involves hormones sensitive to ambient conditions and manipulation, as is well known for the social insects. Hormonal involvement is well-documented by studies of morphological caste determination [2]; and there is increasing evidence for hormonally mediated switches controlling exclusively behavioral caste and task-performance differences [3, 4]. Switch mechanisms may also be influenced by alleles at polymorphic loci, as shown in some ants [5], some stingless bees [6], and worker honeybees [7]. These alleles could either affect the setting of the threshold for switching between alternatives, or could influence the variables (such as individual size) that affect whether or not a threshold is passed. Probably all such switches, however, are influenced by both allelic and environmental factors [1, 6].

The developmental significance of a switch is that it canalizes development or behavior in a particular direction. The evolutionary significance of a switch is that it determines which phenotype is expressed, and subjected to selection. To the degree that alternative phenotypes are independently expressed they are independently subject to selection, and can evolve in different directions [1]. This is the evolutionary principle underlying the divergence of queen and worker phenotypes, as well as the behavioral specializations within the worker caste; and the morphological subcastes in highly eusocial ants and termites. Furthermore, the nature and timing of developmental switch points during ontogeny probably determines the kind and degree of social polymorphism that can be achieved in a given lineage [8, 9].

FLEXIBLE ALTERNATIVES AND MAJOR STEPS IN EVOLUTION

Alternatives controlled by a switch facilitate the elaboration of novel traits sometimes considered "difficult to evolve," such as worker behavior, or parasitic reproduction in species where it is clearly more profitable to be a queen. This is because alternatives are expressed as options: a switch mechanism can be "programmed" by natural selection so that a relatively low-profit alternative is expressed only when it is likely to be advantageous, as indicated by some social or environmental cue (such as dominance status; larval size; or photoperiod, when relative advantage undergoes seasonal change).

In the social insects this regulatory flexibility can make individuals vulnerable to manipulation by others, e.g., of larvae by adult workers [10, 11, 12]. But the protective effect of optional expression has probably facilitated the evolution of most major innovations in the history of the social insects. Major traits that likely originated as alternatives include (a) life in groups, which occurs as an alternative to solitary nesting in most primitively social species (examples in [12]; (b) worker behavior, which is an alternative phenotype in all known social insects and likely originated either as an individually advantageous option or a product of manipulation [10]; (c) particular kinds of morphological worker castes, as a product of pre-existing developmental patterns [8, 9]; and social parasitism, which occurs as an intraspecific alternative in close relatives of species specializing in parasitism as an exclusive way of life [13].

CONCLUSION

Appreciation of the role of developmental and behavioral flexibility in social evolution demands greater attention by evolutionary biologists to studies of regulation, especially to hormone physiology and the genetics of individual differences in phenotype. It also promises to bring the social insects once again into prominence as an exemplary group for testing new generalizations about the evolution of sociality.

1. West-Eberhard, M.J. 1989. Annu.Rev.Ecol.Syst.20:249-78.
2. Wilde,J. and Beetsma,J. 1982. Adv.Ins.Physiol.16:167-246.
3. Röseler, P-F. 1985. In: Experimental Behavioral Ecology, B. Hölldobler and M. Lindauer, Eds. Fischer Verlag, N.Y.
4. Robinson, G.E. 1987. Behav.Ecol.Sociobiol. 20:329-38.
5. Heinze, J. and Buschinger, A. 1987. Insectes Soc. 34(1): 28-43.
6. Kerr, W.E. 1966. Genetics 54(3):859-866.
7. Robinson, G.E. and Page, R.E. 1989. Behav. Ecol. Sociobiol. 24:317-323.
8. Wheeler, D. E. 1986. Amer.Nat. 128(1):13-24.
9. Noirot, C. in press. Ethology Ecol. and Evol.(Florence).
10. Alexander, R.D. 1974. Annu.Rev.Ecol.Syst.4:325-83.
11. Le Moli, F. and Mori, A. 1987. Experientia Suppl. 54:333-363.
12. West-Eberhard, M.J. 1978. J.Kansas Ent.Soc.51(4):832-56.
13. West-Eberhard, M.J. 1986. Proc.Nat.Acad.Sci.USA 83:1388-92.

EFFICIENCY VS. RESILIENCE IN THE ERGONOMICS OF SOCIAL INSECTS

Paul Schmid-Hempel
Zoologisches Institut der Universität
Rheinsprung 9, CH-4051 Basel, Switzerland.

Behavioural strategies of individual workers or the demographic chracteristics of colonies of social insects are selected through their consequences on reproductive success. The study of ergonomics, and of adaptation in particular, contributes to the understanding of the diversity of such traits found in the social insects. However, the difficulty does not reside in accepting this general principle but in analysing the precise way in which a trait of interest contributes to maximum genetic representation in future generations under the prevailing selective forces.

Because populations of social insects are stratified, selection can operate among colonies and among individuals within colonies which has obvious consequences for the evaluation of adaptive hypotheses. Studies of ergonomics have generally assumed that the colony is an unit and ignored the possibility of conflict within. These models have often been successful in explaining various aspects of individual work strategies on the short-term scale. This can be illustrated with the analysis of individual and collective foraging behaviours. For many such problems, the assumption that energy and time is efficiently used by the individual workers may thus provide a convenient approach. However, less success has been met in explaining larger-scale patterns, such as caste structures or division of labour in ants.

Among other things, progress in the field is hampered by the difficulty or neglect of formulating appropriate fitness measures, i.e. to formulate hypotheses about the relevant selective forces. Furthermore, experimental evidence indicates that colonies of social insects are rather resilient against disturbance and show considerable intrinsic variation, such that the effects of environmental factors are often unclear. For example, while many studies have concentrated on the way resources are gathered or processed, parasites may also exert subtle selective pressure, as illustrated by our own studies on bumblebee populations. Those have hitherto largely been ignored.

The way individual behaviours or various forms of organisation may be adapted to particular ecological circumstances is a tantalizing question that is of interest to the study of social organisms in general. To answer this problem, the analysis of small phenomena is only a beginning that my provide insight into constraints that otherwise would go unnoticed. The link of those phenomena to the life history of colonies and eventual reproductive success is still little understood but crucially important. In addition, a proper population biology of social insects that takes into account the various levels of selection and the intricate population genetics of this group, in relation to behavioural and ecological questions is still missing. Some of the future perspectives will be discussed.

ADAPTIVE DIVERSITY IN FUNGIVOROUS INSECTS

Professor T N Ananthakrishnan
Director
Entomology Research Institute
Loyola College
Madras 600 034
I N D I A

Contrary to the understanding of Insect-Plant Interactions, the ecobehavioural, physiological and biochemical aspects of fungivorous insects have not received due recognition. Many detritivorous wood feeding insects exploit fungi for feeding while many species actively culture fungi for feeding. Inspite of the increasing number of taxa that have been recorded from dead and decaying wood, woody and fleshy fungi, forest litter and some plantation crops which offer suitable sites for these insects, adaptive strategies in relation to resource exploitation and consequent fecundity appear important. A few relevant examples of fungivorous insects and their adaptive strategies relating to feeding and reproduction are examined involving such insects as the wood wasps, fungus growing termites, ambrosia beetles such as platypoids and scolytids and mycophagous tubuliferous Thysanoptera.

ENTOMOLOGY IN INDIA : PROGRESS AND PERSPECTIVES

S. JAYARAJ
VICE-CHANCELLOR
TAMIL NADU AGRICULTURAL UNIVERSITY
COIMBATORE-641 003, INDIA

1.0 THE SETTING

People in ancient India were aware of the importance of insects like honeybee, lac insect, silkworm etc. Scientific study of insects was undertaken only from 18th century and since then gradual development in Entomological work has taken place. Indian Entomologists have contributed immensely for the growth of entomology during this century (David and Kumarasamy, 1988). The history of entomology in India was reviwed by Pradhan (1964). The progress made in entomology from 1965 to 1990 is presented here.

2.0 PROGRESS FROM 1965 TO 1990

Entomology is the science of insects affecting plant, animal, man, etc. In a broader sense it encompasses Agricultural, medical including public health, veterinary, forest and urban entomology.

2.1 EDUCATION

Entomology is one of the core courses in the curricula of agricultural Universities at graduate and post-graduate levels. Recently, forest entomology has been included in the curriculum in B.Sc. (Forestry) programmes and the role of forest entomology in enhancing the forest wealth is immense (Nair, 1988). A degree programme on sericulture is offered in some Universities. Some colleges have a post-graduate diploma course in sericulture. Apart from Agricultural Universities, entomology is taught as a special subject in many science colleges. A two year post-graduate diploma course in medical entomology with emphasis on vector control in public health is developing under the guidance of Indian Council of Medical Research (ICMR). Thus, the multifaceted educational programmes have resulted in the development of a strong base with highly qualified entomologists who are competent to take up research in entomology.

2.2 RESEARCH

The number of institutes devoted to entomological research has risen phenomenally. Specializations in biocontrol, biosystematics, ecology, bionomics, sericulture, lac culture, apiculture, pest management, etc., have come to stay. In the past, several national institutes like Indian Lac Research Institute, Namkum, Ranchi, Central sericultural research and training institute, Mysore, Bee Research Institute, Pune, were established. The starting of a national centre for integrated pest management (NCIPM) by the Indian Council of Agricultural Research (ICAR) during 1988 is a right step in this direction.

Research in Entomology includes a) Insect oriented approaches based on individual species or group of insects such white grubs, honey bees etc. b) Crop oriented approaches in field crops like paddy, cotton, sugarcane, wheat, oilseeds, pulses, etc., and Horticultural and Plantation Crops like vegetables, fruits and spices and condiments. Basic and applied research is done in various centres under the auspices of ICAR, ICMR and other national centres like Zoological survey of India, Central Food Technological Research Institute, Bhaba Atomic Research Centre, etc. The science colleges and University Departments also undertake entomological research. A few institutes such as St. John's School of Entomology, Agra and Entomology Research Institute, Loyola College, Madras have contributed substantially to the development of entomological science.

2.3 PROFESSIONALISM

The last 25 years is a landmark in the development of entomology as evidenced by the formation of many professional societies serving the cause of this branch of science. Until late 1960's, the Entomological Society of India was the only major professional organisation. In recent years, several new bodies have come into existence. It is noteworthy to mention the establishment of Indian Society for Biocontrol Advancement, Coimbatore, Association for the advancement of entomology, Thiruvananthapuram, Indian Society for the advancement of insect science, Ludhiana, Entomology Research Association, Udaipur and Indian Academy of Entomology, Madras. The knowledge explosion of the 1970s and 80s led to the starting of new journals such as Journal of Aphidology, Cecidologia internationale, Indian Bee Journal, Entomon, Pesticides, Pestology, Journal of Biological control, Phytophaga, Journal of Entomological Research, Journal of insect science, Indian Journal of Applied Entomology etc. which serve as vehicles for knowledge sharing. These developments have enabled better interactions among scientists and facilitate publication of findings.

The growth of literature has been tremendous as evidenced by the publications of periodicals by many of the learned societies. These are circulated at national and international levels. Accumulation of scientific knowledge in different facets of entomology is also evidenced by increasing number of text books published by Indian authors to suit the needs of the students and scientists under Indian conditions.

3.0 PERSPECTIVES

3.1 EDUCATION

Non-formal education in Agricultural, Medical including public health, veterinary, forest and urban entomology will help to create an awareness among the people since insects are involved in day to day activities of people in one way or other. Specialised courses on apiculture, lacculture, vector biology in relation to public health, high altitude and aquatic entomology, entomology of plantation crops and arid zone and integrated pest management will go a long way in developing human resources and in reducing the losses in Agricultural crops, transmission of diseases of human beings and livestock etc. Urban entomology is an unexplored area and should be given more impetus particularly in cosmopolitan cities which are fast expanding. In the University education, courses on use of Remote sensing for pest management, computer modelling for prediction of pest outbreaks etc., should gain more importance. Biotechnology in entomology might be the panacea for many problems and so entomological biotechnology has to develop to meet the challenges of the 21st century.

3.2 RESEARCH

Insects are one of the important biological constraints in the Third World countries and limit the food production (Odhiambo, 1983). Hence, identification of thrust areas and establishment of centres of excellence for Biosystematics, ecobehaviour, biological control, establishment of a network of projects in different agro-climatic zones for monitoring resistance/resurgence of insects caused by insecticides, establishment of national centres for insect pathogens, parasitoids etc., inception of national research centres for some of the key pests like Heliothis armigera (Hub), Whitefly, brown planthopper, Nilaparvata lugens (Stal), green leafhoppers, Nephotettix spp., bollworm complex in cotton, Spodoptera litura (Fab.) etc., are some of the areas that are worth a thought. Besides there should be a national body to coordinate all sub-sciences of entomology to conduct annual symposia, seminars, and to develop human resources in entomology.

3.3 DEVELOPMENT OF PROFESSIONALS

Entomology can be taken up as a profession by educated unemployed. Women can take up to this profession particularly in public health and urban entomology which will go a long way in reducing the human diseases transmitted by insects (Willis, 1987). Centres to identify the insects as well as for information retrival can be set up by elder and senior entomologists after their retirement from service.

3.4 HOBBY FOR AMATEURS

Entomology should be popularised among the amaturists. As a hobby it can stimulate people's interest. Butterfly gardening will be a satisfying hobby, a self-education route to enjoy and appreciate the nature and when applied on a grand scale can be an educational and conservation device that could reach thousands and measurably enhance local wild life. For this purpose 'ENTO CLUBS' on the lines of ECOCLUBS should be organised in all major cities. The Universities should come forward to motivate people to organise themselves.

3.5 LINKAGE WITH INDUSTRIES

Entomology as applied to pest management has greater scope if a proper linkage with industries is established. The user industries should involve in mass production of parasites, pathogens etc; funds must flow from the industries for undertaking research by individuals and organisations and a better interaction will definitely pay dividends. Private consultancy and firms for pest management must come in a larger measure (Michael et. al., 1989).

3.6 HUMAN RESOURCE DEVELOPMENT

Entomology as a science should bloom by undertaking extensive basic studies which might then form a strong scientific background for future work. For this, harnessing the man power is very important. Entomophobia is one reason why many people are scared at the sight of insects. If this is done away with, more people will tend to learn about insects and take up to this science (Hardy, 1988). The role and involvement of women in entomology should increase (Willis, 1987). Compared to America, the number of women entomologists is less in India. Placement service for entomologists should be made a reality. The openings and avenues should be made known to students. This might attract youngsters to become professionalists.

Looking at the progress made in the past and the perspectives, entomology will be a challenge and an opportunity in the future. It is a challenge because of the innumerable ways in which the insects affect the human life. These tiny creatures threaten the very existence of human life. It is an opportunity, since insects as biological entities offer much scope for solving many unknown mysteries and to attain excellence in scientific pursuits. Ultimately, this will enable society to feed, to shelter and to protect millions of people from hunger, diseases and disasters.

REFERENCES

David, B.V. and T. Kumaraswami. 1988. Elements of Economic Entomology Popular Book Depot. Madras pp. 505.

Judith H. Willis, 1987. Changing status of Women in Academic Entomology, Bulletin of Ent. Soc. of America 33 (1): 17-18.

Michael T. Lambur, Richard F. Kazmerczak, Jr. and Edwin G. Rajotte. 1989. Analysis of private consulting Firms in Integrated Pest Management. Bulletinof Ento. Soc. of America 35 (1): 5-12.

Nair, K.S.S. 1988. Pest Management in Indian Forestry: How to bridge the gap beteen theory and practice In: Integrated Pest Control, Progress and Perspectives. (Eds.) N. Mohandas and George Koshy. 1988. Trivandrum. pp. 32-34.

Odhiambo, R. Thomas, 1983. Biological constraints of food production and on the level and efficient use of chemical inputs. In: Chemistry and World Food Supplies: The new frontiers, Chemrawn II, Perspectives and Recommendations (eds.) G. Bixler and L.W. Shemilt. IRRI, Philippines 65-88.

Pradhan, S. 1964. History of Entomology in India. In: Entomology in India, published by Entomological Society of India. IARI, New Delhi.

Tad N. Hardy. 1988. Entomophobia: The case for Miss Muffet, Bulletin of Ent. Soc. of America 34 (2): 64-69.

Symposium 1

TERMITE BEHAVIOUR AND EVOLUTION

Organizer : **R. H. LEUTHOLD, Switzerland**

INTRODUCTION TO THE SYMPOSIUM

R.H.Leuthold, University of Berne, Zoological Institute, Division of Animal Physiology, Erlachstrasse 9a, 3012 Berne Switzerland

The transition from a cockroach-like solitary insect to the eusocial termite society is a considerable evolutionary step. A further and no less important evolutionary process is the development from a primitive scattered termite colony to the perfectly structured "superorganism" of the higher termites. Innovative elements of behaviour characterise termite evolution to increasingly complex and totally new collective bodies. Termite evolution is still little understood. This symposium will be an opportunity to bring together new and different views and hypotheses on this subject. We shall start with some fundamental questions regarding termite evolution and then gradually focus on the key aspect, which is behaviour. In the second part we shall hear about new findings concerning termite behaviour with respect to social activities and finally we shall deal with links between termite activities and the ecosystem.
A modern view of termite phylogeny is the background for understanding evolution. B.Thorn will deal with the question of the ancestry of termites from the point of view of new taxonomical and paleontological analyses. New taxonomy methods combining morphology and biochemistry used in higher termites are described in the poster by R.K.Bagine et al. What are the conditions and factors that favoured the evolution of termites? Different views and hypotheses will be presented in the 5 following papers. T.Abe presents his view of social evolution, which can be summarised in the following keywords: symbionts, parental resources, neotony, inbreeding, inclusive fitness, altruism. The change of "life-type" which happened with the separation of the foraging area from the nesting site led to a higher degree of specialisation. Here Abe sees a main impetus for the evolution of the true worker caste. R.Rosengaus emphasises monogamy and inbreeding as conditions favouring altruism. Y.Roisin and J.M.Pasteels demonstrate how the loss of neotenic flexibility was promoted by evolutionary specialisation of castes. Causative factors for the evolution of eusociality are claimed in nutrition by S.D.Basalingappa and in nesting behaviour by G.Veeranna. A general discussion of evolution in termites will follow this first part of the symposium.

A.M.Stuart will introduce the second part with an evolutionary interpretation of mechanical and chemical communication as derived from originally defensive functions. New findings on mechanisms of defence and their evolutionary aspects are presented in a poster by S.Basalingappa. M.Kaib will describe a comprehensive analytical study done in his laboratory of the regulation of communicative behaviour by pheromonal information in Schedorhinotermes. Important chemical pheromone identification will be reported by C.Bordereau et al. In higher termites sophisticated foraging strategies have evolved for bringing the food into the nest from the surrounding area. S.Kumar and M.L.Thakur will report on worker caste polyethism during food acquisition by a desert harvester, where minor workers specialise in preparing the foraging holes and the majors in harvesting the food. D.Rajagopal presents his observations of the affect of ecological factors such as moisture, temperature and food availability on foraging behaviour in an Odontotermes species. Two posters will present new information on termite building behaviour. S.Kumar has found an interesting division of labour between two types of building: the nest is repaired mainly by minor workers, whereas spontaneous nest expansion is carried out chiefly by major workers. Australian compass termite mounds are evidence of integrative, adaptive building behaviour. P.Jacklyn's poster shows the adaptive value of nest orientation in these termites and the use of magnetic cues for orientated building behaviour. The evolution of behavioural strategies is understood as an adaptive, competitive process aimed at claiming resources in the long term changing environment. On the other hand, termites are modifying agents for the environment and participate in the development of the ecosystem. This is presented in the review of the extensive, comparative study carried out by E.Garnier-Sillam and co-workers in tropical rain forest. Two different types of termites, a humus feeder and a fungus-grower, claim their niches by different strategies and each helps to modify the soil in its own particular way.
I should be pleased if everyone would think in advance about the points which are going to be raised so that we can have a constructive discussion at the end of the symposium.

EVOLUTION OF THE ORDER ISOPTERA : A REVIEW AND REINTERPRETATION OF ITS PALEOZOIC ROOTS

Barbara L. Thorne
Department of Biology and Museum of Comparative Zoology
Northeastern University, Boston, Ma. 02115 U.S.A.
Harvard University, Cambridge, Ma. 02138 U.S.A.

ABSTRACT.

The phylogeny of the Dictyoptera (termites, cockroaches and mantids), and evolutionary relationships among primitive isopteran families have been re-evaluated using modern principals of systematic thought and techniques of phylogenetic analysis. Morphological, developmental and behavioral character states for these taxa have been compiled and analyzed using cladistic methodology. This approach suggests that the general conception of termites as immediate descendants of a 'Cryptocercus-like' ancestor needs to be reconsidered. Study of the paleozoic history of the termites gives insight into the ecological context under which termites evolved from solitary to social insects.

CUTICULAR HYDROCARBON PROFILES AS A SYSTEMATICAL TOOL: A CASE STUDY IN THE TERMITE GENUS *ODONTOTERMES*

Richard K. Bagine*, Roland Brandl and Manfred Kaib****

* National Museums of Kenya, PO Box 40658, Nairobi, Kenya and ** Lehrstuhl Tierökologie und Tierphysiologie, Univ. Bayreuth, Postfach 101251, D-8580 Bayreuth, Fed. Rep. Germany

Odontotermes is a diverse genus with approximately 30 species in East Africa. So far, the taxonomy of this genus is based on the numerical analysis of morphometrical data. However, there is often little morphological difference between species, some being distinguished only upon size. Consequently, the present taxonomical classification within that genus is uncertain and needs to be tested by additional methods.

DNA-hybridization, allozymes, or metabolic products provide valuable data sets for taxonomical studies. Heritable biosynthetic pathways, for example, lead to species-specific patterns of cuticular hydrocarbons, which are broadly considered to be chemical cues for species recognition by termites. This paper gives evidence, that a numerical analysis of such hydrocarbon profiles leads to congruent systematical relationships between species compared with classical morphometrical data.

Six *Odontotermes* species were collected in Kenya: *O. badius, O. kibarensis, O. stercorivorus, O. nolaensis, O. tanganicus,* and *O. zambesiensis*. For morphometrical characterization 15 distinct parameter were measured from ≥ 15 soldiers. Further analysis of these data proceeds by: (1) log-transformation of raw data; (2) size-correction by regression techniques; (3) discriminant analysis of the residuals; (4) clustering of the group centroids. For chromatographical fingerprinting of complex hydrocarbon profiles ≥ 20 individuals from either caste or morphe were extracted in hexane and separated by GLC. Profiles, based on the 15 most prominent peaks occuring in at least one of the tested samples, were calculated and clustered.

Like previous studies, our morphometrical analysis confirms two groups: the '*badius*' group (*O. badius, O. kibarensis, O. zambesiensis*) and the '*tanganicus*' group (*O. nolaensis, O. stercorivorus, O. tanganicus*). Hydrocarbon profiles of major and minor workers are nearly identical, and those of different species are qualitatively and/or quantitatively distinct. Compared with the differences of profiles within each morphometrical group, the between-group difference is very large. Two peaks, which are not present or only present as a minor constituent (≤ 1% of the total profile) in the '*badius*' group, amount to 10% to 27% each in the '*tanganicus*' group. Both, morphometrical and chemical techniques provide congruent classifications in the genus *Odontotermes*.

EVOLUTION OF WORKER CASTE IN TERMITES

Takuya ABE (Department of Zoology, Kyoto University, Kyoto 606 Japan)

Noirot & Pasteels (1987, 1988) proposed an attractive hypothesis on the evolution of worker caste in termites: the workers have evolved in a polyphyletic fashion several times at the family as well as the subfamily levels. I propose a hypothesis that worker caste of termite society has evolved in relation to the change in life type.

1. Life types in termites

Abe (1984, 1987) proposed the life type to explain the process of colonization of the Krahatau Islands by termites. Based on the information about nesting systems and feeding habits, he distinguished 6 life types, which were grouped into 3 large categories: one piece, intermediate and separate types. In one piece type, a piece of wood serves both a nesting site and food source, while in separate type, which usually nests in the soil, nesting and feeding sites are separate. Intermediate type nests in wood and consumes the wood, but at the same time constructs foraging galleries to consume the other wood.

Dry wood and damp wood termites of Kalotermitidae, Termopsidae and Prorhinotermitinae of Rhinotermitidae belong to one piece type, most species of Termitidae, all species of Hodotermitidae and some species of Rhinotermitidae to separate type, and Mastotermitidae, a few Kalotermitidae, some Termitidae and most Rhinotermitidae to intermediate type.

2. Worker caste in termites

Worker caste of termites has developed independently in Mastotermitidae, Hodotermitidae, Termitidae and at least some groups of Rhinotermitidae, while termites of Termopsidae, Kalotermitidae and Prorhinotermitinae of Rhinotermitidae have developed only pseudergate (Noirot & Pasteels, 1987, 1988). The presence of worker caste corresponds neither to feeding habits nor nesting sites, but corresponds well to life types. Most species with pseudergates are one piece type, while most species with workers are separate or intermediate type. It is notable that only Prorhinoermitinae is one piece type in the Rhinotermitidae and this subfamily has pseudergates only (Roisin, 1988). The evolution of workers from pseuderagates in termite societies seems to have occurred in relation to the change in life types from one piece type to intermediate or separate type.

3. A story on the evolution of worker caste

Although termites and ants, bees and wasps are social insects, termites are different in various ecological aspects from social Hymenoptera. The difference in their food is very important. Most of primitive termites nest in and consume wood. Abe (1989) divided insects largely into 2 groups; cell wall consumers and cytoplasm consumers. Cell wall, a major component of wood, is abundant but low in food quality (nitrogen content is very low), while cytoplasm of both plants and animals is scarce but high in quality. Cellulose, which is a major component of cell wall and the most abundant organic matter on the earth, is difficult for most animals to decompose. Termites (cell wall consumers) and Hymenoptera (cytoplasm consumers) have developed eusociality with very different ecological bases.

Primitive termites have solved the two serious problems through the association with microorganisms; cellulose digestion with protozoa, and nitrogen acquisition with nitrogen-fixing bacteria. This must be prerequisite for ancestral termites to evolve subsocial organization. The logs which termites nest in and feed on are usually longs-lasting and clumptly distributed. This might make it possible for newly-born termites to take either of two strategies to get wood resource; if parental resources are not available, settlement of a new society after alate dispersal, and, if parental resources are available, philopatry to be neotenics and inherit those resources (Myles, 1988). Some offsprings might have delayed maturation to take over their parental resources and have become helpers during their stay in parental nests. They are very pseudergates of most lower termites. The reproductive activities by neotenics lead to inbreeding, which might enhance altruistic behaviour of offsprings (Hamilton, 1972; Bartz, 1979).

When termite societies become large and abundant, and therefore the rate of wood decomposition becomes high, termites with separate life type have competitive advantage over those with one piece type (Abe, 1988), and become dominat in termite communities as is shown in the abundance of Termitidae with separate type in the present world. Then, termites with separate life type must solve another problem; larvae must be fed with food, because feeding and nesting sites are separate. This situation must have promoted the production of specialists of collectiong food and caring for reproductives and larvae, that is,workers. However, the evolution of sterile workers cannot be explained without the selection at colony level, or parental manipulation.

References

Abe, T. 1984. Colonization of the Krakatau Islands by termites (Insecta: Isoptera). Physiol. Ecol., Japan, 21:63-88.

Abe, T. 1987. Evolution of life types in termites. "Evolution and co-adaptation in biotic communities" (Kawano, S., Connell, J.H. & Hidaka, T.), 125-148. University of Tokyo Press.

Abe, T. 1989. Ecology of termites. 156pp. University of Tokyo Press. (in Japanese)

Bartz, S.H. 1979. Evolution of eusociality in termites. Proceedings of the National Academy of Sciences, USA, 76:5764-5768.

Hamilton, W.D. 1972. Altruism related phenomena, mainly in social insects. Ann. Rev. Ecol. Syst., 3:193-232.

Myles, T.G. 1988. Resource inheritance in social evolution from termites to man. "The ecology of social behavior" (Slobodchikoff, C.N. ed.), 379-432. Academic Press.

Noirot, C. & Pasteels, J.M. 1987. Ontogenetic development and evolution of the worker caste in termites. Experientia, 43:851-860.

Noirot, C. & Pasteels, J.M. 1988. The worker caste is polyphyletic in termites. Sociobiology, 14:15-20.

Roisin, Y. 1988. Morphology, development and evolutionary significance of working stages in the caste system of *Prorhinotermes* (Insecta, Isoptera). Zoomorphology, 107:339-347.

STUDIES ON THE EVOLUTION OF MONOGAMY AND EUSOCIALITY IN TERMITES

Rebeca Rosengaus, Dickens No. 80-801, Colonia Polanco, Mexico DF 11560, Mexico

Despite the convergences in social organization between termites and the social Hymenoptera, there are important fundamental differences concerning the mode of colony foundation and development between these two groups as well as the role of the reproductive male in the colony, the female/male sex ratio and the diploid genetic disposition of termites. Two aspects of termite social organization were studied to provide information on the evolution of a monogamous mating system and of eusociality in this order.

Current theories suggest that male monogamy in termites may involve mate guarding and/or mate assistance (1). To better understand the behavioral organization and significance of bisexual colony foundation in termites, we recorded task diversity, time budgets and transition probabilities of acts for male and female primary reproductives in incipient colonies of the camp-wood termite <u>Zootermopsis angusticollis</u>. After being entered into a computer file, all behavior was sequenced and the duration of each act determined.

Repertory size for males and females was identical and the frequencies with which they performed tasks did not vary significantly. Similarly, there were no significant differences in the average time spent by each sex in each behavioral act. The transition probability between sequential behavior shows that males were more likely to continue performing certain acts within the same category, though definite roles could not be established. Thus, sexual polyethism in this species is absent, at least during the incipient stages of colony foundation, which is consistent with the hypothesis that a monogamous mating system could have evolved as a result of mate assistance. This mating system could have been in turn a preadaptation for the evolution of eusociality. Several hypotheses have attempted to explain its origin. From these, Bartz's (2) theoretical framework involving elevated coefficients of relatedness through cycles of inbreeding and outbreeding has had a major impact on the study of eusociality. The effect of sibship of primary reproductives on the survivorship and growth of incipient colonies was studied by periodical census and behavioral observation of colonies. Male and female showed similar mortality patterns, though reproductives paired with non-sibling mates had a higher mortality rate. There were no overall differences in the number of eggs and larvae produced by sibling and non-sibling pairs, suggesting that there are no negative effects of inbreeding. Our results are not consistent with some of Bartz's implicit assumptions on the evolution of eusociality in termites. We believe that inbreeding might have played an important factor in the evolution of sterile castes in higher termites, but that the origin of social behavior was dependent on factors that promoted some degree of inbreeding and not inbreeding itself (3). Factors such as high alate predation during dispersal, high disease risks by outbreeding, low probability of founding incipient colonies owing to unpredictable or scarce resources, and labor being performed mainly by immature insects could have led to a social structure similar to the one observed in lower termites. Inbreeding could have followed, and after selection purged the exposed deleterious alleles, the ancestral prototermite unit could have regained its fitness, only now with a higher degree of relatedness among nestmates that could in turn promote sterile castes.

<u>References</u>

1. Thornhill R. and Alcock J. (1983) The Evolution of Insect Mating Systems, Harvard University Press, Cambridge.
2. Bartz S.H. (1979) Proc.Natl.Acad.Sci. 76: 5764-5768.
3. Myles T.G. and Nutting W.L. (1988) Quart.Rev.Biol. 63: 1-23.

EVOLUTIONARY TRENDS IN NEOTENY AND SECONDARY REPRODUCTION IN TERMITES

Yves Roisin and Jacques M. Pasteels

Laboratoire de Biologie animale et cellulaire, C.P. 160, Université libre de Bruxelles, 50 av. F.D. Roosevelt, 1050 Bruxelles, Belgium

INTRODUCTION

Neotenics, i.e. individuals that reproduce without going through the winged adult stage, are widespread in termites. With dispersal flight and helper (worker or soldier) behaviour, neoteny is one of the three options for a social insect to maximize its inclusive fitness. Neotenics benefit from own reproduction and avoid risky dispersal. Despite its potential advantages, neoteny is of highly variable occurrence among the Isoptera. This paper constitutes an attempt to outline the complexity of this phenomenon and its evolutionary trends.

NEOTENY IN THE LOWER TERMITES

The small societies of the Kalotermitidae probably represent the most primitive level of social organization in extant termites. In such societies, neoteny is the least commonly expressed of the three fitness options : the individuals serve as workers when young; afterwards, most of them proceed to the winged adult or keep working as pseudergates [1]. In most instances, reproduction remains confined to the founding pair : neotenics appear only in a small proportion of the field societies, as replacement reproductives [2].

By contrast with the Kalotermitidae, multiple neotenics are very common in *Mastotermes* and some Termopsidae and Rhinotermitidae, where the differentiation of supplementary neotenics may allow the society to increase its rate of egg production to cope with a higher population, and to restore the reproduction in groups isolated from the rest of the society. Such colonies expand their range and reproduce through satellite calies [2].

EVOLUTIONARY TRENDS IN THE HIGHER TERMITES

In termites without true workers, only minor morphological details allow a distinction between neotenics originating from late larvae, nymphs or pseudergates. With the onset of a distinct worker caste, neotenics split into two clearly separated classes : ergatoids, derived from workers, and nymphoids, derived from nymphs. In orphan nests of *Microcerotermes papuanus* (Termitidae), ergatoids quickly differentiate as replacement reproductives, to be superseded by nymphoids at the seasonal period of late nymph development [3]. Whereas the fertility of the ergatoids seems very low, nymphoid queens become readily physogastric and efficient egg layers.

A marked trend in termite evolution is the loss of the neotenic potential, while imagoes usually acquire the ability to reproduce within their colony of origin. Examples of the occurrence of imaginal replacement reproductives include several Macrotermitinae and Apicotermitinae [4,5]. In *Nasutitermes polygynus*, satellite nests headed by multiple imaginal reproductives are very common and certainly play a major role in the multiplication of colonies [6].

However, exceptions to this pattern are common, and paradoxical situations are sometimes encountered : for instance, in orphan *Nasutitermes novarumhebridarum* societies, replacement reproductives develop from workers but not from nymphs or alates [7], whereas in *N. princeps*, nymphoids are unknown, but their function is assumed by microimagos, resulting from an anticipated imaginal moult [8].

CONCLUSION

The variable fate of neoteny in termite evolution can be explained in part by the diversity of strategies for dispersal and reproduction of colonies. Whereas some species seldom produce replacement reproductives, and only in case of primary queen or king death, others produce large amounts of supplementaries to assume reproduction in satellite calies. However, dispersal strategies do not explain why the roles of replacement or supplementary reproductives, assumed by neotenics in the lower termites, are often taken by imagos in the Termitidae. We suggest that the answer may lie in a quest for plasticity at the individual level : an alate can respond more quickly than a nymph to the vacancy of the reproductive function, since its response is behavioural rather than physiological (e.g., *Astalotermes quietus* [5]). In the higher termites, where tasks are done by a specialized worker caste, nymphs would thus find an advantage in proceeding to the winged adult before choosing between staying and leaving. By contrast, a lower termite nymph going through the imaginal moult must forego its worker functions and the associated inclusive fitness benefits; staying as a pluripotent nymph may thus provide compensations.

ACKNOWLEDGEMENTS

Supported by the Belgian FNRS (postdoctoral fellowship to Y.R.) and FRFC (grant No. 2.4513.90). Contribution No. 201 from King Léopold III Biological Station, Laing Island, Papua New Guinea.

REFERENCES

1. Noirot, C. 1985. In: Caste Differentiation in Social Insects, J.A.L. Watson, B.M. Okot-Kotber and C. Noirot, Eds, Pergamon Press, Oxford, pp. 41-57.
2. Myles, T.G. and W.L. Nutting. 1988. Quarterly Review of Biology, 63: 1-23.
3. Roisin, Y. Submitted.
4. Noirot, C. 1956. Insectes Sociaux, 3: 145-158.
5. Noirot, C. 1985. In: Caste Differentiation in Social Insects, J.A.L. Watson, B.M. Okot-Kotber and C. Noirot, Eds, Pergamon Press, Oxford, pp. 177-186.
6. Roisin, Y. and J.M. Pasteels. 1986. Insectes Sociaux, 33: 149-167.
7. Roisin, Y. and J.M. Pasteels. 1987. Entomologia experimentalis et applicata, 44: 277-287.
8. Roisin, Y. and J.M. Pasteels. 1986: Behavioral Ecology and Sociobiology, 18: 437-442.

SOME FUNCTIONAL AND EVOLUTIONARY ASPECTS OF MECHANICAL AND CHEMICAL COMMUNICATION IN TERMITES

A.M. STUART

UNIVERSITY OF MASSACHUSETTS

AMHERST, MASSACHUSETTS 01003, U.S.A.

FUNCTIONAL ASPECTS.

Mechanical Communication.

The Snapping, jittering, zig-zag locomotory movement complex.

These behaviours while classified separately, blend into each other and all are initiated by an environmental perturbation. The snapping movement in *Zootermopsis nevadensis* and *Z. angusticollis* in both soldiers and larvae consists of: the mandibles opening, a lunge with almost no displacement of the legs, and the jaws opening again when the lunge is fully completed. This is the response to a high intensity stimulus. Jittering resembles the lunge, but without the opening of the mandibles and with a change in the temporal sequence and frequency: a lower grade of response. The Zig-zag movement appears to have its origin in Jittering being superimposed on a locomotory response. Each of these movements can induce further excitation in other individuals: the latter two being, however, essentially communicatory.

Head-banging or vertical vibratory movement.

This movement has been described and analysed in *Zootermopsis* [1, 2] where it is relatively unstructured, and in *Coptotermes lacteus* where there is a structured sequence of 11-12 pulses exhibited by soldiers [2]. In *Zootermopsis* both soldiers and larvae perform the movement.

Chemical Communication.

Sternal gland secretion.

The pheromones from this gland are implicated in sex attraction in *Zootermopsis* [3,4] as well as in *Reticulitermes*, [4] and in trail laying [5].

Frontal gland secretion.

The frontal gland secretion in Nasutitermes acts as an alarm pheromone in addition to its more obvious defensive function [6,7].

EVOLUTIONARY ASPECTS.

Primitively the transmission of excitation causing specific alarm [8] (=alarm recruitment) had as its main cause a perturbation in the immediate environment, and had a defence function; individuals also laid trails. The same system of vibratory movements and a trail later became connected with foraging behaviour.

The main function of head-banging in lower termites is probably modulatory while in higher termites the more structured signal produced by soldiers is coupled with distinct polyethic behaviour in both initiators and recipients: workers leave the site of excitation and soldiers remain. Similar responses are seen in Nasutitermes, but here the initiating signal is a chemical one from the frontal gland pheromone of soldiers. So a chemical initiator has been substituted for the more primitive mechanical one, presumably in association with the reduction of mandibles and the complexity of the frontal gland and nasus.

In terms of the "superorganism" concept, homeostasis of the colony is partly mediated by selective caste polyethism (both in response to the initial stimulus and to the secondary communicatory stimulus), by the physiological state of individuals, and by the importance of competing perturbations at any one time.

REFERENCES

1. Howse, P.E. 1962. Symp. Genet. et biol. it., 11:256-268.

2. Stuart, A.M. 1988. Sociobiology, 14: 49-60.

3. Pasteels, J. 1972. Experientia, 28: 105.

4. Stuart, A.M. 1975. In: Pheromones and Defensive Secretions in insects. C. Noirot, P.E. Howse, G. Le Masne Eds., Dijon, France pp. 219-223.

5. Stuart, A.M. 1961. Nature, 187: 419.

6. Moore, B.P. 1968. J. Insect Physiol., 14: 33-39.

7. Stuart, A.M. 1975. In: Olfaction & Taste V, D. Denton and J. Coghlan, Eds., Academic Press, New York, pp. 343-348.

8. Stuart, A.M. 1963. Physiol. Zool. 36, 69-84.

11th International Congress—IUSSI 1990, India

MULTIPLE FUNCTIONS OF EXOCRINE SECRETIONS IN TERMITE COMMUNICATION : EXEMPLIFIED BY *SCHEDORHINOTERMES LAMANIANUS*

Manfred Kaib

LS Tierphysiologie der Universität Bayreuth, Postfach 101251, D-8580 Bayreuth, Federal Republic of Germany

The social organization of termite colonies depends on an efficient communication system, which is mainly based on chemical signals. These signals are produced by different exocrine glands and trigger or modulate a diversity of behavioural patterns at the level of an individual, and at more integrated levels. This allows a well organized cooperation among a large number of nestmates in a termite colony. Secretions of different glands are not produced for single functions, but are adaptive for multiple functions, depending on the environmental or social context. Furthermore, secretions of different glands may be interactive in terms of their signal function. In addition to the role in intraspecific communication, exocrine secretions also have interspecific functions, e.g. during defense.

Schedorhinotermes lamanianus maintains disperse nest structures: A widely spread network of galleries connect the subterranean main nest with arboreal subsidiary nest divisions, and with foraging sites. Only the agile minor soldiers accompany workers during their extranidal activities while major soldiers remain in the nests. The disperse nest structure puts high demands on the communication system between the individuals of a colony. For extranidal communication, secretions of three exocrine glands play a key role: (1) Sternal gland, which is present in workers and in dimorphic soldiers. (2) Labial glands, which are well developed in workers. (3) Frontal gland, which is present in both soldier morphes, never in workers.

Sternal gland. The sternal gland is located at the 5th abdominal sternite. It produces a multi-component secretion, which is deposited on the ground by individual workers or soldiers, and which is employed during gallery building, orientation between distant nest divisions, and foraging and recruitment. Foraging is only initiated by individually searching minor soldiers. They lay homing trails (low pheromone concentrations), which lead minor soldiers to unprotected foraging areas, but do not act on workers. That caste-specific polyethism is based on a difference in trail-pheromone sensitivity. Minor soldiers respond to secretion approximately two times more sensitively than workers. Compared with workers, minor soldiers can operate more distant from the gallery exits or more laterally from the trail, where pheromone concentrations decrease by distance. Outside the galleries termite workers are always guided and flanked by minor soldiers. Thus intruding ants are confronted first by minor soldiers, which release frontal gland secretion during combat.

Only foraging trails being reinforced by frequent commuting of minor soldiers are followed by workers. They are led to newly discovered food sources by minor soldiers. Recruitment starts, when the first workers return to the nest. On their course back workers lay a recruitment trail, which instantly leads nestmates to the food source for communal exploitation. Minor soldiers on their own are not able to chemically elicit mass recruitment.

Labial glands. In workers the reservoirs of the well developed labial gland extend into the abdomen. Secretion can be released into the oral cavity. Water-soluble constituents of the secretion modulate the exploitation of food sources. It leads towards aggregation of workers who are feeding and, thus, to communal food uptake. By aggregating, workers restrict exploitation of food to limited areas, which can more easily be secured by minor soldiers.

Secretions of the sternal glands and of the labial gland are interactive in behavioural terms. Sternal gland secretion inhibits the feeding aggregations. The interaction of these two chemical signals causes a spacial separation of feeding areas from commuting zones.

Frontal gland. The frontal gland reservoir of both soldier morphes differs in size (major soldiers ≈ 50% of total body weight, minor soldiers ≈ 15%). Based on morphological criteria, we assume the epithelium of minor soldiers secretes more actively than that of major soldiers, which may compensate for the frequent release of secretion by minor soldiers during encounters with predatory ants. The secretion flows into a terminal brush at the labrum. Volatile constituents rapidly evaporate from here.

Chemically the secretions of both soldier morphes differ by the presence of saturated or unsaturated methylketones and ß-keto-aldehyds in minor soldiers.

During agonistic behaviour minor soldiers release frontal gland secretion, which immediately alerts additional soldiers and attrackts them to sites of combat. Attacked ants are disabled by the vapour of the soldiers' secretion and the ants' recruitment behaviour is disrupted. The chemical signal also alerts termite workers and elicits combative behaviour in workers which is then directed towards ants which are disabled by secretion of soldiers.

Alertness or attack behaviour of termite workers only occurs when minor soldiers have released their secretion during combat. Vibratory signals or secretion on its own do not elicit attack behaviour. Workers distinguish clearly between frontal gland secretion of minor and major soldiers. Major soldiers are concentrating only in the nests and do not accompany termite workers into foraging areas. Their secretion does not elicit alarm or combat behaviour.

In addition, frontal gland secretion of minor soldiers inhibits dose-dependent trail-following behaviour at low pheromone concentrations (< three-fold of the threshold concentration), as they occur during the initial phase of foraging outside the galleries. This prevents termite workers from leaving the shelter by the galleries during ant attacks. However, at higher pheromone concentrations, which are typical for well established foraging trails or within the galleries, the vapour of frontal gland secretion does not affect trail-following. Because of their lower threshold concentration, minor soldiers respond more sensitively to the sternal gland secretion than workers. This leads to a spacial separation between fighting soldiers and foraging workers in combat situations .

CIS-3 CIS-6, TRANS-8 DODECATRIEN-1-OL: SEX AND TRIAL FOLLOWING PHEROMONE IN A HIGHER FUNGUS-GROWING TERMITE?

C. BORDEREAU[1], A. ROBERT[2], O. BONNARD[1] and J.L. LE QUERE[3]

1. Université de Bourgogne, Zoologie, UA CNRS 674, 6 bd Gabriel, 21000 Dijon. France.
2. Université des Sciences et Techniques de Masuku, Dpt de Biologie, Franceville. Gabon.
3. INRA, Laboratoire de Recherches sur les Arômes, 17 rue de Sully, 21000 Dijon. France.

The sternal gland of termites is known as the source of trail following pheromone in workers, and sex pheromone in alates. However, some behavioural observations suggested that both pheromones could be one and the same pheromone (1,2). Here are reported the results obtained in *Pseudacanthotermes spiniger*, a fungus-growing termite very abundant in the savannah near Franceville (Gabon).

The sex attraction can be observed in the laboratory if the swarming imagoes of both sexes are gathered together in Petri dishes. The females take up a "calling" position with the abdomen highly raised exposing the sternal gland. The males are attracted by these females and show a great excitation, palpating, licking or even biting the posterior part of their abdomen. There is no tergal gland in imagoes of *P. spiniger*. Sex attraction bioassays show that only the female sternal glands were as attractive as the entire female extracts. Pentanic washes (1h at 4°C) of male and female alates and sternal gland extracts were studied by gas chromatography. One compound found in the sternal gland of both sexes is present in much higher quantities in females than in males (about ten times). After gas chromatography fractionation, only this compound attracts males as well as extracts from female sternal glands do. Combined gas chromatography-mass spectrometry (electronic impact, chemical ionisation), nuclear magnetic resonance, infrared spectrometry and hydrogenation reactions show that this compound is the cis-3, cis-6, trans-8-dodecatrien-1-ol.

Trail-following bioassays show that this compound also is the most active of the sternal compounds on workers. They can detect it at extremely low concentrations (10^{-13} -10^{-14} g/cm), strongly suggesting that this unsatured alcohol is used as trail following pheromone in *P. spiniger*. However, this must be confirmed by chemical investigations on worker sternal glands and natural trails.

Cis-3, cis-6, trans-8-dodecatrien-1-ol was found in several species of *Reticulitermes*. Although it was isolated from entire workers eating infected wood with fungus containing this compound, it was considered to be the trail pheromone (3). Our results demonstrate for the first time that this alcohol does exist in the termite sternal gland. So, this compound plays a major role in the social and sexual biology of termites and the question is whether it may be a basic component conserved in the course of Evolution. Its presence in *Reticulitermes* and *Pseudacanthotermes* reinforces the idea of a phylogenetical proximity between the more advanced lower termites which are the Rhinotermitidae and the Macrotermitinae which represent a particular group within the higher termites (4). Moreover, the results of Kaib et al. (5), showing the non-species-specific trail-following, suggest that the dodecatrienol could be an anonymous signal (6) common to a large number of termite species (Hodotermitidae excepted).

References

1. Leuthold, R.H. and M. Lüscher. 1974. Insectes Sociaux, 21: 335-342.
2. Quennedey, A. and R.H. Leuthold. 1978. Insectes Sociaux, 25: 153-162.
3. Matsumura, F.H., C. Coppel and A. Tai. 1968. Nature, 219: 963-964.
4. Deligne, J. 1985. Actes Colloque Insectes Sociaux, Sect. Française IUSSI, Diepenbeek, 2: 35-42.
5. Kaib, M., O. Bruinsma and R.H. Leuthold. Journal of Chemical Ecology, 8: 1193-1205.
6. Hölldobler, B. and N.F. Carlin. 1987. Journal of Comparative Physiology A, 161: 567-581.

TERMITE SPECIES AND DISTRIBUTION IN CHINA

Li Gui-xiang, Dai Zi-vong
Guangdong Entomological Institute
105 Xingang Road West Guangzhou,
510260 China

According to the latest statistics by December of 1988, there are the 4 Families, repressenting 43 Genera and 385 species of termites in China. All of them are recorded and described as follows:

1. Kalotermitidae	6 genera	66 sp.
2. Hodotermitidae	1 genera	3 sp.
3. Rhinotermitidae	8 genera	151 sp.
4. Termitidae	28 genera	165 sp.

The genera representative of economic importance are: Cryptotermes Reticulitermes, Coptotermes, Odontotermes and Macrotermes, containing about 30 the most destructive species of termites. The termite distribution prevails throughout the 25 provinces, cities and autonomous regions of China except the provinces of Heilongjiang, Jilin, Inner Mongolia, Ningxia, Qinghai and Xingjian.

The north limit of termite distribution is about 40°N (Liaoning: Dandong and Beijing) going through Shanxi (Jiexiu, about 37°N), Shaanxi (Hancheng 35.5°N). Gansu(Wenxian 33°N) to Tibet(Medog 29.5°N and Cona 28°N) Which is an oblique line from the high (northeast) to the low (southwest). The southeast of obloque line is the termite distribution region, but there is no termite distribution in the vast area of the northeast, northwest and Tibet. The area of the termite distribution approximately covers the 40% of total area of the whole country. a). The north limit of Reticulitermes is about 40°N (Liaoning Dandong and Beijing). b). Odontotermes 35°N (Henan: Laoyang). c). Coptotermes 33.5°N (Jiangsu:

Jianhu). d). Cryptotermes 28.4°N (Sichuan: Jiangan) and in parts of the southern Zhejiang and southern Guizhou, and it is mainly distributed in Hainan, Guangdong, Yunnan and Fujian. e). The mound termites are distributed in the province, south of 25°N yunnan (Odontotermes yunnanensis, Macrotermes annandalei and Globitermes sulphureus), Guangxi (Macrotermes annandalei) and Hainan (Microcerotermes remotus and Capritermes wuzhishanensis). The record of the highest latitude for termite distribution is 2442 meters with Reticulitermes (Yunnan: Lijiang Shigu).

MORPHOLOGY OF THE DIGESTIVE TUBE AND SALIVARY GLANDS OF *SERRITERMES SERRIFER* (ISOPTERA : SERRITERMITIDAE)

COSTA-LEONARDO, A.M. and CAMARGO, R.S., Department of Biology Institute of Bioscience, UNESP, 13500 - Rio Claro, SP, Brazil

Serritermes serrifer is the single species from the family Serritermitidae and is found only in Brazil. Workers and soldiers of **Serritermes serrifer** from Brasília (D.F.) had their digestive tubes and salivary glands isolated and analysed by the morphological point of view. Drawings were made with a camera lucida in stereomicroscope and the histological slides were stained with Haematoxylin-eosin. The foregut is not so developed : the termites don't have a enlarged crop and their gizzard is rather reduced, not chitinized and imperceptible in fixed material. The esophageal valve is short and penetrates into the midgut. The midgut doesn't present mixed segment or projections and it is long and thick. The limit between the middle and hindgut is circular as in the Kalotermitidae and looks like a ring-shaped fold. The Malpighian tubules are eight and insert individually and radially at the junction of middle and hindgut. The enteric valve penetrates in the paunch which is enough voluminous and presents folds characteristics. The first segment preceding the enteric valve is short and tube-shaped The firsts folds of the paunch are wide and seems harbour several flagellates, mainly the greatest ones. The colon is long, the rectum is short and muscular and the Malpighian tubules end in its walls. In the soldiers there is only one hypertrophied salivar reservoir that pushes the gut to ventral region. The salivar gland acini are tubular and appear in a great number in workers than soldiers.

(Financial Support CNPq)

USING BAIT TRAPS TO COLLECT SOIL TERMITES IN TABASS OASIS, SOUTHWEST KHORASSAN OF IRAN

HOSSEIN HOOSHMAND AND M. BAGHER SHAHROKHI
Plant Pests & Diseases Research Lab. of Mashad , Mashad 91375 Iran.

In agricultural station of Tabass, Microcerotermes sp. is occuring as a structural and agricultural pest. Many living plants including palm trees are heavily attacked by the termite. Some bioecological studies have been carried out since 1986, a part of this study is the reaction of this termite to bait traps of FRENCH et al (1), the effect of moisture in baits and the proportions of different castes in groups have been determined.

Material and methods

The experiment was carried out in 2 treatments, 15 saturated with water baits and 15 fairly wet baits in cans were used. The traps were examined one month later, no saturated with water bait was occupied, whereas 4 traps with fairly wet baits were occupied by termite groups. The attracted groups can live actively and make aerial bridges in the containers, the proportions of each caste per trap is given in table 1.

Table 1: Numbers of different castes of Microcerotermes sp. attracted to fairly wet bait traps - within a month.

Trap No.	Worker	Nymph	Soldier	Larva	Other artropods
1	2149	7	13	1	-
2	1162	-	3	-	2 pillbugs
3	1708	10	14	-	-
4	4431	488	24	1	-

Reference:

1. French, J.R.J.-P.J.Robinson 1981. Baits for aggregating large numbers of subterranean termites, J.Aust. ent. Soc.20, 75-76

THE TERMITE POPULATION OF THE LAC DE GUIERS REGION IN THE REPUBLIC OF SENEGAL

AGBOGBA C.
Départment de Biologie Animale - Faculté des Sciences
Université C. A. DIOP de Dakar.

This study of the termite population (as a biological indicator) was carried out in a areas influenced by the Lac de Guiers. It represents the soudano-sahelian sector of the Sahel. Termite studies show the predominance of one species Psammotermes hybostoma which especially attacks Balanites aegyptiaca. The damage done to it varing from 33% to 100%. The damage diminishes on Balanites where diversity of species trees increase. Acacia are relatively unscathed.

Key words : Sahel - Vegetation - Termites.

The extended summary has been given under "Additional Paper".

THREE ESTIMATES OF COLONY SIZE FOR *HETEROTERMES AUREUS* IN SOUTHEASTERN ARIZONA (ISOPTERA : RHINOTERMITIDAE)

W. L. Nutting
Professor Emeritus
Department of Entomology
University of Arizona
Tucson, Arizona 85721, USA

A termite control treatment provided a unique opportunity to estimate colony size of this destructive subterranean termite. Subsoil injection of pesticide (probably chlordane, especially noted for its repellency) along the foundation of a prominent church in downtown Tucson flushed over 30,000 termites from their galleries through extensive tubes on a basement wall and out onto the floor where they died. The entire 0.5-x2-m windrow of dead insects was collected and counted. They had obviously died under great stress, for most of the 15,880 workers had evacuated the hindgut and 18,452 soldiers held the mandibles crossed in tetany.

Studies on the ecology of Sonoran Desert termites over some 20 years have included attempts to estimate the size of colonies and natural populations. Herewith is a summary of two previous estimates of H. aureus colony size for comparison with that derived from this mass pesticide kill.

1. By determining long-term forager and colony densities, and using surface/subsurface proportions, Haverty and Nutting [1] calculated H. aureus colony size at ca. 23,000 individuals.
2. By using mark-release-recapture and mass trapping techniques, Jones [3] reported perhaps more realistic estimates of 45,000 to 300,000 with an average of 120,000.
3. Known proportions of soldiers, albeit only for foraging groups, permit easy calculation of group size (colony in this case?). If Jones' [3] figure of 8.6% soldiers is used, then this church colony contained ca. 215,000 termites. This percentage seems high for a subterranean colony; the 1.5% from Haverty and Nutting [2] gives us ca. 1,230,000. Perhaps a more reasonable estimate lies somewhere in between, i.e. ca. 700,000.

While results of the first two approaches differ greatly, the third estimate, based on an actual count of soldiers, provides strong support for the use of the mark-recapture method as used by Jones [3]. I would be interested to hear of similar control-related incidents; the phenomenon might be adapted to flushing termites - or other social insects - from their nests for population estimates.

References

1. Haverty, M. I., W. L. Nutting and J. P. LaFage. 1975. Environmental Entomology, 4(1): 105-109.
2. Haverty, M. I. and W. L. Nutting. 1975. Environmental Entomology, 4(3): 480-486.
3. Jones, S. C. 1990. Southwestern Naturalist, In press.

BIOLOGY OF *COPTOTERMES FORMOSANUS* SHIRAKI IN CHINA

Li Gui-xiang, Dai Zi-vong
Guangdong Entomological Institute
105 Xingang Road West Guangzhou,
510260 China

The genus Coptotermes in China has rich species up to about 30 species. The species that its biology has been studied comparatively clearly is C. formosanus Shiraki.

a). Nests: The nests of C. formosanus are separated into the primary nests and secondary nests, and the sites of buiding nests are under the ground, on the ground surface or inside the tree trunks. The nests of C. formosanus have three external features: 1). Excreta: Common name is "termite soil". 2). Swarming holes: The holes are previded with swaring out for alates. 3). Ventilating holes: Generally thought to air exchange. It is detected by the trailing survey with the isotopes I^{131} and Au^{198} that the radius of the colony activities in the mature could reach 50 meters. The activities are more active at night than by day. The distance between the primary and secondary nests is within 30 meters and the termites move back forth between both nests for connection in 4 hours.

b). Swarm: The nymphs occur when the nests colony develops into the mature stage. According to our results of the indoor and indoor breeding of the paired males and females, it takes 8 years for a nest colony to maturate, and the alate swarming would take place. The swarming often occurs at 7:00-8:00 P.M. at dusk , at RH 85-90% and 21-29°C. The swarming begin in the early April--late June. Each colony swarms two to six times every year.

c). Establishment of the colony at the earlier time: If the

Oxford & IBH Publishing Co. Pvt. Ltd., 66 Janpath, New Delhi 110 001, India

sufficient water and food are given, the paired alate begin to lay eggs after 5 to 7 days under the conditions of 25-30°C and RH 85%. The hatching period is 24 to 32 days..The fast or slow increase of the colony individuals depends mainly on whether the ecological conditions is fully satisfied or not, especially the temperature, humility and food. On 27 April 1973 the alate were paired and put in a incubator at constent temperature 30°C and constent humility 85%, and suitable moisture and food. On 24 July 1974, there were 1039 termites, workers were 915 and soldiers were 124, and a crowd of termite eggs was unable to be ccunted.

d). Formation and development of the colony:

Population quantity in the nests of different years
(The number in parenthesis is the nest number)

Nest year	Range	Average	Workers (average)	Soldiers (average)	Soldiers ratio (%)
0.5	20-102	50.23(19)	42.85	7.38	14.69
1	33-243	158(10)	137.46	20.54	13.00
1.5	177-600	300(4)	267	33	11.00
2	130-956	483(8)	432	51	10.56
2.5	1492	1492(1)	1357	135	9.05
3.5	2506-2967	2731.7(3)	2567.8	163.9	5.99
8*	87041	87041(1)	83215	3801	4.37
20** (estimated)	395766	395766(1)	378738	17028	4.30

* 25 nymphs were found, workers and soldiers of colony were run away

** According to the dissecting investigation of the 80kg subterranean nest dug out on 16 August 1970

FORAGING POLYETHISM IN THE HARVESTER DESERT TERMITE *ANACANTHOTERMES MACROCEPHALUS* (DESNEUX) (ISOPTERA : HODOTERMITIDAE)

Sushil Kumar and M.L.Thakur
Entomology Branch, Forest Research Institute
Dehra Dun (U.P.), India 248006

Anacanthotermes macrocephalus is an important harvester termite which is widely distributed in the desert areas of Gujarat and Rajasthan. Although much is known concerning its biology and ecnomic importance, this species has been studied very little so far as polymorphism and polyethism are concerned. There are two peaks of head-width in workers of A.viarum collected in southern India and Sri Lanka. Preliminary information concerning foraging polyethism is presented here.

MATERIALS AND METHODS

Our observations were made in the second fortnight of December 1989. This species makes epigeous "mounds" in the form of small heaps of soil. Foragers were collected from these epigeous mounds by quickly transferring the heaps into polythene bags. To collect foragers in the open a pebble was placed in the exit hole and returning foragers were picked up with tweezers. Workers open exit holes and close them after foraging. Termites busy opening and closing exit holes were scooped up. Workers' and solders' head-width was measured using an ocular micrometer.

RESULTS

The head-width of workers and soldiers showed two peaks, and they were thus designated as small and large workers (SW and LW) and small and large soldiers. The foragers remain in the epigeous portion of the nest in the early hours of the morning and again in the evening after sun-set; however, they retreat underground into the nest during day-time. Foraging, however, takes place from 10 p.m. to 2-3 a.m. Data show that during opening and closing of exit holes small workers predominate over large ones, while small or large soldiers guard the openings from inside. Small workers are also predominant in the epigeous mound and rarely come out into the open. In the middle of the night large workers predominate (up to 90%) in foraging. No soldiers accompany the workers in the open. Termites carry short bits of dry grass horizontally but long bits longitudinally between their legs. When disturbed, workers often hide between debris or remain motionless.

Foraging takes place in the close vicinity of the exit holes, usually within 20 or 30 cm, but occasionally up to 100 cm or more from the holes, depending upon the availability of the food. Some workers remain at the exit hole and allow the foragers in. But if the blade of grass is long and cannot be carried easily, workers standing inside the hole pull it in. If a foraging worker is carrying a long blade of grass and is unable to enter the hole with the grass, it inserts one end into the hole and the grass is pulled in by workers inside, while the carrier pushes the piece in from the other end.

REFERENCES

1. Ronnwal M.L., (1975) Zeitsch. f. angew. Entomol. 78, 424-440.
2. Watson J.A.L. (1973) Insectes Soc. 20, 1-20.

FORAGING BEHAVIOUR OF *ODONTOTERMES* SPP. (ISOPTERA: TERMITIDAE)

D.Rajagopal
Department of Entomology
University of Agricultural Sciences
GKVK, Bangalore-560 065, INDIA

During the process of food collection, termites forage constantly for cellulose material including grass, crop plants, trees, litter, wooden logs, root stubbles, harvest leftovers, dung pads etc., often causing total destruction of organic matter in the nature. They are known to cover almost all food materials by earthen sheeting and replace them by heaps of soil. Although foraging activity in termites assume a diverse mechanism, the information on the foraging behaviour is fragmentary, especially on the dominant group of Odontotermes spp. in India.

Studies on the foraging behaviour of three mound building species viz., Odontotermes obesus (Rambur), O.redemanni (Wasmann) and O.wallonensis (Wasmann) and two subterranean species viz., O.horni (Wasmann) and O.ceylonicus (Wasmann) were made. The seasonal fluctuations on foraging activity of these species throughout the year in relation to the food sources and weather parameters were recorded.

The results revealed that the foraging behaviour in all the five species of Odontotermes follow a similar pattern where the foraging columns are composed of mainly workers and very few soldiers for guarding. The foraging termites varied with proportion from 35:1 to 90:1 ratio between the workers and soldiers population in foraging sites depending on the nature and availability of food. Foraging column is always found to move in subterranean galleries extended from the nest radially and emerged through exit holes to food sources at varied distances and foraged always under cover of earthen sheetings. They do not forage openly unlike other species such as Macrotermes sp. and Trinervitermes sp. They forage continuously in the same area by opening and closing mechanism of the exit holes at the foraging site. The exit holes varied from 4 to 6 mm in dia. and more than one exit holes are present in the same foraging area.

The subterranean galleries are found distributed from 10 cm to 120 cm depth below the ground level. However, the maximum number of galleries are distributed at 30-60 cm depth. The diameter of the galleries varied from 5 to 9 mm depending on the size of the foraging termites.

The extent of earthen sheeting varied depending on the nature of food availability on the ground from a narrow tubular runways to broader patches of several centimetre diameter on grass to several metres height on the live trees. Even the foraging may be continuous in some places which remained unnoticed by feeding on the roots of crop plants in soil and tunnelling the heartwood from the base of trees and timber in the building without any indication outside.

The results of the observations on the seasonal fluctuations revealed that they forage throughout the year although the intensity of foraging is related to the seasonal weather conditions with the maximum foraging activity during post monsoon period from October to January when there is sufficient ground moisture. Foraging activity increased moderately during the onset of pre monsoon season (April to June) when the occasional heavy rains started in the region. The minimal activity occurs during the rainy season due to more saturation and again from February to middle of April due to dry condition in the soil.

The foraging activity may be diurnal or nocturnal. During summer, the foraging activity is restricted to the early morning and evening hours and sometimes extended the activity even during the entire night. Higher temperature and more sunshine hours of the day reduced the foraging activity (2). The foraging is continuous during the entire day under the shade and also outside during cloudy weather, irrespective of the seasons under Bangalore conditions. However, the optimum temperature and relative humidity for foraging activity was 19° to 25°C and 70 to 80% respectively. These observations have also indicated that they avoid higher temperature (>26°C), cold temperature (<17°C) and continuous rainfall.

Apart from foraging activity above the ground level, the mound building species are found to deposit their food in separate food chambers as in the nest of O.wallonensis (1) and also in the fungus chambers of other species during the post monsoon season.

1. Rajagopal,D. and Veeresh,G.K., 1978. In: Soil Biology and Ecology in India. UAS.Tech.Ser., 23:334-336.

2. Rajagopal,D. and Veeresh,G.K., 1981. J.Soil Biol.Ecol., 1:56-64.

REPAIR VERSUS EXPANSION POLYETHISM IN *ODONTOTERMES OBESUS* (RAMBUR) (ISOPTERA : MACROTERMITINAE)

Sushil Kumar

Entomology Branch, Forest Research Institute
Dehra Dun (U.P.), India 248006

Odontotermes obesus is an economically important mound-building termite species in India, which comprises large and small workers (LS and SW). Prior to this study, it was thought that only one type of worker was predominant in repair and expansion processes as both involve carrying and depositing soil particles. Our findings are presented here in brief.

MATERIALS AND METHODS

In order to study nest repair, a portion of a mound was damaged. Termites engaged in repair work were collected simply by dislodging the portion on to a plastic sheet. One quick sweep was made at a natural expansion site for the nest expansion study. The composition of samples was analysed.

RESULTS AND DISCUSSION

In the case of nest repair, a total of 8,454 termites were collected from 26 samples; they proved to be made up of 21.55% LW, 60.93% SW and 17.51% soldiers. One immature stage (larva) was also found. During the process of nest expansion, a total of 4,952 termites were collected. The composition of different polymorphic forms was 52.68% LW, 33.47% SW and 13.81% soldiers. Two termitophiles were also found. Although both the repair and expansion processes involve the workers' carrying and depositing tiny, moist balls the dominant worker type for each job was significantly different. Soldier population increases far more than its actual proportion in the nest (2 to 5%).

INFLUENCE OF TERMITES ON THE FORMATION OF HUMUS LAYERS IN TROPICAL FORESTS

GARNIER-SILLAM Evelyne

Laboratoire de Biologie des Sols et des Eaux, Université Paris XII, 94 000 Créteil. FRANCE.

The role of termites in soils has attracted the attention of zoologists and a few agromists for some time but it was not until the 1970's that it was first addressed from a combined zoological and agricultural view point (1).

Our Works are a contribution at the knowledge on the role of Termites in the process of humification of the organic matter of the hemi-organic horizons.

By different methods (ethology, chemical studies coupled with transmission electron microscope (TEM) observations, permitting an understanding of the characteristics of the initial food material and its modification during transit through the gut (bolus of food and feces) and pedological analyses of organo-mineral samples linked to studies of the activity of this species which constructs an extensive network of galleries in the soil, thus permitting a better under-standing of the physico-chemical transformations of hemi-organic horizons resulting from the building and feeding behavior of the termites) the action of termites on the transformation of organic debris in a humid tropical forest (Mayombe, Forest, Congo) has been undertaken, comparing several termite species with different feeding regimes

o humivorous termites : *Thoracotermes macrothorax, Cubitermes speciosus Noditermes sp. and Crenetermes albotarsalis.*

o fungus-growing termites : *Macrotermes muelleri*

HUMUS-FEEDING TERMITES

This termites influenced humified horizons extend to 1-3 m from the nest and are thick (3-15cm). With respect to control soils, these horizons are enriched in - organic matter - exchangeable cations (in particular Ca, Mg and K) - phosphorus (2) (3).

It appear that the species particular ethology (building and nutritional behaviour as well as the incorporation of excrements into the horizon) influence both the vegetative organic matter transformation and the biochemistry of exchangeable cations.

The biological activity of humivorous termites favours the development of an active humified horizon around its nest which is rich in mineral and organic colloïds. This leads to the slowing down of the mineralisation of chemical elements while enhancing the humification processes(4).

An electron-microscopic scanning of faeces (2), (5) and (6) emphasizes the building up of organo-mineral aggregates, consisting of spingle-shaped mineral particles which themselves are adsorbed on clearly recognizable vegetal debris, vegetal granules and on bacterial accumulations surrounded by polysaccharide secretions. A certain amount of these bonds persists after akkaline reagent extraction (pyrophosphate + soda), which emphasizes their stability. Thus, a certain amount of organic matter (humin) is included in the faeces, being locked by mineral particles.

FUNGUS-GROWING TERMITES

M. muelleri cuts leaves in the litter into small pieces (confetti) which are transported back to the nest. After a few days storage, they are ingested, passed rapidlythrough the gut where they are broken up physically but subjected to little chemical action, and then they are deposited as mylospheres on the upper surface of the fungus comb (2).

The fungus "*Termitomyces sp* ", exosymbiont of the termite *Macrotermes mülleri* drills the pecto-cellulosique walls of the vegetal cells present in the fungus comb in such a way to get near the dark pigments (polyphenols-proteins). Then, this fungus possessed cellulolytic, pectinolytic and perhaps ligninolytic activities and it degraded in a great extent the brown products. Subsequently, this enzymatic equipment of the termite and of its endosymbionts continued in the digestive tract the degradation of the different plant materials. The faeces which were deposited on the nest floors and on the ground were essentially composed of organic granules and organo-mineral microaggregates. These microaggregates were produced during the intestinal transit by the adsorption of clay particles on to the vegetal biodegradated organic matter. The weakness of the organo-mineral bridges let us to suppose that the organic residue will be easyly mineralised(7).

The humified horizons located in the visinity of the *M. mülleri* nest are thin and are poorer in organic matter than the control horizons. The nest of this species is surrounded by an epigenic soil several centimetres thick characterised by an uniform textural distribution. This results put the workers activity (current or past) of accumulating in it materials lifted from deep horizons. Both the nest walls and the influenced surface horizons show an increase in exchangeable cations; a phenomenon which can be related to the termite's metabolic products present on the built floor, the lifting to the surface of slightly weathered deeply located materials, and by analogy with the hydrology of other Macrotermitinae the recovery of both surface runoff and rising underground water(2). The intensive removal of the litter by this species which exceeds its restoration rate empoverish the humified horizon of organic colloïds. The termites however, influence positively the biochemical cycle of certain cations (Ca, Mg and K) by the huge soil depth in which they work on(8).

Our studies on the physical soil termite properties (9), shows that the specific biological behaviours of the species influence differently the structural stability (Is test) and the permeability (K test) of strata. The soil-feeding termite *Thoracotermes macrothorax*, through the sheer use of a material elaborated from an equilibrated texture (clays/coarse sands) and by means of its own faeces as cement, does improve deeply the structural stability of its area. Whereas the fungus-growing termite *Macrotermes mülleri*, from the very marked upturnings of clays at ground level and from its inability to turn its faeces to use as building materials, locally brings about an imbalance of its own humiferous stratum.

Termites have actually peopled every kind of tropical environments; they have undergone all sorts of dietary conditions to satisfy their feeding needs.

Such diverse trophic strategies - whether from species or genus - have indeed widely differing influences on humifying processes, thus playing a significant role in the fertility of tropical soils (10).

1. Lee K.E. and Wood T.G. 1971. Académic Press, Londres, New York, p.251.
2. Garnier-Sillam E. 1987. Thèse d'état, Université Paris XII, p. 276.
3. Garnier-Sillam E. 1988. Environmental Biogeochemistry, (in press).
4. Garnier-Sillam E., Toutain F., Renoux J. 1989. Soil Biology & Biochemistry, 21 (4) : 499-505.
5. Garnier-Sillam E., Villemin G., Toutain F., Renoux J. 1985. C. R. Acad. Sc. ,Paris, 301, III (5) : 213-218.
6. Garnier-Sillam E., Villemin G., Toutain F., Renoux J. 1987. In : Actes de la VIIème Réunion Internationale de Micromorphol. des Sols, Fédoroff N., Bresson L.M. et Courty M.A. Publishers, Paris, pp. 331-336.
7. Garnier-Sillam E., Villemin G., Toutain F., Renoux J. 1988. Canadian Journal of Microbiology, 34 (11): 1247-1255.
8. Garnier-Sillam E. 1989. Sociobiology, 15 (2) : 181-196.
9. Garnier-Sillam E., Villemin G., Toutain F., Renoux J. 1988. Pedobiologia, 32 : 89-97.
10. Garnier-Sillam E., Renoux J. 1987. In : Coll. Inter. sur le "Maintien de la fertilité des sols tropicaux", Garnier-Sillam E., Kabala M. et Sénéchal J. , Eds., Unesco Publishers, Pointe -Noire, pp. 105-112.

ORIBATID SYMBIONTS OF THE NEST OF *ODONTOTERMES OBESUS*

M.A.HAQ, N.RAMANI and P.NEENA
Division of Acarology
Department of Zoology
University of Calicut
Kerala, 673 635
INDIA

INTRODUCTION : Termites represent the best known group of exclusively eusocial insects of the tropics. The nests of social insects are commonly harboured by a variety of animals including mites. Many mesostigmatid mites lead a phoretic life on social ants while others act as scavengers and ectoparasites(Wheeler,1910; Bernard, 1968). Honey bees also act as hosts for some ectoparasitic Scutacaridae(Morgenthaler, 1968). Uropodid mites serve as scavengers of termites(Hirst, 1927). The present paper reports the association of a few species of oribatid mites with the termite, Odontotermes obesus and their symbiotic role in the nest.

MATERIALS AND METHODS: Oribatid species were recovered from the nest of O.obesus by hand sorting and Berlese funnel extraction. Observation was made on the feeding habits of the mites and gut content analysis of the field collected specimens was performed following appropriate staining procedures.

OBSERVATION: Examination of nest materials, particularly the fungus gardens disclosed the occurrence of a variety of fungal communities like Termitomyces, Aspergillus flavus, Penicillium and Curvularia. 13 species of oribatid mites belonging to 13 genera and 10 families were found inhabiting the fungal gardens of O.obesus. Berlesezetes brazilozetoides, Allonothrus monodactylus and Zetomotrichus lacrimans exhibited comparatively dense populations. Gravid females also contributed to the population density of several species. Gut contents of *Oppia kuhnelti*, *B. brazilozetoides*, *Z. lacrimans* and *A. monodactylus* disclosed the presence of fungal hyphae of *A. flavus*, *Curvularia* sp. and *Fusarium* sp. This was confirmed in the laboratory also where they showed preference to *A. flavus* and *Curvularia* sp.

DISCUSSION: Recovery of 13 genera and 13 species of oribatid mites from termite nest as observed in the present study shows the complexity of oribatid population that can survive even in such restricted environments. All the species with the exception of *Annectacarus mucronatus* and *Xylobates monodactylus* belong to the microphytophagous category(Luxton, 1972; Haq and Prabhoo, 1976; Haq, 1982). Results of laboratory feeding and gut content analysis confirmed their dependence on fungal diets of *A. flavus*, *Curvularia* sp. and *Penicillium* sp. *A. flavus*, a pathogenic fungus of *O. obesus* (Mani, 1982) has been found preferred by 4 species of oribatids thereby excerting an effective check on its growth in the nest. Thus the mites serve to protect the termites from the pathogenic fungus in turn receiving food and shelter.

REFERENCES:

Bernard, F. 1968. Les fourmis (Hymenoptera:Formicidae) d' Europe accidentale et septentrionale. Faune de l'Europe et du Bassin Mediterranean.No.3, Masson et Cie, Paris,p411.

Haq, M.A.1982. Indian Journal of Acarology, 6:39-50.

Haq, M.A. and N.R.Prabhoo.1976. Entomon, 1(2): 133-137.

Hirst, S.1927. In: Insects of Samoa Pt. 8. Terrestrial Arthropoda other than insects, British Museum(Natural History), London, 1: 25-27.

Mani, M.S.1982. General Entomology, Oxford and IBH Publishing Co.Pvt. Ltd., New Delhi, p.912.

Morgenthaler,O.1968. In: Traite' de biologie de l'abeille, R.Chauvin, Ed. Vol.IV: 233-252.

Wheeler, W.M. 1910. Journal of Morphology, 22(2): 307-325.

Symposium 2

EVOLUTION OF SOCIALITY LESSONS FROM PRIMITIVELY EUSOCIAL WASPS

Organizer : **RAGHAVENDRA GADAGKAR, India**

WASPS AND OUR KNOWLEDGE OF INSECT SOCIAL BEHAVIOUR

Charles D. Michener
Snow Entomological Museum, Snow Hall
University of Kansas
Lawrence, Kansas 66045, U.S.A.

Studies of early stages of evolution of sociality involve bees and wasps, since intervening stages between solitary and highly eusocial exist among these insects. Although early studies were with eumenid wasps (Roubaud, 1911), such work has been a favorite activity of melittologists because bees have a far richer array of intermediates between solitary and eusocial. It is as though Polistes contained solitary species and species with various combinations of social attributes up to fully eusocial.

Wasp behavior supports theory-making. Many wasps like Polistes make nests consisting of exposed combs where the wasps are easily counted. In a single population nests can contain one foundress or multiple foundresses, so that the contribution of each toward the reproductive success of the colony can be estimated. Thus wasps invite studies by theorists interested in Hamiltonian aspects of social evolution. Persons working on other groups usually fall back on theory developed with wasps.

Wasps (Polistes) are ideal for studies of individual behavior. For example, although social hierarchies were first recognized by Hoffer (1882, 1883) in bumble bees (long before such hierarchies were studied in vertebrates), they have been better analyzed for wasps (e.g., Pardi, 1948) than for other insects. Other major early findings about sociality also resulted from work with wasps. One was the nutritional control of caste (Marchal, 1897), another, trophallaxis (Roubaud, 1916). Each of these, however, has been more fully investigated subsequently in other groups of social insects.

Nest architecture was early recognized as important for studies of classification and phylogeny of wasps (Ducke, 1914). Wasps continue to be the architecturally most studied group (Wenzel, 1988), at least from the viewpoint of phylogeny, for the nests are usually large and in the open.

Chemically mediated behavior has been studied much more in bees and ants than in wasps. Thus kin or nestmate recognition has been studied intensively in other groups. Gamboa et al. have shown that such recognition occurs in wasps, so it was there to be observed. Of course odor trails are best observed in wingless insects, but are well known in bees. The chemistry of many pheromones and defense substances is known for the other groups; little such work has been done for wasps. Both behavioral and pheromonal aspects of wasp mating behavior have been neglected. I know of no intrinsic reason why chemical investigations and related behavioral work using wasp products have lagged.

While exposed combs have facilitated certain wasp studies, this very feature is disadvantageous for others. There is no narrow nest entrance to guard. Guards in primitively social bees revealed kin and nestmate recognition and recognition pheromones in insects (Greenberg, 1979).

The economic importance of other social insects has led to intensive studies of their sociality. The existence of Apis mellifera provides a comparative basis for behavioral and chemical studies of social bees. Wasps, however, have provided much of the basis for theories of evolution of social behavior. Obviously the whole field will be enhanced when data permit better comparisons.

Differences in the sociality of insect groups are evident when one tries to define social levels; definitions satisfactory for one group may not work for the next. Even within groups, taxonomic units are important. For example, a principal components analysis of bees, using only characteristics relevant to social level, resulted in clusters that were not the social levels (solitary, communal, eusocial, etc.) but were taxonomic groups (Michener, 1974). Thus behavioral features of families or subfamilies overshadowed social information. Clearly history (phylogenetic constraints) is an essential factor in the evolution and origin of behavioral or life history attributes.

PRIMITIVELY EUSOCIAL INSECTS : WHY THEY PLAY A SPECIAL ROLE IN STUDIES OF SOCIAL EVOLUTION

M. J. West-Eberhard

Smithsonian Tropical Research Institute
(Correspondence: Escuela de Biologia, Universidad de Costa Rica, Ciudad Universitaria, Costa Rica, Centroamerica)

One reason sometimes given for the key role of primitively eusocial insects in studies of social evolution is their intermediate level of sociality, considered "transitional" between casteless group life and the occurrance of highly eusocial colonies with morphologically distinct castes. Another is the observability of their relatively small groups, which in the case of primitively eusocial wasps often inhabit unenclosed nests in easily reached places. A third reason, and the one I will emphasize here, is the great variety of their social behaviors. This variation -- both within and between species -- facilitates comparative study and the establishment of generalizations regarding the evolution of social behavior.

Comparative study has long been the mainstay of evolutionary biology. To understand the importance of primitively eusocial insects for evolutionary thought, it is helpful to reflect on the nature of comparative studies and the kinds of hypotheses they test.

COMPARATIVE STUDY AS AN EXPERIMENTAL SCIENCE

Laboratory experimentation involves comparison, of states with and without a particular manipulation. Comparative studies of organisms embody the same principle, except that they start with a character state, and then undertake comparisons designed to reveal the evolutionary/historical manipulations (conditions) that have produced that result under natural selection. Phylogenetic analysis may help determine the direction of change (polarity) between two states in a particular group. But what I will call "causal transition hypotheses" can be profitably examined without reference to phylogeny. A causal transition hypothesis describes the circumstances under which natural selection may have produced a particular kind of character. It attempts to answer the question "in what circumstances does this trait evolve?" Data in support of a causal transition hypothesis for the evolution of a social trait, or favoring a contradictory alternative hypothesis, can come from any social taxon where the trait is seen. It is the formulation and testing of causal-transition hypotheses that can lead to the discovery of cross-phyletic generalizations about social behavior, applying alike to wasps, bees, primates and birds.

Causal transition hypotheses are best examined in variable groups, where the trait of interest is present in some species and absent in others; or where it is present in some individuals (or at some times or places) and not in others of the same species. Comparative study then examines the correlates, or conditions, of presence and absence of the trait. The success of this method depends on the validity of the assumption that the correlation of trait and situation arises from or is maintained by selection, rather than being a vestige of selection on an ancestor which persists regardless of circumstances. Intraspecific comparisons, taking advantage of plasticity in the expression of the trait, are especially informative, because they offer immediate clues about the context in which the trait is employed in place of alternative traits. Primitively eusocial insects are particularly useful in this regard, because adults show a high degree of phenotypic plasticity in social behavior. In the highly eusocial insects, by contrast, pre-adult canalization of adult phenotype in effect restricts the variety of natural experiments that can be observed by comparative science.

AN EXAMPLE: VARIETY AND FLEXIBILITY IN THE SOCIAL BEHAVIOR OF POLISTES

Polistes wasps, like their melliferous counterparts the halictine bees, show a seemingly inexhaustable variety of social behaviors. The competitive reproductive tactics of females include solitary and social nest founding, nest usurpation, surreptitious oviposition, worker behavior, differential egg-eating, mock foraging behavior, and even idleness. Aggressiveness can take the form of ritualized display, overt fighting, and even mortal battle. Pattern in the occurrance of these traits conforms to the predictions of causal transition hypotheses applicable to social insects in general (e.g., see [1] on the occurrance of social parasitism; and [2] on the evolution of ritualized display). The relation between intraspecific phenotypic flexibility and evolutionary lability and species diversity in social behavior [3] suggests that the two kinds of variability should co-exist within taxa, as they do in Polistes.

1. Wcislo, W.T. 1987. Biol. Rev. 62:515-543.
2. West-Eberhard, M.J. 1979. Proc. Am. Phil. Soc. 123:222-223.
3. West-Eberhard, M.J. 1989. Ann. Rev. Ecol. Syst. 20:249-78.

MATERIAL HANDLING AND THE EVOLUTION OF SPECIALIZATION

Robert L. Jeanne
Department of Entomology
University of Wisconsin
1630 Linden Drive
Madison, Wisconsin 53706
U.S.A.

Much of the specialization that characterizes social wasp colonies is manifested through the collection and handling of the various materials required by the colony. The flow of materials into and through the colony is central to colony success. To understand the evolution of specialization, we need to understand the evolution of how each material is handled.

Within the Vespidae there is a range of behavior with respect to how materials are handled. At one extreme, in the solitary wasps there is no division of labor; the solitary female must carry out unaided and in the proper sequence all the tasks associated with feeding herself and constructing and provisioning a nest. At the other extreme are the swarm-founding wasps. Among workers of Polybia occidentalis, for example, there is a well-defined temporal division of labor accompanied by complete task partitioning. That is, all four kinds of material (pulp, water, prey, nectar) brought in by the foragers (older individuals) are transferred to nest workers (younger individuals) for distribution/utilization in the colony [1,2]. How did this form of division of labor evolve from the solitary condition? For an answer we must look to the intermediate stages.

In the primitively eusocial wasps and in the vespines we encounter a variety of intermediate states in which the partitioning of materials-handling tasks into foraging and utilization at the nest is less complete. Among the Stenogastrinae and the genera Ropalidia, Mischocyttarus, and Polistes prey and nectar loads are usually shared by the forager with one or more adults on the nest; less frequently, the forager feeds the entire load to larvae herself or gives the entire load to a nestmate [2]. The same is true in the vespines, although in some species prey loads are usually passed entire to a nest wasp [3]. The situation is quite different for water and pulp. These two materials are typically utilized at the nest by the forager that brings them in, even in the vespines [2,3]. In other words, there is little if any task partitioning with respect to these materials in these species. Surprisingly, when sharing of pulp does occur it is in the primitively eusocial species and not in the vespines, even though the latter have much larger colonies and are socially more advanced.

These patterns are difficult to understand in terms of colony-level selection, which ought to favor colony efficiency. This might lead us to expect the degree of specialization to correlate roughly with the level of social organization or colony size. At the very least, we would expect within a given taxon to see a level of specialization with respect to food materials (prey and nectar) that parallels that with respect to nest materials (pulp and water).

A better understanding of the patterns in the primitively eusocial species comes if we interpret them within an individual-selection framework. West-Eberhard has argued that social behavior among primitively eusocial wasps is best understood in terms of selection acting on individuals via their inclusive fitness [4]. By this reasoning, the role a female adopts in the colony constitutes a conditional strategy set by relative reproductive success attainable by different routes: direct reproduction on the parental nest, helping on the parental nest (inclusive fitness), or direct reproduction via founding her own nest [2]. As a female offspring in a queenright colony ages, the likelihood that she will realize any fitness by reproducing directly diminishes relative to the (inclusive) fitness she can realize by helping. Thus a young female 'hedges her bets' and does not engage in risky or costly tasks that would compromise her chances of personal reproduction. As the relative value of the helping option increases, however, the female engages in increasingly 'altruistic' tasks, proceeding as she does so through the normal temporal sequence of worker roles. Thus her own self-interest comes increasingly to converge on that of the queen and her behavior becomes increasingly 'cooperative.'

Each of the materials collected by foragers is, of course, vital to colony success. Because each female realizes her own fitness through the successful reproduction of the colony, each material has a value to each individual. But the relative values may vary according to the reproductive status of the individual, and these differences are reflected in the differing thresholds among individuals for the handling of each material. Prey and nectar are nutrients utilized by adults as well as larvae and have value in enhancing personal fitness. Thus they are valued highly by the queen and high-ranking workers, and consequently they tend to flow upward through the dominance hierarchy. Pulp and water do not directly enhance personal fitness, but are more directly means of investing in the brood in the nest, i.e. inclusive fitness. These materials do not tend to move upward in the hierarchy. An exception to this appears to be the pulp utilized by the queen in some species of primitively eusocial wasps in which the queen forcibly removes the load from the forager and uses it to initiate a new cell in which she subsequently oviposits [5,6].

REFERENCES

1. Jeanne, R.L. 1986. Monitore Zool. Ital. (N.S.), 20(2): 119-133.
2. Jeanne, R.L. In: The Social Biology of Wasps, K.G. Ross and R.W. Matthews, Eds., Cornell University Press, Ithaca, New York. (In press)
3. Akre, R.D., W.B. Garnett, J.F. MacDonald, A. Greene, and P.J. Landolt. 1976. J. Kansas Ent. Soc., 49: 63-84.
4. West-Eberhard, M.J. 1981. In: Natural Selection and Social Behavior. Recent Research and New Theory, R.D. Alexander and D.W. Tinkle, Eds., Chiron, New York, pp. 3-17.
5. Dew, H.E. 1983. Z. Tierpsychol., 61: 127-140.
6. West-Eberhard, M.J. 1969. Misc. Publ. Museum Zool., Univ. Michigan, 140: 1-101.

RELATEDNESS IN PRIMITIVELY EUSOCIAL WASPS

Colin R. Hughes, Joan E. Strassmann and David C. Queller
Department of Ecology and Evolutionary Biology, Rice University,
P. O. Box 1892, Houston, Texas, 77251, U.S.A.

Relatedness has been a primary focus of attention in attempts to explain the evolution of eusociality in hymenoptera since Hamilton proposed his kin selection theory and pointed out that the haplodiploid sex determination system can lead to very high relatedness among female nestmates[1,2]. The haplodiploid hypothesis proposes that eusociality has arisen particularly often in the Hymenoptera because full sisters are related by 3/4 so a female should prefer to help raise full sisters rather than produce her own offspring to which she would be related by 1/2.

Recent advances have made it relatively easy to measure genetic relatedness. We collected females of 14 species of *Polistes* and *Mischocyttarus*, two genera which lack morphological castes. The wasps were kept frozen until they were analyzed using starch gel electrophoresis of proteins; individuals were scored at all polymorphic loci (from 1-6 per species) and these data were used to generate an estimate of relatedness using the method of Queller and Goodnight[3].

Relatedness was generally high among female nestmates; the mean value for all populations was 0.54, very close to the value for parent offspring relatedness[4]. Relatedness values in three species were close to the full sister value of 3/4 and were significantly above the $r = 1/2$ expected between females and their offspring. Full sister relatedness was included in the 95% confidence limits of 8 species, while the confidence intervals of 11 species included the parent-offspring value of 1/2. In only one species was the estimate of relatedness significantly below 1/2. These results suggest that the evolution of sociality in the primitively eusocial wasps has not generally been driven by female relatedness values above that expected between parents and offspring. However, the general expectation of high relatedness among primitively eusocial insects is met and provides a stark contrast to those highly eusocial ants where relatedness can be extremely low[5].

Another prediction of kin selection theory is that the degree of helping should be positively correlated with relatedness between the altruists and beneficiaries. The nest founding stage in *Polistes* provides an ideal arena for investigating this prediction: in some species most females start their own nests in the spring while in others many females function as subordinate helpers to a relative. We tested the prediction that the frequency of spring helpers is positively correlated with relatedness among the reproductive female nestmates in the autumn (because these are the females that may join together the following spring[6]). By considering only an ecologically similar set of congeners, 8 populations of*Polistes* in Texas, we minimized the chance that other differences among species would obscure the pattern of association. In these populations the frequency of subordinate helpers in the population varied from 1.7% in *P. metricus* to 77.6% in *P. annularis,* though four populations had rather similar frequencies near 50%.

The prediction of a positive correlation between frequency of helping and degree of relatedness was not met. The actual correlation coefficient was negative (Pearsons $r = -0.58$) though it was not significantly different from zero. A reasonable explanation for this result is that the effect on costs and benefits of differences in the ecology of the seven species are more important than relatedness in explaining the patterns of variation in helping.

The failure of these predictions might be explained if females could identify and choose to aid their closest relatives within their colony but we found no evidence for such discrimination. *Polistes annularis* foundresses emerging from hibernation begin new colonies in the spring with some of their natal nestmates. However, their relatedness to the natal nestmates they joined was not higher than relatedness to the natal nestmates that they did not join[7].

Primitively eusocial wasps are generally quite closely related, as expected. But they are often not more closely related to their colony mates than they would be to their own offspring. Moreover, differences between species in within-colony relatedness do not explain differences in the extent of helping. Thus while relatedness remains central to explaining primitively eusocial wasp societies, knowledge about variation in relatedness alone is insufficient to explain variation in behavior.

1. Hamilton, W. D. 1964. J. Theor. Biol. 7: 1-52.
2. Hamilton, W. D. 1972. Ann. Rev. Ecol. Syst. 3, 193-232.
3. Queller, D. C. and K. F. Goodnight. 1989. Evolution, 43, 258-275.
4. Strassmann, J. E., D. C. Queller, C. R. Hughes, S. Turillazzi, R. Cervo, S. K. Davis, and K. F. Goodnight. 1989. Nature 342: 268-270.
5. Pamilo, P. 1982. Heredity 48: 95-106.
6. West-Eberhard, M. J. 1969. Misc. Publ. Mus. Zool. Univ. Mich. 140: 1-101.
7. D. C. Queller, C. R. Hughes and J. E. Strassmann. submitted.

LESSON FROM AUSTRALIAN, SOUTHEAST ASIAN AND JAPANESE *ROPALIDIA*

Yosiaki Itô

Laboratory of Applied Entomology and Nematology, Nagoya University, Chikusa, Nagoya 464-01 Japan

Ropalidia is a unique genus in eusocial wasps, because it contains both of 2 levels of eusociality, with and without morphologically distinct castes. In the largest subgenus Icariola, 7 species so far studied (including 2 Indian species) have no distinct castes, and establish their colonies by pleometrotic independent founding. On the other hand, at least 2 species of the subgenus Icarielia, R. montana and R. romandi found their nests (with envelope) by swarming, and have many morphologically distinct queens. In addition, there is the third group. R. (Anthreneida) sumatorae has no distinct castes and makes non-enveloped nests in cavities, but colonies had many egg-laying, inseminated females (Yamane et al.,1983). Large colonies of R. (Icariola) socialistica (Hook and Evans, 1986 and my personal obs.), which construct their non-enveloped nests in cabities and have no distinct castes, may also have many egg-layers (see also description on R. bambusae in Spradbery and Kojima, 1989).

In Icariola, dominance interactions and numbers of functional queens are variable. In R. fasciata, some colonies are polygynous (Itô, 1987a), and first brood females can leave reproductive progeny (Itô and Yamane, 1985), but in other colonies a single foundress ate eggs laid by subordinates (Iwahashi and Yamane, 1989). Progeny females sometimes supersede foundresses (Turillazzi and Turillazzi,1985; Iwahashi and Yamane, 1989). In R. revolutionalis, only one foundress per colony can oviposit despite lack of dominance interactions during pre-emergence period, but post-emergence colonies have multiple egg-layers (Itô, 1987b). In R. variegata jacobsoni, polygynous pre-emergence colonies change to monogyny after emergence of progeny, but large post-emergence colonies again become polygyny (Yamane, 1986). In R. cristata, only one female in each colony is inseminated and had developed ovaries (Spradbery and Kojima, 1989).

Although Carpenter (1989) reached an opinion that long-term polygyny was always derived from monogyny and did not evolve from rudimentary-caste-polygyny, the data for Ropalidia suggest the latter possibility. Large colonies of independent founding Ropalidia are polygynous, and they often reconstruct nests by foundress-progeny groups,the situation similar to swarm-founding.

This facultative polygyny may easily change to swarm-founding permanent polygyny in R. sumatorae before evolution of morphological castes. Lack of monogynous species with morphological castes in Ropalidia suggests that Icarielia type polygyny has evolved from social sytems like R. sumatorae.

Species	Stage	No. found-resses	% pleo-metro-sis	Mean No. of egg-layers	% of polyg-yny	Aggres-sion	Swarm-ing?	Ref.
(Icariola)								
fasciata	pre-	2.5(614)	60	2(4)	75(4)	M-FS	no	4,5
	post-	-	-	3(9)	57(23)	W-FS	-	
revolu-tionalis	pre-	3.4(17)	88	1(7)	0(7)	R	no	6
	post-	-	-	3(9)	63(9)	FS	-	
sp.nr. variegata	pre-	3(6)	83	1.7(3)	66(3)	no	no	3,4
	post-	-	-	1.3(3)	33(3)	FS	-	
gregaria	pre-	6(3)	100	5(1)		FS	no	3,4
	post-	-	-	many		FS	-	
variegata jacobsoni	pre-	2.3(10)	90	1.5(2)	50(2)	M	no	12
	post-	-	-	1(3)	0(3)	M	-	
	post-*	-	-	3.7(3)	100(3)	FS	-	
socialistica				many?	100?	?	yes?	
(Anthreneida)								
sumatorae				329(2)	100(2)	?	yes	SY
(Icarielia)								
montana				30(2)	100(2)	?	yes	13
romandi				many(3)	100(3)	?	yes	SY

R:rare,M:mediocore,FS:frequent/strong. *later post-emergence stage.
SY:S. Yamane, personal communication.

References

1.Carpenter, J.M.1989. Cladistics, 5:131-144.
2.Hook, A.W. and H.E.Evans 1982. J. Aust. Ent. Soc., 21:271-275.
3.Itô,Y.1986a. J. Aust. Ent. Soc., 25:309-314.
4.Itô,Y.1986b.Social Evolution in Wasps. Tokaidaigaku Syuppankai.
5.Itô,Y. 1987a. In:Animal Societies:Theories and Facts,Itô et al. Eds., Japan Sci. Soc. Press, Tokyo, pp.17-34.
6.Itô,Y. 1987b. J. Etholoigy, 5:115-124.
7.Itô,Y. and Sk. Yamane 1985.Insectes Sociaux, 32:403-410.
8.Iwahashi,O. and S.Yamane 1989. Societies of Ropalidia. Tôkaidaigaku Syuppankai, Tokyo.
9.Richards, O.W.1978. Aust.J.Zool.Suppl.Ser.,No.61:1-132.
10.Spradbery, J. P. and J. Kojima 1989. Kontyû, 57:632-653
11.Turillazzi,S.and C.M.Turillazzi 1985. Monit. Zool.Ital.,19:219-230.
12.Yamane, S. 1986. Monit. Zool.Ital., 20:135-161.
13.Yamane,S.,J.Kojima and Sk.Yamane 1983. Insectes Sociaux, 30:416-422.

EVOLUTION OF EUSOCIALITY : LESSONS FORM THE MECHANISM OF NESTMATE DISCRIMINATION IN THE PRIMITIVELY EUSOCIAL WASP *ROPALIDIA MARGINATA*

ARUN VENKATARAMAN AND RAGHAVENDRA GADAGKAR
Centre for Ecological Sciences, Indian Institute of Science, Bangalore-560 012, INDIA.

The multiple origins of eusociality in the Hymenoptera have been ascribed to the genetic asymmetry created by haplodiploidy such that full-sisters are more closely related to each other than a mother is to her daughters (1). However, many Hymenopteran queens are known to mate multiply and simultaneously use sperm from different males and thus produce different patrilines of daughters. This destroys the asymmetry created by haplodiploidy unless full- and half-sisters can be discriminated and preferential aid given to the former (2).

Female wasps of the primitively eusocial wasp *Ropalidia marginata* (Lep.) (Hymenoptera : Vespidae) can discriminate nestmates from non-nestmates outside the context of their nest. In order to do so however, it is essential that both the discriminating animals and the discriminated animals have been exposed to a fragment of their nests and a subset of their nestmates. This suggest that both labels and templates used in recognition are acquired by the wasps from such common sources outside their bodies as the nest or their nestmates. It appears unlikely therefore that full- and half-sisters or other genetic differences within a colony will be recognised. One implication of these results is that kin recognition is unlikely to restore the genetic asymmetries created by haplodiploidy but broken down by multiple mating (3) and serial polygyny (4) in this species. Eusociality is thus unlikely to be selected for by haplodiploidy alone in *R.marginata* (5). In general, we suggest that the multiple origins of eusociality in the Hymenoptera should be ascribed to haplodiploidy with caution (6).

A second consequence of our results is that acceptance and tolerance of conspecifics are likely to depend not merely on genetic relatedness but on other factors too. Here we describe results of experiments designed to explore this possibility. Pairs of nests were collected from localities separated from each other by at least 10 km to ensure that wasps eclosing from the same nest would be more closely related to each other than they would be to the wasps eclosing from the other nests in the pair. One nest was cut into three parts. One part was fixed in a cage and allowed to be regenerated by the adults present at the time of collection. All animals eclosing on the second fragment of the nest were allowed to remain on that fragment for a period of 10-20 days and were thus exposed to a fragment of their nests and a subset of their nestmates. These constituted 'Exposed' relatives of the animals on the regenerated nests. From the third fragment animals were removed from their pupal cases prior to their natural eclosion and were thus not exposed to their natal nests and nestmates. These constituted 'Isolated' relatives of the animals in the regenerated nest. The second nest in each pair was cut into two halves, which were used to obtain 'Exposed' non-relatives and 'Isolated' non-relatives of the animals on the regenerated nest. Four wasps (which were at least one month old) of each of the four categories namely 'Exposed' relatives, 'Isolated' relatives, 'Exposed' non-relatives and 'Isolated' non-relatives were released into the cage containing the regenerated nest. Behavioural interactions between the nest animals and the introduced animals were recorded for three weeks after introduction.

In the three repetitions of such an experiment, a total of 46 animals were released. However, no animal irrespective of category, was accepted onto the colonies. But by means of tolerance indices used for quantifying tolerance of the nest animals towards the foreign animals, we found that, in the vicinity of the nest, nest animals were significantly more tolerant towards 'Exposed' relatives than 'Exposed' non-relatives. Away from the nest however, no category of introduced animals was treated more tolerantly than any other category by the nest animals. In fact, as if in accordance with the latter result, a satellite nest initiated by four foreign animals was joined by three nest animals. These findings seem to indicate that complex context-dependent rules govern tolerance and acceptance of foreign animals and that genetically unrelated individuals can sometimes cooperatively found nests.

Because genetic asymmetries created by haplodiploidy are unlikely to be sufficient to explain the presence of worker behaviour in *R.marginata*, we have begun to explore models of mutualistic interactions that might facilitate social evolution (5). The observations that factors other than genetic relatedness modulate acceptance of and tolerance towards foreign animals and that genetically unrelated individuals can cooperatively found new nests are clearly conducive to the development of mutualistic interactions that may facilitate the evolution of eusociality (7)

1. Hamilton, W.D. 1964. J.Theor.Biol., 7:1-52.
2. Gadagkar,R. 1985. Proc. Indian Acad.Sci. (Anim.Sci.), 94:587-621.
3. Muralidharan, K., Shaila, M.S. and R.Gadagkar. 1986. J.Genet., 65:153-158.
4. Gadagkar, R., Chandrashekara, K., Swarnalatha Chandran and Seetha Bhagavan. *this volume.*
5. Gadagkar, R. In : Social Biology of Wasps, K.G.Ross and R.W.Matthews, Eds., Cornell University Press, New York, *in press.*
6. Venkataraman,A.B., Swarnalatha, V.B., Nair,P. and R.Gadagkar. 1988. Behav.Ecol.Sociobiol., 23:271-279.
7. Lin, N. and Michener, C.D. 1972. Q.Rev.Biol., 47:131-159.

EVOLUTION OF EUSOCIALITY : LESSONS FROM SOCIAL ORGANIZATION IN *ROPALIDIA MARGINATA* (LEP.) (HYMENOPTERA : VESPIDAE)

K.CHANDRASHEKARA AND RAGHAVENDRA GADAGKAR
Centre for Ecological Sciences, Indian Institute of Science, Bangalore-560 012, INDIA.

In the primitively eusocial wasp *Ropalidia marginata* which lacks morphological caste differentiation, all adult females on a nest can be assigned to one of the three behavioural castes namely, Sitters, Fighters and Foragers (1). Here we explore the manner in which social organization and division of labour are achieved through such a behavioural caste differentiation. To do this, we examine the behavioural, morphological and anatomical correlates of behavioural castes in 12 post-emergence colonies of *R. marginata* from Bangalore and Mysore, India. Our sampling methodology and the variables used in the study are described elsewhere (1,2). Analysis of time-activity budgets of wasps as described before (1) showed the presence of three behavioural castes namely Sitters, Fighters and Foragers in all colonies. Logistic regression analysis revealed that the risky task of foraging for food is performed largely by Foragers. Having very poorly developed ovaries, Foragers have the least chance of becoming queens in the future. On the other hand intra-nidal tasks such as feeding larvae and nest building are shared by Sitters and Fighters. Fighters also showed dominance signficantly more often than either Sitters or Foragers. Both Sitters and Fighters have equally well developed ovaries and can be treated as hopeful queens. No morphological differences were seen between the three behavioural castes. A complex network of dominance-subordinate relationships rather than a simple linear hierarchy was evident in most colonies. The queens were not necessarily the most dominant animals in their colonies; in fact the queens did not show any dominance behaviour in three colonies. The frequency of dominance behaviour was correlated with the frequency of such behaviours as Feed larva, Extend walls of cells, and Build new cells, suggesting that individuals showing dominance behaviour also perform several intra-nidal tasks.

In all 12 colonies only one animal was ever seen laying eggs and she was thus designated as the queen. Queens in 11 out of 12 colonies were Sitters. A particularly striking result was that in five out of 11 colonies the queens were unmated. This was true inspite of a mated female being present in two of these colonies. Behavioural caste differentiation into Sitters, Fighters and Foragers was seen in all colonies whether or not their queens were mated. In addition, we failed to detect any quantitative differences between the two types of colonies inspite of comparing them with respect to several variables using different statistical methods. Our results suggest that mating is not essential for the development of a female's ovaries and that unmated females can become queens, prevent nestmates from laying eggs and maintain normal social organization.

R. marginata is characterized by perennial indeterminate colony cycles and undergoes frequent queen replacements. This provides opportunities for other individuals to become queens. Such opportunities for future reproduction are expected to result in behaviours that are moulded by selection both at the colony level and at the individual level. Indeed, we suggest that behavioural caste differentiation into Sitters, Fighters and Foragers is a manifestation of selection acting at both kin or colony level and at the individual level. Foragers who appear to have the least chance of reproduction take up the risky task of foraging and enhance colony fitness and thereby also maximize their own inclusive fitness. Sitters and Fighters on the other hand share intra-nidal tasks and thereby enhance colony fitness and their own inclusive fitness. But by being dominant, Fighters appear to increase their chances of becoming queens in the future. By avoiding foraging and dominant acts, Sitters appear to conserve energy and thus pursue an alternate strategy of enhancing their chances of becoming queens in the future. Most individuals thus try to maximize their own chances of future reproduction without hurting the colony's interests too much. When an opportunity for direct reproduction arises due to the death or ageing of a queen however, we might expect individuals to sacrifice the interests of the colony to become queens and gain direct fitness. That such a scenario is possible is suggested by the fact that unmated females can become queens even in the presence of mated females who may make better queens for the colony.

Worker-brood relatedness is expected to be low in *R. marginata* both because of multiple mating by the queen (3) and frequent queen replacements (4). Owing to these reasons it has been argued (5) that indirect fitness gained by workers is unlikely to maintain group life in this species and that mutualistic interactions among hopeful reproductives are more important for the evolution of group life. Such an arguement is consistent with results presented here which show that division of labour and social organization are achieved through behavioural caste differentiation and not, as in many species studied, through a dominance hierarchy led by a despotic queen suppressing all her nestmates into worker roles.

1. Gadagkar, R. and N.V. Joshi. 1983. *Anim. Behav.* 31:26-31.
2. Chandrashekara, K. and R. Gadagkar. 1987. Natl. Symp. Social Insects, Univ. Agril. Sciences, Bangalore, India.
3. Muralidharan, K., Shaila, M. S. and R. Gadagkar. 1986. *J. Genet.* 65:153-158.
4. Gadagkar R., Chandrashekara, K., Swarnalatha Chandran, and Seetha Bhagavan. *this volume.*
5. Gadagkar, R. In: Social Biology of Wasps, K.G. Ross, and R.W. Matthews, Eds., Cornell University Press, New York. *in Press.*

ANISCHNOGASTER: FROM LITTLE TO LESS SOCIALITY (STENOGASTRINAE) (HYMENOPTERA, VESPIDAE)

M.H. Hansell* and S. Turillazzi**

* Dept. of Zoology, Glasgow University, Glasgow, Scotland, U.K.
** Dip. di Biologia Animale e Genetica, Universita di Firenze, Italy

The current state of our knowledge of the colony biology and social organisation of wasps of the subfamily Stenogastrinae comes mainly from observations on three of the oriental genera (*Parischnogaster*, *Liostenogaster*, *Eustenogaster*) [cfr. 7,2,1]. Only a little information on social life is available for the fourth oriental genus (*Metischnogaster*) [3,4,9] and for the two genera living exclusively in New Guinea (*Stenogaster* and *Anischnogaster*) [5,6,9]. With the aim of completing, at any rate in outline, our picture of the level of social organisation and the nature of the colony biology for the subfamily as a whole, we studied colonies of two species of *Anischnogaster* in the region around Madang (Madang Province, Papua New Guinea) in October-November of 1989. *A.iridipennis* (or a new species very similar to it) builds nests with up to 20 cells scattered or clustered (depending from suspension length) along rootlets or plant stems. *A.laticeps* has a very compact nest architecture with a limited number of cells which open ventrally, all nearly on the same level. Nest material is entirely composed of very fine plant fragments. In both species colony size ranges from 1 to 2 females. Only two out of eight active nests of *A.iridipennis* were seen with two females and then for only 17 per cent of the study period; while only one of 25 collected nests of *A.laticeps* had two rather than a single female. The level of social organisation for both species is thus very modest, however observations on a two-female colony of *A.iridipennis* showed that the relationship between the two wasps was fairly typical of that know for other stenogastrine species: social interactions were simple but effective, and division of labour was extreme in the period (19 days) of nest sharing with the oldest female remaining almost always on the nest and the other being most of the time absent in foraging and pulp collecting trips. For *A.laticeps* the size of the nest and brood suggest a nest designed for care by one female. Dissection of the wasps in the two colonies with two females showed that one of the pair was probably just newly emerged with undeveloped ovaries and empty spermatheca.

The two egg laying sequences observed in *A.iridipennis* show that, unlike previously studied stenogastrinae, pap secretion after the placing of the egg were absent in both and that the secretion before the laying of the egg was absent in one. No egg deposition was observed in *A.laticeps* but none of the eggs (or larvae) removed from collected nests had any trace of pap secretion on them, so production for pap secretion is as yet unrecorded in this species. Both species, however, still possess a large Dufour's gland, the gland responsible for the production of the secretion.

Social life in *Anischnogaster* seems to be an extreme low for the subfamily and can be explained as an evolutionary trend towards a less social life in this genus. This essentially completes our picture for the subfamily which is of small colony sizes but sharply differentiated behavioural roles whenever more than one female occurs on the nest. The contrast between this and the vastly more complex societies to be found among the Polistinae and Vespinae shows again that some feature common to all Stenogastrinae must account for their failure to achieve anything but very small colony size in any of their species.

References

1. Carpenter, J.M. 1988. The phylogenetic system of the Stenogastrinae (Hymenoptera: Vespidae). J.N.Y. Entomol. Soc. 96: 140-175.
2. Hansell, M.H. 1987. Nest building as a facilitating and limiting factor in the evolution of eusociality in the Hymenoptera. Oxford Surveys in Evolutionary Biology. P.H. Harvey and L. Partridge (edits). 4: 155-181.
3. Pagden, H.T. 1958. Some Malayan social wasps. Malay Nat. J. 12: 131-148.
4. Pagden, H.T. 1962. More about *Stenogaster*. Maly Nat. J. 16: 95-102.
5. Spradbery, J.P. 1975. The biology of *Stenogaster concinna* Van der Vecht with comments on the phylogeny of the Stenoastrinae (Hymenoptera Vespidae). J. Austr. Entomol. Soc. 14: 309-318.
6. Spradbery, J.P. 1989. The nesting of *Anischnogaster iridipennis* (Smith) (Hymenoptera: Vespidae) in New Guinea. J. Austr. Entomol. Soc. 28: 225-228.
7. Turillazzi, S. 1986. Les Stenogastrinae: un groupe cle pour l'etude de l'evolution du comportement social chez les guepes. Actes coll. Insectes Soc. 3: 7-32.
8. Van der Vecht, J. 1972. A review of the new genus *Anischnogaster* in the Papuan Region (Hymenoptera Vespidae). Zool. Meded. Leiden 47: 240-256.
9. Van der Vecht, J. 1977. Studies of oriental Stenogastrinae (Hymenoptera: Vespidae). Tijschrift Entomol. 120: 55-75.

PRIMITIVE EUSOCIALITY : COMPARISONS BETWEEN HYMENOPTERA AND VERTEBRATES

H. Jane Brockmann
Department of Zoology, University of Florida, Gainesville, FL 32611 USA

Theories on the evolution of insect eusociality have developed in some isolation from theories on the evolution of sociality in other organisms. Primitively eusocial groups are made up of adults (usually overlapping generations) who cooperatively rear young that are not direct descendants. Reproductive dominance, with one individual doing most of the breeding (egg laying), is maintained through dominance or other asymmetries. Defined in this way, there are many species of birds and mammals that show primitive eusociality or cooperative breeding.

Explanations for the evolution of eusociality are somewhat different from those used to explain cooperative breeding. In particular, relatively little attention has been paid to the ecological constraints hypothesis. Cooperatively breeding vertebrates are thought to be living at the maximum possible population size for the available habitat ("saturated"), with intense competition among conspecifics for breeding opportunities. Among those studying helpers-at-the-nest (birds) or den (mammals), it is generally agreed that delayed breeding occurs when the gain (in inclusive fitness) from helping is greater than the gain from independent breeding. This typically occurs when there is some constraint which either prevents some individuals from attaining breeding status, such as a shortage of territories, or which raises the costs of independent breeding to prohibitive levels, such as when one pair cannot bring in adequate food for their young or when more than two individuals are required to defend resources or ward off predators. When the ecological requirements of a species are specialized to the extent that they cannot find suitable marginal habitats, or if the quality of the environment deteriorates over the season, or if there are marked unpredictable changes in the quality of the environment for initiating breeding, then younger individuals are at a disadvantage relative to established breeders. This can favor helping over dispersal and independent breeding.

Cooperative breeding in primitively eusocial Hymenoptera may be viewed in much the same way. Social insects may also be living at saturated population levels, with few breeding opportunities. This has not been explored in detail, but many argue that social competition is intense. What are the constraints on independent nesting? Are there correlations between the degree of difficulty in becoming established as a breeder and the frequency of eusociality? If there is seasonal deterioration of the environment for independent breeding, then it means that the first individuals to emerge in the season may gain a considerable advantage, and thus overwintering conditions are crucial. This would be particularly likely in areas where there is a short nesting season. These and other comparisons will be made, showing the similarities and differences between hymenopteran and vertebrate cooperative breeding.

SOCIAL ORGANISATION IN LABORATORY COLONIES OF *ROPALIDIA MARGINATA*

SWARNALATHA CHANDRAN AND RAGHAVENDRA GADAGKAR
Centre for Ecological Sciences, Indian Institute of Science, Bangalore-560 012, INDIA.

Ropalidia marginata (Lep.) (Hymenoptera : Vespidae) is a primitively eusocial wasp which lacks morphological caste differentiation. By constructing time activity budgets for individually identified wasps and subjecting these to multivariate statistical analysis, we have shown that adult female wasps can be classified into one of the three behavioural castes namely Sitters, Fighters and Foragers (1). In recent times we have often found it necessary to use laboratory colonies for a variety of experiments. It is important therefore to see if social organisation of the laboratory colonies is similar to that in natural colonies. This report describes the results of 306 hrs of observations of 2 laboratory colonies that were established by marking and releasing all adults present on two natural colonies.

Analysis of time activity budgets of wasps in such laboratory cages as described else where (1) showed that three behavioural castes corresponding to the Sitters, Fighters and Foragers of natural colonies were present in both the colonies. An interesting difference however was that a fourth cluster of wasps, unlike anything seen in natural colonies, was present in both colonies. The wasps in this cluster spent almost all their time away from their nest and appeared to be completely excluded from the social organisation on the nests. Several wasps leave natural colonies from time to time and we suspect that wasps in the laboratory cages which would have left if they could, constituted this cluster. A second difference concerned the behaviour of feeding larvae. In natural colonies the foragers that bring food often pass it on to others on the nest and seldom feed the larvae themselves (2). This probably ensures efficient division of labour as the foragers are free to go back and bring more food. In contrast, foragers in laboratory cages that brought food often fed the larvae themselves before they went back to bring more food. They appeared to have learnt that the food in the cage will always be available. Apart from such differences which were obviously in response to laboratory conditions, we found no significant differences between natural and laboratory colonies in social organisation. These studies thus demonstrate the ability of the wasps to adjust to laboratory conditions and also justify the use of laboratory colonies where necessary.

1. Gadagkar, R. and N.V.Joshi. 1983. Anim.Behav. 31:26-31.
2. Chandrashekara, K and R.Gadagkar. *this volume*.

THE DETERMINANTS OF DOMINANCE IN A PRIMITIVELY EUSOCIAL WASP

PADMINI NAIR, PARTHIBA BOSE AND RAGHAVENDRA GADAGKAR
Centre for Ecological Sciences, Indian Institute of Science, Bangalore-560 012, INDIA.

In many species of primitively eusocial wasps the position of an animal in the dominance hierarchy determines its role in the colony. *Polistes gallicus*, in which dominance hierarchies play an important role in social organisation, is the only species where an attempt has been made to study the determinants of dominance (1). Although an animal's position in the dominance hierarchy has been postulated to be less important in *Ropalidia marginata* (2), dominance hierarchies do exist and dominant-subordinate interactions are quite common (3). It may hence be of interest to study the determinants of dominance in this species. We have therefore begun to study the effects of body size, age, social experience and other factors in making an animal dominant or subordinate. We have maintained pairs of female *R.marginata* in laboratory cages and made behavioural observations to identify the dominant member and the egg-layer of each pair.

The following conclusions emerge from the first set of such experiments :

1. There is a positive correlation between behavioural dominance and egg laying ability. In 12 out of 14 experiments, the wasp which was dominant laid eggs. ($p = 0.012$, Sign test).

2. Body size is unlikely to be a determinant of dominance because in only 8 out of 17 experiments the larger animal was dominant.

3. Similarily body size is unlikely to influence an animal's ability to become an egg-layer because in only 6 out of 14 experiments the larger of the two wasps became the egg-layer.

4. The age of an animal may influence its chances of becoming dominant because in 16 out of 22 experiments the older of the two animals became dominant. ($p = 0.05$, Sign test).

1. Roseler, P.-F., Roseler, I., Strambi, A. and R.Augier. 1984. Behav. Ecol. Sociobiol., 15:133-142.
2. Chandrashekara, K. and R.Gadagkar. *this volume.*
3. Premnath, S., Chandrashekara, K., Chandran, S. and R.Gadagkar. *this volume*

CONSTRUCTING DOMINANCE HIERARCHIES IN A PRIMITIVELY EUSOCIAL WASP

SUDHA PREMNATH, K.CHANDRASHEKARA, SWARNALATHA CHANDRAN AND RAGHAVENDRA GADAGKAR
Centre for Ecological Sciences, Indian Institute of Science, Bangalore-560 012, INDIA.

It is well known that dominance-subordinate interactions form an important part of social organisation in primitively eusocial wasps (1). In many studies of social insects, the rank of an animal is based merely on the frequency with which it shows dominance behaviour. This is misleading because a high frequency of dominance behaviour may be accompanied by an equally high frequency of subordinate behaviour. Moreover showing dominance over a high ranking individual should be weighed differently from showing dominance over a low ranking individual. For these reasons, we have used a modified version of the index of fighting success that was developed for red deer to solve similar problems (2). Thus we compute an index of dominance for each animal in a colony of n individuals as:

$$\frac{\sum_{i=1}^{n} B_i + \sum_{j=1}^{m} \sum_{i=1}^{n} b_{ji} + 1}{\sum_{i=1}^{n} L_i + \sum_{j=1}^{p} \sum_{i=1}^{n} l_{ji} + 1}$$

where ΣB_i measures the rate at which the subject shows dominance behaviour towards colony members and Σb_{ji} measures the sum of the rates at which all animals dominated by the subject in turn show dominance behaviour towards colony members. 1 to m are thus the individuals towards whom the subject shows dominance. Similarly ΣL_i measures the rate at which the subject shows subordinate behaviour towards colony members. Σl_{ji} measures the sum of the rates at which those animals towards whom the subject show subordinate behaviour in turn show subordinate behaviour towards colony members. 1 to p are thus the individuals towards whom the subject shows subordinate behaviour. Indices of dominance calculated in this fashion permit the construction of a simple linear hierarchy with relatively few ties. Animals which do not participate in dominance-subordinate interactions are of course all tied with a dominance index of 1.0.

1. Pardi, L. 1948. Physiol.Zool., 21:1-13.
2 Clutton-Brock, T.H., Albon, S.D., Gibson, R.M. and Guinness, F.E. 1979. Anim.Behav., 27:211-225.

PERENNIAL INDETERMINATE COLONY CYCLE IN A PRIMITIVELY EUSOCIAL WASP

K. CHANDRASHEKARA, SEETHA BHAGAVAN, SWARNALATHA CHANDRAN, PADMINI NAIR AND RAGHAVENDRA GADAGKAR
Centre for Ecological Sciences, Indian Institute of Science, Bangalore-560 012, INDIA.

The colony cycle of primitively eusocial wasps consists of three phases: the pre-emergence phase, the post-emergence phase and the declining phase. Most species of wasps in temperate regions follow a seasonal colony cycle. Being initiated synchronously in Spring by overwintered females, colonies grow through summer and are abandoned in Fall, after producing reproductives. Females produced in Fall mate and overwinter while the remaining individuals, including the males, die (1). In the tropics, however, colonies are aseasonal and may be initiated throughout the year. Colony cycles in most species, either temperate or tropical, may thus be termed determinate since they are abandoned after a fixed time after initiation.

Ropalidia marginata follows a colony cycle which often encompasses multiple repeats of a typical determinate colony cycle (2). We therefore call it an indeterminate colony cycle. A major portion of the brood and cells are destroyed and a large fraction of adults leave the nest during the declining phase of each unit of the colony cycle but a small number of females may stay back on the nest and begin a new unit of the cycle. This leads to two interesting questions: why is there a decline if the colony is to continue? and why is the decline not complete as in the determinate colony cycle. We propose two alternate hypotheses. One is that the colony cycle is a response to predation by *Vespa tropica*. It may be adaptive to issue 'swarms' of dispersing wasps periodically to found new nests, before all is lost to the predator. This hypothesis predicts that queen replacements need not necessarily coincide with the beginning of every unit of the colony cycle. The second hypothesis is that the queen produces reproductives and dies at the end of each unit of the colony cycles. One of her daughters may however stay on and use her natal nest to produce her brood. This hypothesis predicts that queen replacements should necessarily coincide with the beginning of every unit of the colony cycle. Our present results are incapable of distinguishing between these hypotheses. In some colonies queen replacements always coincide with new units of the colony cycle but in others, this is not so. Further studies of such perennial indeterminate colony cycles are bound to be rewarding.

1. Jeanne, R. L. 1980. Ann.Rev.Entomol. 25:371-396.
2. Gadagkar, R. In: Social Biology of Wasps, K.G.Ross and R.W. Matthews, Eds., Cornell University press, New York, *in press*.

MT-DNA LENGTH POLYMORPHISM IN SOCIAL WASPS (VESPINAE)

J. Schmitz and R.F.A. Moritz
Bayerische Landesanstalt für Bienenzucht Erlangen, Friedrich-Alexander-Universität Erlangen-Nürnberg, Burgbergstraße 70,
D-8520 Erlangen (Federal Republic of Germany)

The genetic relationship of six European social wasps (*Vespula vulgaris*, *Vespula germanica*, *Vespula rufa*, *Dolichovespula saxonica*, *Dolichovespula media*, *Vespa crabro*) was studied by restriction fragment length polymorphisms (RFLP) of mitochondrial DNA. The RFLP method is a useful tool to estimate the relationship between species which are relatively closely related, i.e. between species belonging to the same genus. To estimate the relationship between species which are very divergent, i.e. *V. crabro* and *Paravespula* species, the fragment method is not very accurate and the restriction site method estimates with a higher precision [1]. Nevertheless, our data nicely fit into some of the previously published models of wasp phylogeny but not into others.
The results support the hypothesis of monophyly of the genera *Vespula* and *Dolichovespula* [2]. In contrast to models which devide the genus *Vespula* in *Paravespula* (*P. vulgaris* and *P. germanica*) and *Vespula* (*V. rufa*), we found that *V. germanica* was the closest to *V. rufa*. The position of the genus *Vespa* remained uncertain due to the precision limits of the RFLP technique used in this study. Nevertheless, distances were estimated that resulted in a closer relationschip to *Dolichovespula* than to *Vespula*.

1. Nei, M. 1987. Molecular Evolutionary Genetics. Columbia University Press, New York, 512pp.

2. Carpenter, J.M. 1987. Syst. Ent., 12: 413-431.

Symposium 3

ANT - PLANT ASSOCIATIONS

Organizers : **A. RAMAN, India,**
K.N. GANESHAIAH, India

THE CHOICE OF THE NESTING SITE BY *OECOPHYLLA LONGINODA* : ROLES OF IMPRINTING AND SELECTIVE ATTRACTION TO PLANTS

Dejean A.*° & Djieto C.°

* Laboratoire d'Ethologie et Sociobiologie, URA CNRS 667, Université Paris XIII, F-93430 Villetaneuse, FRANCE.
° Laboratoire de Zoologie, Faculté des Sciences, Yaoundé, ÇAMEROON.

The *Oecophylla* build their nest from tree leaves which they bind together by means of the silk secreted by last stage larvae which they manipulate like shuttles. Fields observations have enabled to determine that in a given locality, the *Oecophylla* prefentially occupy one type of plant and that this "preference" varies from one locality to the other. We have supposed that the modalities that lead to the choice of a site may depend on many factors of which one may be due to imprinting to the plant from which individuals of the new society are originated (foundation by isolated queen or by colony fission are possible in *Oecophylla*). Jaisson (1980) showed that *Camponotus* bred in contact with a plant which is normally repulsive, are henceforth attracted to this plant under the effect of an imprinting.

MATERIAL AND METHODS

We have tested the attraction of four plants of economic interest on the *Oecophylla* : mango tree (Anacardiaceae), orange tree (Rutaceae), cocoa tree (Sterculiceae) and guava tree (Myrtaceae). For this, we confront sets of workers plus brood comb with choice between two kinds of tubes of roled leaves serving as a refuge. In order to test the influence of an imprinting, workers originating from a plant A have a choice between tubes of leaves from this plant A and tubes of leaves from a plant B. Reciprocally, workers originating from a plant B are confronted to the same choice. This enables a cross statistical test, the six possible cases being considered.

In order to test spontaneous attraction of the different plant species tested on the *Oecophylla*, we used individuals from societies settled on *Cardia aurentica* (Boraginaceae). They are subjected to the same types of tests as described previously. Cross tests being impossible in this case, the results obtained are compared to a random theoretical distribution.

RESULTS

1. Influence of the original surroundings on the choice of a refuge :

In all cases, the workers statistically take refuge in tubes made up of leaves of the same species of plant from which they originate. The differences varying from P>0.01 for the comparison between orange and cocoa trees to P>0.001 for the five other cases. Theses results tend to confirm imprinting to the plant from which the workers are picked.

2. Study of spontaneous attraction :

In all cases studied, we obtained some significant differences (mango tree/orange tree: P> 0.02; orange tree/cocoa tree : P> 0.02; guava tree/cocoa tree : P >0.05) and some highly or very highly significant differences (mango tree/ guava tree : P>10^{-3}; mango tree/ cocoa tree : P> 10^{-5}; orange tree/ guava tree : P> 10^{-5}).

Based on these results, a statistical comparison enables to show that the mango tree is the most attractive plant. It is followed by the orange tree, then guava tree and lastly the cocoa tree.

DISCUSSION

Besides intra- and inter-specific agressivity, the settling of an *Oecophylla* society on a tree depends on two factors. One innate factor linked with selective attraction that different plants exert on the *Oecophylla* and one acquired factor by imprinting by contact with the leaves of the plant that make up the nest, during larval stages and imago period where the workers do not go out from the nest. This explains ecological surveys that show differences in "prefered" plant species from one locality to the other.

We may also suppose that selective attraction tends towards a hierarchical system whereby plants compete with each other to attract ants for their protection. It is interesting to note that the same study carried out on *Tetramorium aculeatum* (nest constructed with leaves and cardboard) leads to an hierarchy between plants which is the opposite of the one encountered in the present study (DEJEAN et al., 1990).

These results besides their interest in applied entomology may be compared to obligatory associations between plants and ants (*Acacia-Pseudomyrmex* in tropical America) or semi-obligatory associations (*Cecropia-Azteca* in America and *Barteria-Pachysima* in Africa) where the plant furnishes shelter and food (nectar, Beltian and Mullerian bodies) in exchange of its protection.

REFERENCES

DEJEAN A., NGOKAM S.& DJIETO C. 1991. Le choix du site de nidification chez *Tetramorium aculeatum* (Formicidae, Myrmicinae). Actes Coll. Ins. Soc., 7, Paris, 1990.

JAISSON P. 1980. Environmental preference induced experimentally in ants (Hymenoptera-Formicidae). Nature, 286, 338-339.

LONG-TERM VARIATION IN A HIGH-ELEVATION TEMPERATE REGION ANT-PLANT MUTUALISM

Dr. David W. Inouye (*) and Dr. Orley R. Taylor, Jr. (**). Rocky Mountain Biological Laboratory, Crested Butte, Colorado, 81224, USA (*Mountain Research Station, University of Colorado; **Department of Entomology, University of Kansas).

Although the classic examples of ant-plant mutualisms came from the tropics, mutualistic ant-plant associations are also common in temperate regions. The temperate associations do not appear to be obligate, the way some of the tropical ones are, and one explanation for this difference could be related to the greater variability in environmental and biological parameters affecting temperate associations. We consider data on variation in flowering from a 16-year study of *Helianthella quinquenervis* (Asteraceae), the aspen sunflower, and the implications that this variation has for the mutualism the sunflower has with ants.

This study was conducted at the Rocky Mountain Biological Laboratory, at an elevation of 2,900 m in the West Elk range of the Colorado Rocky Mountains, USA. Inouye and Taylor (1979, Ecology 60:1-7) documented a mutualistic relationship between *H.quinquenervis* and several species of ants that occur in the same montane meadows. Extrafloral nectar secreted from the involucral bracts of the developing flower heads is rich in amino acids and sugars, and attracts ants (e.g., *Formica obscuripes*) that collect it both day and night. In the process of patrolling the flower heads the ants encounter flies (Tephritidae and Agromyzidae) that are attempting to oviposit in them. The ants interrupt or deter oviposition, and significantly reduce levels of seed predation by fly larvae that develop in the flower heads.

We established a permanent plot in 1974 (450 m^2) and another in 1975 (365 m^2) for annual counts of the number of flower heads produced by *H. quinquenervis* that develop to the stage of ripening seeds. Numbers of flower heads (Fig. 1) have ranged from 3450 (in 1982) to 3 (1985) in the larger plot, and 1235 (in 1984) to 1 (1976) in the smaller plot. These plants are long-lived perennials, may not begin flowering until about 10 or more years of age, and can probably live longer than 50 years (Inouye and Taylor, unpublished). In years when flower numbers are low, the vegetative parts of the plants are still present. The years with very low numbers of flower heads are usually the consequence of a random environmental event, late (2nd or third week in June) hard frosts (e.g., temperatures of -7° C), sometimes accompanied by snow. These temperatures result in damage to or the death of developing flower heads, but do not appear to result in damage to the vegetation. These frost-kills are most common following years with low snowpack, when plant development begins earlier than usual.

The loss of flower heads does not appear to have had severe consequences for the plant populations, which appear to have remained stable over the length of this study. The ants that would normally feed on extrafloral nectar from the flower heads are opportunistic feeders, and do not appear to suffer greatly from the absence of this resource. The flies that normally lay their eggs in the flower heads are perhaps most greatly affected, since they do not appear to have alternate host plants. The loss of flower heads constitutes a reproductive bottleneck for these flies, and the few flowers that survive the frosts are usually heavily damaged by fly larvae. The year following a frost, however, fly populations are usually greatly reduced, while flowering may be two orders or magnitude greater. In such years seed production is often very high, and two years after a frost there is usually a large increase in the number of seedlings (there is no seed dormancy).

The random, yet relatively frequent (3-5 year) occurrence of events such as these late frosts therefore has significant and long-term consequences for the demography of the plant, the ant-plant mutualism, and the fly populations. Without the long-term perspective that provides information on the frequency of these events, and their consequences, we could have a very distorted understanding of these interactions. The fact that these disruptions occur with such frequency has probably selected against the evolution of an obligate relationship between the ants and plants.

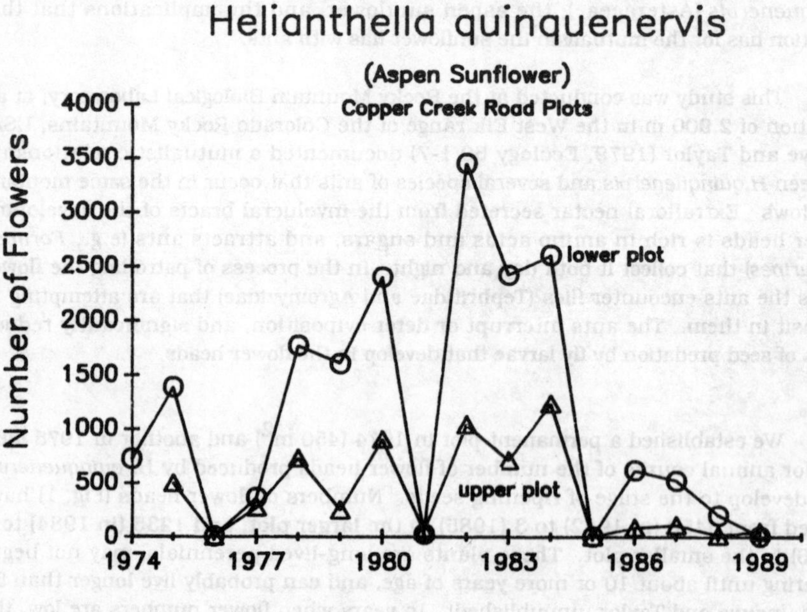

ANT-CROTON MUTUALISM FOR SEED DISPERSAL : SPECIFIC ADAPTATIONS OF THE HOST PLANT AND THE ANTS

Ganeshaiah, K.N., Department of Plant Genetics and Breeding, Agril. College,
Uma Shaanker, R., Department of Crop Physiology, Agril. College, UAS, GKVK.
Veena, T., Department of Veterinary Physiology, Vety. College, UAS Hebbal.
Bangalore 560065

Croton bonplandianum Baill. attracts a set of ant species to the extrafloral nectaries borne on the pedicel of female flowers (Veena et al. 1989). Few of these ant species help in dispersing the seeds from under the plant (Ganeshaiah and Uma Shaanker, 1989). We studied the evolutionary modifications in the structure and behaviour of the ants and plant that specifically strengthen this mutualistic relation.

The nectaries exhibit temporal specificity in their function; they become active only when the fruits mature and when the seeds are splitting. The number of ants attracted to an inflorescence (benefit to the plant) increased non-linearly with increase in sugar content per nectary or number of nectaries per inflorscence (costs to the plant). From this relation, using the graphical method (Smith and Fretwel, 1974; Ganeshaiah and Veena, 1989), we estimated the optimal levels of sugar content and nectaries per inflorescence at which the benefit to cost ratio is maximised. The plants appeared to produce these optimum levels during the periods when the ants are most needed for seed dispersal and during their peak reproductive stages.

After feeding on the nectar, on their way back to the nest, ants carry the seeds from the ground. However, the effective removal of seeds was restricted to a radius of 45-60 cm around the stalk (nectar influence zone). Since croton occurs in dense colonies, this behaviour of the ant serves as an essential evolutionary feed back such that only those genotypes that provide the nectar to the ants are favoured in the course of selection. Interestingly, plants also exhibit unique maturity and branching pattern that facilitates the placement of seeds within this nectar influence zone; the inflorescences are borne on polychasially branching pattern and the fruits that are towards the axial side (parent axis) mature first such that the splitting seeds always face obstruction and fall within the nectar influence zone.

Among several species of nectarivorous ants found on the ground in the croton habitats, few were preferentially attracted by the croton plants. For instance, few such as Pheidole sps which help in seed dispersal were attrracted through the chemical signals of the plant. This was evident by the significantly more number of them attracted to the plant extract with sugar than to sugar alone.

Thus both plants and ants exhibited several specific modifications that strengthen the mutualisic relation between them.

References:

Ganeshaiah K.N. & R. Uma Shaanker. (1988). Evolution of a unique maturity pattern strengthens ant-plant mutualism for seed dispersal in Croton bonplandianum Baill.. Oecologia, 77, 130-135.

Ganeshaiah,K.N. & T.Veena (1988). Ant-Plant mutualism : Selective forces and adaptive changes. In Ananthakrishnan,T.N. & Raman (Eds.), Insect plant interactions. New Delhi: Oxford .

Smith, C.C. and S.D. Fretwell (1974). The optimal balance between size and number of offspring. Am. Nat., 108, 499-506.

Veena, T., A.R.V.Kumar and K.N. Ganeshaiah (1989). Factors affecting ant (Formicidae: Hymenoptera) visits to the extrafloral nectaries of Croton bonplandianum Baill. Proc. Indian. Acad. Sci (Anim. Sci) , 98, 57-64.

MYRMECOCHORY BY *TETRAMORIUM CAESPITUM* IN IRELAND

ANNE O'BRIEN and JOHN BREEN

Thomond College of Education, Plassey, Limerick, Ireland

Three kinds of seeds were found in *Tetramorium caespitum* nests at a site in southeastern Ireland [1], viz., *Thymus praecox*, *Cerastium* spp. (the two most common species) and *Euphrasia* sp. Relative abundance varied according to season, perhaps due to availability (figure 1).

Cerastium and *Thymus* seeds were reported previously from *T. caespitum* nests in Britain [2,3], as were seeds of several heath plants (*Calluna vulgaris*, *Molinia caerulea*, *Ulex minor* and *Agrostis setacea* [3]). This report seems to be the first for *Euphrasia*. *Euphrasia* seeds were found in July both with and without a netted seed covering, presumably as a result of ant activity.

We did not find seeds of *Polygala vulgaris* and *Viola tricolor* ssp. *curtisii* recently reported from *T. caespitum* nests in the Netherlands [4], although these plants occur at the study site.

ACKNOWLEDGEMENTS.
This work was supported by a grant from the British Ecological Society.

REFERENCES

1. Breen, J. and A. O'Brien. 1990. This Congress Proceedings.
2. Donisthorpe, H.St.J.K. 1927. British Ants. Routledge, London.
3. Brian, M.V., J. Hibble and D.J. Stradling. 1965. Journal of Animal Ecology, 34: 545 - 555.
4. Oostermeijer, J.G.B. 1989. Oecologia, Berl. 78: 302 - 31.

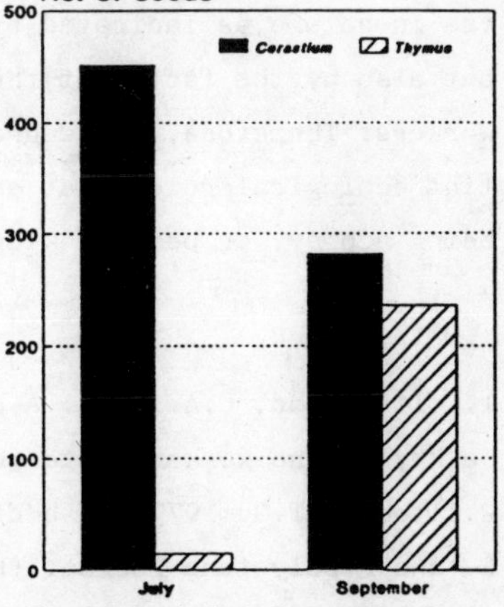

Figure 1. Seeds from *T. caespitum* nests.

A MORPHOLOGICAL COMPARISON BETWEEN PREY AND KLEPTOPARASITIC ANTS (*FORMICA FUSCA* GROUP) OF A CARNIVOROUS PLANT (*PINGUICOLA NEVADENSE*, (LINB) CASPER, 1962). HYMENOPTERA; LENTIBULARIACEA)

ZAMORA,R & TINAUT,A.

Dpto.BIOLOGIA ANIMAL, ECOLOGIA Y GENETICA.

UNIVERSIDAD DE GRANADA.

18071. GRANADA. SPAIN.

During a study about the feeding ecology of **P.nevadense**, an endemic carnivorous plant of Sierra Nevada (Southern Spain), we observed that captured ants appeared on the rosettes closest to the ant nest. We also observed that ants of the same species showed kleptoparasitic behaviour stealing prey captured by the plant, without being captured. To determine the possible differences between the prey and the kleptoparasitic ants, we collected individuals of both types to examine later in the laboratory. Acording to (1), both samples corresponded to **F. fusca** Linné, 1758 and **F. lemani** Bondroit, 1917 in a proportion of 69.86% and 30.13% for the kleptoparasitic ants (n=73) and 60.0% and 24.4% for the captured ants, respectively. In the latter we found 15.5% secretergate individuals, whereas within the 73 kleptoparasitic ants no secretergate individual appeared.

These results show that the high mortality rate of the secretergates in nature,as indicated by (2), would be caused not only by disease but also by the fact that they are easier for predators than the normal workers. Therefore, the degree of vitality seems to determine the distint ecological roles that ants of the same spaecies will play, such as being a prey, or being a kleptoparasite of a carnivorous plant.

1.Collingwood, C.A. 1978. A provisional list of Iberian Formicidae with a key to the worker caste (**Hym. Aculeata**). EOS, 52: 65-95.

2.Elton, E.T.G. 1977. On a disease of the labial glands in **Formica rufa** L. and **F.polyctena** Foerst. (**Hym. Formicidae**). **Proceedings** of the VIII International Congress I.U.S.S.I. Wageningen: 138-139.

ROLE OF ANTS IN THE PEST MANAGEMENT OF FINGER MILLET (*ELEUSINE CORACANA* GAERTNER)

T.M. MUSTHAK ALI
Department of Entomology, University of Agricultural Sciences, G.K.V.K. Campus, Bangalore 560 065, INDIA

Ants are one of the important components of the agro-ecosystem. Presence and absence of certain species of ants in an ecosystem determines the pest composition of a cropping system (Lesten, 1970). In the present study the role of most commonly encountered ant species (Table) in determining the insect pest complex of finger millet, was studied.

The finger millet (FM) crop was grown in three blocks of 750 m² each with an isolation distance of 50 m between the blocks. The crop matures in 120-130 days. On the 30th day after sowing (when generally 80% of the insects appear on the crop) treatments were imposed. The block-I was maintained as control, where the insect fauna including ants were not disturbed. In the block-II the ants, *Camponotus compressus* and *Formica* sp. which are closely associated with FM root aphid, *Forda orientalis* were selectively eliminated by treating their nests with insecticides (@ 10 ml Aldrin in 1 lt of water/nest). Similarly in the block-III all the visible species of ants were eliminated from the vicinity; in addition, BHC 10% dust @ 15 kg/ha was applied in a band of 1 m, round the plot to check the insurgence of ants from the neighbouring fields. Three weeks after the treatment, incidence of different pests was recorded in all these blocks, following suitable sampling techniques.

The selective elimination though had no significant effect on overall pest status; it did reduce significantly the infestation of root aphid, the 'targetted pest'. Though the indiscriminate (complete) elimination of ants in block-III also brought down significantly the infestation level of root aphid, it resulted in a drastic increase in the levels of other pests. For instance, the stem borers (Lepidoptera 2 spp.) and defoliators (grasshoppers 2 spp., *Myllocerus* 3 spp., caterpillars 5 spp.) significantly increased in block-III over control block. Similar effects were also found in earhead pests, (caterpillars 3 spp.) and the incidence of termites (1 sp.). Sucking insects (Aphid 1 sp., leaf hoppers 5 spp., Pseudococcid 1 sp.) in the canopy were not affected (Figure).

Table: Ants categorised based on their interactions with other insect fauna in finger millet ecosystem

Categories* (mean nest density per 100 sq.m) P	phsn	phsn RA	pHns ra	pS
Diacamma rugosum (0.8796	*Monomorium indicum* (1.4709)	*Camponotus compressus* (1.8577)		*Pheidole* sp. (1.224)
Leptogenys processionalis (0.7071)	*Meranoplus bicolor* (1.8577)	*Myrmica* (0.8796)	*Tapinoma melanocephalum* (1.0521)	*Monomorium scabriceps* (1.5871)
L. chinensis (1.0520)	*Solenopsis geminata* (1.2241)			
L. diminuta group (0.8990)				

*P = Predatory; H = Honey-dew; RA = Root aphid; S = Seed harvestor; N = Nectar; Lower case : Facultative

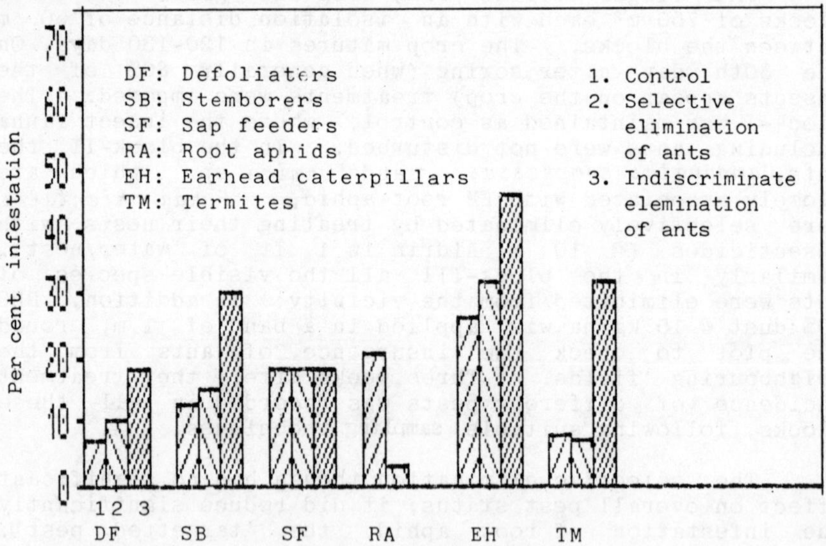

FIGURE: INFLUENCE OF ANTS ON THE INFESTATION LEVEL OF FINGER MILLET PESTS

Thus, these results demonstrate that management by selective elimination of ants is the most appropriate strategy to regulate the pest incidence without causing ecological imbalance of an agro ecosystem.

Leston, D., 1970, Entomology of the cocoa farm. *Ann. Rev. Entomol.*, 15:273-294.

THE EFFECTIVENESS OF ANTS ON THE PLANT PESTS IN KHORASSAN

MOHAMMAD BAGHER SHAHROKHI
Plant Pests & Diseases Research Lab.
Mashad IRAN

Abstract-in studying the pests of different plants specially fruit trees and pastures it was observd that many pests were attacked by Ants as a strong predacious Arthropods. Among these pests, codling moth (Laspeyresia Pomonella), Ovina leaf defoliator (Zamacra flabellaria) and termites (Anacanthotermes spp.) are hunted more than the others. Based on field investigation and observation some tests have been done and it has been known that ants can reduce the population of these pests noticeably.

1. Introduction:

Ants are one of important groups of insects which have various activities in building, pastures farms and orchards. In agricuiture ants have two roles. Beneficial and harmful.By the first role ants are a strong enemy against many pests and by the other they attack some of plants and crops as a pests.

2. Materials and Methods:

1-2. The particularities of the main hosts.

Codling moth is a pest of many fruit trees specially apple which eat of fruits in larval stage.

Ovina leaf defoliator is a species of LEPIDOPTERA which act on many desert bushes chiefly Frula Ovina and, termite act on buildings and deserts and eat of woody facilites and dried bushes.

2-2. The study area:

Ants are spread in all regions of Khorassan and in different conditions. Some tests have been done in an orchards of Mashad that is a cold region and in a pasture of Kashmar with warm and semi desert condition.

3-2. Testing:

The tests were based on field investigation and observation as following:

A- In 1978 in an orchard of Mashad 3 trees of apple were chosen and three different number of larvae of codling moth were released on their trunks. After a few minutes ants began to find and notice them, First one or two ants reached to a larva and attacked it. Bit by bit the number of ants increased and tried to kill and eat the larvae. After two hours the result of the test was cleared as table 1.

Table 1: The effect of ants on the larvae of codling moth

Total	Eaten	Hurt	Sound	Mortality(%)
15	10	2	3	73
10	8	1	1	80
12	10	-	2	83
37	28	3	6	78.7

B- In the pasture of Khorassan in 1988, the activity of ant on ovina leaf defoliator was investigated. By sampling the predatism capability of ants has been determined (Table 2).

Table 2: The range of predatism of ants on Ovina leaf defoliator

Date		Mortality(%) of pupae in cocons					
		1	2	3	4	Total	Mean
12-5	(1987)	21	17	28	25	91	23
22-7		46	48	60	54	208	52
12-9		75	82	82	79	318	79
19-10		65	80	71	76	292	73
18-12		94	85	91	89	359	90

C- Ants are strong enemies against harvester termites. In 1988 by searching the nests of ants a mass of termite carcases have been observed.

Symposium 4

ECOLOGY AND EVOLUTION OF HONEY BEE BEHAVIOUR

Organizer : **FRED C. DYER, U.S.A.**

COMPARATIVE ANALYSES OF THE HONEY BEE DANCE LANGUAGE : PHYLOGENY, FUNCTION, MECHANISM

Fred C. Dyer
Department of Zoology
Michigan State University
East Lansing, Michigan 48824 USA

Most studies of the honey bee dance language have concerned the European species Apis mellifera, and have focused on how bees acquire and communicate information about the direction and distance of flight to food. The several species of Asian honey bees offer a comparative perspective which, especially during the past decade, has led to a new phase of discovery into the phylogeny, adaptive design, and underlying mechanisms of this remarkable behavior. This paper reviews recent comparative investigations on each of these levels of analysis.

PHYLOGENY

According to Lindauer's [1] original hypothesis about the phylogenetic history of the dance language, a primitive ancestor similar to A. florea (open-nesting, direction communicated in dances oriented to celestial cues) gave rise to a bee with A. dorsata's characteristics (open-nesting, dances oriented to celestial cues or to gravity) and then to the ancestor of A. cerana and A. mellifera (cavity-nesting, dances oriented to gravity or to celestial cues when present). Although the ancestral Apis may well have nested in the open and oriented dances to celestial cues, it now appears that subsequent changes were more complicated than previously supposed. Apis florea orients to celestial cues in a way that is fundamentally different from the other Apis species [2,3]. This suggests either that A. florea's way of using celestial cues is derived, or that the incorporation of gravity as a reference for the other species was accompanied by a change in the way dancers orient to celestial cues.

FUNCTION

Interspecific comparisons have also been used for the most direct test so far of the hypothesis that "dialects" in the distance code of the dance language are adaptively tuned to flight range [1]. In Thailand, the distance dialects of three species hardly differ, but the flight ranges, as inferred from forage maps, differ substantially [4].

Thus, the relationship between ecology and distance code may be more subtle than previously thought, if it exists at all.

MECHANISM

The least appreciated outcome of interspecific comparisons is the potential for understanding the details of behavioral processes shared by different species; features which are obvious in one species may lie hidden in another. For example, the open-nesting Asian honey bee Apis florea uses visual cues but not gravity as a reference for dance communication; on overcast days dancers, lacking celestial cues, refer to landmarks surrounding the nest [2]. Because A. mellifera dancers can use gravity in the absence of celestial cues, there was little reason to suspect that they should have similar abilities, but experiments patterned after those used with A. florea now suggest that they do [5]. This may offer an opportunity to use A. mellifera's dancers to study landmark orientation in insects, just as they have been used to study celestial orientation [e.g., 6].

1. Lindauer, M. 1956. Zeitschrift für Vergleichende Physiologie, 38: 521-557.
2. Dyer, F.C. 1985. Journal of Comparative Physiology A, 157: 183-198.
3. Dyer, F.C. Submitted.
4. Dyer, F.C. and T.D. Seeley. Submitted.
5. Dyer, F.C. In preparation.
6. Rossell, S. and R. Wehner. 1986. Nature, 323: 128-131.

EVOLUTION OF REPRODUCTIVE BEHAVIOR IN HONEY BEES

G.Koeniger and N.Koeniger

Institut für Bienenkunde (Polytechnische Gesellschaft)
Fachbereich Biologie der J.W.Goethe-Universität Frankfurt
Karl-von-Frisch-Weg 2, D-6370 OBERURSEL, Fed.Rep.Germany

Queens of Apis mellifera (A.m.) and Apis cerana (A.c.) are polyandrous, they mate with 10 to 20 drones. In contrast drones are monogamous. During mating a queen collects sperm of many drones in her oviducts. Later the queen transfers less than 1/10 of the sperm of each drone into the spermatheca and discharges the rest (1).

Queens of Apis florea (A.f.) also mate with several drones. But in this species drones transfer the sperm directly into the spermatheca (2). While the endophalli of A.m. and A.c. are thick at the end, the endophallus of A.f. ends in a fine tip which might enter the spermaduct (fig.1).

Thus basically two different principles of reproductive system occur. The anatomy of the endophallus is adapted to the different mechanism of spermtransfer (fig.1).

Sperm numbers in queens and drones prove polyandry for many Asian species (3) (tab 1). Sperm numbers of A.koschevnikovi (A.k.) (4) are unknown.

Figure 1: Endophalli of Apis species

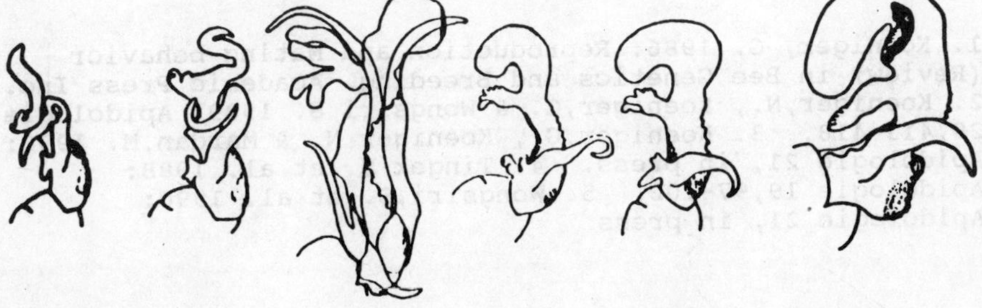

A.a.(5) A.f. A.d. A.k. A.c. A.m.

Table 1: number of spermatozoa in queens and drones
mechanism of spermtransfer

	A.m.	A.c.	A.d.	A.f.	A.a.
No spermatozoa (in million)					
ves.sem.	11	1,2	2,5	0,4	0,1
spermatheca	5	2,5	3,7	1,3	1,0
sperm transf	Ovid.	Ovid.	?	sptheca	sptheca

We want to suggest an evolutionary hypothesis of mating behavior: In the ancestral Apis drones produce a medium number of spermatozoa and deposit them into the queen`s oviducts (general type of Apoidea). The sperm is transferred partially into the spermatheca.

From this basic situation two different strategies may have developed:

1. Drones increase individual fitness by producing more spermatozoa. This will reduce the number of drones succeeding to transfer sperm into the spermatheca. Small drone numbers cause adverse effects on colony fitness. So, the drone's strategy is counterbalanced by the queen's transfer of a small percentage of sperm into the spermatheca. <u>A.m.</u> may represent this stage.

2. Drones increase their individual fitness by injecting their sperm portion in a "better position" (for example near the orifice of the ductus spermaticus). In a final stage drones "develop" an endophallus which enables the transfer directly into the spermatheca. Again, one or small numbers of "fathers" reduce colony fitness. So drones are favoured which produce smaller numbers of spermatozoa and leave room for additional drones. <u>A.andreniformis</u> (<u>A.a.</u>) represents this stage.

According to these ideas A.m. and A.a. are the two ends of a dichotomous evolution.

1. Koeniger, G. 1986: Reproduction and Mating behavior (Review) in Bee Genetics and Breeding. Academic Press Inc. 2. Koeniger,N., Koeniger,G. & Wongsiri S. 1989: Apidologie 20,413-418. 3. Koeniger,G., Koeniger N. & Mardan,M. 1990: Apidologie 21, in press. 4. Tingek S. et al. 1988: Apidologie 19,97-102. 5. Wongsiri,S. et al. 1990: Apidologie 21, in press

OPTIMIZING STRATEGIES IN CHOICE BEHAVIOUR IN THE HONEY BEE

U. Greggers and R. Menzel
Freie Universität Berlin, Institut für Neurobiologie,
Königin-Luise-Str. 28-30, D - 1000 Berlin 33, FRG

Single worker bees (Apis mellifera) were trained to feed exclusively from a single patch of four arteficial flowers. The color, position and continuous reward rate of each flower was adjustable for each experiment, and the time-course and frequency of four major behavioral components - licking, inter-flower tube returns, retries in flight and inter-flower flights - were calculated from the inter-response times detected by a computer.

In a first series of experiments, all flowers were adjusted to equal reward rates. The inter-flower flight time increased when the overall patch reward decreased, and a symmetrical flight directionality distribution was observed, as already reported by Schmidt-Hempel (1985) Physiol. Entomol. 10: 439-442.

In a second series, reward rates were adjusted to the ratio 1:2:2:8. Color signals and flower positions did not change in both experiments, and inter-flower flights now showed a small preference to the higher reward. In contrast, retries in flight showed a stereotype time-course, and their frequency matched the rate of reward. Retries in flight were dominant in optimizing choice behavior.

The foraging strategy of the honey bee in a single patch is interpreted as "moment-to-moment" decisions. The proposed mechanism for optimizing choice is sensitization, which is dependent on the strength of the US. The bees were conditioned to color and position, as further experiments demonstrated.

DIVERSITY OF *APIS* IN SOUTHEAST ASIA

Gard W. Otis
Department of Environmental Biology
University of Guelph
Guelph, Ontario, Canada N1G 2W1

The genus *Apis* has recently been recognized as being far more diverse than was previously believed to be the case. Most researchers in the recent years prior to 1980 recognized only four species: two cavity-nesting species, *Apis mellifera* and *A. cerana*, and two open-nesting species, the tiny dwarf honey bee, *A. florea*, and the giant or rock honey bee, *A. dorsata* [1, 2]. In contrast, Maa [3] recognized 24 "species," many of which are clearly only geographical races or subspecies. Contained within his taxa, however, are at least three good biological species which have been described in the 1980's.

The *dorsata* complex consists of at least 2 and possibly as many as 4-5 species. *A. laboriosa* from the Himalayan region is the largest honey bee [4]. It appears to be ecologically distinct from *dorsata* [5], although the male endophalli of the two taxa are indistinguishable [6] and the mechanisms of reproductive isolation are currently unknown. Other distinct forms are found in the Philippines (*A. breviligula*), on Sulawesi and the Sula Islands (*A. binghami*) [2], and possibly in the Andaman Islands [7]. Their status is still uncertain, with some authors considering them as subspecies [2] and others considering them sufficiently distinct to be recognized as full species [8, 9].

Two dwarf honey bees, *A. florea* and *A. andreniformis*, were separated in 1987 [10], with more recent confirmation of species status based on male genitalia and worker wing venation [11]. Maa [3] presented additional information. Where the two species are sympatric in southern China, Assam, and much of Indochina, *florea* is the lowland species which is replaced at higher elevations by *andreniformis* [3, 7]. However, where only one species is found, it can occur from sea level up to 1300 m (e.g., *florea* in S. India, *andreniformis* in the Malay Peninsula and Sarawak, Borneo), with highland bees being larger in size [7]. The food-storage comb differs substantially, with cells on the dance platform either pointing upwards (*florea*) or nearly horizontal (*andreniformis*). Multivariate analyses of morphometric measurements separate the species into two adjacent and non-overlapping clusters of points [12]. *A. andreniformis* shares the mite *Euvarroa sinhai* with *A. florea* [7, 13]. Virtually nothing is known about isolating mechanisms, ecological differences, and similarity in behavior of the two dwarf honey bees.

A reddish species of cavity-nesting honey bee, *A. koschevnikovi* (= *A. vechti* Maa), was recognized in 1988 [14, 15] from Borneo and Sumatra [16]. It typically forms small colonies that swarm and abscond frequently [7]. Drone flights of *koschevnikovi* occur later than those of *cerana* and earlier than those of *dorsata*, thereby providing reproductive isolation [17]. *A. koschevnikovi* can be separated by multivariate analyses from the other cavity-nesting *Apis* species [16]. *Varroa jacobsoni* parasitizes drone brood of both *koschevnikovi* and *cerana*, although they are larger when reared on the larger host species. No detailed studies of the behavior or ecology of this species have been conducted.

Allozymes of the three cavity-nesting species, two dwarf bees, *dorsata*, and *binghami* have recently been analyzed [18]. Variation was found between species for esterase, malate DH (dehydrogenase), alpha glycerophosphate DH, fumarase, glucose DH, and succinate DH. These latter three enzymes have not been found to be polymorphic in *Apis* before. All taxa except *dorsata* and *binghami* differed substantially at one or more loci, supporting their status as distinct species. However, allozyme data do not appear to be very useful for constructing phylogenies of *Apis* [18, 19]. The extreme similarity of enzymes and male genitalia [7] of *binghami* and *dorsata* suggest that *binghami* should not be given species status.

The recognition of several new species of *Apis* increases the opportunities for comparative research on their ecology and behavior. Given how much we know about *Apis mellifera* and comparatively how little we know about most of the Asian honey bees, there is a great need for further research in this geographic region if we are to understand the evolution of the honey bees.

REFERENCES

1. Michener, C.D. 1974. The Social Behavior of the Bees. Harvard University Press, Cambridge, MA, p. 404.
2. Ruttner, F. 1988. Biogeography and Taxonomy of Honeybees. Springer Verlag, Berlin, p.284.
3. Maa, T.C. 1953. Treubia 21: 525-640.
4. Sakagami, S.F., T. Matsumura, and K. Ho. 1980. Insecta Matsumurana 19:47-77.
5. Roubik, D.W., S.F. Sakagami, and I. Kudo. 1985. Journal of the Kansas Entomological Society 58: 746-749.
6. McEvoy, M.V. and B.A. Underwood. 1988. Journal of the Kansas Entomological Society 61: 246-249.
7. Otis, G.W. Personal observations.
8. Starr, C.K., P.J. Schmidt, and J.O. Schmidt. 1987. Pan-Pacific Entomologist 63: 37-42.
9. Michener, C.D. 1989. Personal communication.
10. Wu, Y. and B. Kuang. 1987. Bee World 68: 153-155.
11. Wongsiri, S., K. Limbipichai, P. Tangkanasing, M. Mardan, T. Rinderer, H.A. Sylvester, G. Koeniger, and G. Otis. In press. Apidologie.
12. Ruttner, F. 1989. Personal communication.
13. Delfinado-Baker, M. 1989. Personal communication.
14. Mathew, S. and K. Mathew. 1988. IBRA Newsletter No. 12: 10.
15. Tingek, S., M. Mardan, T.E. Rinderer, N. Koeniger, and G. Koeniger. 1988. Apidologie 19: 97-102.
16. Ruttner, F., D. Kauhausen, and N. Koeniger. In press. Apidologie.
17. Koeniger, N., G. Koeniger, S. Tingek, M. Mardan, and T.E. Rinderer. 1988. Apidologie 19: 103-106.
18. Gan, Y.Y., G. Otis, M. Mardan, and S.G. Tan. 1989. Proceedings, Isozyme Conference, 1989.
19. Sheppard, W.S. and S.H. Berlocher. In press. Apidologie.

ABSCONDING BEHAVIOUR OF *APIS CERANA* IN SRI LANKA

R.W.K.Punchihewa, N.Koeniger[1] and D.Howpage[2]

Sri Lanka Dept. of Agriculture, Agriculture Research Station, Makandura, Gonawila (NWP), Sri Lanka.
[1]**Institut fur Bienenkunde, Karl-von-Frisch-Weg 2, 6370 Oberursel 1, Federal Republic of Germany.**
[2]**Sri Lanka Dept. of Agriculture, Apiculture Development Centre, Bindunuwewa, Bandarawela, Sri Lanka.**

Absconding could be called the most difficult behavioural problem in managing Apis cerana in Sri Lanka. The intensity of this problem seem to have a close relation to the eco-climatology of this island. It is most acute in Low Country Dry and Low County Intermediate climatic zones while less acute in Low Country Wet climatic zone. Beekeepers in higher altitudes(above 300m MSL) encounter this problem to a lesser extend. In areas where the problem is most acute observational evidence strongly suggest it is a problem under artificial conditions (eg. colonies in movable frame hives) rather than on natural colonies nesting in natural nesting sites.

One of the distinct characters of a colony in preparation for absconding is the reduced number of returning pollen forages. An index developed on observing the quantitative aspects of this behavioural characteristic is called the **Colony Performance Index or CPI** which could be used to predict absconding. The CPI is based on the relative abundance of pollen carriers returning to the nest in relation to observational period and to the total incoming bees.

$$CPI = \frac{(F_p)^2}{T_{obs} \cdot F_t} \times 100$$

Where, F_p = Total number of foragers returned with pollen loads
T_{obs} = Observational period in seconds
F_t = Total number of returning bees

Preparation for absconding is not an instantaneous decision by the colony but a process that shows a gradual build-up phase which culminates in abandoning the existing nest site. This process could be observed externally by a progressive reduction of daily/weekly CPI value of a colony where on the day of absconding the CPI falls to 0.00. CPI of a colony could be reduced by artificial means such as constant disturbance, smoking, high temperature etc. so that the **Absconding Impulse or**

I is eventually activated. Once a colony gets the **AI** it shows a eculiar behaviour where the colony does not defend itself gainst intruders such as wax moth, ants etc. and cease to rear rood in spite of the fact the queen continues to lay. Therefore ne wax moth infestations of combs, broodlessness in combs and ood shortages which become apparent in colonies in preparation or absconding seems like secondary events rather than primary uses of absconding.

endency for a colony to get **AI** is much greater in movable frame ves compared to natural colonies existing in the same or milar eco-climatological conditions. Natural colonies kept der observation for the past 18 months neither absconded nor ney encounter any pest incidence while all the hived colonies ept in same locations one time or the other wanted to abscond.

ce a colony get the **AI** firmly established (when CPI is <0.25) is often difficult to prevent it from leaving the existing st site even by offering it a frame of brood with good stores food. In many occasions it can only delay the absconding by a w days where the bees would wait until all the sealed brood to tch out to take with them. One way of not loosing a hived lony if it gets the **AI** is by allowing it to abscond under ntrolled conditions when the **CPI** is <0.10 and giving it a nest te after the pseudo-absconding.

e absconding tendency in hived colonies seem to be higher ring the eco-cilimatological stress periods while natural lonies are not so prone. The empirical evidence are highly ggestive of implications of the micro-climatic conditions thin the nest site as a primary cause in activating the **AI**. Of ese micro-climatic factors temperature and humidity seem to ay a key role in activating the **AI**. In our experimental lonies we have been able to elevate the **CPI** by temperature and midity regulation so as to prevent the colony getting in to the wer **CPI** regimes which activate the **AI**.

MATING BEHAVIOUR OF *APIS CERANA* IN SRI LANKA

R.W.K.Punchihewa, N.Koeniger[1] and G.Koeniger[1]

Sri Lanka Dept. of Agriculture, Agriculture Research Station, Makandura, Gonawila (NWP), Sri Lanka.
[1]Institut fur Bienenkunde, Karl von Frisch Weg 2, 6370 Oberursel 1, Federal Republic of Germany.

Honeybee drones leave the colony and practice their mating flight activity during the afternoon hours of each day. Most active drone flight period occurred during 15:30 hours to 17:00 hours. Young virgin queens leave the hive on the mating flight usually on the 2nd or the 3rd day after emergence and they fly out on the mating flight between 16:15 hours to 16:55 hours. Success of the mating flight was determined by examining the presence of the mating sign in the vagina of the returning queens. In all our observations mating flight of the queens lasted for 9.3 ± 1.7 minutes. Individual drones flew between 10 to 16 minutes and returned to the hive for a 2 to 4 minute period for rest or for food.

With the use of a hydrogen filled balloon live queens or 9-OD pheromone capsules were floated between 3 to 6 meter heights to detect the drone congregation areas. The drone congregation areas (**DCA**) were formed within 250 meters from the drone colonies and often more than one congregation area occurred for a group of drone colonies. The **DCA**s were found among the canopies of trees. The space in which the drones were flying had distinct boarders demarcated horizontally by the canopies of trees on the periphery and vertically by canopies of taller trees above. With the use of insect nets tied to tall bamboo poles flying drones were caught at **DCA** and marked with quick drying paints to determine the flight range of drones. The rate of recovery after 24 hours in the nearby drone colonies were 92% indicating a short flight range.

The **DCA**s of Apis cerana in Sri Lanka has some features common in with Apis mellifera in Europe but in many aspects A. cerana **DCA**s are distinctly different.

CONSTRAINTS TO QUEEN REARING DISCRIMINATION IN THE HONEY BEE

Francis L. W. Ratnieks

Department of Entomology, University of California, Berkeley, CA 94720, USA

Workers in polyandrous and polygynous insect societies have a potential indirect reproductive strategy or rearing queens more related to self. In polyandrous societies empirical support of this discrimination is provided by studies of the honey bee [1, 2, 3] However, in these studies the extent of biased queen rearing, or discriminatory feeding, was relatively weak. This paper considers two constraints on discriminatory queen rearing, which apply both to the honey bee and generally to polyandrous social Hymenoptera.

THE ACCURACY OF FULL-SISTER VERSUS HALF-SISTER RECOGNITION

Kin recognition requires information for assigning conspecifics to various kin classes. In different contexts different sources of variation can provide relevant information. For example, nest derived odor cues can provide information for distinguishing nestmates from non-nestmates (4). Discrimination between queen- and worker-laid male eggs may use cues of egg maternal caste origin [5, 6]. However, for full- versus half-sister recognition in a polyandrous society neither environmental nor maternal cues provide useful information. The evidence for queen rearing discrimination, therefore, indicates the use of a phenotype matching system [7] to compare heritable odor cues between self and potential queens.

The accuracy of this system of discrimination depends greatly on the genetic variance of heritable odors [8, 9, 10]. A single multiallelic locus, in which alleles code for distinct odors, permits recognition no better than shown in Table 1. (Note that total errors, E1 + E2, descrease as the number of alleles increases.)

Table 1. Minimum probabilities* of mistakenly assigning a half-sister as a full-sister (E1), or a full-sister as a half-sister (E2), for self-referent phenotype matching between hymenopteran sisters using a single polymorphic locus with a variable number of alleles at equal frequency as information source.

	1 or 2 alleles---consider a sister as a full-sister if she has in common with you--2 alleles			
# alleles	E1	E2	E1	E2
2	0.91	0	0.44	0.25
5	0.79	0	0.14	0.40
20	0.59	0	0.03	0.45
infinite	0.50	0	0	0.50

* population mean mistake probabilities are independent of queen mating frequency, although the variance of mistakes between colonies of different mating types is affected (i.e., mistake probabilities vary between colonies according to the genotypes of the queen and her mates, and the number of mates).

The use of heritable odor cues in discrimination, therefore, raises the important question of the effect of discrimination on the underlying genetic variance. In some situations discrimination increases genetic variance (e.g., the pollen-stigma compatibility system in plants [11]; gamete compatibility in sea squirts [12]), whereas in other situations it may decrease variance (e.g., genotype mediated aggression in marine invertebrates [13]). Analysis of queen rearing discrimination [14] suggests that in species with queens mated to three or more males queen rearing discrimination will reduce genetic odor cue diversity to a single allele, and for species with double mated queens to two alleles.

Table 2. Effect of discrimination on the genotypes of queens produced in the colony.

colony #	queen	queen's mates	worker and queen genotypes	disfavored queen genotype, allele
1	*aa*	*a, a, A*	*aa, aa, aA*	*aA, A*
2	*AA*	*a, A, A*	*aA, AA, AA*	*aA, a*

The rationale behind the treble mating result is shown in outline in Table 2. A and a are two alleles coding for different odors which are used in discrimination. Mating types 1) and 2) are equivalent, but with the "name" of the allele changed. In 1) aA queens will be discriminated against relative to aa queens, so that the final frequency of A in queens produced in these colonies will be below the initial frequency. Similarly, in 2) a will be reduced. The discrimination in these two cases should be equal (because the pattern of worker genotypes is the same), so that the gain in frequency of a in one type 1) colony equals the gain in A in one type 2) colony. However, over the whole population type 1) colonies will be more frequent than type 2) colonies if a is more frequent than A . (Other paired mating types occur, with the same result.) Therefore, when $a > A$, a increases in frequency and vice versa. This is a form of frequency dependent selection and leads to fixation of the common allele, with loss of genetic variance.

The importance of this result is that queen rearing discrimination reduces the source of variation on which it depends, so providing a possible reason for the weak discrimination in the honey bee. However, given that discrimination does occur, where does the necessary genetic variance come from? Nestmate versus non-nestmate recognition, if partly a result of heritable odor cues, could increase genetic variance [14]. However, in the honey bee environmentally acquired odors appear to be of primary importance in distinguishing nestmates from non-nestmates [14], although definitive experiments remain to be done. Other loci, which are polymorphic for other reasons (e.g., overdominance, frequency dependent selection, drift) could also be a source of information. The highly polymorphic honey bee sex determination locus also carries a large amount of potential information [15], which is maintained by frequency dependent selection. Whether this information is usable, that is can be detected, is unknown [14, 15].

COSTS OF DISCRIMINATION

If queen rearing discrimination has zero cost to the colony then it should be universally favored; but what if discrimination reduces colony productivity or survival? For example, if discriminatory workers are less effective in other duties, as occurs when workers have a high probability of direct reproduction [16].

This situation has been examined using a population genetics model to investigate the likelihood that a rare discriminatory or non-discriminatory allele can invade (i.e, an allele which affects the behavior of workers to make them discriminators or non-discriminators) [16]. The general result for polyandrous Hymenoptera is that a rare discriminatory or non-discriminatory allele will invade if the increased work efficiency of rare strategy workers relative to common strategy workers is $> (1 - b)/(1 + b/n)$, where n is queen mating frequency and b is the final proportion of queens of the rare strategy patriline when the allele is inherited paternally divided by n (i.e., when $b = 1$, there is no discrimination).

The biasing effect of discrimination, b, is reduced as the frequency of recognition errors increases, and importantly is also affected by n. In particular, the bias against a rare non-discriminatory allele reduces as n increases, because the many patrilines of discriminatory workers effectively cancel each other out [16]. As a result a small increase in work efficiency of non-discriminators can permit the formation of a mixed polymorphism of discriminators and non-discriminators. This conclusion is supported by empirical evidence for the presence of both discriminators and non-discriminators in the honey bee [3]. For $n = 20$, with zero kin recognition mistakes (to show the most restrictive case), and half the queens removed by discrimination, a rare non-discriminator will invade if workers are only 4% more efficient than discriminators.

1. Visscher, P. K. 1986. Behavioral Ecology and Sociobiology 18: 453-460.
2. Noonan, K. C. 1986. Ethology 76: 295-306.
3. Page, E. E., G. E. Robinson and M. K. Fondrk. 1989. Nature (Lond.) 338: 576-579.
4. Gamboa, G. J., H. K. Reeve and D. W. Pfennig. 1986. Annu. Rev. Entomol. 31: 431-454.
5. Ratnieks, F. L. W. and P. K. Visscher. 1989. Nature (Lond.) 342: 796-797.
6. Ratnieks, F. L. W. 1988. American Naturalist 132: 217-236.
7. Sherman, P. W. and W. G. Holmes. 1985. In: Experimental Behavioral Ecology and Sociobiology, B. Hölldobler and M. Lindauer, Eds, Sinauer Associates, Sunderland, Massachussetts.
8. Getz, W. M. 1981. Journal of Theoretical Biology 92: 209-226.
9. Getz, W. M. 1982. Journal of Theoretical Biology 99: 585-597.
10. Lacy, R. C. and P. W. Sherman. 1983. American Naturalist 121: 489-512.
11. de Nettancourt, D. 1977. Incompatibility in Angiosperms. Springer-Verlag, Berlin.
12. Scofield, V. L., J. M. Schlumpberger, L. A. West and I. L. Weisman. 1982. Nature (Lond.) 295: 449-502.
13. Crozier, R. H. 1986. Evolution 40: 1100-1101.
14. Ratnieks, F. L. W. 1990. American Naturalist (in press).
15. Crozier, R. H. 1988. In: Invertebrate Historecognition, R. K. Grosberg, D. Hedgecock and K. Nelson, Eds, Plenum Press, New York.
16. Ratnieks, F. L. W. and H. K. Reeve. 1990. Journal of Evolutionary Biology (in press).

11th International Congress—IUSSI 1990, India

HOW REAL ARE TEMPERATE-TROPICAL HONEY BEE DIFFERENCES?

David W. Roubik
Smithsonian Tropical Research Institute
Balboa, Panama (APO Miami 34002-0011, USA)

Recent work suggests latutudinal differences in honeybee behavior, and the general model that best fits the data involves either *r* vs *K* selection or *"rate maximizer* vs. *efficiency maximizer"*. While some of the predictions may turn out to be true, the original field data on which they were based seem much less certain. Examples can be found in appraisal of differences between African and European honeybees (both in the Americas!) that relate to reproduction, foraging, and cold tolerance. These show that while helping to direct studies at underlying mechanisms, such as metabolic rate [1], the broadening biological data do not match some original expectations or studies.

COLD TOLERANCE

The thermoregulatory ability of honeybee colonies was thought to differ in some fundamental way between Africanized and European honeybees [2], but the behavioral and physiological mechanisms used by clusters to stay warm are very similar [3]. Both bees elevate metabolic activity and retain this added heat within a cluster [3]. Differences in survival ability in cold climates seem more related to nest site, honey storage, brood rearing during cold weather, and the maximum longevity of an inactive worker. Thermoregulation by Africanized honeybee clusters is considerably more efficient at warm temperatures (23°C) than is that of European honeybees. The two ecotypes are perhaps more similar than we think, but different in other ways that we still do not recognize.

REPRODUCTION

The Africanized honeybee of the Neotropical lowlands was thought to reproduce more or less continually during the year, potentially giving rise to 8 to 12 swarms [4]. This reproductive activity, measured in apiary hives, was two or three times greater than that of European honeybees in temperate areas. It was thought to reflect a greater risk of colony mortality due to density-independent factors such as predators [4,5,6]. However, seven years of field data from a broad transect of lowlands in central Panama show that the average colony goes through one yearly reproductive cycle [7]. Three swarms per year are probably produced by an Africanized honeybee colony, roughly the same number produced by a temperate honeybee colony in the temperate zone. Reproduction took place during the major yearly flowering episodes—one or two peaks from the dry season to the first half of the wet season. They seemed to coincide with the reproduction of apiary colonies of European honeybees. As for the effect of enemies or other natural factors on these ecotypes, and their evolutionary ecology, there are virtually no comparative data. Indeed, a study in Botswana revealed that *A. mellifera scutellata* , in natural sites, is not particularly defensive [8].

FORAGING ECOLOGY

During certain times and in certain rather poorly defined settings, there are differences between European and Africanized honeybees that seem related to the need to reproduce rapidly by Africanized honeybees and the need of European honeybees to survive a long period with no forage [6]. In two cross-fostering studies, where bees of one ecotype were introduced into colonies of the other, the adopted bees foraged as their hosts. The cues perceived and produced by nest mates are clearly essential in regulating the process. As shown by Seeley [9], food-storing bees are central to determining colony foraging activity. If the age-related tasks of Africanized honeybees are shifted toward younger ages, and if forager mortality is relatively high (a trait related to foraging activity itself), then there may be enhancement of the proportion of food-storers in the nest (beyond the 20% of European honeybees [9]), so that the colony is always more receptive toward forage than European bees. Specific environmental cues—the pollen and nectar species and quantities available—are potentially important determinants of whether colonies forage intensively or do not. While there are apparently some statistical differences between the amount of pollen collected by colonies of the two ecotypes [10], their general performance is very similar, but the location and kinds of forage they accept or prefer has not been studied well, nor related to foraging performance. Thus the tendencies of these bee ecotypes to seek or forage in certain times or places, within a single habitat or among seasons, are not known. With a scant empirical basis and incomplete theory, the reasons for ecological differences among honeybees [11] continue to be elusive.

REFERENCES

1. Heinrich, B. 1979. Journal of Experimental Biology, 80:217-229.
2. Seeley, T. D. 1985. Honeybee Ecology, Princeton University Press. p. 201.
3. Southwick, E. E., D. W. Roubik and J. M Williams. in press. Comparative Biochemistry and Physiology.
4. Winston, M. L, O. R. Taylor and G. W. Otis. 1983. Bee World, 64:12-21.
5. Lee, P. C. and M. L. Winston. 1987. Ecological Entomology, 12:187-195.
6. Roubik, D. W. 1989. Ecology and Natural History of Tropical Bees, Cambridge University Press, New York. p. 514.
7. Roubik, D. W. and M. M. Boreham. in press. Interciencia [Caracas].
8. Schneider, S. and R. Blyther. 1988. Insectes Sociaux, 35:167-181.
9. Seeley, T. D. 1989. Behavioral Ecology and Sociobiology, 24:181-199.
10. Danka, R. G., R. L. Hellmich II, T. E. Rinderer and A. M. Collins. 1987. Animal Behavior, 35:1858-1863.
11. Ruttner, F. 1988. Biogeography and Taxonomy of Honeybees. Springer-Verlag, Berlin. p. 284.

MITOCHONDRIAL DNA AND BIOGEOGRAPHY OF *APIS* SPECIES

Deborah Roan Smith, Museum of Zoology, Insect Division, University of Michigan, Ann Arbor, MI 4810S
USA

Mitochondrial DNA (mtDNA) restriction site polymorphisms are used in a study of the biogeography and subspecies differentiation of European and Asian honey bees.
The following populations of Apis mellifera were sampled: Apis mellifera mellifera from Norway, Sweden, Denmark, and France; A. m. carnica from Austria and Yugoslavia; A. m. ligustica from Italy; A. m. iberica from Spain; A. m. lamarckii from Egypt and A. m. unicolor from Madagascar (collected by S. Goodman, Chicago Field Museum, Chicago, IL, USA); A. m. scutellata and A. m. capensis from South Africa (collected by R. Crewe, University of the Witswatersrand, Johannesburg, RSA).
Samples of Asian honey bee populations were provided by F. Dyer, Michigan State University, East Lansing, MI, USA (Indian samples); H. Pechhacker, Institut fur Bienenkunde, Lunz-am-See, Austria (Thai samples); and G. Otis, University of Guelph, Guelph, Ont., Canada (Thailand and all other sites). The following species and populations were sampled:
Apis cerana indica (as defined by Ruttner [1]) from Bangalore, India; Chang Mai, Pitsanulok and Bangkok, Thailand; peninsular Malaysia; Sabah (north Borneo), Malaysia; southern Sulawesi, Indonesia; and the island of Luzon, the Philippines.
Apis koschevnikovi from Sabah (north Borneo), Malaysia.
Apis dorsata dorsata (as defined by Ruttner [1]) from Bangalore, India; Thailand; peninsular Malaysia; and Sabah (north Borneo), Malaysia.
Apis dorsata binghami (as defined by Ruttner [1]) from Sulawesi.
Apis florea from Bangalore, India and Thailand;
Apis andreniformis from peninsular Malaysia and Sabah (north Borneo), Malaysia

Mitochondrial DNA was prepared from workers of each hive sampled [2,3]. Aliquots of each mtDNA were digested with the 6-base restriction enzymes AccI, AflII, AvaI, BclI, BglII, Eco0109, EcoRI, EcoRV, HindII, HindIII, NdeI, PstI, PvuII, SpeI, XbaI and XhoI. The resulting DNA fragments were radioactively end-labeled with 32P nucleosides and separated by electrophoresis on 1% agarose gels and 4% polyacrylamide gels [4,5] and visualized on X-ray film by autoradiography. Cleavage maps showing the number and location of the restriction sites generated by these enzymes, were constructed for each subspecies or population sampled [2,3,6,7]. Percent sequence divergence among mitochondrial genomes were estimated from these maps [8].

Three main types of mitochondrial DNA were found in A. mellifera, indicating that the subspecies examined belong to three main lineages within the species. These are a western Mediterranean group, which includes A. m. mellifera and A. m. iberica from northern Spain; an eastern Mediterranean group, which includes A. m. ligustica, A. m. carnica and A. m. lamarckii; and an African group which includes A. m. iberica from southern Spain, A. m. intermissa (based on data from J.-M. Cornuet, INRA-CNRS, Bures sur Yvette, France), A. m. scutellata and A. m. capensis. Percent sequence divergence among the three types or lineages of mtDNA ranged from 2.3% to 4% [2]. The three lineages indicated by mtDNA may owe their existance to the colonization of Europe, the middle East and Africa by three lines of Apis mellifera, as proposed by Ruttner [9].

Preliminary work on populations of Apis cerana showed approximately 2% sequence divergence among the following 3 populations: southern India; Thailand; peninsular Malaysia and Sabah. The samples from Luzon, Philippines, were strikingly different from the other A. cerana samples; estimates of percent sequence divergence between the Luzon population and each of the other three populations is approximately 8%. These results make sense in light of the recent geological history of southeast Asia. Islands on the shallow continental Sunda shelf (e.g., Borneo) were joined to the Asian mainland during the late Pleistocene. Luzon, however, was never attached to the Asian mainland [10]. The geological history of the Sunda shelf and the present distribution of the dwarf bee A. florea suggests that the island of Palawan was probably colonized by A. florea no later than the late middle Pleistocene (approximately 160,000 years ago). A. florea occurs on the Asian mainland and on those islands of the Sunda shelf that were at one time connected to the mainland. It does not appear to have crossed even short distances across the sea: it occurs on the Philippine island of Palawan, but not on Luzon. The channel between Borneo and Palawan is 145 m deep, and sea levels have not been low enough for the two islands to be joined by land since the late middle Pleistocene [10].

Completed work on southeast Asian A. cerana and comparative results for the dwarf and giant honey bees will be presented at the XIth International Congress of IUSSI at Bangalore.

REFERENCES

1. Ruttner, F. Biogeography and Taxonomy of Honeybees, Springer-Verlag, New York.
2. Smith, D. R., O. R. Taylor and W. M. Brown. 1989. Nature, 339: 213-215.
3. Smith, D. R. and W. M. Brown. 1990. Annals of the Entomological society of America, 83 (1): 81-88.
4. Brown, W. M. 1980. Proceedings of the National Academy of Science, U.S.A., 77: 3605-3609.
5. Wright, J. W., C. Spolsky and W. M. Brown. 1983. Herpetologica, 39: 410-416.
6. Smith, D. R. 1988. In: Africanized Honey Bees and Bee Mites, G. R. Needham, R. E. Page Jr., M. Delfinado-Baker and C. E. Bowman, eds., Ellis Horwood, Ltd., Chichester. pp.303-312.
7. Smith, D. R. and W. M. Brown. 1988. Experientia, 44: 257-260.
8. Nei, M. and F. Tajima. 1983. Genetics 105: 207-217.
9. Ruttner, F., L. Tassencourt and J. Louveaux. 1978. Apidologie 9 (4): 363-381.
10. Heaney, L. R. 1986. Biological Journal of the Linnean Society, 28: 127-165.

11th International Congress—IUSSI 1990, India

MATING AND GENE FLOW AMONG HONEY BEE POPULATIONS

Orley R. Taylor
Dept. of Entomology
7005 Haworth Hall
University of Kansas
Lawrence, KS 66045 USA

Spatial aspects of the honey bee mating system are not well known primarily because matings occur in mid-air at a height and speed which make direct observations extremely difficult. Owing to the highly dynamic nature of queen and drone flight, the genetic consequences of natural matings have also remained unpredictable. Neutral or null models, based on assumptions and conditions similar to those of mixed mating models for plant populations may help us to better understand the relationships between the mating system and genetics in honey bees.

Two primary assumptions of a neutral mating model are (1) that reproductives (drones and queens) of various genotypes do not differ in characteristics which affect mating success (for example, flight performance, copulatory behavior, or fertility), and (2) mating is random with respect to genotype.

With these simplifying assumptions, the honey bee mating system is simulated as a surface topography of relative frequencies of drones and queens over a given area. Field studies, using aerial drone traps and mark-and-recapture techniques indicated that drones occur about their home colonies in patterns which approximate a bivariate normal distribution with standard deviations ranging from 800 to 1500 meters. By modelling distributions of drones from two or more sources, the expected frequencies of drones, encountering queens at any point in space, is calculated. These expected values provide a null hypothesis against which observed frequencies may be tested.

A hypothetical queen flight distribution is modelled with a peak frequency at a distance (ranging from 0 to 5 km) away from the queen source location. Because the mode of this queen distribution occurs at a distance away from the origin, flight distributions of queens could be described as hyper-dispersed.

Tests of the model show that it is possible to obtain predetermined frequencies of drones of two or more types at specified distances by altering the distance between drone sources and the number of drones at each source. Matings by genetically marked queens within linearly structured drone populations indicated that most queens mate at distances of 1 - 1.5 km from their colony. Frequency distributions for mating flights of both queens and drones provide a basis for measuring gene flow among honey bee populations via the mating system. When combined with information on the dispersal distances of swarms we should be able to establish the conditions that lead to genetic heterogeneity among isolated and semi-isolated feral and managed honey bee populations.

THE SEASONAL CYCLE OF NESTING AND MIGRATION BY THE HIMALAYAN HONEY BEE, *APIS LABORIOSA*

Benjamin A. Underwood
Department of Entomology
Cornell University
Ithaca, NY 14853

ABSTRACT

The Himalayan honey bee, Apis laboriosa,nests in the open high on cliffs at altitudes of 1200 to 3500 m. In an effort to determine how colonies cope with the relatively harsh weather conditions at such altitudes, the seasonal nesting cycle of A. laboriosa was studied in western Nepal. One of the keys to high-altitude survival is seasonal migration. Nest sites within the subalpine zone (above about 2800 m) are occupied for a maximum of only four months in summer (June through September), while those within the warm-temperate zone (1200 to 2000 m) may be used for as much as ten months of the year (February through November). Predation, especially by humans, serves as the impetus for some migratory movements, but a combination of poor foraging conditions and low colony stores also leads laboriosa colonies to migrate. In the steep, narrow valleys inhabited by these bees, the seasonal abundance and distribution of floral resources vary greatly within short distances and colonies need not move more than about 10 to 20 km in migrating between cliff sites separated by as much as 2000 m in altitude. Late in November or in early December, laboriosa colonies abandon their nests on the cliffs and move into the forest at altitudes below 2000 m. There they survive the cold winter months of December and January by huddling in energy-efficient combless clusters placed in concealed locations near the ground. Colonies return to the cliffs in early February to begin the yearly cycle again.

GENETIC DIVERSITY AND CONSERVATION OF *APIS CERANA* IN HINDU KUSH-HIMALAYAS

L. R. VERMA
International Centre for Integrated Mountain Development (ICIMOD), Kathmandu, Nepal.

Hindu Kush-Himalayan region is rich in bee resources.There are at present four or more species of honeybees in this region. Among these, Asian hive bee, Apis cerana, is eqivalent of the European hive bee, Apis mellifera. Apis cerana has meny valuable characteristics of biological and economic importance. These include docile nature, less prone to attacks of wasps and high level of resistance to parasitic Asian mites (Varroa jacobsoni and Tropilaelaps clarae that plague Apis mellifera. Apis cerana can co-exist with other native bee species and requires least chemical treatment of colonies to control epidemics. However, yet this native bee species has not become very popular amongst the beekeepers because of its several bahavioral characteristics. These include frequent swarming and absconding, proness to robbing, production of large number of laying workers and lower honey yields.

There is a current movement in Asia to import allopatric Apis mellifera for commercial exploitation. Many such importations of the exotic Apis mellifera have proved disastrous because of its allopatric nature and the introduction of new diseases and parasitic mites. There is now apprehension that importation of Apis mellifera would lead to the decline of Apis cerana population in its native habitat to a level that threatens its extinction as a genetic resource. Apis cerana has already become a rare species in Japan and parts of China. Before this happens in Hindu Kush-Himalayan region, a conservation strategy through development and pormotion of beekeeping with this native bee species need to be adopted to help maintain its genetic diversity. Such stategies first require the exploration and evaluation of different sub species/races/ecotypes of this native bee species and then improvement of best of them through selective breeding, appropriate apiary management practices and biotechnological research. The genetic diversity of *Apis mellifera* has been organized into 24 sub-species having varied economic usefulness. So far only four sub-species of *Apis cerana* are recognized. The northern and high altitude sub species/ecotypes of *Apis cerana* is likely to yield valuable honeybee germplasm which may have commercial applications not only throughout Asia, but also in western hemisphere where beekeeping with *Apis mellifera* is threatened with parasitic Asian bee mites and spread of aggressive Africanized bees.

Discriminant analysis results show that *Apis cerana cerana* in the north-west Himalayas comprises of two separate geographic populations, that have been arbitrarily named *Himachali* and *Kashmiri*. *Apis cerana* from the north-east Himalayas, form a separate cluster from the bees of the north-west Himalayas, and are a separate race named as *Apis cerana himalaya*. Multivariate discriminant analysis results show that the following 3 geographic populations of *Apis cerana himalaya* are distinguished: (1) The foot hills of the Himalayas, (2) the Brahmaputra valley and Khasi hills, and (3) the Naga and Mizo hills (Mattu and Verma, 1983, 1984 a,b; Singh *et al* 1989).

References :

1. Mattu, V.K. and L.R. Verma 1983. Comparative morphometric studies on Indian honeybee of the north-west Himalayas. 1 Tongue and antenna. J.Apic.Res. 22:79-85.

2. Mattu, V.K. and L.R. Verma, 1984 a. Comparative morphometric studies on Indian honeybee of the north-west Himalayas. 2. Wings. J. Apic. Res. 23:3-10.

3. Mattu, V.K. and L.R. Verma 1984 b. Comparative morphometric studies on Indian honeybee of the north-west Himalayas. 3 Hind legs, tergites and sternites. J. Apic.Res. 23:117-122.

4. Singh, M.P., L. R. Verma and Howell V. Daly. 1989. Morphometric analysis of the Indian honeybee in north-east Himalayan region. J. Apic. Res. (in press).

A PHYLOGENETIC ANALYSIS OF HONEY BEES (HYMENOPTERA :APIDAE : *APIS*)

Byron Alexander, Snow Entomological Museum, Snow Hall, University of Kansas, Lawrence, Kansas 66045 U.S.A.

A quantitative cladistic analysis of 6 species and species groups in the genus Apis using 21 characters of adult morphology and behavior found a single most parsimonious cladogram with a consistency index of 0.94 (autapomorphies not included in calculation). Detailed discussion of this analysis is presented elsewhere (1), but synapomorphies for each clade shown in Figure 1 are listed below:

Genus Apis: compound eyes hairy in all adults, meeting dorsally in males; angles formed by intersection of forewing veins 2 rs-m and 3 rs-m with vein M less than 45°; queens with numerous ovarioles in each ovary; workers with barbed sting shaft; sting sheath in queens and workers unpigmented, with extremely short, inconspicuous setae; venter of metasomal segment 8 in queens and workers a conspicuous membranous, interiorly setulose bulb surrounding the base of the sting shaft; proboscis of males and queens reduced; male endophallus enormously enlarged.

florea + andreniformis: loss of distal abscissa of hindwing vein M (also occurs in mellifera); adult males with a thumblike process bearing a dense pad of stiff bristlelike setae on its inner surface; flagellum of male antennae conspicuously shortened.

dorsata group + (mellifera + (cerana + koschevnikovi)): ventral gonocoxite desclerotized or absent; dorsal gonocoxites reduced, widely separated mesally; gonobase of genital capsule lost; metageotaxis (2).

dorsata group (= Apis dorsata, A. laboriosa, A. binghami, A. breviligula): males with dense pads of frondlike setae on ventral face of middle and hind tarsi.

mellifera + (cerana + koschevnikovi): further reduction of dorsal gonocoxites to scale-like sclerites less than half as long as penis valves; male metasomal tergum 8 with vertical arm much longer than horizontal arm; male metasomal sterna 7 and 8 fused and greatly attenuated mesally.

cerana + koschevnikovi: capping of drone cells with a central pore.

The above characters are those for which polarity decisions based on outgroup analysis were straightforward and unequivocal. Other features exhibiting variation within Apis, such as aspects of the dance language and the morphology of the male endophallus, were not included in the analysis because they are unique to Apis and therefore polarities cannot be unequivocally established. Nevertheless, the cladogram supported by this analysis is perfectly consistent with the phylogeny of Apis proposed by Lindauer (3) and von Frisch (4) on the basis of their comparative analyses of the dance language of four species of Apis. It therefore provides independent confirmation of their phylogenetic hypothesis.

References

1. Alexander, B. ms. submitted to Annals of the Entomological Society of America.
2. Jander, R. and U. Jander. 1970. Zeitschrift für Vergleichende Physiologie, 66: 355-368.
3. Lindauer, M. 1956. Zeitschrift für Vergleichende Physiologie, 38: 521-557.
4. Frisch, K. von. 1967. The Dance Language and Orientation of Bees; Harvard University Press, Cambridge, Massachusetts.

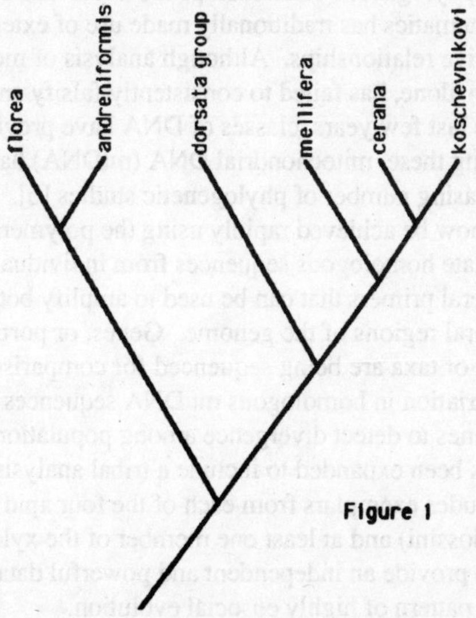

Figure 1

MOLECULAR PERSPECTIVES ON THE PHYLOGENY OF THE APIDAE

S. A. Cameron, Biology Department, Washington University, St. Louis, Missouri 63130, USA

Investigations of insect social evolution often make use of the methods of comparative social behavior. Although these investigations are critical to our understanding of the patterns and processes of social evolution, many descriptive hypotheses have been generated that are not easily subject to rigorous testing [but see 1]. Often, behavioral investigations are made without regard to a phylogenetic framework. In fact, phylogenetic patterns are the best testing grounds for evolutionary scenarios, for homologizing behaviors, and for making decisions about what is primitive and what is derived.

The origin of highly eusocial behavior within the bees remains an exciting and controversial question. There are two highly eusocial bee taxa, the stingless honey bees (Meliponinae) and the true honey bees (Apinae). Both of these groups are found within the family Apidae. Several competing hypotheses have been advanced to explain the origin of high eusociality. One hypothesis advances a single origin, with highly eusocial behavior arising once from a common, highly eusocial ancestor [2]. A second suggests a single origin from a common, primitively eusocial ancestor [3]. A third suggests two independent parallel origins for the highly eusocial systems [4, 5].

Methods to assess phylogenetic relationships are numerous and continue to develop. Hymenopteran systematics has traditionally made use of external and internal morphology to derive relationships. Although analysis of morphology is important, this approach taken alone, has failed to consistently falsify any of the above competing hypotheses. In the last few years, classes of DNA have provided alternative characters for analysis. Among these, mitochondrial DNA (mtDNA) has become the molecule of choice for an increasing number of phylogenetic studies [6]. Because selective gene amplification can now be achieved rapidly using the polymerase chain reaction (PCR), it is feasible to isolate homologous sequences from individual insects. Our laboratory has developed several primers that can be used to amplify both vertebrate and insect mtDNA from several regions of the genome. Genes, or portions of genes, from an increasing number of taxa are being sequenced for comparison. In *Bombus*, I have found sufficient variation in homologous mt DNA sequences from the large and small ribosomal RNA genes to detect divergence among populations, species and subgenera [7]. This study has been expanded to include a tribal analysis of the Apidae. The entire analysis includes exemplars from each of the four apid tribes (Apini, Meliponini, Bombini, and Euglossini) and at least one member of the xylocopid outgroup. The mtDNA sequences provide an independent and powerful data set for inferring phylogeny and the pattern of highly eusocial evolution.

REFERENCES

1. Carpenter, J. M. 1989. Cladistics, 5: 131-144.
2. Michener, C.D. 1944. Bull. Am. Mus. Nat. Hist., 82: 151-326.
3. Prentice, M. and H.V. Daly. In preparation.
4. Winston, M.L. and C.D. Michener. 1977. Proc. Natl. Acad. Sci. USA, 74: 1135-1137.
5. Kimsey, L.S. 1984. Syst. Entomol., 9: 435-441.
6. Harrison, R.G. 1989. TREE, 4: 6-11.
7. Cameron, S.A. In preparation.

This study was supported by a postdoctoral fellowship from the National Institutes of Health.

SCANNING ELECTRON MICROSCOPIC STUDIES OF THE PROVENTRICULUS OF HONEYBEE WORKERS, DRONES AND QUEENS-CORRELATION WITH FEEDING BEHAVIOR

K. Crailsheim and M. A. Pabst
Institut für Zoologie and Institute für Histologie und Embryologie
Karl-Franzens-Universität, A-8010 Graz, Austria.

In the honeybee (Apis mellifera L.) the proventriculus (pv) regulates the flow of food from the crop into the midgut. It can separate liquid and solid nutrients by gulping movements of the 4 partially hairy pv folds (Fig. 1 a,b). This function is esepecially important for workers who consume much more pollen than the other castes.

Studies were done on the pv of 4 and 6-7 day old worker pupae, just emerged workers, foragers, flying drones, and queens that were at least half a year old.

In the youngest pupae the organ has not yet reached its original size and differentiation (Fig. 1c) The final size is reached about two days before emergence. The pv of all three castes has the same principal structure. About two thirds of the folds reach into the crop. The length of the pv in the crop differs; it is 25% and 28% smaller in queens and drones than in workers respectively.

The size difference of the organ between the workers at the one hand and the drones and the queens at the other hand is even greater than the measured one if the larger body size of the latter is considered. The larger pv of the caste which has to digest pollen for other members of the colony (especially for the brood), and has to separate foraged nectar from pollen in the crop, might be an adaptation to this larger need of food-flow regulation. Supported by the Fonds zur Förderung der wissenschaftlichen Forschung.

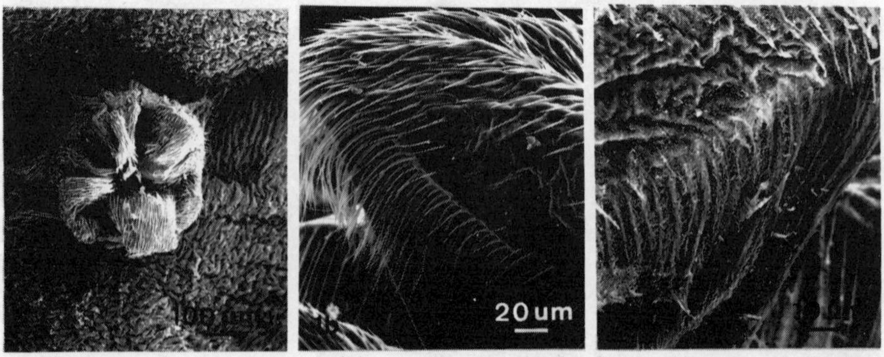

Fig. 1: Proventriculus of a queen (a), tip of a pv fold (b,c) from a worker with filiform hairs split at their ends (b) and from a 4 day old pupa with differentiating cuticle and hairs still stuck together (c).

ABSCONDING DANCE IN *APIS CERANA JAPONICA:* A SLOW AND LONG TAIL-WAGGLING MOTIVATES THE WHOLE COLONY TO TRANSLOCATE

Masami Sasaki
Laboratory of Entomology, Faculty of Agriculture, Tamagawa University, Machida, Tokyo, 194 Japan

DESCRIPTION OF COLONY CONDITION BEFORE AND AFTER ABSCONDING

Sequential behavioral events leading to the absconding were recorded with the Japanese honeybee, A. cerana japonica colony which had been kept in a glass observation hive equipped with two standard Langstroth frames in 27 C, LD-controlled room. Actual absconding occurred on August 17th, 1989.
Day -10: The colony was overpopulated; ca 8,000 workers/3,200 cm comb surface space. There were few stored honey and brood but nothing special in behavioral repartoire.
Day -8 to -3: Number of foragers decreased gradually to 5/min. Guard bees at hive entrance also reduced from ca 10 to none.
Day -2: Very limited activity at entrance and the "absconding dance" began on brood area (no actual eggs and larvae). Many workers inserted their head and body into empty cells in the upper comb and no oviposition by the queen was noticed.
Day -1: The dance continued at low frequency.
Day 0: The dancing bees and colony performance were video-recorded. Number of the slow-dancing bees increased and the absconding occurred at 13:50. All bees left the hive within 4 min and made an absconding swarm under a tree branch (Cryptomeria japonica) of ca 10 m above ground and ca 40 m apart from the mother hive. Number of bees remained was only 75 and all of which were within several hours after emergence. Remained sealed brood were 234 (on the day of emergence, 79; day -1, 45; -2, 40; -3, 21; -4, 8 and -5, 1). The swarm was caught and re-introduced into a 3-frame observation hive where the ample space and stored honey were available.
Day +3: Bees were quite relaxed and no more absconding dance was observed.

CHARACTERISTICS AND SPECIFICITY OF "ABSCONDING DANCE"

1. The absconding dance seemed to be first performed by a few specific workers and the number of dancers increased during few hours before the colony leave. The dance was performed within the hive (not in swarm) and the majority (maybe all) of bees danced without scouting.

2. The dance was composed of repetitive (but nonrhythmic) slow and long tail-waggling runs for 8 to more than 60 s (average 16 s) without return runs like in foraging dance (Fig. 1). The dancer moved around in various parts of the upper comb but never in the usual foraging dance area located close to the entrance.

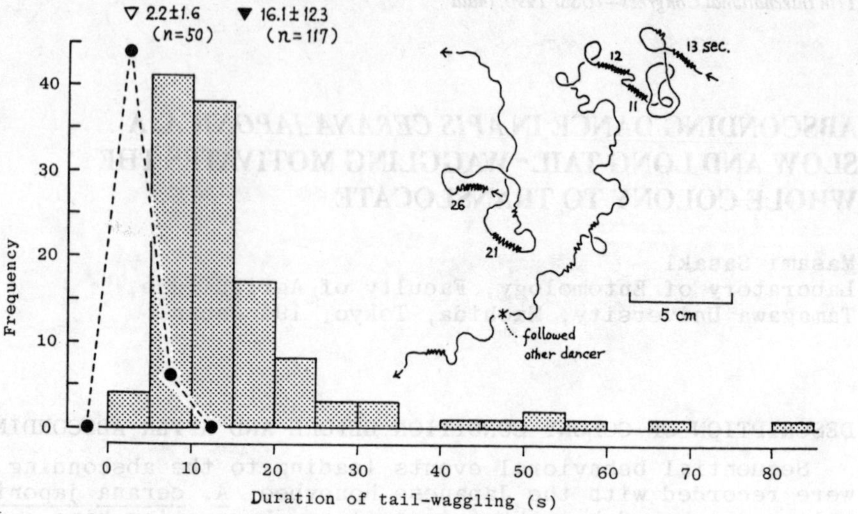

Fig. 1. Comparison of the frequency distribution of the duration of tail-waggling phase between absconding dance (av. 16.1± 12.3 s) and usual foraging dance (2.2±1.6 s, dotted line). The trace shows the movement of two dancers during 5 min (1 h before absconding).

3. Duration of waggling varied even in sequential ones by single bee. Interval time between waggling phases was also irregular. Sixteen seconds (average) of waggling duration is equivalent to ca 6 km if it was a communication for foraging site, and the figure is far beyond the estimated foraging limit (3 km) in A. cerana japonica (based on our unpublished data).

4. Direction (angle to gravity axis) of waggling run tended to be constant not only in sequential performance in single bee but also in all dancing bees. It changed in counter-clock direction (ca 15°/hr) to compensate sun's movement.

5. There was no apparent difference in the behavioral pattern of dance followers (5.5 ± 1.3 bees at once) except that they stayed in the hive.

6. It was so quiet and no other colony activities during several hours prior to the pandemonium of absconding flush.

FUNCTIONAL ASPECTS

The function of this significantly-modified dance system seems to "motivate" whole colony member to leave the mother nest and translocate. Probably the new nesting site is not yet specified at this phase and also, slow tempo does not necessarily mean long distance migration. Recently, Ratnieks (personal communication) found that the scout bees of Africanized swarms perform an interesting slow dance before long distance migration (presumably to an unspecified goal). In cerana absconding dance, however, the dancer performed "within" the hive. Unity among individuals and compensation with the lapse of time in the angle of waggling run appears to be remaining fragment of the originally evolved system and to have no functional meaning.

MATING SUCCESS OF BOTH NATIVE *APIS CERANA JAPONICA* AND INTRODUCED *A MELLIFERA* IN SYMPATRIC CONDITION

Tadaharu Yoshida and Jûro Saito
Institute of Honeybee Science,
Tamagawa University, Machida-shi,
Tokyo, 194 Japan

Difference in the mating flight time in sympatric Apis species is thought to be a key factor of interspecific reproductive isolation (Koeniger and Wijayagunasekera, 1976; Koeniger et al., 1988). Intraspecific variation of flight time in different races and lines of A. mellifera was also reported (Rowell et al., 1986). Ruttner (1988) mentioned about reproductive barrier between native mellifera and introduced A. cerana indica in Europe. However, there have been no detailed information on the interaction between native cerana and introduced mellifera in Asia. The mating flights in both queens and drones of the two species were compared in the same biotope in Tokyo.

Materials and Methods

This study was done in the normal mating season for both honeybee species, from middle May to middle June in 1989. Queen mating flights were observed with the aid of a specially designed queen excluder fitted to the entrance of nucleus hive (Koeniger, 1984). When the queen appeared at the entrance, the excluder was taken away and let her fly away. The returned queen was carefully inspected for a mating sign and then allowed to enter. The time and duration were recorded for all flights. Drone flights were observed using modified UC Davis-type observation hives (Gary, 1976) placed in 26± 2°C, LD12:12 rooms and normal hives placed outside.

Results and Discussion

Total 39 flights of 7 queens for cerana and 48 flights of 8 queens for mellifera were observed (Fig.1). Cerana queens departed on flights between 1315 and 1715 with the peak at 1530. Four queens returned with the mating sign from 1445 to 1600. While mellifera queens departed on flights between 1230 and 1500 with the peak flight at 1330. Five queens returned with the mating sign from 1330 to 1430.

The flight duration of cerana queens ranged from 0.5 to 37 min with the average 27.9 min. Mating succeeded in the relatively long flight between 16.5 and 37.4 min. Orientation flight was supposed to be within 15 min. The success-

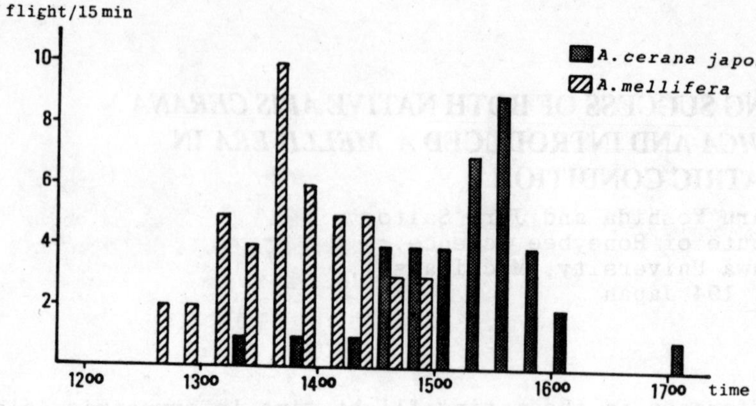

Fig.1 Comparison of the queen flight time between *A. cerana japonica* and *A. mellifera*

ful mating flights were between 5 and 18.2 min with the average of 11.4 min, which is significantly shorter than that in cerana. Orientation flight was estimated as short as 5 min or less.

Judged from the presence or absence of a mating sign, cerana queens flew 1 to max 5 times, during which more than one mating was made, to fill their spermathecae (no. of observed queens = 4). Mellifera queens, on the other hand, flew 2 to 3 times to complete matings (n = 5). In queens of both species, mating flight occurred at 6 to 11 days old.

Drone flight time was well synchronized with that of queen in both species: ranged from 1315 to 1615 with the peak 1430 to 1530 in cerana and 1215 to 1530 with the peak 1230 to 1300 in mellifera, respectively.

Considering from these time distribution, there seems to be chronological reproductive isolation between the two species, before 1430 for mellifera and after 1430 for cerana. This might be the major reason allowing co-existence of the native and introduced honeybees in Japan.

References

Gary, N. E. 1976. How to construct and maintain an observation bee hive, Division of Agricultural Science, Univ. California. Leaflet 2853.

Koeniger, G. 1984. Apidologie 15:198-204.

Koeniger, N. and H.N.P. Wijayagunasekera. 1976. J. Apic. Res. 15:67-71.

Koeniger, N., G. Koeniger, S. Tingek, M. Mardan and T. E. Rinderer. 1988. Apidologie 19:103-106.

Ruttner, F. 1988. Biogeography and taxonomy of honeybees, Springer Verlag, Berlin. P. 284.

THE MEDIAN AND LATERAL OCELLI CONTROL THE FEEDING FLIGHT IN THE HONEYBEE

Gerald KASTBERGER
Institute of Zoology, Dept. of Neuroethology,
University of Graz, A 8010 Graz, Austria.

The ocelli are known to play some role in orientation [1]. There is, however, strong evidence now from ethometric experiments of free-flying honeybees that the ocelli are more than a simple device for phototaxis control. They apparently have the job of keeping irritation low by determining whether phototactic orienting or pattern-induced orienting behaviour is more important in a particular state of motivation. In this study [2] the role of the lateral ocelli versus median ocellus in the control of the flight course was analysed. The bees under test were trained to cross an arena in free flight to get to the feeding place. They were stimulated artificially by side light switches, and it was observed by video how they compensated for flight course errors. 8 groups of bees (fully sighted and bees with the ocelli occluded) were compared, measuring temporal and spatial parameters of course and yawing responses to light stimuli in the time domain of hundreds of milliseconds after side light on- and off-switching.

The resulting 32 patterns of responses (course/yawing, on/off, 8 conditions of ocellar occlusion) were outlined in rate plots and mean curves. They can be plausibly interpreted by two minimal model networks (Fig.1) postulating 5 subsystems (the ocelli as an illumination detector (A), the median (B) and lateral (C) ocelli and the compound eyes (D) as event detectors, the median ocellus (E) inhibiting off-effects) and applying 4 rules (the ocelli reveal dark activity (a), occlusion of the ipsilateral ocellus inverts the responses (b), the subsystems are (c) multiplicatively joined and (d) controlled by gating) which reveals a hierarchical order in the event-detector subsystems.

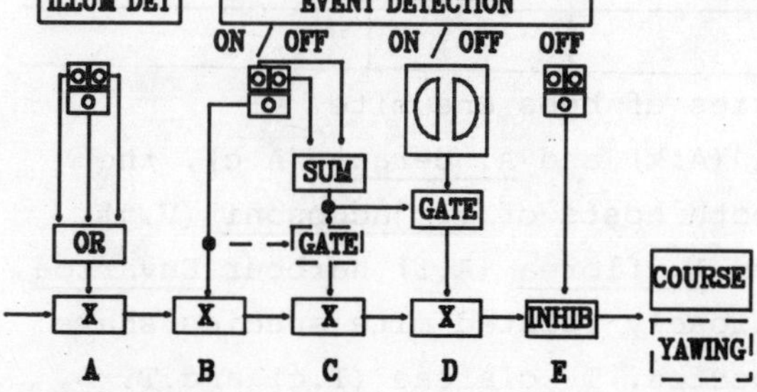

Fig.1. Minimal model networks for visual control of course and yawing after on or off light switch. Yawing control differs only in an additional gating instance (dotted lines). Here the median ocellus can be considered as a master instance gating the lateral ocelli.

1. Kastberger, G. 1990. The ocelli control the flight course in honeybees; Physiol. Ent. (in print).
2. Kastberger, G. 1990. The role of the median and the lateral ocelli of free-flying honeybees in yawing and course control; Zool. Jb. Physiol. (in print).

COEVOLUTION OF ASIAN HONEY BEES AND THEIR PARASITIC MITES

N. Koeniger

Institut für Bienenkunde (Polytechnische Gesellschaft)

Fachbereich Biologie der J.W. Goethe-Universität

Karl-von-Frisch-Weg 2, D 637 Oberursel, F.R. Germany

The western honey bee *Apis mellifera* was originally allopatric to the other Apis species, which are summerized as 'Asian' in this paper. When *A. mellifera* is brought into contact with the Asian honey bee species, the natural balance among the sympatric bee species and their parasites is disturbed. For example, two Asian mite species (*Varroa jacobsoni*, *Tropilaelaps clareae*) have recently changed over to *A. mellifera*. Therefore, I have only considered reports from countries and areas in Asia where *A. mellifera* is not yet present.

ASSOCIATION OF BEES AND THEIR PARASITIC MITES

Tab. 1 Natural association of mites and bees

	V.u	V.j	Ev.s	T.c	T.k
A.a			+		
A.f			+		
A.c	+	+			
A.k		+			
A.d				+	+
A.d.l				+	+

Some closely related species of bees one mite species. *A. koschevnikovi* (A.k) and *A. cerana* (A.c), the cave dwelling bees, are both hosts of *V. jacobsoni* (V.j). *A. andreniformis* (A.a) and *A. florea* (A.f) harbour *Euvarroa sinhai* (Ev.s). Further, closely related mite species share the same honeybee host species. *T. clareae* (T.c) and *T. koenigerum* (T.k) are reported from *A. dorsata* and *A. dorsata laboriosa*. V. *underwoodi* and *V. jacobsoni* are found in *A. cerana*. So, the natural distribution of the mites corresponds well to the systematic position of their honeybee host species. This seems to indicate a long coevolution between bees and their parasitic mite species.

COEVOLUTION OF PARASITIC MITES AND THEIR BEES

Several behavioural reactions (grooming etc.) of the bees suppress the mite population. But, some characteristics of the mite also limit their multiplication in the colony (reproduction only on drone brood etc.). The natural balance between the parasite and its host is a result of multiple regulations, in which both the mite and the bee seem to cooperate.

The Varroaidae and the Tropilaelapidae seem to have entered the Apis colony independently from each other. The TRopilaelapidae are less adapted to parasitism of honey bees ('complete' males, no parasitims of adult bees). Thus, they may have come more recently compared to the Varroidae, perhaps, after the separation of the A. dorsata complex from the rest of Apis.

PHYLOGENY OF PARASITE-HOST RELATIONSHIP OF VARROIDAE-APIS.

It seems certain that Apis and Varroidae are monophyletic groups. Hence, the common ancester of Apis must have had a parasitic ancestral mite. From this origin A. cerana with V. jacobsoni on one side and on the other side A. Florea with Ev. sinhai have evolved.

The degree of drone brood preference differs significantly between Ev. sinhai and V. jacobsoni. It can be suggested that evolution went over three stages:

1. A mite reproducing on worker and drone brood causes damages to the host colony which in turn reduces the fitness of the mite (like V. jacobsoni and A. mellifera).

2.The mite enters both types of brood but reproduction is limited to drone brood (like V. jacobsoni and A. cerana). During the time the mite spends in a worker brood cell (without reproduction), it might miss a drone brood cell which allows reproduction.

3. In the final stage the female mite can distinguish between drone and worker larvae and does not enter worker brood cells any more. According to this idea the Ev.sinhai/A. florea relation has reached the most advandes stage.

FEEEDING OF JELLY TO ADULT BEES (*APIS MELLIFERA* L.)

K. Crailsheim
Institut für Zoologie, Karl-Franzens-Universität
Universitätsplatz 2, A-8010 Graz, Austria

Trophallactic behavior is known to be an important factor of sociality. Foragers feed other foragers and younger hivemates with nectar; nurses supply jelly, honey and pollen.

MATERIAL AND METHODS

^{14}C-phenylalanine was injected into nurse bees (^{14}C-nurses). They had emerged from a comb in an incubator 8 days before, were marked with small spots of dyes or numbered plates, and were then kept in two small hives with three combs each. One and a half to two hours after injection some of them were assayed for ^{14}C incorporated into the protein fractions of their head, their thorax and their abdomen. The rest of the injected bees was reintroduced into the hive for one night. Some free-flying foragers of each hive had been marked a few days before and together with the ^{14}C-nurses some young drones and workers were introduced. In this way, the age or the function of a part of the population was known: 1 day old workers (1d), 1 day old drones, nurses at an age of 8 days, foragers (usually more then 14 days old) and the queen. The following morning all bees were killed rapidly with CO_2 snow and kept frozen at -20°C. Each bee was assayed for content of ^{14}C. The nurses that had been injected with ^{14}C phenylalanine (^{14}C-nurses) were divided into their three body segments before analysis.

RESULTS AND DISCUSSION

About a quarter of the injected ^{14}C could be detected in the protein fraction of the head of nurses after 1.5 hours. In laboratory experiments these nurses distribute jelly to workers of different ages and to drones (1). The results presented here show the same trend; all age classes investigated were fed by the ^{14}C-nurses (Table 1).

	Number of workers	number of 14C-nurses	relation of fed marked bees to total marked bees 1d	nurses	foragers	drones	% of fed 14C jelly fed to workers	% of workers fed
colony 1	161	3	4/6	5/5	8/10	3/3	28.0	13.7
colony 2	618	5	2/5	3/8	3/23	3/3	13.4	12.0

Table 1
Survey of population of two colonies, the relation of age-defined bees that received jelly from the 14C-nurses to those who did not, percentage of fed jelly received by workers, and percentage of fed workers of all workers in the hive.

The ^{14}C that was injected in the evening minus the ^{14}C found in the morning in the nurses was calculated as being fed; fed amount as percentage of total ^{14}C injected was 26.5% and 33.4% in colony 1 and 2 respectively; exhalation as $^{14}CO_2$ or excretion was not considered. More than 0.01% of the ^{14}C fed by the nurses had to be found in another bee in order to be identified as having been fed. Most fed workers (about 2/3) received less than 0.2% of the distributed jelly but some more than 1%; queens received 0.19% and 0.25% respectively.

Results show that in free-flying colonies jelly is not only fed to the brood and to the queen, but also to a considerable number of workers of any age and to young drones; the latter were already discussed in the literature (2) as having been fed in this way. After emergence, young bees increase their protein content and therefore require protein-rich food. Foragers have a quick turnover of protein (3). Both groups have at the same time a lower level of intestinal proteases as compared to nurses (4). A digestion of protein from the pollen, which is quite difficult to digest, by the nurses and its subsequent feeding as jelly - to bees with other duties - can be interpreted as a special feature of the well known division of labor in the honeybee colony.

REFERENCES

1. Crailsheim, K. 1990. Zoologische Jahrbücher. in press
2. Mindt, B. 1962. Zeitschrift für Bienenforschung, 6: 9-33.
3. Crailsheim, K. 1986. Journal of Insect Physiology, 32(7): 629-634.
4. Moritz, B. and Crailsheim, K. 1987. Journal of Insect Physiology, 33(12): 923-931.

POLYPLOIDIZATION OF FAT BODY TROPHOCYTES IN *APIS MELLIFERA*

Bitondi, M.M.G. (*), Z.L. Paulino-Simões(*) and R.B.R. de Almeida(**).
(*) Faculdade de Filosofia Ciências e Letras de Ribeirão Preto, Univ.São Paulo, Brazil. (**) Faculdade de Medicina de Ribeirão Preto, Univ.São Paulo, Brazil.

We have used the fat body of Apis mellifera to study polyploidy and the effect of juvenile hormone (JH) on DNA replication. The dorsal fat body under the last tergite was extracted, transferred to a gelatinized microscope slide, gentle squashed and fixed in ethanol-acetic acid. Hydrolisis was performed in 5N HCl for 20 min at 30 C. The preparation was stained for 40 min in Schiff's reagent. Microspectrophotometric DNA determination at 570 nm was done with a Zeiss scanning microphotometer. Each nucleus was automatically scanned in 1 um steps under a 40x objective. Measurements of the trophocyte nuclei were carried out on recently ecloded workers and drones and worker pupae (14 days old) treated or not at 10 days (prepupae) with 1 ug of JH. Trophocytes of queens that did not receive treatment were measured just after eclosion. The results show that:

- the distribution of Feulgen stained trophocyte nuclei from adult queens shifts toward higher values when compared with the distribution of untreated adult workers. The difference between these distributions was significant (Kolmogorov-Smirnov test);
- the trophocytes of untreated adult workers can be separated into two classes of nuclei with respect to DNA content. The observed peaks suggest replication of the genome. The distribution of nuclei after JH treatment was not significantly different from the control;
- the detected difference between distributions of trophocyte nuclei from treated and untreated worker pupae was not significant. The same occurred with treated and untreated adult drones.

TRANSFER OF PATERNAL MITOCHONDRIAL DNA IN FERTILIZATION OF HONEYBEE (*APIS MELLIFERA* L.) EGGS

Michael S. Meusel and Robin F.A. Moritz

Bayerische Landesanstalt fur Bienenzucht, Universitat Erlangen/Nurnberg, Burgbergstr.70, D-8520 Erlangen (Federal Republic of Germany)

Mitochondria are inherited exclusively via the female in most eucaryotic organisms. During fertilization only very few paternal mitochondria enter the egg cytoplasm. In honeybees polyspermic fertilization and complete penetration is the rule and many sperms including their mitochondria rich flagellum penetrate in the egg cytoplasm. Hybrid brood of different developmental stages of two subspecies of honeybees (*Apis mellifera carnica, A. m. capensis*) were tested using restriction fragment length polymorphisms and a diagnostic mitochondrial DNA (mtDNA) probe. Densitograms of autoradiographs indicated that the male contribution can represent up to 10% of the total mitochondria in the fertilized egg. Irrespective of this initial high male contribution, the paternal mtDNA was not found in larval stages of the same hybrids. This indicates that the paternal mtDNA is not replicated and degenerates during larval development.

SPECIFIC DISTINCTIVENESS OF *BOMBUS NEVADENSIS* CRESSON AND *B. AURICOMUS* ROBERTSON (HYMENOPTERA : APIDAE)- ENZYME ELECTROPHORETIC DATA

A. Scholl, R.E. Owen*, R.W. Thorp** and E. Obrecht. Universitat Bern, Zoologisches Institut, Baltzerstr. 3, CH-3012 Bern, Switzerland, *The University of Calgary, Dept. of Biological Sciences, Calgary, Alberta, Canada T2N 1N4 & Mount Royal College, 4825 Richard Rd. S.W., Calgary, Alberta, Canada T3E 6K6 and **University of California, Dept. of Entomology, Davis, California 95616, USA.

This contribution is part of a project investigating the systematics of bumble bees by enzyme electrophoretic techniques. The Subgenus Bombias is restricted to North America and, according to present views, consists of two subspecies (Krombein, K.V., Hurd, P.D. Jr., Smith, D.R. and Burks, B.D. 1979. Smithsonian Inst. Press.), B. n. nevadensis and B. n. auricomus. Bombus n. nevadensis occurs principally from transmontane California E to 100°W longitude, where it is replaced by auricomus.

Bombus n. auricomus was orginally described as a separate species, but Milliron (1961. J. Kansas Entomol. Soc. 34:49-61) found that this taxon is only subspecifically different from nevadensis and he stated that all intergradations between the two forms are to be found. Intergradation of the two forms of nevadensis between 102° and 98° W longitude in Nebraska was subsequently observed by LaBerge and Webb (1962. Univ. Nebr. Agr. Exp. Sta., Res. Bull. 205, 38 p.).

In fact, intergradation was only inferred from coat color variation in areas of overlap of both taxa in the mid-western and northwestern United States and in Alberta. However, coat color variation is quite typical for bumble bees and in particular many North American bumble bee species show gradation from one color form to another, resulting in covergence toward local Mullerian mimicry groups (Plowright, R.C. and Owen, R.E. 1980. Evolution 34:622-637; Thorp, R.W., Horning, D.S. Jr. and Dunning, L.L. 1983. Bull. Calif. Insect Surv. 23. 79 pp). Therefore one might alternatively speculate that nevadensis and auricomus are specifically distinct, but that one or the other or both species might converge at the western or eastern limits of their distribution to the eastern or Great Basin color pattern respectively.

We have tested his hypothesis by enzyme electrophoretic investigations, using specimens of both taxa from a wide area of their geographical distribution, including in particular areas of apparent overlap (as suggested by the distribution maps in Milliron, H.E. 1971. Mem. Entomol. Soc. Can. 82:1-80) in southern Alberta, Montana, Nebraska and Iowa. The results are that both taxa are genetically highly differentiated. There is no evidence of intergradation as judged by these genetic data. We conclude that the taxa nevadensis and auricomus are specifically distinct.

CEPHALIC NEUROENDOCRINE SYSTEM OF THE ROCK HONEY-BEE, *APIS DORSATA* (F.) WORKER

D.B. Tembhare and G.N. Paliwal
Department of Zoology, Nagpur University Campus
Nagpur 440 010, INDIA

INTRODUCTION

Besides a good deal of work on the neuroendocrine system, especialy aimed at manifesting hormonal basis of caste differentiation in the honey-bees, no information is, however, available on the rock honey-bee, Apis dorsata [1-3]. The present work has therefore been undertaken on A. dorsata and this report encapsulates an information on structural organization of the cephalic neuroendocrine system (CNS) of the worker rock bee.

MATERIAL AND METHODS

The workers of A. dorsata were collected from the nests hanging on the toll and unaccessible trees in the forest of Vidarbha region (Maharashtra/India) after seeking help from tribal honey-hunters. The cephalic neuroendocrine organs were dissected out in saline solution and were fixed in aqeous Bouin's fixative. After thorough dehydration, tissues were cleared in xylene and embedded in paraffin wax at 58-60°C. The serial paraffin sections were cut at 4-5 µm thickness and stained with Bargmann's Chrome Hematoxylin-Phloxine and Cameron and Steele's paraldehyde fuchsin (PF) staining techniques [4].

OBSERVATIONS AND DISCUSSION

The CNS of the worker rock bee consists of the neurosecretory cells (NSC) in the brain, a pair of corpora cardiaca (CC) and that of corpora allata (CA). The medial neurosecretory cells (MNC), lateral neurosecretory cells (LNC) and posterior neurosecretory cells (PNC) are located in the pars intercerebralis medialis, pars lateralis and mid-posterior regions of the protocerebrum of the brain of A. dorsata, respectively representing almost identical distribution with that of A. mellifera [5]. Althogh Breed [6] proclaimed the presence of two pairs of LNC groups in A. mellifera, only a single pair of LNC could be observed in A. dorsata.

Besides above mentioned three groups, present study interpolates existence of the ventral neurosecretory cells (VNC) in the tritocerebral lobes and the optic neurosecretory cells (ONC) in the internal medullary region of the optic lobes of the worker rock-bee. On the basis of various cytomorphological features and tinctorial properties, the NSC of various groups are classified into AI, A2, B and C cells. The neurosecretory pathways arising from various groups of NSC and leading to the CC are well evident.

The CC consist of cerebral axonal fibres loaded with variable amount of neurosecretory material (NSM) and the intrinsic neurosecretory cells (INC). The histomorphological structure of CC and CA of A. dorsata resembles to a great extent with that of A. mellifera [7] and A. cerana indica [3]. A short nerve (NCA) lies in-between the CC and CA and contains fine granules of NSM. Adjoining part of external wall of the CA is intimately fused with the outer muscle layer of oesophagus, supporting the earlier observations on A. mellifera [7]. The CA are devoid of NSM.

ACKNOWLEDGEMENTS

Award of JRF from DST (New Delhi) under the project of Centre of Science for Villages, Wardha (India) to one of us (GNP) is gratefully acknowledged.

REFERENCES

1 Rembold, H. and G. Ulrich. 1982. In : The Biology of Social Insects, M.D. Breed, C.D. Michener and H.E. Evans, Eds. Boulder Co. USA : Westview Press, pp. 370-374.

2 Ulrich, S. and H. Rembold. 1983. Cell. Tiss. Res. 230 : 49-55.

3 Mishra, R.C. and G.S. Dogra. 1983. Proc. II. Internat. Conf. on Apic. in Tropical Climate. ICAR (India) pp. : 289-300.

4 Panov, A.A. 1980. In : Neuroanatomical Techniques, N.J. Strausfeld and T.A. Miller, Eds., Springer-Verlag, New York, pp. 25-50.

5 Ritcey, G.M. and S.E. Dixon. 1969. Proc. Entomol. Soc. Ontario, 100 : 124-138.

6 Breed, M.D. 1983. J. Apic. Res. 22 (1) : 9-16.

7 Laere, O. Van. 1970. J. Apic. Res. 9 (1) 3-8.

KIN RECOGNITION AND NEPOTISM IN *APIS MELLIFERA*

P. Kirk Visscher

Department of Entomology

University of California, Riverside

Riverside, California, USA 92521

This presentation concerns itself with the question: when rearing queen bees, do honey bee workers skew their investment toward more closely related individuals? There might be an inclusive fitness benefit from doing so, since full sisters (coefficient of relatedness 0.75) are 3 times as closely related to a given worker as half sisters (0.25), and the entire accumulated resources of the colony are left to one or a few new queens at swarming.

Two hurdles must be crossed in answering this question. First, one must distinguish the offspring of different drones (subfamilies) in the colony. Second, to analyze the queens produced (i.e. the product of nepotism), though not behavior toward developing queens (the process), there must be a difference between the distribution of subfamilies among the larvae being reared and that among the adult workers doing the rearing. If this difference is absent, the expected result of nepotistic rearing does not differ from non-nepotistic rearing, since each would result in an expected distribution of adult queens identical to that among workers and queen nurses.

I here evaluate evidence from three studies which have reported detecting nepotism [1,2,3] Visscher reported that workers reared larvae a foreign colony headed by a close relative of their own queen in greater numbers than the larvae they reared from completely unrelated foreign colonies. Noonan reported that in a colony where two subfamilies were visually distinguishable on the basis of the genetic marker cordovan, workers more frequently visited the cells of, and more frequently fed the larvae of the same phenotype (and subfamily) as themselves. Page et al. reported that in a colony in which subfamilies could be distinguished by their phenotype at MDH, that queens were more likely to be reared from larvae of the subfamily which was most overrepresented among the queen nurses, relative to subfamily frequencies among the larvae.

These studies are the strongest evidence for adult-queen larva recognition in honey bees, and used widely varying techniques, all pointing to the same conclusions: that there is a tendency for worker bees to rear more closely related larvae as queens, but that this tendency is quite weak. However, each of these studies suffers from difficulties in interpretation. Visscher's experiments involved two quite unnatural circumstances: transferring larvae between colonies, and a choice between larvae slightly less closely related than half sisters (0.22) and completely unrelated larvae. Also, his results could be explained as a response by the bees to genetically determined odors of the queens which laid the eggs, which would not distinguish eggs and larvae in the natural choice between full and half sisters. Noonan's experiments involved a colony with only two subfamilies, and cordovan phenotypic markers; both these may make discrimination easier. Visscher and Seeley (unpublished) showed that bees which differed from each other in cordovan phenotype, but not in relatedness to groups of test bees, elicited different levels of aggressive behavior from the test bees, raising the possibility that Noonan's results were an artifact of the cordovan genetic markers. Page et al.'s study was closest to a natural context, in that it used three subfamilies, and a marker (MDH) which seems less likely to have direct behavioral effects. However, as presented, Page et al's study involved a statistical analysis which contained a systematic bias toward the reported effect. I have reanalyzed these data, removing the systematic bias, and the nepotism effect is much weaker, but is still statistically significant (p=0.03 rather than 0.001). With the corroboration provided by the above and other studies, it seems that a small but real kinship bias exists in queen rearing.

A model I constructed of rearing decisions in early phases of queen rearing suggests that a bias of only a few percent by workers in their choices of which larvae to feed could lead to the nepotism biases reported. There are several explanations of why nepotistic biases in queen rearing are small. Nepotism may impose a price in colony efficiency (e.g. see [3]) so that colony-level selection disfavors it. There may ambiguities in recognition mechanisms that constrain recognition precision and result in many "mistakes". There may be mechanisms by which nepotists among the workers are restrained by their half sisters, similar to the presence of "policing" mechanisms in worker drone production [4].

Future work on the question of nepotism in honey bee queen rearing should further address the question of just how much of a bias workers show, whether this is dependent on different social contexts (e.g. see [6]), and whether it has effects in normal colonies, and why it is so small when the payoffs to nepotism superficially seem so great.

REFERENCES

1. Visscher, P.K. 1986. Kinship discrimination in queen rearing by honey bees (Apis mellifera). Behavioral Ecology and Sociobiology, 18: 453-460.

2. Noonan, K. C. 1986. Recognition of queen larvae by worker honey bees (Apis mellifera). Ethology, 73, 295-306.

3. Page, R.E. 1989. Genetic specialists, kin recognition, and nepotism in honey bee colonies. Nature 338:576-579.

4. Hillesheim. E. N. Koeniger, and R.F.A. Moritz. 1989. Colony performance in honeybees (Apis mellifera capensis Esch.) depends on the proportion of subordinate and dominant workers. Behavioral Ecology and Sociobiology, 24:291-296.

5. Ratnieks, F.L.W. and P. K. Visscher. 1989. Worker policing in the honeybee. Nature 342:796-797.

6. Hogendoorn, K. and H.H.W. Velthuis. 1988. Influence of multiple mating on kin recognition by worker honeybees. Naturwissenschaften 75:412-413.

A NUPTIAL DISADVANTAGE FOR EUROPEAN HONEY BEES IN VENEZUELA

R. L. Hellmich II [1] and A. M. Collins [2]

[1] USDA,ARS Honey Bee Breeding, Genetics & Physiology Laboratory, 1157 Ben Hur Road, Baton Rouge, Louisiana 70820

[2] USDA, ARS Honey Bee Laboratory, 2413 East Highway 83 Building 204, Weslaco, Texas 78596

INTRODUCTION

The objective of this paper is to test the hypothesis that temperately-evolved honey bees, *Apis mellifera,* have nuptial traits that are maladaptive in the tropics.

MATERIALS AND METHODS

The research was conducted near Acarigua, Portuguesa in Venezuela. For 7 days in March European queens and drones and Africanized queens and drones were observed exiting their colonies. The time queens returned and whether they returned with a mating sign was noted. Five people observed the drone source colonies in a round robin type of arrangement. Each observer counted the number of drones leaving a colony for 90 seconds then moved to another colony. On average each colony was observed every five minutes. A similar arrangement was used for queen observations. Another five co-workers observed mating colony entrances by making continuous rounds. A queen was prevented from leaving a colony by a queen excluder cage. When a queen was found at the entrance a gate was opened and she was allowed to leave. Only data from queens which successfully mated and were laying eggs one week after observations were concluded are presented. A total of 15 drone source colonies and 42 mating colonies were observed.

RESULTS AND DISCUSSION

Africanized queens and drones were well synchronized; mean departure times for queens and drones was 5:10 and 5:14, respectively. European reproductives, on the other hand, were less synchronous. European queens left their colonies 48 minutes before their drones (queen $\bar{x}$ = 4:09; drone $\bar{x}$ = 4:57). This asynchrony might explain why European queens took more (Afr. $\bar{x}$ = 1.7;

Eur. $\bar{x}$ = 3.1; $P < 0.03$) and longer (Afr. $\bar{x}$ = 18.4 min; Eur. $\bar{x}$ = 26.3 min; $P < 0.003$) mating flights.

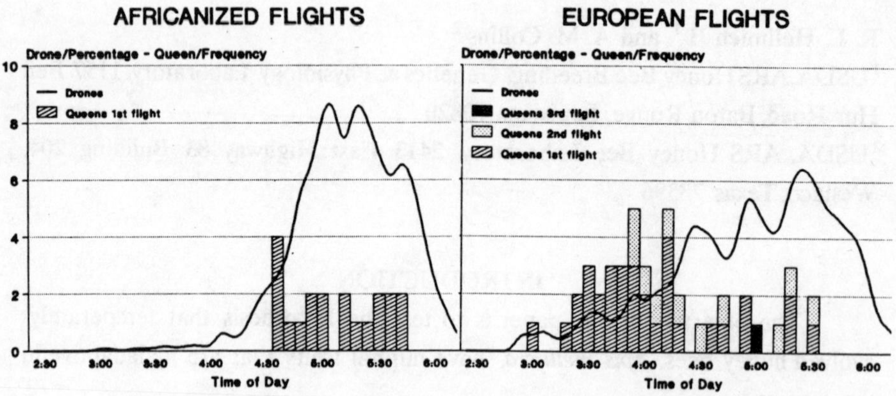

Figure. Time Africanized and European reproductives exited colonies.

The European asynchrony also might have resulted in less efficient matings. European queens had many more unsuccessful flights than the Africanized queens. Every Africanized queen that returned with a mating sign made only one mating flight that day. The Europeans, on the other hand, usually took two and sometimes three flights before they mated. The total time that each European queen was away from her colony averaged 116 minutes. Each Africanized queen averaged much less time away from her colony -- only 42 minutes. Consequently, European queens were exposed to environmental hazards nearly three times longer than Africanized queens. These traits might have contributed to the demise of European honey bees and the remarkable success of Africanized honey bees in the neotropics.

Symposium 5

EVOLUTION AND SPECIATION IN SOCIAL PARASITES

Organizers : **ALFRED BUSCHINGER, F.R. Germany**
STEFANO TURILLAZZI, Italy

EVOLUTIONARY TRANSITIONS BETWEEN TYPES OF SOCIAL PARASITISM IN ANTS, HYPOTHESES AND EVIDENCE

Alfred Buschinger

Institut für Zoologie der Technischen Hochschule Darmstadt
Schnittspahnstr. 3, D-6100 Darmstadt, FRG

Parasitic ants living in dependence of other ant species exhibit a variety of more or less well-defined types of life habits: Xenobiosis (guest ants), temporary parasitism (dependent colony foundation, then independent colonies), dulosis (slave-making), and inquilinism (permanent parasitism, usually workerless species coexisting with the host colony queens). A few workerless species kill the host queens, or invade already orphaned host colonies.

Several hypotheses on the phylogenetic relations among these types have been put forward. They are summarized in a diagram in Wilson (1971) according to which three pathways, from dulosis, temporary parasitism, and xenobiosis, all eventually lead to inquilinism. Hölldobler and Wilson (1990) have modified this scheme in that a fourth pathway from polygynous colony organization might directly lead to inquilinism, polygyny being a precursor also of dulosis and temporary parasitism.

Evidence from our recent field and laboratory studies suggests that transitions between types of social parasitism in fact do occur. Thus, the genus Epimyrma comprises active slave-makers, species with reduced workerforce and slave-raiding, and workerless species. The latter, however, kill the host colony queens as do their dulotic relatives (Buschinger 1989). The genus Chalepoxenus also comprises a number of slave-making and one workerless species, all killing the host queens when invading a colony (Buschinger et al. 1988). Among the parasites of Leptothorax acervorum, a slave-maker (Harpagoxenus sublaevis), a "murder-parasite" killing the host queens (Doronomyrmex goesswaldi), and one or two true inquilines (D. kutteri, perhaps D. pacis) are found. Despite an evidently close relationship among these parasites, however, it is not possible to arrange them in a convincing evolutionary order (Buschinger 1990).

Looking through the list of the about 200 parasitic ants in Wilson

(1971), there are in fact only four clusters of species (related to Formica, Tetramorium, Leptothorax subgenera Leptothorax and Myrafant, resp.) comprising parasites of two or three different types each. An evolutionary transition from the slave-makers or temporary parasites to the true inquilines can hardly be substantiated in any of these groups, whereas some of the workerless "murder-parasites" very probably originated from slave-makers of the respective cluster.

The remaining parasites in Wilson's list are arranged in 15 systematically scattered groups containing exclusively inquilines, and five others with only temporary parasites. Moreover, in some of these groups inquilinism presumably has evolved several times convergently, e.g., nine times in tropical Pheidole species (Wilson 1984). Since in all these groups no slave-makers or guest ants are known which might be the precursors of the inquilines, and no derived inquilines in the groups with temporary parasitism, I hypothesize that most probably the various forms of slavery, temporary parasitism, and inquilinism have evolved independently from one another, inquilinism being the most frequent form. Transition from one type of social parasitism to another has occurred only very rarely, from slavery to a derived form of permanent parasitism without workers, which differs from inquilinism, however, in that the parasites do not coexist with the host colony queens.

References:

Buschinger, A., 1989: Evolution, speciation, and inbreeding in the parasitic ant genus Epimyrma (Hymenoptera, Formicidae). J. evol. Biol. 2, 265-283

Buschinger, A., 1990: Sympatric speciation and radiative evolution of socially parasitic ants - heretic hypotheses and their factual background. Z. Zool. Syst. Evolutionsforschung, in press

Buschinger, A., Cagniant, H., Ehrhardt, W., Heinze, J., 1988: Chalepoxenus brunneus, a workerless "degenerate slave-maker" ant (Hymenoptera; Formicidae). Psyche 95, 253-263

Hölldobler, B. and Wilson, E.O., 1990: The Ants. In press

Wilson, E.O., 1971: The Insect Societies. The Belknap Press of Harvard University Press, Cambridge, Mass.

Wilson, E.O., 1984: Tropical social parasites in the ant genus Pheidole, with an analysis of the anatomical parasitic syndrome (Hymenoptera: Formicidae). Insectes Soc. 31, 316-334

11th International Congress—IUSSI 1990, India

MORPHOLOGY OF THE PARASITIC MYRMICINE ANT

Per Douwes
Department of Zoology, Helgonav. 3, S-223 62 Lund, Sweden.

Social parasites among ants and other Hymenoptera are assumed to be closely related to their hosts. If so, the difference in morphology between a social parasite and the host species or the genus to which the host belongs (Table 1) should reflect an adaptation to a parasitic mode of life. Table 1 shows the material studied. The characters chosen for this study are alitrunk length (=body size) and (relative to body size) head length and width, antenna length, alitrunk height and width, femur and tibia length, and petiolus and postpetiolus length, height and width. Three kinds of social parasites are recognized: slavemakers, euinquilines (coexist with the host queen(s)), and pseudoinquilines (constitute the only queen of the colony).

Table 1. Myrmicine social parasites, hosts and supposed closest free-living relatives analysed in this study.

	Parasite	Host	Closest relative
Sla-	*Epimyrma algeriana*	*Myrafant spinosus*[1]	*Myrafant* spp.
ve-	" *bernardi*	" *gredosi*	"
ma-	" *ravouxi*	" *unifasciatus*	"
ker	*Myrmoxenus gordiagini*	" *lichtensteini*	"
	Chalepoxenus muellerianus	*Temnothorax recedens*	" or *T. recedens*
	Harpagoxenus sublaevis	*Leptothorax acervorum*	*L. acervorum*
Pse-	*Epimyrma adlerzi*	*Myrafant exilis*	*Myrafant* spp.
udo-	" *corsica*	" "	
inqu-	" *kraussei*	*Temnothorax recedens*	"
line	*Doronomyrmex goesswaldi*	*Leptothorax acervorum*	*L. acervorum*
	Anergates atratulus	*Tetramorium caespitum*	*Tetramorium* spp.
Eu-	*Myrmica "microgyna"*	*Myrmica rubra*	*Myrmica rubra*
inqu-	" *hirsuta*	" *sabuleti*	" *sabuleti*
line	*Sifolinia karavajevi*	" *rugulosa*	*Myrmica* spp.
	Doronomyrmex kutteri	*Leptothorax acervorum*	*L. acervorum*
	Teleutomyrmex schneideri	*Tetramorium caespitum*	*Tetramorium* spp.

[1] Not available for this study.

Table 2 lists the eight most important morphological differences. Reduced body size is the most significant parasitic adaptation in queens (except in *Chalepoxenus* and *Harpagoxenus)* and in both female casts heightening and widening of the postpetiolus and to a lesser extent the petiolus have occurred. Queens have elongated antennae and

Table 2. The eight most significant differences between parasite and closest free-living relative.

In queens		In workers	
Alitr. length	−	Postpet.width	+
Petiol. height	+	Postpet.height	+
Postpet. height	+	Petiol.width	+
Postpet. width	+	Eye length	+
Anten. lenght	+	Anten. lenght	−
Head length	+	Petiol.height	+
Head width	+	Petiol.length	−
Petiol. width	+	Alitr.height	+

enlarged head, while slavemaker workers have enlarged eyes and shortened antennae. There is no obvious difference between different kinds of parasitic queens except that slavemaker queens have large eyes (as their workers). The adaptive significance of small size in queens is obvious (no extra resources for independent nest founding) as are the short antennae in slavemaker workers (fighting; there is also a tendency for shorter legs). The large eyes are probably also associatedwith slave raiding (orientation). The pronounced changes in postpetiolus/petiolus proportions still awaits an explanation. Large head and long antennae in queens could be a consequence of small size (allometry).

Table 3. Sum of differences (absolute values) in body proportions between parasitic species and closest free-living relative.

	Species	Slave-makers	Inquilines Pseudo-	Eu-
Tetramorium group	*Teleutomyrmex schneideri*			336
	Anergates atratulus		296	
Myrafant/	5 species	167		
Temnothorax group	3 species		180	
Leptothorax group	*Harpagoxenus sublaevis*	127		
	Doronomyrmex goesswaldi		79	
	" *kutteri*			77
Myrmica group	3 species			86

There is a variation in the degree of morphological specialization among taxonomic groups and kinds of parasites (Table 3). *Tetramorium* parasites are the most specialized ones, probably because they evolved from very large queens and in *Teleutomyrmex* also because the very special habit of climbing on the host queens back evolved. Among the rest the following pattern emerges. Slavemakers are the most and euinqulines are the least specialized ones. Obviously slavemaking requires a specialized morphology, while in euinquilines, which have to obey the rules of the host colony, selection should favour a high similarity to the host (= closest relative). In pseudoinquilines, which should be more in control of the host nest, selection for similarity should be less if any. Thus in the *Myrafant/Teleutomyrmex* group, where pseudoinquilines are as specialized as slavemakers, the former might have a slavemaker origin (degenerate slavemakers), whereas the little specialized *L. goesswaldi* probably has evolved directly from its host.

The size of euinquiline queens (excl. *Teleutomyrmex)* is strongly correlated with host worker size ($r = 0.98$, $n = 4$), the queens being 0.93–1.08 the size of a worker, supporting the idea that such queens mimic host workers . There is also a high correlation between slavemaker workers and host workers ($r = 0.98$, $n = 5$), the slavemakers being 1.00–1.20 times as big as the host.

THE EVOLUTION OF INQUILINE ANT PARASITES : THE INTERSPECIFIC VERSUS THE INTRASPECIFIC HYPOTHESIS

N.R. Franks*, A.F.G. Bourke**

* School of Biological Sciences, University of Bath, BA2 7AY, U.K.
** Department of Zoology, University of Cambridge, CB2 3EJ, U.K.

In ants there are four kinds of obligate, interspecific social parasites. These are (a) temporary social parasites, (b) queen-intolerant workerless parasites, (c) inquilines (queen-tolerant workerless parasites), and (d) slave-makers (1,2). Inquilines, as their designation indicates, are forms which lack a worker caste and whose queens enter the nests of other ant species and produce further parasitic queens and males in the presence of the host queen. They are the most species-rich parasite type, and have evolved many times independently (3).

Almost all inquilines share the following traits: (a) each species is specific to a single host species; (b) inquiline queens and males are smaller than the sexuals of their host species; (c) inquilines are polygynous and parasitize polygynous host species; (d) inquilines are close phylogenetic relatives of their host species. This last trait is known as Emery's rule (1,4). We distinguish between a loose form of Emery's rule (inquilines and their hosts are closely related) and a strict form (inquilines are more closely related to their host species than to any other nonparasitic species) (5).

Controversy surrounds the route by which inquiline ants evolve (3,5). There are two principal hypotheses (6). The first (the interspecific hypothesis) is that one nonparasitic species evolves into an inquiline parasite of another, closely related nonparasitic species (e.g. 1). The second (the intraspecific hypothesis) is that inquiline species arise when intraspecific inquilines speciate from their host stock (e.g. 3).

It is hard to discriminate between these two hypotheses because both can account for the shared traits of inquiline species that we outline above. However, in our view the intraspecific hypothesis more plausibly explains inquiline/host species pairs following the strict form of Emery's rule. This is because phylogenetic constraints on host choice do not seem severe enough to account for the strict Emery's rule under the interspecific hypothesis (6).

But the intraspecific hypothesis in turn raises the difficulty of explaining the sympatric reproductive isolation of the parasites. Such isolation conceivably results from divergent breeding behaviour in the parasites associated with their small size. However, more work on inquiline phylogeny, genetics, behaviour and mating biology is required to discriminate between the two hypotheses with certainty. A solution to the enigma of inquiline evolution is especially desirable, as a verified intraspecific route to inquilinism would strongly support West-Eberhard's (7) important "alternative adaptation" hypothesis for the origin of species diversity.

REFERENCES

1. Wilson, E.O. 1971. The Insect Societies, Belknap Press, Cambridge, Massachusetts.
2. Sudd, J.H. and N.R. Franks. 1987. The Behavioural Ecology of Ants. Blackie, Glasgow.
3. Buschinger, A. 1986. Trends in Ecology and Evolution, 1: 155 - 160.
4. Emery, C. 1909. Biologisches Centralblatt, 29: 352 - 362.
5. Ward, P.S. 1989. In: The Genetics of Social Evolution, M.D. Breed and R.E. Page, Eds., Westview Press, Boulder, pp. 123 - 148.
6. Bourke, A.F.G. and N.R. Franks. 1990. Submitted manuscript.
7. West-Eberhard, M.J. 1986. Proceedings of the National Academy of Sciences, U.S.A., 83: 1388 - 1392.

DOMINANCE INTERACTIONS AND SOCIAL PARASITISM IN ANTS

J. Heinze
Zool. Inst. (II), Röntgenring 10, D8700 Würzburg, F.R.G.

Two major pathways have been proposed to explain the evolution of social parasitism and slavery in ants (see [1]). According to the territoriality hypothesis, slave-raiding evolved from territorial contests between free-living species. However, it does not explain the dependent colony founding behavior of the parasite queen. In the second hypothesis, the invasion of host colonies by founding females is thought to be derived from facultative polygyny [2]. The adoption of young inseminated females by conspecific colonies in facultatively polygynous species has in fact been called "parasitism" [3] or "autoparasitism" [4]. Since in many cases of secondary polygyny young females usually are adopted only by a colony if they are closely related to its members, neither the resident queens nor the workers suffer reduced fitness. Polygyny thus does not appear to involve "parasitism" of established colonies by additional queens [5], at least as long as the young females do not try to dominate or replace the old queens.

Dominance interactions and fighting between females in established colonies might characterize an intermediate colony organization between polygyny and monogyny, and therefore be helpful to explain the change from apparent ignorance of each other in polygynous nests of, e.g. *Leptothorax (Myrafant) curvispinosus* [6], to bellicose colony founding and regicide in strictly monogynous parasites, such as *Protomognathus americanus*. Though dominance interactions and usurpation have been reported from cofoundresses of numerous ant species, aggression between queens has been thought to be absent from secondarily polygynous nests.

The behavior of inseminated females has been studied in two undescribed non-parasitic species of *Leptothorax* (s.str.) from North America - one functionally monogynous, the other polygynous. First data suggest that ritualized and openly aggressive behavior may play a role in regulating the fertility of individuals in mature colonies.

In *Leptothorax* sp.A, young females may return into their mothers nests after mating, where they spend several breeding cycles without becoming fertile [7]. After hibernation, females engage in ritualized and openly aggressive interactions which may lead to colony fission or to the emigration of some single females. The responses of individuals during encounters with nestmates apparently reflect the existence of linear dominance hierarchies. In each colony, only the highest ranking individual lays eggs, in most colonies this female presumably is the mother of the other females. Indirect evidence suggests that in some cases a young female might dominate and replace its mother. In one laboratory colony for example, the old queen was found dead about five weeks after the end of hibernation, the antennae had apparently been torn off, and another female had started to become fertile.

Similar aggressive interactions have also been observed in a closely related facultatively polygynous species, *Leptothorax* sp.B (= *L. canadensis* ?).

Leptothorax (s.str.) species are parasitized by the slave-makers *Harpagoxenus sublaevis* and *H. canadensis*, and by several workerless parasites of the genera *Doronomyrmex* and *Leptothorax*. Queens of *Harpagoxenus* [8, 9], and *Doronomyrmex goesswaldi* [10], and most probably also of *L. wilsoni* (unpubl. res.) kill the host queens during colony foundation. Their fighting techniques, though far more expert, are quite similar to the mandible fights among females in *Leptothorax* sp.A.

References

1. Buschinger, A. 1986. Evolution of social parasitism in ants. TREE, 1: 155-160.
2. Buschinger, A. 1970. Neue Vorstellungen zur Evolution des Sozialparasitismus und der Dulosis bei Ameisen (Hym., Formicidae). Biol. Zentralbl., 89: 273-299.
3. Elmes, G.W. 1973. Observations on the density of queens in natural colonies of *Myrmica rubra* L. (Hymenoptera, Formicidae). J. Anim. Ecol., 42: 761-771.
4. Bolton, B. 1986. Apterous females and the shift of dispersal strategy in the *Monomorium salomonis* - group (Hymenoptera: Formicidae). J. Nat. Hist., 20: 267-272.
5. Nonacs, P. 1988. Queen number in colonies of social hymenoptera as a kin-selected adaptation. Evolution, 42: 566-580.
6. Wilson, E.O. 1974. Aversive behavior and competition within colonies of the ant *Leptothorax curvispinosus*. Ann Entomol Soc Am, 67: 777-780.
7. Heinze, J. and Buschinger, A. 1988. Polygyny and functional monogyny in *Leptothorax* ants (Hymenoptera: Formicidae). Psyche, 95: 309-325.
8. Viehmeyer, H. 1921. Die mitteleuropäischen Beobachtungen von *Harpagoxenus sublaevis*. Biol. Zentralbl., 41: 269-278.
9. Buschinger, A. 1968. Untersuchungen an *Harpagoxenus sublaevis* Nyl. (Hymenoptera, Formicidae). III. Kopula, Koloniegründung, Raubzüge. Ins. Soc., 15: 89-104.
10. Buschinger, A. and Klump, B. 1988. Novel strategy of host-colony exploitation in a permanently parasitic ant, *Doronomyrmex goesswaldi*. Naturwissenschaften, 75: 577-578.

AGGRESSION, COMPETITION AND ASYMMETRIES IN THE EVOLUTION OF ANT SLAVERY

Robin J. Stuart, Department of Entomology, University of California, Davis CA 95616 USA

Slavery in ants is an unusual and highly complex phenomenon and debate continues over how it might evolve (1-4). All obligatory slavemakers perform two specialized tasks: a) nonindependent colony foundation in which the parasite queen invades a host nest and usurps the role of reproductive; and b) slave raids in which parasite workers attack neighboring nests, steal brood, and rear it to produce slave workers. This paper will discuss some ideas on the evolution of these behaviors and the general ecological circumstances underwhich obligatory slavery might evolve, especially in leptothoracine ants.

Slave raids are aggressive intercolonial interactions which are provoked by slavemaker scouts and which proceed in a relatively stereotyped manner. Comparisons of the raiding behavior of certain leptothoracine slavemakers and the kind of aggressive territorial behavior which can be elicited among similar free-living species leave little doubt that the latter behavior patterns formed the evolutionary substrate upon which specialized slave raiding evolved in this group of ants (1,4).

The evolution of parasitic colony foundation among leptothoracine ants is not well explained but is probably quite independent of slave raiding *per se*. Queen adoption by secondarily polygynous or functionally monogynous colonies was probably involved (2) and perhaps polydomy (4). Recent discoveries of dominance behavior among queens in certain functionally monogynous species (J. Heinze, pers. com.) might provide the necessary aggressive component which is so characteristic of parasitic colony foundation.

Ecologically, I hypothesize that ant slavery results from intensive competition between closely-related sympatric species, which normally occur at high densities, are polydomous, reproduce by budding (at least facultatively), are similar enough to form successful mixed colonies with one another (2) but which differ enough for one species to evolve into a highly aggressive and invasive parasite of the other. The evolving parasite should be less numerous than the evolving host species and under some degree of stress, perhaps as a direct result of its more successful competitor. Further, suppose that the evolving host is functionally less aggressive, territorial, or discriminating in terms of nestmate recognition and colony defence than the evolving parasite, traits which might relate to the relative success of these species under particular ecological circumstances. Also, suppose that the evolving parasite is functionally monogynous, adopts conspecific queens, and displays aggressive dominance interactions among its queens to the extent that young queens compete aggressively for the dominant role. This might form the basis for parasitic colony foundation and easily evolve into true monogyny so common among slavemakers. The evolving host might have a similar queen status or be facultatively polygynous since this is so common among host species, and either might facilitate the integration of the parasite queen and the formation of genetically diverse colonies (5). The characteristics I have portrayed occur among some sympatric ants to some extent and together these kinds of asymmetries between evolving host and parasite might provoke the evolution of obligatory slavery.

1. Stuart, R. J. and T. M. Alloway. 1983. Behaviour 85: 58-90.
2. Buschinger, A. 1986. TREE 1: 155-160.
3. Buschinger, A. and E. Pfeifer. 1988. Ins. Soc. 35: 61-69.
4. Pollock, G. B. and S. W. Rissing. 1989. Amer. Nat. 133: 61-70.
5. Stuart, R. J. 1988. Proc. Natl. Acad. Sci. (USA) 85: 4572-4575.

GEOGRAPHIC VARIATION IN THE DEVELOPMENT OF PARASITISM IN BEES

W.T. Wcislo
Department of Entomology, Snow Hall
University of Kansas
Lawrence, Kansas 66045-2119, U.S.A.

For 114 surveys of the bee fauna in numerous localities the linear correlation between {the percentage of parasitic bee species} and {the latitude of the area sampled} is $\underline{r} = 0.66$ [1]. Among bees, social parasites and cleptoparasites are quantitatively more abundant in regions at higher latitudes. This cline in abundance of brood parasites apparently also holds for ants and wasps, yet it is opposite to that observed in brood parasitic birds (quantitative data for ants, wasps, and birds are lacking) [1,2,3]. These distributions occur despite the fact that facultative parasitic behavior is observed frequently in all localities, based on available information [1,4]. Together these observations suggest that the fixation of parasitic behavior occurs more frequently under certain environmental conditions.

The ecological and social conditions which may favor the evolution of parasitism include highly synchronized nest-founding within potential host populations [refs. in 1], and high levels of social competition [1,5]. Synchronized nest-founding is ubiquitous in temperate (seasonal) areas, probably due to strong stabilizing selection. Such selection is less likely to occur with the same intensity or constancy in many tropical regions. [In the tropics, strongly seasonal areas should be the better places to look for new parasitic species.] In regions where seasonality is pronounced, any losers in social competition (e.g., due to nest failure) are less likely to be successful as free-living individuals because of the limited time available for independent nesting. This limitation is negligible in aseasonal areas. Due to temporal considerations, the <u>relative</u> success of an alternative mode of reproduction as a parasite is likely to be greater in areas with pronounced seasonality. These findings help explain why facultative parasitic behavior has gone to fixation more frequently at higher latitudes.

Comparisons among facultative and obligate parasites also demonstrate an important feedback linkage between behavior and evolution (i.e., between phenotypes and genotypes). They show how phenotypic changes in behavior alter the conditions under which populations evolve, sometimes selecting for genetic changes in morphology, behavior, physiology, etc. [refs. in 6].

Little is known about the ontogeny of parasitic forms [1]. In many respects structural features of parasitic bees represent the "masculinization" of the body plan of a free-living species, including features not obviously related to parasitism [7]. Scant data suggest that morphological aspects of a parasitic phenotype occur at low frequencies within populations of free-living species. Such structural aberrations occur as developmental anomalies, and can be induced, to varying degrees, by pathogens or the topical application of hormones [8,9]. It is possible that ecological aspects mentioned above may facilitate the expression of these anomalies.

REFERENCES

1. Wcislo, W.T. 1987. Biological Reviews, 62 : 515-543.
2. Wilson, E.O. 1971. The Insect Societies, Harvard University Press, Cambridge.
3. Payne, R. 1978. Annual Review of Ecology and Systematics, 8 : 1-28.
4. Yom-Tov, Y. 1980. Biological Reviews, 55 : 93-108.
5. West-Eberhard, M.J. 1986. Proceedings of the National Academy of Sciences, U.S.A., 83 : 1388-1392.
6. Wcislo, W.T. 1989. Annual Review of Ecology and Systematics, 20 : 137-169.
7. Pérez, J. 1884. Actes de la Société Linnéenne de Bordeaux, 1884 : 1-63.
8. Salt, G. 1927. Journal of Experimental Zoology, 48 : 223-331.
9. Campos, L.A.O. 1978. Journal of the Kansas Entomological Society, 51 : 228-234.

A CHEMICAL BASIS FOR HOST RECOGNITION IN CUCKOO BUMBLE BEES

R.M. Fisher and D.R. Greenwood*

Department of Biology, Acadia University, Wolfville, Nova Scotia, Canada, & DSIR Biotechnology, Department of Scientific and Industrial Research, Palmerston North, New Zealand.

Cuckoo bumble bees (*Psithyrus* spp.) are obligate social parasites of bumble bees (*Bombus* spp.). *Psithyrus* queens lay their eggs in the nests of their hosts and effectively commandeer the host worker force to raise their offspring. Nest and trail odours of host species provide olfactory cues by which *Psithyrus* queens locate nests of their respective host species. The basis for host recognition has been examined for *P. vestalis*, which parasitises *B. terrestris*.

A bioassay was constructed to determine the source of olfactory cues for host attraction. Pentane extracts of body parts and organs from queens of *B. terrestris* were presented to *P. vestalis* queens and their antennal responses were recorded. Of the regions and organs tested, only the Dufour gland and the sixth tergite with its associated epidermal glands were active in the bioassay. From capillary GC-MS examination of these active extracts, compounds common to these two sources and not present in adjacent tergite and sternite regions were identified. Of the 61 compounds characterised in total, several aliphatic esters and branched hydrocarbons were unique to these two gland sources.

These results demonstrate that specific chemicals may be used by *Psithyrus* queens to discriminate host species and that such compounds arise from exudates of specific glands rather than from a general body aroma. That these compounds are active in isolation, rather than in admixture with other volatile compounds, has yet to be confirmed by bioassay.

These studies have implications for nestmate recognition. Such a generic approach examining intra-species recognition, may enable elimination of the compounds from the plethora of all compounds potentially involved, thereby simplifying kin-based recognition studies.

ON THE EVOLUTION OF SOCIAL PARASITISM IN *POLISTES* WASPS (HYMENOPTERA, VESPIDAE)

Rita Cervo and Stefano Turillazzi

Dipartimento di Biologia Animale e Genetica. Università di Firenze. Via Romana 17. 50125 Firenze, Italy.

Taylor [1] proposed four stages in the evolution of social parasitism in wasps. Only the first (facultative, temporary intraspecific parasitism) and the fourth [obligatory, permanent and interspecific parasitism (inquilinism)] are found in the Polistinae - but whilst intraspecific usurpation seems more common [2,3], only the three known species of the *Sulcopolistes* genus, which invade colonies of four species of *Polistes* distributed prevalently round the Mediterranean [4], practise inquilinism.

Usurpation of host colonies by female *Sulcopolistes* occurs slightly after the *Polistes* foundresses have started activity at the beginning of the season, and a similar delay occurs between intraspecific usurpations in *P. b. bimaculatus* and nest foundation, but in this case because the usurper herself has lost her own nest.

Whether the type of strategy used by the obligate parasite to invade the host colony is violent or non-violent, seems to depend on the need to keep the resident foundress on the nest, but *Polistes* females, which usurp a conspecific nest and seldom depend on contribution from the foundress for rearing their brood, always adopt a violent strategy.

Some behaviours presented by the parasites following their arrival on the host colony, such as nest stroking and destruction of part of the immature brood, are common to both to the obligate and facultative parasites, as well as to *P. gallicus* females which have been presented experimentally to a nest of a conspecific female [5]. Nest stroking using the abdominal tergites, which has already been observed in other vespine obligate social parasites [6,7], could be used by the parasite females to apply a pheromone on the nest. Its function could be to induce the workers about to emerge to recognise the intruders as nest mates and consider them as relatives.

The mechanisms by which an alien female penetrates the social system of a colony and takes over control seems similar in interspecific and intraspecific usurpation.

The shorter the colonial cycle, the more advantageous usurpation would be, since the chances of success of a newly founded colony would be very low. Such situation is found in species such as *P. b. bimaculatus* which nidificates in mountain areas where the season favourable for brood rearing is extremely brief.

This seems to confirm that in *Polistes* wasps obligatory, permanent social parasitism has evolved in species in which intraspecific usurpation provides foundresses with a good behavioural alternative if they lose their nest early in the season.

REFERENCES

1] Taylor, L.H. 1939. Ann. Ent. Soc. of America 32: 304-315.

2] Klahn, J.E. 1988. Behav. Ecol. Sociobiol. 23: 1-8.

3] Makino, S. 1989. Insectes Soc. 36: 116-128.

4] Scheven, J. 1958. Insectes soc. 5: 409-438.

5] Cervo, R. and S. Turillazzi. 1989. Ethol. Ecol. Evol. 1: 185-193

6] Jeanne, R.L. 1977. J. Kansas Ent. Soc. 50: 541-557.

7] Reed, H.C. and R.D. Akre. 1983. Insectes Soc. 30: 259-273.

EVOLUTION OF SOCIAL PARASITISM IN YELLOW JACKETS AND HORNETS (HYMENOPTERA : VESPIDAE : VESPINAE)

Hal C. Reed, Biology Department, Oral Roberts University, Tulsa, OK 74171

Roger D. Akre, Entomology Department, Washington State University, Pullman, WA 99164-6432

Vespine social parasites include facultative, temporary parasites with a worker caste (e.g. Vespula squamosa) as well as workerless, obligatory, permanent inquilines (e.g. V. austriaca and Dolichovespula arctica) (1). The widespread occurrence of intra- and interspecific usurpations among yellowjacket and hornet queens (1) has been postulated as the evolutionary origin of vespine social parasitism (2). Successful colony takeover by a conspecific queen has been documented in incipient V. maculifrons colonies (3). Multiple usurpation attempts by conspecific queens of early vespine colonies also occur as several researchers have discovered dead queens in entrance tunnels to subterranean nests (1). Successful interspecific usurpation resulting in a mixed colony of workers of two nonparasitic species has been reported in V. pensylvanica and V. vulgaris (4), and such interspecific queen competition probably occurs among other sympatric vespines.

Differences in the usurpation strategy are seen in the North American parasitic vespines. The workerless, obligate parasite D. arctica passively invades a host colony and coexists for a time with the host queen (5,6). In contrast, V. austriaca parasites aggressively invade host colonies with no evidence of a peaceful coexistence period with the host queen (7), a usurpation strategy much like nonparasitic vespine queens. This inquiline, as well as most vespine social parasites, is active later and for a longer time during the spring and early summer than the host queen (7). D. arctica and V. austriaca exhibited similar behavior during colony occupations, but the latter exhibited more frequent aggression toward host workers (8). The facultative social parasite, V. squamosa, has more reproductive options than the obligatory inquilines and thus serves as an example for discussing selective pressures leading to social parasitism in the Vespinae. V. squamosa queens may initiate and build their own colonies, aggressively invade host colonies, or in subtropical climates, join post-reproductive annual conspecific colonies, thereby creating polygynous, perennial colonies (9). Investigations of quéen-queen interactions in such polygynous colonies could offer insight into queen competition, usurpation, and the development of obligatory parasitism in some species. We hypothesize that queen-queen competition during times of waning queen control and asynchrony in queen activity periods lead to intra- or interspecific usurpations, queen adoption, or coexisting queens (i.e. polygynous associations). The climatic situation (e.g. tropical or temperate), colony stage, and differential queen control in competing queens will affect whether encounters between queens will result in successful usurpation, peaceful coexistence, or a polygynous colony. Social parasitism could then be viewed, along with polygynous vespine colonies, as alternate yet successful reproductive options among the usually monogynous species of yellowjackets and hornets.

REFERENCES

1. Akre, R. D. 1982. Social wasps. In: Social Insects, H. R. Hermann, Ed. Academic Press, New York, vol. IV, pp. 1-105.

2. Taylor, L. H. 1939. Annals of the Entomological Society of America, 32: 304-315.

3. MacDonald, J. F. and R. W. Matthews. 1981. Journal of the Kansas Entomological Society, 54(3): 433-457.

4. Akre, R. D., C. F. Roush, and P. J. Landolt. 1977. Environmental Entomology, 6(4): 525-526.

5. Jeanne, R. L. 1977. Journal of the Kansas Entomological Society, 50(4): 541-557.

6. Greene, A., R. D. Akre, and P. J. Landolt. 1978. Melanderia, 29: 1-28.

7. Reed, H. C. and R. D. Akre. 1983. American Midland Naturalist, 110(2): 419-432.

8. Reed, H. C. and R. D. Akre. 1983. Insectes Sociaux, 30(3): 259-273.

9. MacDonald, J. F. and R. W. Matthews. 1984. Journal of the Kansas Entomological Society, 57(1): 134-151.

PARALLEL SIZE-RELATED ALTERNATIVES IN MALES AND FEMALES AND THE QUESTION OF SYMPATRIC SPECIATION IN HYMENOPTERAN SOCIAL PARASITES

M. J. West-Eberhard
Smithsonian Tropical Research Institute
(Correspondence: Escuela de Biologia, Universidad de Costa Rica, Ciudad Universitaria, Costa Rica, Centroamerica)

Intraspecific competition and phenotypic plasticity commonly give rise to alternative adaptations in both male and female insects [1, 2]. In many cases derived (or secondary) female and male tactics (including surreptitious oviposition or intraspecific social parasitism in females; and alternative mating sites or behaviors in males) are associated with losing out in social competition; and this, in turn, is often associated with small size in both sexes. Positive assortative mating by size is also a fairly common phenomenon in insects [1]. Could assortative mating by small males and females having different expressed phenotypes or "alternative adaptations" (e.g., small parasitic females and small males mating within the nest) have given rise to social parasite species via sympatric speciation? If so, it would seem to explain "Emery's Rule," which holds that socially parasitic species are more closely related to their host species than to any other.

The ALTERNATIVE-ADAPTATION HYPOTHESIS of social-parasite speciation proposes that parasitic behavior and the beginnings of distinctive male mating patterns originated prior to speciation, in the form of alternative reproductive tactics [2, 3]. Selection favoring contrasting alternatives can lead to considerable behavioral and morphological divergence between alternatives, without reproductive isolation of forms -- a process of intraspecific divergence like that producing strikingly different workers and queens in the social insects. These adaptations, originally alternatives, then could come to characterize a reproductively isolated lineage via a process of "phenotype fixation" in which other (non-parasitic) options were lost [2, 3]. In theory, this process (of phenotype fixation and reproductive isolation) could occur in allopatry, with females originally disposed to intraspecific parasitism beginning to parasitize other (especially, closely related) species, and this behavior coming to characterize all members of the species. Or it could occur in sympatry, e.g. via assortative mating of males and females having parallel (size-related) alternative adaptations including parasitism. In the sympatric case, genetic divergence between a parasitic and non-parasitic form would require (in lieu of geographic isolation) that there be a genetic component to the determination of alternatives (e.g., affecting individual size, and hence adoption of complementary behavioral alternatives and assortative mating in both sexes). The linkage of both parasitism and mating behavior to size accomplishes the simultaneous divergence of both, often considered a problem for such (sympatric speciation) hypotheses [e.g. 4]. This hypothesis differs from previous ones [e.g. 4] in having extensive divergence occur prior to, and independently of, reproductive isolation (assortative mating).

Emery's Rule, if strictly true, would support the sympatric version of this hypothesis; but in fact only an approximate Emery's Rule applies in most cases (e.g., even in workerless inquiline ants). Parasites are often so derived as to make assignment of relationships difficult; or the taxonomy and phylogeny of the group concerned is often too poorly known to allow the precise statement of relationship implied by the Rule (sensu strictu). The sympatric version of the alternative-adaptation hypothesis is supported by evidence of recurrent mutations affecting size, eg. in microgyne ants [5]. Sympatric speciation of ant social parasites may sometimes have occurred via rapid host transfer involving closely related species pairs, rather than in situ speciation within host colonies. There is evidence for this in the genera Leptothorax, Pogonomyrmex, and Pseudomyrmex [6]. Divergence in the form of alternative adaptations, as well as rapid host transfer, may give the impression that intranidal sympatric speciation has occurred when it has not.

1. Thornhill, R. and J. Alcock. 1983. The Evolution of Insect Mating Systems. Harvard Univ. Press, Cambridge, Mass.
2. West-Eberhard, M. J. 1989. Annu. Rev. Ecol. Syst. 20:249-278.
3. West-Eberhard, M. J. 1986. Proc. Natl. Acad. Sci. USA, 83:1388-1392.
4. Maynard Smith, J. 1966. Amer. Nat. 100:637-50.
5. Cammaerts, M. C., R. Cammaerts and H. Bruge. 1987. Annls Soc. r. zool. Belg., 117: 147-158.
6. West-Eberhard, M. J. in prep.

BEHAVIOUR OF A NEW SOCIALLY PARASITIC INDIAN BEE, *BRAUNSAPIS KALIAGO*, IN NESTS OF ITS HOST, *B. MIXTA*. (HYMENOPTERA : XYLOCOPINAE)

Suzanne W. T. Batra, USDA-ARS, Beneficial Insects Laboratory, Beltsville, MD 20705, USA: Yasuo Maeta, Faculty of Agriculture, Shimane University, Matsue 690, Japan; Shoichi F. Sakagami, Institute of Low Temperature Science, Hokkaido University, Sapporo 060, Japan.

Both sexes of Brausapis kaliago (Reyes and Sakagami, 1990) live in nests of B. mixta in northwestern India. Direct intranest observations revealed that female B. kaliago may functionally replace host queens, much as do Psithyrus in Bombus nests. Parasites usually avoided encounters with hosts by cringing behavior and by quietly resting near the brood. Dominance was sometimes asserted by means of oophagy, and by several specialized agonistic behaviors. Established females became unable to fly effectively, and if pushed out of nests by hosts, they could not return. One parasite evidently was killed by hosts. They relied on host food stores deposited in nests or on larvae, and also aggressively solicited food from hosts by trophallaxis. Both hosts and cleptoparasites groomed the brood and removed the chorion during hatching. Larvae were fed by the ventral deposition of provisions by both hosts and parasites. Adult-larval trophallaxis was also seen. These nonforaging cleptoparasites may reduce the population of foraging, and pollinating, host bees.

INTRASPECIFIC NEST USURPATION IN THE LARGE CARPENTER BEE *XYLOCOPA SULCATIPES* MAA (APOIDEA : ANTHOPHORIDAE)

Roland E. Stark
Albert - Ludwigs - Universität Freiburg, Institut für Biologie I, Zoologie, Albertstraße 21 a, D - 7800 Freiburg i.Br., West Germany.

Introduction: The large carpenter bee *X. sulcatipes* typically nests in twigs of dead branches where the female excavates a straight tunnel (Gerling et al., 1983). Females also use pre - existing cavities like in Arundo, Ferula and other substrates occurring naturally in the habitat. Thereafter she builds a linear array of cells. Each cell is provisioned with pollen and nectar, an egg is deposited upon the bee bread and the cell is sealed. Both solitary and social nests occur. Social nests may be formed in several ways. Associations of overwintering females occur in old nests at the onset of the breeding season. Females disperse from and also join these hibernating assemblages. In such a nest association in which breeding subsequently occurs quasisociality may result from incomplete dispersal of females, producing two-female associations. New associations of 2 females were also formed in empty nests throughout the breeding season. On rare occasions a nest association was formed by a female joining an established two -foundress association where brood cells had already been produced. In these nests one of the two original cofoundresses deserted the nest, leading to a new two-female association. Quasisociality was generally of a temporary nature, because dominance relations developed leading to a semisocial association. Other females remained solitary, and some associations dissolved before breeding began, giving rise to solitary nests. Later in the season, after maturation and emergence of the first brood, one or more of the young daughters stayed in the nest, giving rise to a matrifilial group. In these nests the individuals show a primitive reproductive division of labour. The mother produces a second brood while one of her daughters acted as a guard. If the mother died or was expelled from the nest and two daughters remained, a semisocial nest was formed (Stark, et al., 1990 and in manuscript).

Usurpation of single - founded nests: Solitary nests were taken over frequently. In 1986 I could observe 19 successful nest take overs which represents 52 % of the cases of solitary nest initiations. In 1987, though, only 7 % of the 31 solitary nests were taken over. The probability of nest usurpation was strongly affected by the stage of nest development. Most successful nest usurpation attempts generally happened when the nest owning female was foraging, hence, the nest was left unguarded. When returning from the foraging trip the female then was faced with the usurper positioned at the nest entrance. Usually extremely agonistic interactions followed, including head to head pushing, biting of legs or antennae, and stinging attempts by the contestants. In two cases it was observed that a potential usurper started to fight a resident female while the latter was guarding at the nest entrance. In all cases where the nest has been taken over successfully the brood present got destroyed immediately or several hours later. The cell partitionings were removed and evicted from the nest as well as the brood. Eggs usually get devoured by the usurper. Where the brood in a cell had reached a larval stage, the remaining provisions were reused by the usurper to provision a new brood cell (Blom and Velthuis, 1988).

Usurpation of social nests: Nests that have reached the matrifilial stage (18 in 1986, 12 in 1987; the imagines have just emerged from the brood cells) were successfully taken over (n = 2) only when the mother was out collecting food for the freshly emerged bees. Usually it took an average of 6 days (5 till 11) after emergence of the young bees until an active guard was present at the nest entrance. It also happened that a guarding daughter was thrown out by an alien bee while the mother was absent. When returning to the nest the mother fought the alien female. After a short fight, though, the mother gave in. In the two successful cases the imagines and the remaining pupae where either pushed out of the nest or adopted by the usurper and a de facto matrifilial nestassociation got established.

Nests presenting the semisocial (n = 8) and quasisocial (n = 28) state were never usurped successfully. In all those cases where an alien bee tried to usurp a nest she was either successfully defeated by the guarding bee or the fight was intercepted by the return of the foraging female.

Conclusion:

Social Nests are protected from becoming usurped, while solitary nests are frequently taken over by an alien bee.

Literature

Blom, J. van der & Velthuis, H. H. W. 1988. Social behaviour of the carpenter bee *Xylocopa pubescens* (Spinola). Ethology. 79: 281 - 294.

Gerling, D., Hurd, P. D., Jr. & Hefetz, A. 1983. Comparative behavioral ecology of two Middle East species of carpenter bees (Xylocopa Latreille) (Hymenoptera: Apoidea). Smithsonian contributions to Zoology, No. 369: 1 - 28.

Stark, R., Gerling, D., Velthuis, H. H. W. & Hefetz, A. 1990. Reproductive competition involving oophagy in the socially nesting bee *Xylocopa sulcatipes* Maa (Apoidea: Anthophoridae). Naturwissenschaften, in press.

Symposium 6

CASTE DIFFERENTIATION IN SOCIAL INSECTS

Organizer : **KLAUS HARTFELDER, F.R. Germany**

THE DEVELOPMENTAL BASIS AND EVOLUTION OF WORKER POLYMORPHISM IN ANTS

Diana E. Wheeler, Department of Entomology, University of Arizona 85721 USA

A physical worker caste can be defined as a set of workers that develops under the same developmental program, a program that includes both growth parameters and the setting of adult size. The number of physical castes characteristic of a species is the number of sets of growth rules used to derive the entire worker caste system.

The diversity of worker castes in ants can be accounted for by three aspects of development: the setting of adult size, the setting of growth parameters such as exponential growth rates, and the reprogramming of adult size and growth parameters.

Two descriptive methods are available for assessing worker castes. First, double-log plots show the relationship between two body dimensions. The form of the plot can be used to estimate growth parameters and to suggest that these were reprogrammed. Second, size frequency distributions, preferably of pupae, show the range and frequency of worker sizes produced under a given set of environmental conditions. The form of the size frequency distribution indicates whether adult size was reprogrammed in some individuals. Confirmation of a revision of adult size can be obtained through laboratory observation and manipulation of larval growth.

All patterns of worker polymorphism examined so far can be accounted for by serial reprogramming. In serial programming, developing larvae face a succession of choices between (a) metamorphosing according to the current developmental rules or (b) revising the rules. Revision may involve resetting adult size alone or resetting both adult size and growth parameters.

Atta and *Pheidologeton* represent caste systems that appear to involve successive resetting of adult size without alteration of growth parameters. Double-log plots are linear and allometric while the distribution of worker size has as many as four modes, representing up to four physical castes. The soldier caste in *Eciton*, with its hook-shaped mandibles, is clearly differentiated from other workers. The *Eciton* caste system could be generated by two sequential reprogrammings of adult size to generate three size modes of workers and one final reprogramming of both adult size and growth parameters resulting in the distinct soldier caste.

Worker caste complexity can evolve by the addition of reprogramming events, by the modification of growth parameters associated with the program for each physical caste, and by the adjustment of threshold values for the decision to revise. Modification of growth parameters would adjust the shape of workers in a caste, such as soldiers in *Eciton*. Adjustment of threshold requirements would shift the proportion of workers in each caste.

Do ecological or developmental factors constrain the evolution of worker castes? The maximum number of physical worker castes in species examined to date is four. No worker caste includes workers of similar sizes but different shapes, which would require that the two worker pathways diverge early. Yet early divergence is possible, since it is found in some queen vs. worker developmental lines. The disadvantage of early divergence as well as castes added by serial reprogramming could be the increased time required to realign perturbed caste ratios. The advantage of increased ergonomic efficiency contributed by additional physical specialization may not offset the disadvantage of decreased flexibility in adjusting caste ratios to environmental and demographic variability.

THE GENETICS OF QUEEN POLYMORPHISM IN LEPTOTHORACINE ANTS

J. Heinze[1], A. Buschinger[2], and U. Winter[3]

[1] Zool. Inst. II, Universität Würzburg, FRG
[2] Zool. Inst., TH Darmstadt, FRG
[3] FB II (Biologie), Universität Bremen, FRG

In some species of ants, two morphologically different types of functional queens occur. Queens may either be gynomorphic, as in most species (i.e., primarily winged and having a thorax consisting of distinct sclerites), or intermorphic (i.e., more or less workerlike, wingless, and having widely fused thoracic sclerites).
In the myrmicine tribe Leptothoracini, queen polymorphism is known in the xenobiotic genus *Formicoxenus* [1], the slave-making ant *Harpagoxenus sublaevis* [2], and in two non-parasitic species, *Leptothorax sphagnicolus* [3] and *L.* sp.A [4]. In all these species, intermorphic females may be inseminated and fertile; the presence of a spermatheca easily distinguishes them from ergatomorphic workers. Intermorphs have been found to be the only fertile queens in approximately 99 percent of over 1100 *Harpagoxenus* colonies from southern Germany (Buschinger, unpubl. res.). In *L.* sp.A, the frequency of intermorphic and gynomorphic queens varies from population to population and appears to be influenced by ecological qualities of the habitat [5].

Breeding experiments have shown that in *H. sublaevis* and *L.* sp.A queen morph is genetically mediated, apparently by one single diallelic locus, E/e. Brood homozygous in e may become either winged female sexuals or workers; EE- and Ee-larvae develop into wingless intermorphic females and workers [5, 6, 7]. The E/e system strongly affects caste determination in both species [5, 8]. *H. sublaevis* colonies with an ee-female inseminated by an e-male, and colonies with an EE-female inseminated by an E-male, produced equal numbers of female offspring during four breeding periods. However, the ratio between workers and females differed significantly: whereas in ee x e colonies 60.3 percent of all female larvae developed into sexuals and 39.7 percent into workers, only 18.8 percent of female offspring in EE x E - colonies were sexuals and 71.2 percent were workers. The E/e-alleles seem to act on the caste-ratio in at least two ways: in *H. sublaevis*, eggs of different genotypes are equally queen-biased, but the development times of ee- and EE-larvae are different. The allele E seems to slow down the development of female brood and thus might extend periods during which the larvae are sensitive to epigenetic caste determining influences. Second, the inhibitory potential of intermorphic queens seems to be larger than that of gynomorphs [8].

Dissection results from *L.* sp.A workers of different genotypes suggest that the E-allele somehow prevents the development of a spermatheca in a certain percentage of larvae that have been epigenetically predetermined to become young sexuals. The ovaries of about 15 percent of all EE-workers consisted of three to six ovarioles instead of the two typical for ee-workers [5].

A two allele single locus control of wing dimorphism is rather common in non social insects and it has been suggested that altered hormone profiles curtail or prevent the normal pattern of wing development (for a review see [9]). Hormones are known to be important factors during the development of castes in social insects; the E/e-system in leptothoracine ants thus might influence caste-ratios by affecting hormone titers, too.

References

1. Francoeur, A., R. Loiselle, and A. Buschinger. 1985. Biosystématique de la tribu Leptothoracini (Formicidae, Hymenoptera). 1. Le genre *Formicoxenus* dans la région holarctique. Naturaliste can. (Rev. Écol. Syst.), 112: 343-403.
2. Buschinger, A. and U. Winter. 1975. Der Polymorphismus der sklavenhaltenden Ameise *Harpagoxenus sublaevis* (Nyl.). Ins. Soc. 22 (4): 333-362.
3. Francoeur, A. 1986. Deux nouvelles fourmis néarctiques: *Leptothorax retractus* et *L. sphagnicolus* (Formicidae, Hymenoptera). Can. Ent., 118: 1151-1164.
4. Heinze, J. and A. Buschinger. 1987. Queen polymorphism in a non-parasitic *Leptothorax* species (Hymenoptera: Formicidae). Ins. Soc., 34 (1): 28-43.
5. Heinze, J. and A. Buschinger. 1989. Queen polymorphism in *Leptothorax* spec.A: its genetic and ecological background (Hymenoptera: Formicidae). Ins. Soc., 36 (2): 139-155.
6. Buschinger, A. 1975. Eine genetische Komponente im Polymorphismus der dulotischen Ameise *Harpagoxenus sublaevis*. Naturwissenschaften, 62 (5): 239.
7. Buschinger, A. 1976. Genetisch bedingte Entstehung geflügelter Weibchen bei der sklavenhaltenden Ameise *Harpagoxenus sublaevis* (Nyl.) (Hym., Form.). Ins. Soc., 25 (2): 163-172.
8. Winter, U. and A. Buschinger. 1986. Genetically mediated queen polymorphism and caste determination in the slave-making ant, *Harpagoxenus sublaevis* (Hymenoptera: Formicidae). Entomol. Gener., 11 (3/4): 125-137.
9. Roff, D.A. 1986. The evolution of wing dimorphism in insects. Evolution, 40: 1009-1020.

A NEW MODEL FOR THE GENETIC DETERMINATION OF CASTE IN *MELIPONA*

H.H.W. Velthuis and M.J. Sommeijer, Department of Comparative Physiology, Utrecht University, P.O.Box 80.086, 3508 TB Utrecht, Netherlands.

Kerr et al. modified in 1975 [1] earlier models for a genetic program regulating queen determination in the stingless bee genus *Melipona*. Our poster on this subject explains another modification, in which next to a genetic preadaptation for the development of queens also preadaptations for worker development are considered. For that purpose the genetic mechanism is translated into physiological mechanisms.

The genetic system A regulates development in the young larva. This could be related to the rate of food intake, either high or low, e.g. as the result of differences in the quality or amount of an enzyme. The enzyme system is less efficient if alleles are homozygous (A^-) than heterozygous (A^+). The heterozygous condition promotes queen development, the homozygous condition worker development. However, both queen and worker disposition are not irreversibly fixed, but can still be modified later in life.

The genetic system B operates at a later larval phase. It regulates the production of JH at either a high or a low level. This gene again is in the same way more effective in heterozygous (B^+) form than if homozygous (B^-). Two alleles are envisaged. The physiological translation of this genetic system can have various forms.

Two genes operating in this way leads to four classes of animals: A^+/B^+, A^+/B^-, A^-/B^- and A^-/B^+. In connection with the further development of these four classes a sensor mechanism is assumed, which measures the amount or the quality of food ingested in the course of larval development as bein high (H) or low (L). The combination of two sets of genes and two classes of food leads to 8 categories. A typical combination of factors for a Melipona queen is $A^+/B^+/H$, for a worker bee $A^-/B^-/L$.

A genetic programming for a lower level of JH is considered to be an adaptation to inferior food conditions. Higher hormonal activity might increase metabolic activity which would in turn lead to reduced adult body weight; this could be disadvantageous if not fatal ($A^-/B^+/L$). If food amounts are higher ($A^-/B^-/H$) a larger worker could develop. The category $A^-/B^+/H$ represents the larvae that are the nearest to queen development, although they still will develop into a worker only. A^+ animals are predestined to become queens, but B^- levels preclude this. All A^+/B^- animals become workers. The $A^+/B^+/L$ will become workers too, so only the $A^+/B^+/H$ type will develop into queens.

The action of B^+ is designed to be the most influential, because the artificial application of JH has stronger effects than those of increasing the amount of food.

The various categories may respond differently to experimental treatments. Therefore, in addition to inducing queen development, the effect of JH application on mortality should also be considered.

REFERENCES

1. Kerr, W. E., Y. Akahira and C. A. de Camargo. 1975. Sex determination in bees. IV. Genetic control of juvenile hormone production in Melipona quadrifasciata (Apidae). Genetics 81: 749-756.
2. Velthuis, H.H.W. and Sommeijer, M.J. (1990) Social and hormonal regulation of caste dimorphism in the stingless bees. In: Morphogenetic Hormones of Arthropods, Vol III, Ed. A.P. Gupta, Rutgers Univ. Press, (in press).

ROLE OF OVERWINTERING AND QUEEN CONTROL IN CASTE DETERMINATION IN THE ARGENTINE ANT, *IRIDOMYRMEX HUMILIS*

[1]Edward L. Vargo and [2]Luc Passera

[1]Brackenridge Field Laboratory and Department of Zoology, University of Texas at Austin, Austin, TX 78712-1064, U.S.A.

[2]Laboratoire d'Entomologie, Universite Paul-Sabatier, CNRS URA 333, 118 route de Narbonne, 31062 Toulouse CEDEX, France

Social insects are characterized by the existence of a reproductive caste (queens) and a more or less sterile caste (workers). For the vast majority of species, the factors regulating the determination of caste are poorly known. This is certainly true of the subfamily Dolichoderinae, which has received little study in this regard.

To help fill this gap, we conducted studies of caste determination in the Argentine ant, Iridomyrmex humilis, a member of the Dolichoderinae. Observations in the field have suggested that both overwintering and queen influence regulate caste determination in this species. Our studies therefore focused on these two factors.

EFFECT OF OVERWINTERING

Overwintering was found to stimulate gyne development, as demonstrated by the fact that queenless colony fragments (ca. 1200 workers and 200-300 larvae of all stages) comprised of freshly overwintered workers and brood produced more gynes (which appeared more rapidly) than those comprised of individuals collected at other times of the year, although gyne potential larvae are always present. Results of experiments in which the production of gynes was compared in colony fragments consisting of different combinations of freshly overwintered and non overwintered workers and brood suggest that overwintering increases both the capability of larvae to develop into gynes and the tendency of workers to rear larvae as gynes.

QUEEN CONTROL

Corpses of queens were used to test the hypothesis that the inbhibitory queen influence is pheromonal. The addition of a fresh queen corpse was as effective as a living queen in inhibiting gyne development, suggesting that a queen pheromone is involved in caste determination. This inhibitory influence was removed when queen corpses were rinsed in pentane, lending further support to the pheromonal hypothesis. Apparently, only mated queens produce the pheromone; living virgin queens of known age were not inhibitory, whereas young mated queens of the same age were.

The introduction of living functional queens or their corpses to queenless colony fragments containing gyne larvae caused workers to execute many of these within hours, demonstrating that this inhibitory pheromone can act retroactively after female larvae become sexualized. In addition, behavioral observations revealed that queens may directly participate in the execution of gyne larvae by joining workers in attacking them. This appears to be the first report of an ant queen exerting control over larval development by direct attack.

CONCLUSION

These results indicate that queens of I. humilis produce a pheromone which inhibits the production of gynes. This is only the second ant species (the other is the myrmicine, Solenopsis invicta [1,2]) for which strong evidence of pheromonal queen control over the production of sexuals has been obtained. Most likely, the pheromone inhibits sexualization of female larvae mainly by limiting the quantity and/or quality of food given them by workers. It can also cause workers to execute larvae which have sexualized. In addition, queens may sometimes join in this execution of gyne larvae.

The finding of a stimulatory effect of overwintering and an inhibitory influence of queen pheromones in this polygyne species accord well with the timing of the production of gynes in the field. Gynes are produced from overwintered larvae only in spring just after a sharp reduction in queen number (and inhibitory queen pheromone) resulting from the execution of nearly all functional queens by the workers [3,4]. These newly-reared gynes then mate in the nest, thereby restoring queen number and its associated inhibitory influence to their previous high levels. Thereafter, gyne production is inhibited until these new queens are executed the following spring.

REFERENCES

1. Vargo, E.L. and D.J.C. Fletcher. 1986. Journal of Comparative Physiology A, 159: 741-749
2. Vargo, E.L. 1988. Insectes Sociaux, 35: 382-392
3. Markin, G.P. 1970. Ann. Entomol. Soc. Am., 63: 1238-1242
4. Keller, L., L. Passera and J.P. Suzzoni. 1989. Physiol. Ent., 14: 157-163

DOES OVARIAN ACTIVITY INFLUENCE THE DOMINANCE STATUS OF BUMBLE BEE (*BOMBUS TERRESTRIS*) WORKERS ?

Dr. Adriaan van Doorn
Koppert B.V., Veilingweg 17
NL - 2651 BE Berkel en Rodenrijs

Among the members of bumble bee (*Bombus terrestris*) colonies there is a hierarchical order, which is connected with the activity of their ovaries. The queen dominates all workers and initially is the only egglayer. At a certain stage of colony development, however, the most dominant workers start egg laying as well. Moreover, they may become aggressive at that time. Dominant workers usually stay within the nest. Subordinate workers go out to forage [1,2,3].
To obtain more insight into the physiological basis of dominance and the division of labour, workers of queenright colonies and of queenless groups were either treated with juvenile hormone (JH) or their ovaries were ectomized [4,5]. As expected, injection of JH induced oogenesis. In queenright colonies it did not influence the workers' dominance or the division of labour, but in queenless groups JH treatment slightly favoured a worker to become dominant. Here I will focus upon the role of the ovaries.

MATERIALS AND METHODS

Freshly emerged workers of *Bombus terrestris* were ovariectomized by making two incisions in the membrane between the fourth and fifth sternite. The ovaries were pulled out with forceps and cut off. In a first experiment the ovaries of all workers from a queenright colony were ectomized. In another colony the operation was performed on only half of the emerging workers. In a second experiment twenty-seven groups of four freshly emerged, ovariectomized workers were confined to wooden boxes. In another fifteen groups one randomly chosen worker retained its ovaries, but two incisions were made in the intersegmental membrane. The other three workers were ovariectomized. In all cases the establishment of the dominance order was studied, using methods described earlier [2,5].

RESULTS AND CONCLUSIONS

Comparance of the results obtained from both queenright colonies (with all or half of the workers ovariectomized) with the results of control colonies, containing merely non-ovariectomized workers, did not reveal any significant differences. No difference was found with respect to the establishment of the hierarchical order, the performance of behaviours which are usually associated with egg laying (a.o. opening/closing of egg cells, and eating eggs) and of aggression by dominant workers, the proportion of the workers which became foragers, the age at which they started foraging and the foragers' position in the hierarchy.
Merely in the colony with half of the workers ovariectomized the most dominant ovariectomized workers tended to be somewhat less active and less dominant towards subordinate workers than the most dominant non-ovariectomized workers.
Temporal and behavioural characteristics associated with the establishment of the dominance order in the small queenless groups of four ovariectomized workers did not differ from those in groups of four untreated workers. This was also the case with respect to the groups of which only three out of four workers were ovariectomized. In six out of fifteen groups the worker with intact ovaries became dominant. This was slightly more than expected (25%), but the difference was not significant (x^2 = 1.8).
It is concluded that ovarian activity has to be considered as a consequence of dominance rather than a cause. Task allocation is not dictated by the ovarian activity of the workers.

REFERENCES

1. Honk, C.G.J. van & P. Hogeweg. 1981. Behavioural Ecology and Sociobiology, 9: 111-119.
2. Doorn, A. van & J. Heringa. 1986. Insectes Sociaux, 33: 3-25.
3. Duchateau, M.J. & H.H.W. Velthuis. 1989. Entomologia Experimentalis et Applicata, 51: 199-213.
4. Doorn, A. van. 1987. Netherlands Journal of Zoology, 37: 255-276.
5. Doorn, A. van. 1989. Physiological Entomology, 14: 211-221.

AN EXPERIMENTAL STUDY OF THE REGULATION OF BROODCELL PROVISIONING IN *MELIPONA* BEES

Sommeijer, M.J. and D. Koedam
Utrecht University, Bee Research Dept., P.O.Box 80.086, 3508 TB Utrecht, Netherlands.

INTRODUCTION

The 'Provisioning and Oviposition Process (POP)' of *Melipona* is characterized by the cooperative filling of the cell with larval food by a small group of workers in a typical and short behavioural sequence, which leads to the oviposition of the queen and immediate cell closure [1,2,3]. Various authors assume a central role for the queen in stimulating workers to start this provisioning process [3]. The regulating mechanism for this process, by which a specific amount of liquid larval food is brought into the cell, is still unknown. In *Melipona*, queens and workers are reared in the same type of broodcell and variations in food quantity and quality are considered to be of importance for queen determination [4,5]. Since in normal colonies a high percentage of females develop into queens, it is of interest to study the regulation of the provisioning of the broodcell.

MATERIAL AND METHODS

Colonies of *Melipona rufiventris* ,*Melipona favosa*, *M. beecheii* and *M. fasciata* were studied in observation hives to investigate the regulation of broodcell provisioning behaviour of workers during POP. To study the induction and the continuation of worker regurgitations, larval food was experimentally taken out or injected into broodcells. The effect of the typical drumming by the queen was investigated by the use of a queen-drumming simulator, imitating the antennal drumming of the queen on the dorsal bodyparts of workers.

RESULTS

During a POP in *M. rufiventris* the cell is filled by 11.8 worker-discharges (n=74). The quantity of larval food regurgitated by successively discharging workers diminishes during the course of the provisioning sequence. Just before oviposition a cell contains about 118 mg. larval food.

Experimentally injecting pure larval food in a cell leads to discharges by workers also when the queen is not near the cell. A small amount of larval food (mean: 3 mg.) resulted in 11.8 discharges, a large amount (mean: 100 mg.) led to only 3.2 discharges before oviposition. Watery-dilutions of larval food also triggered provisioning, but the more diluted the larval food, the smaller the chance of a discharging respons. Pure water did not stimulate workers to discharge. Artificial drummings on body-inserting workers also could initiate a discharging respons by workers of all species.

REFERENCES

1. Sakagami, S. F., M. J. Montenegro and W. E. Kerr. 1965. Behavior studies of the stingless bee, with special reference to the oviposition process. V. Melipona quadrifasciata anthidioides Lepeletier. J. Fac. Sci, Hokkaido Univ. Ser VI, Zoology 15: 578-607.
2. Sakagami, S. F. 1982. Stingless bees, pp. 361-423. In Hermann, H. R. (ed.). Social Insects , vol. III, Academic Press, New York.
3. Sommeijer, M.J. (1985). The social behavior of Melipona favosa (F): Some aspects of the activity of the queen in the nest. J. Kansas Entom. Soc. 58(3): 386-396
4. Kerr, W. E., Y. Akahira and C. A. de Camargo. 1975. Sex determination in bees. IV. Genetic control of juvenile hormone production in Melipona quadrifasciata (Apidae). Genetics 81: 749-756.
5. Darchen, R. and B. Delage Darchen. 1975. Contribution à l'étude d'une abeille du Mexique Melipona beecheii B. (Hymenoptère: Apide). Apidologie 6: 295-339.

HALICTINE CASTE DETERMINATION

Gerd Knerer
Department of Zoology, University of Toronto, 25 Harbord Street, Toronto M5S 1A1, Ontario, Canada

The Halictinae are a species-rich group of primitive bees found worldwide and showing a great diversity of shape, colour and nest architecture. Above all they exhibit a wide spectrum of behavioral patterns that embrace solitary, communal and social ways of life, thereby providing excellent paradigms for studying the evolution of social behaviour and caste systems.

Halictine bees, unlike honeybees, show little anatomical caste differences, apart from the sometimes quite obvious size discrepancies although most species show considerable overlap between the castes. E. marginatus is unique in its complete absence of a physical caste difference. Despite the common occurrence of the two female types, size is not an absolute factor for determining a caste, since in every species studied so far, worker-sized queens have been observed with queen-sized individuals that functioned as worker. Ill-defined castes as those found in Halictus confusus and Evylaeus calceatus are linked to a primitve social level with small nest populations and quite a few males in the first brood. Highly social species as E. malachurus have bimodal castes, large nest populations and very few males in the first brood. Changes in daylengths throughout the season provide the cues that trigger the switch from the excavation of small cells to the large cells in which queens are reared. These females possess substantial fat bodies apparently responsible for the ovarian diapause that require an extended period of inactivity after eclosion. Summer females (workers) lack the fat bodies and commence activity immediately after leaving their cell.

Even when gynes have been inseminated and have overwintered, they do not become queens automatically. In several species, spring associations of females are common that cooperate in the rearing of the first brood of workers. In _E_. _linearis_, for example, less than 40% of overwintered gynes become fully reproductive queens, while the rest perform worker-like duties in the polygynous nests.

Finally, caste changes are possible through queen replacement. Auxilliaries in polygynous associations can substitute the principal egg layer on her death or disappearance, as can workers in matrifilial societies during the summer. In both cases, a time factor comes into play since replacements can only occur in the early stages of polygynous nests and is more frequent in summer societies if the queen had disappeared before the eclosion of her daughters.

The summary above clearly suggests that caste determination in halictine bees is a complex phenomenon under the control of intrinsic and environmental factors. The interplay of seasonal parameters with intranidal conditions set the stage for caste determination in almost all cases but _E_. _marginatus_, where insemination alone decides queen status.

CASTE AND METAMORPHOSIS: MODULATION OF ECDYSTEROID TITER, JUVENILE HORMONE TITER, AND CORPORA ALLATA ACTIVITY IN HONEY BEE LARVAE

Klaus Hartfelder[§], Anna Rachinsky[§], Colette Strambi[*], Alain Strambi[*]

[§]LS Entwicklungsphysiologie, Universität Tübingen, Auf der Morgenstelle 28, D-7400 Tübingen, Fed. Rep. Germany

[*]C.N.R.S., Laboratoire de Neurobiologie, B.P. 71, F-13402 Marseille Cedex 9, France

In the last larval instar, the nutritional history of a honey bee larva is transformed into endocrine signals, which subsequently control caste differentiation as an element integrated into the process of metamorphosis. In this paper we report the results of ecdysteroid and juvenile hormone (JH) titer analyses of haemolymph extracts from fourth and fifth instar queen and worker larvae, and we discuss the observed JH titer fluctuations in relation to JH synthesis by the corpora allata (CA).

MATERIAL AND METHODS

Larvae and pupae were taken from Apis mellifera carnica colonies. Two highly specific radioimmunoassays (1,2) were used for hormone titer analysis. The composition of ecdysteroids in larval haemolymph was determined by HPLC and subsequent RIA analysis. JH synthesis in CA of queen and worker larvae was studied in vitro by a radiochemical assay (3).

RESULTS AND DISCUSSION

In queen and worker larvae, the ecdysteroid haemolymph titer remains low during the larval feeding phase, and only one maximum was found in the prepupal phase (queens 100.2 pmol/ml haemolymph; workers 47.7 pmol/ml). Formation of this peak starts earlier in queens. Shortly before pupal ecdysis, ecdysteroids are cleared from haemolymph in both castes. The ecdysteroid titer, thus, is caste-specifically modulated with respect to duration and height of the peak in spinning larvae and prepupae. HPLC separation of haemolymph extracts from queen prepupae showed that makisterone A is the predominant ecdysteroid compound in larval haemolymph (4). This corroborated earlier results obtained for pupal haemolymph (5). In distinction to pupae, however, 20-hydroxyecdysone also makes an important contribution to the ecdysteroid titer in larval stages.

Curves for JH titer of queen and worker larvae showed a remarkable caste difference at the beginning of the fifth larval instar (387.8 pmol/ml haemolymph in queens; 27.5 pmol/ml in workers). In prepupae, a smaller titer maximum was found in both castes. It is being established simultaneously with the ecdysteroid peak in the middle of prepupal development (4). Our data for JH haemolymph titer are almost identical with data on JH-III content in whole body extracts of honey bee larvae (6), indicating that haemolymph may be the only body compartment where JH can be detected in significant amounts.

For most of the last larval stadium, fluctuations in JH titer are paralleled by changes in JH release from the CA. Accordingly, highest values of JH release (2.9 pmol JH/h/pair CA) were observed in early fifth instar queen larvae (7). JH release in worker larvae remains at a low level (0.1 - 0.16 pmol JH/h/pair CA) until the middle of the spinning phase. A smaller maximum of JH release (queens 0.9 pmol JH/h/pair CA; workers 0.3 pmol) was found in early prepupal development in both castes.

The profiles of JH and ecdysteroid titer in Apis mellifera queen larvae compare well with the conditions generally required for post-embryonic developmental regulation in holometabolous insects (8). Such conditions were analyzed in detail for Manduca sexta. Worker larvae deviate from this model species, especially with respect to JH titer during the larval feeding stage. Our results indicate that from an endocrinological point of view it is the reproductive caste which best represents the general traits of a holometabolous insect.

REFERENCES

1. Strambi C., A. Strambi, M.L. de Reggi, M.H. Hirn, M.A. Delaage. 1981. Eur. J. Biochem. 118: 401-406.
2. de Reggi M.L., M.H. Hirn, M.A. Delaage. 1975. Biochem. Biophys. Res. Commun. 66: 1307-1315.
3. Pratt G.E., S.S. Tobe. 1974. Life Sci. 14: 575-586.
4. Rachinsky A., C. Strambi, A. Strambi, K. Hartfelder. 1989. Gen. Comp. Endocrinol. in press.
5. Feldlaufer M.F., E.W. Herbert, J.A. Svoboda, M.J. Thompson, W.R. Lusby. 1985. Insect Biochem. 15: 597-600.
6. Rembold H. 1987. Insect Biochem. 17: 1003-1006.
7. Rachinsky A. and K. Hartfelder. 1990. J. Insect Physiol. in press.
8. Bollenbacher W.E. 1988. J. Insect Physiol. 34: 941-947.

CASTE DIFFERENTIATION OF THE HONEY-BEE, *APIS MELLIFERA* L. CYCLIC CHANGES OF CAMP TITERS DURING LARVAL DEVELOPMENT

Ch. Czoppelt
Max-Planck-Institut für Biochemie
D-8033 Martinsried, FRG

Introduction

Hormone- and cyclic nucleotide-titers are closely tied together in animal systems. Hormones regulate intercellular, cyclic nucleotides intracellular communication. Hormone binding promotes an increase of the intracellular level of cyclic AMP (cAMP) which operates as "second messenger" by mediating the effect of hormone action in several tissues (1).
One of the hormones which regulate postembryonic growth of the honeybee is juvenile hormone (JH) III. Particular steps to caste formation are regulated by caste-specific modulation of the JH-III-titer (3). It is possible that the level of cAMP is modulated during bee-development in a similar way. Therefore eggs and larvae from both the castes were investigated for their total endogenous amount of cAMP.

Materials and Methods

Dated eggs and larvae of different instars were collected from outdoor colonies of Carnica bees. Particular larval instars were also characterized and classified by measurement of head capsule width and by individual weight (4). Queen larvae originated from swarm or from queen rearing cells by use of modified procedures (2, 7). Both eggs and larvae were weighed, subsequently frozen in liquid nitrogen and stored at -25 °C for not more than one month before processing. Determination of cAMP was carried out with a radioimmunoassay (RIA, 5).

Results and Discussion

Both the castes show a similar pattern of rather high cAMP pools in early and middle-aged larval instars (L1 - L4; 72-144 h after egg-laying, EL) and of relatively low titers in L5 (168 h after EL, Table 1). The cAMP-titer in queen larvae is enhanced by 2 to 10 times compared with the level in worker larvae. The titer is constant at the beginning of postembryonic phase before it decreases (worker) and increases (queen), resp., in L3. The cAMP-rate of 3835 pmol/g body-weight (BW) in queen L4 is remarkably high as compared with the rate of 408 pmol/g BW in worker L4.

It seems to correspond with the period of maximum synthesis and succeeding release of stainable material of the neurosecretory cells in the brain (6). cAMP, therefore, might act as second messenger for the hormones in the bee. Histological studies on larvae (6) have demonstrated that the first step to queen bee formation has taken place within 120 h after EL. The activation of corpora allata (CA) by neurohormonal signals leads to the enlargement of their volume and to the increase of JH-synthesis. On the other hand CA of worker larvae remain smaller in their volume and obtain their highest activity in L5-stage.

Table 1 cAMP-Titer during larval growth of honeybee

Stage	Time after egg-laying (h)	Total number of individuals (n)	Average body-weight (mg)	cAMP-rate/g body-weight (pmol)
		Worker		
E	48-72	1988	0.09	464.1
L1	72	100	0.09	396.6
L2	96	780	0.33	670.3
L3	120	117	2.19	375.1
L4	144	37	10.5	408.5
L5	168	9	101.2	34.4
		Queen		
L3	120	53	2.89	770.1
L4	144	7	21.9	3835.2
L5	168	7	127.6	672.3

An increase of the cAMP-titer parallels increment of the JH-III-level during larval growth (3). Both the JH-III- and cAMP-titers in the young larvae (L1, L2) increase and achieve a maximum in L3 and L4, but decrease to a minimum in L5. This decrease in L5 is attributed to low pools of both JH-III and cAMP in worker and queen larvae.

Literature

1. Bodnaryk, R.P. 1983. In: Endocrinology of Insects - Invertebrate Endocrinology Vol. 1, R.G.H. Downer, H. Laufer, Eds., A.R. Liss Inc. New York, pp. 567-614.
2. Jenter, K. 1988. In: Varroa jetzt biologisch im Griff. Stuttgart, pp. 72-106.
3. Rembold, H. 1986. Advances in Invertebrate Reproduc tion, 4:59-68.
4. Rembold, H., J.P. Kremer and G.M. Ulrich. 1980. Apidologie, 11(1):29-38.
5. Steiner, A.L., A.S. Pagliara, L.R. Chase and D.M. Kipnis. 1972. Journal of Biological Chemistry, 247: 1114-1120.
6. Ulrich, G.M. and H. Rembold. 1983. Cell and Tissue Research, 230(1):49-55.
7. Weiss, K. 1971. Apidologie, 2(1):3-47.

EFFECT OF JUVENILE HORMONE (JH) AND PROTEIN KINASE C ACTIVATOR ON HONEYBEE WORKERS

H. Sasagawa[1], Y. Kuwahara[1], T. Kusano[2] and M. Sasaki[3]
1) Inst. of Appl. Biochem., Univ. of Tsukuba, Tsukuba Ibaraki 305, Japan. 2) Inst. of Agr. & Forestry, Univ. of Tsukuba, Tsukuba, Ibaraki 305, Japan. 3) Lab. of Entomol., Fac. of Agr., Tamagawa Univ., Machida, Tokyo 194, Japan.

In the honeybee society, division of labor (age polyethism) is a characteristic phenomena, which was regulated by JH increase in haemolymph with age. JH-III titer is influenced by age and season; it increased with age in all seasons and the titer in summer bee were higher than the corresponding bees in winter and in rainy seasons, during the time the titer remained at low level. Positive correlation was demonstrated among JH titer, age and kinds of labor in summer and winter bee.

When bees were treated (both of injection and topical application) with methoprene (JH analog) and JHs at day-0 during summer to autumn seasons, guards and pollen-collecting foragers appeared sooner than the control bees. This change was restricted not only in behavioral level, but also in physiological state. For example, α-glucosidase activity of the hypopharyngeal gland (the major source of bee milk) responsible for honey production was enhanced by in vitro (JHA; 0.1-10μg) and also in vivo treatments of methoprene (JHA; 0.001-10μg). These results indicate that the physiological phenomena related to task was governed by a direct action of JH with age.

To elucidate the mechanism of division of labor affected by JH, following experiments were carried out. A colony consisted of a queen and 1200 workers of "isochronological age" was used and parts of workers were treated topically with methoprene (0.1, 1, 10 μg/bee, total 150 bees, respectively), 1-oleoyl-2-acetyl glycerol (OAG, the protein kinase C activator; 0.4, 4, 40 μg/bee, total 150 bees respectively) and two kinds of controls (intact and 0.5μl-acetone treatment, each 150 bees). The colony was kept in a observation hive and behavioral effects were examined hereafter, and also bees were sampled at appropriate intervals for investigating physiological changes. Induction of age polyethism by both treatments was demonstrated; OAG treatment accelerated behaviorally the division of labor just like methoprene's (JHA's) and haemolymphal JH titer determined by a micro-HPLC system changed correspondingly against treated dose of both agents. Following equipments and conditions were used for the micro-HPLC: a Micropak column (1.5mm i.d. x 250mm in length) packed with μs-Finepak SIL CN (JASCO); a JASCO 880-PO HPLC pump with an ML-425 micro-injection system (1μl sampling loop); a JASCO 875-UV detector (1μl flow cell) at 217 nm; a JASCO 802-SC system controller and a HITACHI 056 recorder at 1mV full scale; an n-hexane-n-butanol mixture (100:0.58) was used as the developing solvent at 0.1ml/min flow rate and 30-31 kg/cm^2 pressure.

Effects of OAG on α-glucosidase activity in the hypopharyngeal gland <u>in vivo</u> were similar with those of methoprene treatment. Calcium ion is essential for activating the protein kinase C, and therefore, haemolymphal calcium titer of individual bee was determined by an atomic absorption spectroscopy (HITACHI 180-70 Polarized Zeeman Atomic Absorption Spectrophotometer). Using the same individual haemolymph, JH-titer was also determined. Results indicated that calcium levels increased with age until 12 days (64.3 ppm) after emergence (7.0-39.5 ppm) and then decreased to low level (day-14; 42.4 ppm, day-18; 52.3 ppm, day-21; 48.5 ppm, day-28; 27.4 ppm, day-40; 13.2 ppm). The worker bee in hive was in average 39.4 ppm, guard 43.8 ppm, forager 41.7 ppm, the queen bee 110 ppm, and the drone 190 ppm, respectively. Namely, calcium level of the forager was higher than hive ones. Calcium level was fluctuated even in the bees of similar JH-titer.

These results explain division of labor, depending upon age, among age-group and also age-composition of colony, and these plasticity of labor-shift mediated by JH and calcium titer which activated the protein kinase C.

THE EFFECT OF JUVENILE HORMONE (JH) ON ESTERASE ACTIVITY AND CASTE DETERMINATION IN *MELIPONA*

Bonetti,A.M.(*), Z.L. Paulino-Simoes(**), M.M.G. Bitondi(**) and M.A. Bezerra(*).
(*) Faculdade de Medicina de Ribeirão Preto,Univ.São Paulo, Brazil; (**) Faculdade de Filosofia Ciências e Letras de Ribeirão Preto, Univ.São Paulo, Brazil.

Melipona has a well defined caste system. Caste determination follows a sequence which always terminates in stimulation of the endocrine system, controlled by hormones during development. The queens of Apis mellifera have a much higher hormone level than the workers (Rembold and Hagenguth,1981). Control of JH has a dominant role in the process of caste determinaion.

While for Apis a different type of food is sufficient to turn all larvae into queen, in Melipona, besides the change in food, according to Kerr's hypothesis, it is also necessary that the larvae be heterozygote for a pair of caste determining genes. However the application of JH to these larvae induces the appearance of nearly 100% queens, suggesting that JH could be a codntrolling agent on the feminizimg genes in Melipona.

The queens obtained through treatment with JH are very similar to natural queens, in terms of morphology and weight of the ovaries, spermatheca volume, and the occurrence of tergal glands.

An analysis of the electrophoretic profile of the esterases, which are partially response for the maintenance of JH levels in the hemolymph, indicate that workers natural and treated queens have:

- a pattern wich varies during development, genic activation and repression, which is expressed through the appearance or not of certain forms of esterase;
- in the prepupa, a decisive phase in caste determination, a more anodic form, which appears to be characteristic of this stage;
- from L1 to prepupae, there is a constant pattern, and soon after there occur patterns characteristic of pupae, while those characteristic for young larvae become weaker;
- beginning at the black eyed, brown bodied pupal stage, the larval patterns begin to reappear;
- in the treated group, it is apparent that JH promotes an acceleration in development, when the esterases are considered; so that we see in some phases, patterns characteristic of the next phase.

ANALYTICAL AND HISTOLOGICAL INVESTIGATIONS ON THE ACTIVITY OF CORPORA ALLATA DURING CASTE DIFFERENTIATION OF *FORMICA POLYCTENA* (FOERSTER) (HYMENOPTERA, FORMICIDAE)

G. H. SCHMIDT and D. TSATI

Department of Zoology-Entomology, University of Hannover, Herrenhäuser Str. 2, D - 3000 Hannover 21, F.R.G.

In Formica polyctena caste determination takes place at the end of the first larval instar. Normally sexuals develop from winter eggs and workers from summer eggs (Schmidt, 1974). Larval development and metamorphosis is under the control of juvenile hormone (JH) which is produced by the corpora allata (CA). Using GC-MS-MIS technique (Rembold & Lackner, 1985) JH-III could be found in significant amounts in the larvae of both sexuals and workers. JH-III was present in 5-7 days old winter egg (0,1 pmol/g), but not in freshly deposited w-eggs. In the 1. instar of sexual larvae 0,4 pmol/g could be registered; prepupae of sexuals and workers contained 3,9 and 3,8 pmol/g, respectively. From these results it can be followed that JH will not transfer from the queens to their eggs.

The activities of the CA can be demonstrated histologically in serial sections (Schmidt, 1964). There may be a direct relationship between intraindividual size of CA and their activities.

In our case we used Bouin solution for fixation and the Azan method for staining the sections from various larval stages of sexuals and workers. For the determination of the size of the CA the largest section was used and the volume was calculated by the formula,

$$V = \frac{4\pi}{3} ab^2.$$

The CA of Formica are rounded and compact. The volume of the right and left gland may be different and this is independent of their activities. The cells are basophilic. No mitosis and other cell divisions could be observed during the larval development in both castes. But an increase in the size of the cell and the nucleus took place presumably by polyploidisation. The number of cells per C. allatum amounted to 10-18 cells in both castes which is independent of the instar.

On the other hand, very high changes in the activities of CA could be observed. Besides changes in the volume, a growth of the glands could be stated independent of the physiological stage. The values of the relation CA volume/ body weight (W) decreased in both castes at the end of the larval period (7,866x10^3 μ^3/mg for sexual and 5,468x10^3 μ^3/mg for worker larvae). The CA of the sexuals grow faster than those of the workers. But the variations in the size of CA were implied by rhythmic and cyclic changes of the glands. A correlation between the volume of CA and that of the nucleus was also observed. In the larvae of both castes four cycles of activities could be stated.

Secretion activities of CA were different in both castes. The largest volume could be observed in sexuals. In contrary to the worker larvae the CA activity of the sexuals went on increasing in each successive instar. The highest peak was found after the 3rd ecdysis. A growth factor, without regarding the activity phases, was calculated which demonstrates an increase in CA volume. It was 18 in sexuals and only 3 in worker larvae. The weight of the latter increased 350 times, where as in sexuals 540 times.

At the end of the 4th instar the CA volumina of both castes approached nearer. We received for sexuals 8,452x10^3 μ^3/mg and workers 8,534x10^3 μ^3/mg (CA volume/W). When compared the JH amount in these stages we may suppose that the same JH concentration is present in the prepupae of both castes .

From our investigations it could be followed that neither JH nor the size of CA take part in caste determination, none the less it is important for caste differentiation.

Rembold, H. and Lackner, B. (1985) Convenient method for the determination of picomol amounts of juvenile hormone. - J. Chromatogr. 323: 355-361.

Schmidt, G.H. (1964) Aktivitätsphasen bekannter Hormondrüsen während der Metamorphose von Formica polyctena Foerst. (Hym. Ins.). - Insects socianx 11: 41-58.

Schmidt, G.H. (1974) Steuerung der Kastenbildung und Geschlechtsregulation im Waldameisenstaat, p. 404-512, in Schmidt, G.H. (ed.) Sozialpolymorphismus bei Insekten. Probleme der Kastenbildung im Tierreich. - Wiss. Verlagsges., Stuttgart, 974 pp.

TERMITE COMPETENCE FOR SOLDIER DIFFERENTIATION

Deborah Ann Waller, Department of Biological Sciences,
Old Dominion University, Norfolk, Virginia 23529 USA

Jeffery Paul La Fage, Entomology Department,
Louisiana State University, Baton Rouge, Louisiana 70803 USA

Many termite species produce soldiers seasonally, often prior to the reproductive flights. Seasonal caste generation may be related to increased competence of older and/or heavier workers for differentiation into soldiers. In the present study, groups of workers from different colonies of the Formosan termite, Coptotermes formosanus Shiraki, were collected four times over a year and maintained in the laboratory for 1) 9 months, 2) 6 months, 3) 3-4 months and 4) 3 days. Workers were then tested for soldier differentiation under controlled conditions. Within a collection period, worker dry weights varied significantly, with some colonies with workers almost twice as heavy as workers from other colonies. However, there was no association between worker dry weight and soldier production. Workers kept longest in the laboratory weighed significantly more than those from newly collected colonies, and they also had higher survivorship and produced significantly more soldiers than newly collected workers. The significance of different factors in termite competence for soldier differentiation is reviewed and discussed.

CASTE PECULIARITIES OF THE TERMITE SALIVARY GLANDS (*ANACANTHOTERMES AHNGERIANUS* JACOBSON)

Mednikova T.K.
Department of Entomology, Biology Faculty,
Moscow University, Moscow, 119899, USSR .

The social nature of termites' life, division of their functions in colony, their caste differentiation influence on structure and functions of termites' salivary glands. Due to complex investigations of ultrastructural cytochemical and biochemical gland peculiarities a conclusion is made that the salivary glands of different termite castes have a partial morpho-functional specialization. The workers have a higher digestive gland activity. The antibacterial barrier function is mach higher in the soldiers. The elder workers, alates and young initiators of a new nest have the best cement properties of salivary secret; the reproductives have the more developed excretory and exchange processes in the salivary glands. Such a partial gland specialization increases the potential secretory abilities of the whole termite colony. But this functional specialization of neuter termite castes is not connected with any obvious ultrastructural modifications of acinary cells. Some changes detected in glandular cells of reproductive individuals do not fully conform known functions. This changes are probably related to some poorly known types of termite gland activity (e.g. hormone or pheromone production or allotrophic role). The salivary glands play an important and various role in termite life. Being directly involved in stomadeal trophallaxis, they form the element of complex regulation system of termite colony. The investigation of these gland functions may be useful for enrichment of knowledge on regulatory mechanism of termite caste differentiation which is very poor for today.

THE DOUBLE MEANING OF THE TERM CASTE

Christian Peeters

School of Biological Science, University of New South Wales, Sydney, **Australia**

Reproductive division of labour is the essence of insect eusociality - adult females either reproduce or are sterile and perform helper roles. It is thus necessary to describe unambiguously which individuals are mated and laying eggs. However there is an additional important characteristic found in highly-eusocial species (stingless bees, honey bees, vespine wasps and ants), i.e. two different morphological classes of female adults. This differentiation into queens and workers occurs during larval development, and it represents morphological specialization for the performance of contrasting roles. "Caste" has been used to describe these distinct female phenotypes. Increased social complexity has been made possible by accentuating the differences between queens and workers, through changes in the underlying developmental programs [1]. The morphological concept of caste makes an implied reference to function, but in some species there is no longer correspondence between morphology and function as a consequence of secondary modifications, e.g. mated workers reproduce instead of queens in some ponerine ants [2].

"Caste" has also become a functional concept, stemming from E.O. Wilson's definition of age groups as being equivalent to morphological castes [3]. Thus "caste" serves to describe the partitioning of non-reproductive activities among workers, and it has also become a synonym for the separation of reproductive and sterile roles. This leads to ambiguity in the literature, e.g. "caste differentiation" refers alternatively to role separation or to morphological dimorphism.

Reproductive division of labour, and the occurrence of (physical) castes, are two distinct phenomena associated with eusociality - the former can occur without the latter . "Caste" should be restricted to the description of morphological differentiation which is initiated in the larval stage, while "role" can adequately refer to the behavioural (and physiological) differences which are the outcome of interactions among adults [2]. This restricted use of "caste" emphasises morphology, which is phylogenetically less labile than function.

REFERENCES

1. Wheeler, D.E., 1986. American Naturalist, 128: 13-34.
2. Peeters, C. and Crozier, R.H., 1988. Psyche, 95: 283-288.
3. Wilson, E.O., 1968. American Naturalist, 102: 41-66.

Symposium 7

THE ROLE OF LEARNING AND MEMORY IN THE ORIENTATION OF SOCIAL INSECTS

Organizer : **R. MENZEL, F.R. Germany**

FUNCTIONAL COMPONENTS OF LEARNING AND MEMORY IN HONEY BEES

R. Menzel, M. Hammer, M. Sugawa, S. Wittstock and G. Braun
Freie Universität Berlin, Institut für Neurobiologie,
Königin-Luise-Str. 28-30, D - 1000 Berlin 33, FRG

Bees associate an olfactory stimulus (CS) quickly with a sucrose reward (US). A single associative learning trial, lasting just a few seconds, establishes a stable memory lasting for more than 24 hours, even under conditions in which the brain is exposed for electrophysiological recordings or for pharmacological treatments (Menzel, 1987). We have used this convenient learning paradigm to examine which neurotransmitters and modulators are involved in memory formation, memory retrieval, sensory processing and motor control (Bicker and Menzel, 1989). Local injections of small quantities (2 - 5 nl) of drugs before or after the single conditioning trial allow to manipulate selectively the processes involved in the formation of the memory trace or those involved in the retrieval of an existing memory trace. Those dependencies and time courses of action were tested for several putative transmitters/modulators predominantly of those like biogenic amines, for which the distribution in the bee brain is known from immunocytochemical studies (see Bicker et al. 1987 for review). In addition, a sample of receptor agonists and antagonists were examined. The limitations of this behavioral-pharmacological approach are quite obvious, and we realized that the procedure allows to distinguish between selective effects on the memory trace (storage, retrieval) and the sensory and motor components only under favorable conditions. Such favorable conditions were found for the action of injected dopamine (DA, 10^{-6} M) into the mid-protocerebrum. DA inhibits the retrieval from an existing memory but does not interfere with the recognition of the chemosensory stimuli, the motor programs of proboscis extension, licking and sucking, and the establishment of a new memory trace (Michelsen 1989, Menzel et al. 1988). More general effects were found for octopmaine (OA), noradrenaline (NA) and 5-hydroxytryptamine (5-HT). Both NA and OA enhance the storage and retrieval processes, but have also more general facilitatory effects on the motor activity and the sensitivity to chemosensory stimuli. NA appears to modulate more specifically the formation and retrieval of the memory trace when it is injected into the mushroom body, whereas OA causes arousal-like effects when it is injected in several neuropils of the brain. 5-HT injections (10^{-8} M) into the mushroom bodies inhibits the storage and retrieval of memory without any obvious effects on the sensory or motor components. The mushroom bodies in the honey bee are known to be an important neuropil for the consolidation of an early sensitive form of memory into a late stable form of memory (Erber et al., 1980). These pharmacological results collaborate these results and indicate an antagonistic relationship between 5-HT and the two amines NA and OA.

In a next step we searched for the role of a second messenger, cAMP, in memory formation. In-vitro incubation experiments reveal that the adenylate cyclase in homogenates of the brain tissue is stimulated by several transmitters/modulators including NA, OA and 5-HT, and most strongly by Forskolin (Sugawa et al. 1989). Dibutyryl-cAMP injected into the mushroom body enhances memory formation. Forskolin injected in relative large quantities into the mid-protocerebrum causes no significant change in memory formation or other behavioral parameters. We tentatively interpret these results to indicate that both the facilitatory and the inhibitory actions of OA, NA and 5-HT respectively are mediated by cAMP. A strong stimulation of the adenylate cyclase in larger areas of the brain as in the case of Forskolin injections causes both facilitation and inhibition of cellular processes related to memory formation, whereas the localised injection of dibutyryl-cAMP into the mushroom bodies stimulates selectively the facilitatory processes.

The role of de-novo protein-synthesis for memory formation was tested with the same olfactory conditioning paradigm and the application of anisomycin or cycloheximide as protein synthesis inhibitors. Solutions of the inhibitors at various concentrations (including saturated solutions) were either fed, injected in the haemolymph, injected directly into the brain or placed in cristaline form directly into the upper-median part of the brain. Incubation experiments with 35-S-methionine determined a very high rate of protein synthesis inhibition. Most surprisingly, the animals survived all these tratments perfectly and showed no change in overall behavior or the sensory and motor components of the proboscis extension reflex. Also the olfactory conditioning revealed no differences to sham treated control groups. The animals were trained either by a single conditioning trial or by 5 learning trials. The tests were carried out minutes, hours or days after conditioning. It is known from earlier experiments (Menzel, 1968) that 3 or more learning trials establish a memory trace which lasts for the lifetime of the animal. We conclude from these results that the formation of a long lasting and stable olfactory memory trace in the honey bee does not need de-novo protein synthesis. Further experiments are carried out to test whether imprinting - like learning in early life time is sensitive to protein synthesis inhibition.

References

1. Bicker, G., Menzel, R. 1989. Nature 337: 33-39.
2. Bicker, G., Schäfer, S., Rehder, V. 1987. In: Neurobiology and Behavior of Honeybees, R. Menzel, A. Mercer, Eds. Springer-Verlag, Heidelberg-New York, pp. 202-224.
3. Erber, J., Masuhr, Th., Menzel, R. 1980. Physiological Entomology 5: 343-358.
4. Menzel, R., Michelsen, B., Rüffer, P., Sugawa, M. 1988. In: Modulation of Synaptic Transmission and Plasticity in Nervous Systems, G. Herting, H.-C. Spatz, Eds., Springer-Verlag, Berlin-Heidelberg.
5. Sugawa, M., Sher, B., Menzel, R., Dudai, Y. 1989. In: Dynamik und Plastizität neuronaler Systeme, W. Singer, N. Elsner, Eds., G. Thieme Verlag, Stuttgart.

HOW HONEY BEES LEARN ABOUT A LANDSCAPE

Fred C. Dyer
Department of Zoology
Michigan State University
East Lansing, Michigan 48824 USA

The ability of honey bees and other social insects to use familiar landmarks to navigate between the nest and distant feeding sites raises compelling questions about how spatial features of the landscape are stored in memory. A recent study [1] suggested that honey bees, like some vertebrates, form "cognitive maps," representations in memory which encode the geometrical relationships among different sites. Experiments showed that bees could head toward a familiar feeding site even if displaced to an arbitrary location other than the hive; this was taken as evidence that they computed a novel route to an unseen goal using a previously learned map of the landscape. I have repeated these experiments [2] and observed that the behavior predicted by the map hypothesis only when the connecting route was familiar to the bees or they could see landmarks associated with a different previously traveled route to the goal. Deprived of such information, but with no other differences in their experience, bees flew in other directions. Hence bees can use spatial information learned on previously traveled routes but cannot abstract from it a more generalized knowledge of the landscape.

Studies of bees in reproductive swarms extend these results and place them in a natural context. Since swarms generally re-nest less than 500 m from the parent colony [reviewed in 3], bees foraging from the new nest will be confronted by landmarks that previously guided their foraging flights from the parental nest. In natural and artificially created swarms, most experienced bees reorient immediately to their new nesting site. Deprived of their new nest, however, they return to their parental nest rather than to other nearby nests, suggesting that the location of the parental nest had been retained in memory [4]. Also, after moving with a swarm, foragers fly preferentially to feeding sites visited previously from the parental nest [5], indicating, in contrast to previous suggestions [6], that the spatial information used to find familiar feeding sites is stored independently of the location of the nest.

1. Gould, J.L. 1986. Science, 232: 861-863.

2. Dyer, F.C. Submitted.

3. Seeley, T.D. 1985. Honeybee Ecology. Princeton University Press, Princeton.

4. Robinson, G.E. and F.C. Dyer. In preparation.

5. Dyer, F.C. In preparation.

6. Cartwright, B.A. and T.S. Collett. 1987. Biological Cybernetics, 57: 85-93.

LEARNED AND SPONTANEOUS REACTIONS TO AIRBORNE SOUND IN HONEYBEES

W.H. Kirchner
Zoologisches Institut II der Universität,
Röntgenring 10, D-8700 Würzburg, FRG

In the dance language successful forager bees inform their nestmates of distance, direction and profitability of food sources. For a long time it was completely unknown, how these informations are transmitted in the darkness of the hive. Substrateborne vibrations of the comb, which bees are known to be sensitive to, and which are used in other communication processes in honeybees [1,2], have been ruled out, they are not produced by dancing bees [2]. Airborne sound signals on the other hand have been shown to be emitted with high intensities in round dances [3] and waggle dances [4]. Simulation of bee dances using an artificial dancer [5] showed that those sound signals are at least a neccessary part of the dance information. On the other hand bees have been said to be deaf and several attempts to train honeybees to respond to airborne sound failed. Recently it could be shown, that bees respond to airborne sound in an aversive conditioning paradigm, using a training procedure which paired a sound signal with a mild electric shock [6].

Using an operant conditioning paradigm it was now possible to determine frequency range and amplitude thresholds of hearing in bees. Bees were trained to visit a feeder, where they had to walk through a Y-shaped gallery. A small amount of sucrose was presented on one side and a continuous sound on the same or in other experiments on the other side. Left and right side were changed in a pseudorandom order. Bees learn in this situation to walk to the rewarded side with sound frequencies up to 500 Hz ($n=514$ bees, $p<0.001$). The threshold of hearing in terms of velocities of air particle oscillation is independent of frequency over a wide range of frequencies 200 mm/s peak-to-peak. Stimulus-transfer experiments and experiments in which bees had a simultaneous choice between 2 different frequencies indicate only poorly developed frequency discriminination in bees.

Spontaneous reactions of bees to airborne sound have been found when single bees were placed in a standing-wave tube at the sites of maximum air-particle movement. Within the bees' nest a so far unknown spontaneous response of worker bees to airborne sound was found. Bees react spontaneously to pulses of airborne sound by folding the wings. This reaction is different and independent of the freezing response, which bees exhibit in response to substrateborne vibrations. The wing folding reaction was observed at frequencies up to 500 Hz, the thresholds are nearly exactly the same as in the conditioning experiments.

The amplitude of sound signals emitted in bee dances is by a factor of 5 to 10 higher than the thresholds of hearing determined in the present study. So bees are able to hear the airborne sound signals emitted in the dances. Ablation experiments indicate that the sense organ is located on the antenna. Using videomicroscopy it was possible to study the sound-induced vibrations of freely moving antennae in live bees. The antenna has a narrow banded resonance at 265 Hz. This is exactly the frequency of the sound signals emitted in bee dances. This indicates that Johnston's organ in the antenna could be the used to detect airborne sound in honeybees.

Recordings of antennal nerve activity in the scape of the antenna confirm that the antenna is involved in hearing. Microelectrode recordings of auditory neurons in the brain are currently in progress.

References:

1. Michelsen,A., Kirchner,W.H., Andersen,B.B. and Lindauer,M. 1986. J comp Physiol A, 158: 605-611.
2. Michelsen,A., Kirchner,W.H. and Lindauer,M. 1986. Behav Ecol Sociobiol, 18: 207-212.
3. Kirchner,W.H., Lindauer,M. and Michelsen,A. 1988. Naturwissenschaften, 75: 629-630.
4. Michelsen,A., Towne,W.F., Kirchner,W.H. and Kryger,P. 1987. J comp Physiol A, 161: 633-643.
5. Michelsen,A., Andersen,B.B., Kirchner,W.H. and Lindauer,M. 1989. Naturwissenschaften, 76: 277-280.
6. Towne,W.F. and Kirchner,W.H. 1989. Science, 244: 686-688.

BEHAVIORAL MECHANISMS OF ASSOCIATIVE CONDITIONING IN THE HONEY BEE, *APIS MELLIFERA*; INDIVIDUAL DIFFERENCES IN LEARNING PERFORMANCE AND POTENTIAL ROLES IN COLONY MAINTENANCE

Brian. H. Smith, Department of Entomology, 1735 Neil Ave., Ohio State University, Columbus, OH 43210

Identification and location of food resources that are needed by the colony is an important task of foraging worker honey bees. The stimuli that allow foragers to predict the spatial and temporal arrangement of these resources change on a rapid scale relative to an average bees' lifetime. This variability necessitates enough phenotypic plasticity to quickly learn new stimulus correlations as they arise and also to forget old correlations as they cease to be of predictive value. Such a learning phenotype can arise from a combination of learning mechanisms that operate on different time scales or according to different rules; mechanisms include habituation, sensitization, alpha-, associative, and operant components. Exactly which combinations of these learning mechanisms is responsible for giving rise to a learning phenotype can determine how bees respond over short- and long-term time spans after the conditioning experience, how selection acts on the behavioral phenotype, and how bees react to changing environmental circumstances. Therefore it is necessary to understand how different learning mechanisms in bees interact to produce a behavioral phenotype.

Free-flying honey bees can learn the association of a variety of visual, olfactory, and tactile stimuli with the nectar and pollen rewards of flowers, as well as predict the distribution of these stimuli in space. Such studies are an invaluable part of any effort to understand learning in an ethological context. However, one limitation to studying free-flying bees is the lack of control over important variables that need to be accurately controlled in order to establish which mechanisms are involved. For example, the point at which flying bees perceive a visual or olfactory stimulus cannot be controlled, and thus the temporal relationships among conditioned and unconditioned stimuli cannot be held constant or accurately manipulated. Intertrial interval varies with the time it takes the subject to freely return to the conditioning situation.

To circumvent this lack of control and supplement studies with free-flying subjects, honey bees restrained individually in harnesses can be conditioned to extend their proboscises in a controlled laboratory environment. After one or more forward pairings of an odor (conditioned stimulus, CS) with a sugar water reward (unconditioned stimulus, US) ca. 40-90% of the bees will extend their proboscises (conditioned

Enhancement of a background proboscis extension to odor is specific to forward pairing and is sensitive to latency of onset of the reward relative to the onset of odor; thus the mechanism is in large part associative, although it contains non-associative components for a short period after conditioning.

In an attempt to develop a more diverse set of proboscis extension conditioning paradigms, our recent experiments show that aversive conditioning stimuli such as shock can also be conditioned. Individual bees learn to avoid shock during differential conditioning where one odor is always followed by a sucrose reward and another odor is followed by sucrose and a 10 volt A.C. shock delivered on an omission schedule (that is, delivered to the proboscis only if the bees extend their proboscises during stimulation of the sucrose taste receptors on the antennae). Subjects learn quickly to extend their proboscises to the odor followed by feeding and not to the odor followed by shock. They also learn within a few trials to withhold proboscis extension (i.e., the unconditioned response) even during stimulation of the sucrose sensory receptors on the antennae; they respond strongly to sucrose after the odor that predicts feeding but significantly fewer of the same bees respond to sucrose after odor that predicts shock. As of yet we do not know if the mechanism is Pavlovian, operant, or a mixture of the two. Experiments designed to determine the necessity of an omission contingency and determine the efficacy of visual and olfactory conditioning stimuli in the aversive paradigm are presently be undertaken.

There are considerable differences in learning performance among individuals on the aversive conditioning task; individual differences in responding are much more pronounced on the aversive task in comparison to discrimination conditioning in which simply a lack of reward is used as a punishment. Subjects differ, among other things, in when they switch from non-responding to responding, whether or not they respond spontaneously to the odor, and whether they stop responding to sucrose.

Shock is clearly an artificial stimulus from an ethological perspective, although bees must certainly learn to avoid natural stimuli that signal lack of a reward or presence of a predator, for example. In order to bring this problem into a more ethological perspective we are at present developing other stimulus procedures for both restrained and free-flying bees to uncover operant components in learning patterns from honey bees. However, regardless of it's artificiality, the shock paradigm uncovers a variety of response strategies (i.e., learning phenotypes) in bees from the same colony. Our present goal is to study how these learning strategies reflect different developmental states that bees go through (i.e., task specializations); for example, are bees that show good learning performance those that are specialized to perform complex learning tasks as foragers? If differences in learning performance do reflect task specialization, then are different castes specialized for solving different learning problems and how do different learning phenotypes help to integrate colony functioning?

THE ROLE OF MEMORY IN THE ORIENTATION OF THREE PONERINE ANTS : *ODONTOMACHUS TROGLODYTES*, *PACHYCONDYLA SOROR* AND *BRACHYPONERA SENAARENSIS* (HYMENOPTERA, FORMICIDAE)

Dejean A. & Corbara B. Laboratoire d'Ethologie, URA CNRS 667, Université Paris XIII, F-93430 Villetaneuse, FRANCE.

In certain species of Ponerinae, the workers forage alone and when they encounter a group of termites they do not recruit nestmates if the society is regularly provisioned. Workers which discovered the termites undertake a series of returns between the prey and the nest. Their exploratory path, very sinuous, is different from the other very direct path, different from one to the other. This enables to establish that no trails are used and that workers are capable of directing themselves, partly due to their memory.

Material and methods

Ten societies of *Odontomachus troglodytes*, six of *Pachycondyla soror* and *Brachyponera senaarensis* were bred in boxes of 6.5 cm in diameter, placed in hunting areas of 120 X 110 cm. Groups of 15 workers *Cubitermes* freshly killed were placed at 90 cm from the ant nest. When a foraging worker finds the prey, the nest is displaced 25 cm perpendicular by to the axis nest-prey. The operation is carried out with 5 to 7 different workers per nest.

Results and Discussion :

- Typical case (A) : the worker burdened with its prey directs itself towards the original situation of the nest. On reaching this point out, overshooting it by a few centimeters, its displacement becomes very sinuous, it leaves its prey and undertakes a new exploration following a rosaceous form of path whose centre goes through the prey. The diameter of the loops tends to increase. If the nest is not located within 5 to 7 minutes, the path takes on a "random" aspect. When the nest is reached, the worker return directly to the prey and carries it back. After having left this prey in the nest, the worker then returns directly towards the group of prey. It was therefore capable of correcting the direction.
- Variants : during the searching of the nest, the worker does not leave its prey (B); the worker, when it reaches the nest, does not return to the prey that it left behind, but, goes directly towards the group of prey (C); the worker returns to find the prey that it left comes back to the nest where it remains (D); the worker comes to a stand still at the corner of the hunting area (F).
- Repartition of the cases :

	A	B	C	D	E	F	
Odontomachus	75.4%	3.3%	6.5%	-	13.1%	1.6%	(61 cases)
Pachycondyla	53.3%	30.0%	-	3.3%	10.0%	3.3%	(30cases)
Brachyponera	58.0%	22.6%	-	-	19.3%	-	(31 cases)

Therefore, case A is the most frequent in theses three species. This implies that the workers do not depend on trails or tracks for their orientation. Their memory intervenes on four levels : for the direction of the return path to the nest after discovering the group of prey; for the distance separating the group of prey and the nest; for seeking the prey left behind during the search for the new situation of the nest; for returning directly to the group of prey whilst carrying out the correction.

Symposium 8

SOCIAL INSECTS IN ECOSYSTEMS

Organizer : **MICHEL G. LEPAGE, France**

TERMITE COMMUNITY IN THE GRASSLAND OF KENYA WITH SPECIAL REFERENCE TO THEIR FEEDING HABITS

Takaya ABE (Department of Zoology, Kyoto University, Kyoto 606, Japan).

Some intensive studies have been carried out on the distribution, abundance and ecological role of termites in African savannas as reviewed by Wood & Sands (1978) and Josens (1985), but information on subterranean termites has been scarce. The present paper deals with distribution, abundance, feeding habits and predator-prey interactions of subterranean termites in the grassland of Kenya.

STUDY SITE AND METHODS

The survey was carried out at bushed grassland dominated by *Acasia tortilis* around the Kajiado Field Station of ICIPE (International Centre of Insect Physiology and Ecology: 1°50"S and 36°48"E, alt., 1700m, mean annual precipitation, 511mm) located 80 km south of Nairobi, Kenya in 1981. An outline of the study was given in Abe & Darlington (1985).

An area of 10m x 20m containing no mound was selected and divided into 200 quadrats of 1m x 1m, all of which were dug to a depth of 70cm from July to October. When soil was dug, subterranean nests and fungus combs were located, and all termites of *Macrotermes* and *Odontotermes* in the nests, fungus combs and subterranean galleries were collected and counted. Other subterranean termites were sampled from soils of 4 small quadrats (50cm x 50cm in size and 70cm in depth). They were extracted from every 10cm layer of soil by hand sorting. The feeding habits of termites were observed in the night as well as when soil was dug.

RESULTS

At least 9 species of termites were found in an area of 10m x 20m; *Macrotermes michaelseni*, *Odontotermes* sp., *Microtermes* sp. , *Synacanthotermes* sp. , *Microcerotermes* sp. , *Amitermes* sp., *Pericapritermes* sp., *Cubitermes* sp. and *Anoplotermes* sp. Around the area *Hodotermes mossambicus* was common. The nests of dominant termites are shown in Fig. 1. *Macrotermes subhyalinus* was found at ca. 30 km south of the study site.

Food contents of termites were largely divided into 5 categories: living grasses, dead plant materials on the ground surface, dead roots of grasses in the soil, soil and dungs. Living grasses were sometimes consumed by *M. michaelseni* and *H. mossambicus* , dead plant materials on the ground surface by *Macrotermes*, *Odontotermes*, *Microtermes* and *Synacanthotermes*, dead roots in the soil by *Amitermes* and *Microcerotermes*, soil by *Cubitermes*. *Pericapritermes* and *Anoplotermes*. The species with similar food requirements tended to segregate their vertical distribution in the soil (Fig. 1). The population density was tentatively estimated as $1329/m^2$. Dead plant feeders on the ground surface (fungus-growing termites) were 741/m²: *M. michaelseni*, 356, *Odontotermes* spp., 8, *Microtermes* sp., 143, and *Synacanthotermes* sp., 234. Dead root feeders were 233/m²: *Amitermes* sp., 95, *Microcerotermes* sp., 138. Soil feeders were 355/m².

Dominant predators were ants. Many species including *Megaponera foetens*, *Pachycondyla* sp. and *Ophthalmopone berthoudi* attacked foraging termites on the ground surface, but two species of *Dorylus* were the most important predators. A subterranean doryline ant, *Dorylus juvenculus* (reddish brown in colouration) attacked *M. michaelseni* through the foraging galleries located 10-20 cm deep in the soil of *M. michaelseni*, while *Dorylus* sp. (yellowish brown in colouration) attacked other subterranean termites located deeper in the soil. Thus they segregated their food items completely.

The distribution pattern of large and small colonies of *Macrotermes michaelseni* was examined in relation to the intra-specific territoriality and predation pressure by Abe & Darlington (1985) and compared with that of *M. bellicosus* studied by Collins (1981). It is probable that the intraspecific territoriality and peculiar foraging pattern of dominant termite species make it possible for subordinate species to coexist in the territories of dominant species.

REFERENCES

Abe, T. & Darlington, J.P.E.C., 1985. Distribution and abundance of a mound-building termite, *Macrotermes michaelseni*, with special reference to subterranean colonies and ant predators. Physiol. Ecol. Japan., 59-74.

Collins, N. M., 1981, Population, age structure and survivorship of colonies of *Macrotermes bellicosus* (Isoptera: Macrotermitinae). J. Anim. Ecol., 50: 293-311.

Josens, G., 1985. The soil fauna of tropical savanna III. The Termites. "Ecosystems of the World 13: Tropical Savanna" (ed. Bourliere, F.), 505-524. Elsevier.

Wood, T. G. & Sands, W. A., 1978. The role of termites in ecosystems. "Production Ecology of Ants and Termites" (ed. Brian, M.V.), 292-292. Cambridge University Press.

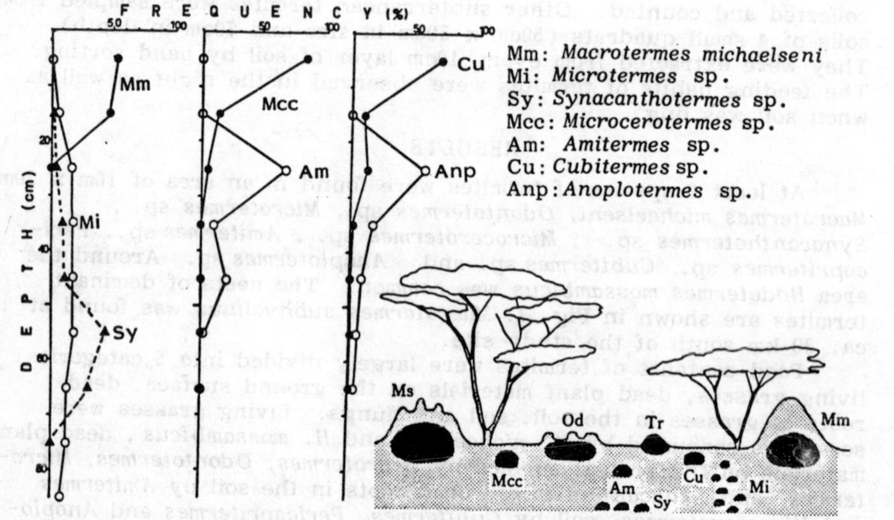

Fig. 1. Vertical distribution of termites at Kajiado, Kenya.

THE INFLUENCE OF GEOPHAGOUS TERMITES ON SOILS OF INUNDATION FORESTS IN AMAZONIA-FIRST RESULTS

Christopher Martius

Max-Planck-Institut für Limnologie, Postfach 165, D-2320 Plön, F.R.G., and Instituto Nacional de Pesquisas da Amazônia, c.p. 478, 69.011 Manaus/AM, Brazil

In seasonally flooded riverine forests of the Amazon near Manaus, Brazil, the most abundant termites, apart from wood-feeding Nasutitermes spp., belong to a geophagous species of soldierless Apicotermitinae, provisionally named Anoplotermes sp. A [1], which builds epigeous earth nests at the base of tree trunks. When these are abandoned or destroyed (see below) they fall to the bottom of the tree where relatively large heaps of old nest material accumulate, rich in organic matter and densely rooted. The following results arise from a study aimed at a first assessment of the role of these termites in the turnover of organic matter and in soil formation.

Alimentary Biology. Nests walls are composed of myriads of dark grey fecal pellets (diameter 0.25-0.50 mm). In the worker gut are mineral soil particles as well as small fragments of leaf litter. Their mandibles are typical of humus feeders. This species has never been observed in dead wood. It is therefore a secondary decomposer feeding mainly on detritus and other organic components of the soil. The feces is steadily plastered on to the nest walls.

Populations. Tab.1 shows the nest densities in various parts of the Ilha de Marchantaria, an island in the Amazon. Nests occur only on the higher, less flooded areas of the forest. Variations in population density appear to be due mainly to the periods of exposure to flooding, but this remains to be studied. The average heights (h) and diameters (d) of adult nests are 4.8 and 0.2 m, respectively, from which a mean nest volume of 0.075 m^3 can be calculated ($V = 0.5\pi h \cdot d/2$).

Soil Properties. Three months after the flood, the top soil, which is covered only by a thin (1-2 cm) litter layer, consists of a 10 cm horizon of grey, heavy, compact and finely textured clay. Below this are 10-20 cm of red mottled clay. The clay content in both horizons is about 80%. The soil under the heaps of old nest material has the same characteristics, but the heaps themselves consist of a thick (≤ 25 cm), dark, crumbly, organic-rich and densely rooted horizon. Below it is a 10 cm mixing zone separating this horizon from the underlying clay. The size of these accumulations of termitogenic soil depends mainly on both the size and age of the related nest and the age of the heap itself. Older heaps are flatter and poorer in organic matter. Studies on the physical and chemical properties of soils, nests and heaps show that the heap material has a considerably lower bulk density than the surrounding soil (0.6 vs. 1.0 g/cm^3), and that the water intake rate is greatly (x 27.5) enhanced. Organic matter (total C) is at least double in nests and heaps by comparison with the soil uninfluenced by termites, and nitrogen content is 1.6-4.0 times higher.

Turnover. Nests subjected to rising water dissolve fast, and their inhabitants tend to concentrate at the top, often increasing the nest by new construction. Many colonies survive the flood in fragmentary nests above the water level, but these are likely to fall down after water level has dropped. Even so, the inhabitants are able to recover in the dry phase and rebuild the nests. (Later in this period, new nests arise on previously uncolonized trees, indicating that nuptial flights have taken place.) As a result, population turnover is highly variable, depending largely on the water level peak and the duration of the flood. The 1989 inundation, the third largest since 1903, led to an almost complete collapse of the populations (plot C, tab.1). However, much soil is stored in the nests (as their volume indicates, tab.1), and if we allow for a mean annual turnover of 25 % of the nests, the mass of soil cycled through the big populations is consistently high.

Conclusions. These first results show that Anoplotermes sp. A has marked effects on the physical and chemical properties of the soils. Particularly important seems to be the formation of soil aggregates due to the mixing of organic and mineral soil particles in the termite gut. Plant growth is likely to be positively affected. Further taxonomic, biological, pedological and chemical analyses are to be made for a better understanding of the role of these termites in Amazonian ecosystems. Particular interest will be on factors controlling their distribution.

TABLE 1. POPULATIONS OF ANOPLOTERMES SP. A

PLOTS	HEIGHT ABOVE SEA LEVEL (M)	AREA (M^2) STUDIED	DENSITY (NESTS/HA)	TOTAL NEST VOLUME (M^3)
P	25-27	1050	219	16.43
P/2	25-27	1452	62	4.65
Z	24-25	6000	3	0.25
C BEFORE FLOOD	24-25	6000	30-50	2.25-3.75
C AFTER FLOOD	24-25	6000	5	0.38

Acknowledgments. I am greatly indebted to Anthony Smith for correcting the English of the first draft of the manuscript.

REFERENCES

1. Martius, C. 1989. Untersuchungen zur Ökologie des Holzabbaus durch Termiten (Isoptera) in zentralamazonischen Überschwemmungswäldern (Várzea). Dissertation, Univ. of Göttingen. Afra-Verlag Frankfurt, 285 p.

IMPORTANCE OF EPIGEOUS TERMITE NESTS IN THE FUNCTIONING OF A SUDANIAN SAVANNA IN COTE D'IVOIRE

Y. Tano

Laboratoire de Zoologie, Faculté des Sciences et Techniques
22 BP 582 Abidjan 22, Côte d'Ivoire

INTRODUCTION

The work was done within a drainage basin of the Sudanian savanna, North-West of Côte d'Ivoire. Termite fauna appeared to be highly diversified and divided into main trophic groups: foraging, fungus-growing, xylophagous and humivorous species. The results will focus on epigeous nests which importánce in the hydrology and soil dynamics of the drainage basin has been studied.

METHODS AND MATERIAL

Abundances, mound size (state of degradation, high, basal area) were determined within transects along the topography. Transects were checked at different intervals during the study. Environmental parameters, such as soil type, vegetation, slope were collected together with the parameters related to termitaria. Previous articles have outlined the main characteristics of termite populations and their environnement [1], [2], [3].

DENSITIES

Results are expressed in relation with topography, soil type and vegetation. Maximum values reached 60 mounds ha^{-1} for <u>Macrotermes bellicosus</u> dead nests and 7 ha^{-1} for alive nests, 150 ha^{-1} for <u>Cubitermes</u> spp. and 60 ha^{-1} for <u>Trinervitermes</u> sp. Table 1 shows the densities of termitaria according to the vegetation.

Table 1. Mound densities in vegetation types (N ha^{-1})

Vegetation	Macrotermes alive	Macrotermes dead	Cubitermes	Trinervitermes
Riverine forest	0	0	0	0
Hydromorphic	0	10	0	25
Grass savanna	0	34	138	37
Fields	0	25	38	25
Recent fallows	0	16	60	40
Shrub savanna	0	34	130	10
Dense shrub savanna	4	46	97	23
Tree savanna	13	43	67	17
Hard pan vegetation	0	20	70	30

VOLUMES AND BASAL AREAS

Volumes involved in mound building activities and total area covered were estimated. Since Macrotermes mounds are killed by the peasants while cropping, there is a considerable increase of the mean basal area as a consequence of erosion on the dead nest: 8.8 m^2 per unit termitarium in crop fields and 14.6 m^2 in recent fallows (1-2 yrs). This lead to important consequences on the hydrodynamic characteristics. Total figures obtained stressed the importance of the termites mounds in this ecosystem, as seen from the table 2.

Table 2. Volumes (m^3 ha^{-1}) and basal areas (m^2 ha^{-1}) of termitaria on different soil types.

Soil type	Volumes	Basal areas
Sand hydromorphic	113.2	278.4
Clay hydromorphic	477.2	1563.6
Brown ferrugineous	122.4	542.4
Yellow ferrugineous	56.8	248.0
Red ferrallitic	50.8	338.4
Ferrallitic indurated	102.0	466.4

ROLE OF TERMITE MOUNDS IN THE DYNAMICS OF THE ECOSYSTEM

From the results obtained, one can attempt to picture the importance of the epigeous termite nests in the functioning of the drainage basin considered.

a- Mounds and erosion. Due to their basal areas (up to 15% of the whole biotope in some cases) termites have a great influence on the hydrodynamic characteristics of the surface soils: their action is of particular importance in fields and recent fallows and in indurated soils where species (Trinervitermes) could increase dramatically water run-off.

b- Mounds and soil. The large building structures and quantities of soil rehandled by termites have large effects on soil characterictics. Volumes involved amounted to nearly 500 m^3 ha^{-1} in some biotopes. Furthermore, termites have been shown to occur down to 5-8 m unter indurated hard-pan plateau (E. Fritsch, pers. comm.). They are therefore able to modify the soil profiles, increasing the alteration of the horizons.

c- Mounds and water infiltration. Through their feeding activities, termites increase water penetration within their foraging areas. Total foraging holes are equivalent to an opening several meters in diameter per ha.

REFERENCES

[1] Lepage, M., & Tano, Y., 1986. Actes Coll. Ins. Soc., 3: 133-142.
[2] Tano, Y. & Lepage M., 1986. In: Chemistry & Biology of Social Insects, J. Eder & H. Rembold eds, Verlag J. Peperny, München: 613-614.
[3] Lepage, M. & Tano, Y., 1988. Actes Coll. Ins. Soc., 4:341-344.

IMPACT OF FUNGUS-GROWING TERMITE SPECIES ON LITTER INCORPORATION IN A PREFOREST SAVANNA (COTE D' IVOIRE)

C. Rouland (1), F. Lenoir (1), L. Abbadie (2), P. Kouassi (3) & M. Lepage (2)

(1) Laboratoire de Biologie des Populations, Université Paris Val de Marne, 94010 Créteil Cedex, France
(2) Laboratoire d'Ecologie, URA CNRS 258, E.N.S., 46 rue d'Ulm, 75230 Paris Cedex 05, France
(3) Laboratoire de Zoologie, Faculté des Sciences et Techniques, 22 BP 582 Abidjan 22, Côte d'Ivoire

INTRODUCTION

Macrotermitinae species have a great impact on litter dynamics and organic matter cycling of most forest and savanna ecosystems in Africa and Asia. Many works have been devoted to assess their importance in food collection. But few of these studies deal with the quality of the food ingested. The degradation of plant material by Macrotermitinae depend upon their double symbiosis: endosymbiosis with gut microflora and ectosymbiosis with a fungus from the genus Termitomyces. The fungus grows on structures ("fungus-combs") built by termites from fecal plant material proceeded through the termite digestive tube.

MATERIAL AND METHODS

The study was done in a preforest savanna situated in the Guinean zone of Côte d'Ivoire (Lamto: 5°02' 1W; 6°13'1N). Underground nests were sampled along trenches examined every two months. Fungus-comb chambers were plotted, measured and combs weighted. Animals were collected from their nests, frozen for transport and storage. Enzyme essays were done in the laboratory using different substrates from plant material. Specific activity was expressed as the amount of glucose equivalent (µM) liberated per mn (=unit) per mg of protein.

FOOD STORED IN UNDERGROUND NESTS

Quantities of food (fungus-combs) stored in termite nests were highly variable according to the vegetation and to the season. In annually burned savannna 212 kg ha^{-1} were found during the rainy season but only 21 kg ha^{-1} during the dry season when food reserves were depleted. While in savanna protected from fire, fungus-combs weight amounted to 361 kg ha^{-1} in rainy season and 915 kg ha^{-1} in dry season [1].

DIGESTIVE OSIDASIC SET AND NUTRITION MODE

Previous works on the digestive osidases of different termite species have shown that there is a good correlation between the digestive osidasic set and the nutrition mode [2]. Then, the digestive osidases from four termite species have been determined to specify the nature of their primary aliment. To precise the role of each organism (termite, fungus, microflora) in the digestion, the enzymatic activities of the midgut (termite and fungus activities), of the hindgut (termite, microflora and fungus activities) and of the mycotêtes (fungus activity) have been determined. Four species have been chosen: <u>Ancistrotermes cavithorax</u>, <u>Macrotermes bellicosus</u>, <u>Pseudacanthotermes militaris</u> and <u>Odontotermes nr. pauperans</u>.

The results obtained showed that, for the four species, the midgut was the most efficient part of the digestive tract for the saccharids degradation. The hindgut presented enzymatic activities generally lower than in the midgut. The fungi presented an original metabolism: it highly degraded some polysaccharids whereas it had very few activities on several oligosaccharids. Globally, our results showed that the four Macrotermitinae species studied exhibited a large range of osidasic set. High cellulolytic activities, due to endocellulase (active on CMC), exocellulase (active on avicellulose) and β-glucosidase (or cellobiase) in termites digestive tract allowed them to use cellulose as nutritive substrate.

CONCLUSIONS

Very clear differences in the set of osidases were noticed among the four Macrotermitinae species studied. This result has brought evidence for the probability of different food sources. The differences in diet composition from plant material could be an important factor in the coexistence of these species in the same biotope. It will be interesting to expand this new kind of study to other termite species which diet is not known, to get informations explaining why some species are depleted or pullulated in cultivated fields.

REFERENCES

[1] Kouassi, Ph., 1987. <u>Master Thesis</u>, University of Abidjan, 129 pp.

[2] Rouland, C., Chararas C. & Renoux, J., 1986. <u>C.R. Acad. Sci., Paris</u>, 9: 341-345.

ROLE OF TERMITES IN ORGANIC MATTER DYNAMICS IN AFRICAN SAVANNAS, WITH SPECIAL EMPHASIS ON NITROGEN CYCLING

M. Lepage & L. Abbadie

Laboratoire d'Ecologie, URA CNRS 258, E.N.S., 46 rue d'Ulm
75230 Paris Cedex 05, France

INTRODUCTION

Termites populations and nests are a main component in the ecology of most African savannas [1]. Of special importance are the termites of the Macrotermitinae subfamily, since their exo-symbiosis with a fungus implies the processing of large quantities of dead plant material withdraw from the decomposition cycle and stored in their nest. One can question the significance of such activities at the level of the ecosystem dynamics. Some workers have outlined their detrimental effect on soil organic matter. The approach outlined in this paper will somewhat moderate this statement.

MATERIAL AND METHODS

The study was done in a guinean savanna, Côte d'Ivoire (5°02' 1W; 6°13' 1N). Species studied belong to Macrotermitinae: 1-large epigeous mounds scattered in the savanna (probably old Macrotermes nests); 2-underground nests composed of subunits ("fungus-comb chambers") and belonging to several genus.
Mapping methods, sampling procedure and size measurements were applied to the termitaria. Influence of termites structures on soil metabolism was tested by respirometry of aliquotes samples in the laboratory under controlled humidity and temperature. Emphasis was given on nitrogen mineralization using Kjeldahl method and Kjeltec apparatus.

IMPORTANCE OF TERMITARIA

Average density of termitaria reached 12.9 ha^{-1}. Mean high was 0.90 m and 70.3 m^2 as basal area. All the mounds covered 9% of the savanna and represented 300 m^3 ha^{-1} of soil. More trees were found on the termitaria (1627 ha^{-1} as compared with 624 ha^{-1} outside). This as a consequence on the dynamics of the tree species in the ecosystem. Soil analysis showed that termitaria are richer in finer particles. As a consequence, exchangeable cations are 4 times more in soil rehandled by termites. Termitaria are clearly sites of organic matter accumulation with 3 times more carbon and nitrogen.
Higher densities of soil fauna are found within the mounds, as compared to the adjacent soil. Of particular interest is

the colonization by underground species of Macrotermitinae, as shown on the table 1.

Table 1. Densities of fungus-combs in termitaria and control soil [2] (number N and dry biomass B in g).

Species	Fungus-combss			
	in Termitaria		in Control soil	
	N m^{-2}	B m^{-2}	N m^{-2}	B m^{-2}
<u>Ancistrotermes</u> sp.	12.4	67.8	3.5	13.2
<u>Microtermes</u> sp.	3.1	1.7	3.3	2.2
<u>Odontotermes</u> sp.	16.9	122.8	0.07	0.4
<u>Pseudacanthotermes</u> sp.	6.2	48.8	1.0	0.6
Total	38.6	241.1	7.9	16.4

According to fungus-combs turnover [2], the total corresponds to a food consumption of 1610 g m^{-2} yr^{-1}.

IMPORTANCE OF UNDERGROUND MACROTERMITINAE NESTS

Experiments conducted in the laboratory showed a significant higher soil metabolism induced by termite activity. Carbon concentration did not differ significantly between soil rehandled and control soil but respiration rates differed (from inside outwards), as shown in table 2.

Table 2. Respiration rates of soil from fungus-comb layers (0-2 mm and 2-5 mm thickness) compared with control soil, expressed in mg of CO^2 day^{-1} per 15 g of soil.

Days	1	2	4	7	10
Control soil	1.73	1.18	1.16	0.32	0.13
2-5 mm layer	4.88	2.04	0.94	0.20	0.30
0-2 mm layer	6.82	2.83	1.28	0.39	0.54

Nitrogen mineralization was measured on the same incubated samples and was 2-3 times more in the soil closely rehandled by termites [3]

CONCLUSION

Two antagonistic processes have been evidenced: nutrient storage in large termitaria and nutrient release in underground structures. Through their feeding and building activities, termites appeared as regulators of nitrogen cycling in the ecosystem studied.

REFERENCES

[1] Lee & Wood, T.G., 1971. <u>Termites and soils</u>. Academic Press, 251 p.
[2] Josens, G., 1972. <u>Doctoral Thesis</u>, Univ. of Brussels.
[3] Abbadie, L. & Lepage, M., 1989. <u>Soil Biol. Biochem</u>., 21: 1067-1071.

THE FORMICIDAE OF THE RAIN FOREST IN PANGUANA, PERU : THE MOST DIVERSE LOCAL ANT FAUNA EVER RECORDED

M. Verhaagh, Staatliches Museum für Naturkunde Karlsruhe, Erbprinzenstr. 13, D-7500 Karlsruhe, Fed. Rep. of Germany

The ant fauna of the biological station "Panguana" in Peru was studied. Panguana (9°37'S, 74°56'W) is situated about 220 m a.s.l. at the Rio Yuyapichis, an affluent of the Rio Pachitea. The region belongs to the preandine hylaea, and is covered naturally by an evergreen seasonal rain forest as defined by Ellenberg [1]. Inundation vegetation is found in few parts. Besides natural ecosystems, there are also different anthropogenic habitats, such as pastures, plantations, and secondary forests. Mean annual temperature at the station is about 24-25°C. The rainy season lasts from the end of October to April, and supplies 80% of the whole annual precipitation of about 2400 mm.

Field work was carried out between May 1983 and Juli 1985. Ants were caught by forceps or aspirator when running around, by breaking up their nests, by pitfall and light traps, in tree eclectors, by means of berlese funnels from litter and epiphyte humus, and by baiting them with tuna. Arboreal species were also collected from cut trees, and by climbing up a big tree. The total area sampled, including forest and other habitats, was about 10 km².

So far, more than 500 ant species belonging to 6 subfamilies and 78 genera were recognized from the vicinity of Panguana. This is the most diverse local ant fauna ever recorded. The number is equal with about 25% of the total number of species known from the neotropical region [2] and greater than the number of ant species known from the whole of Europe (ca. 370 species from 52 genera [3,4]).

About 10% of the species collected are likely to be new to science.

Some notable examples of species diversity in Panguana are:

1. 114 species (41 genera) from 13 pitfall traps in a forest area of 10 x 10 m within 14 weeks; 2. 44 species (21 genera) from tree eclectors at one big tree within 8 weeks; 3. 38 species (18 genera) from one big tree hand-sampled in 20 m height. The two last figures mean

that ant species richness on a single big tree in Panguana might be as high as in the entire native ant fauna of the British Isles which consists of 46 species from 16 genera [5]. Interestingly, Wilson [6] found similar data on a single big tree after insectizidal fogging in the Tambopata Reserved Zone, Peru.

The genera richest in species numbers are Pheidole (Myrmicinae = M) and Camponotus (Formicinae) with more than 50, each. Together with the species of Pseudomyrmex (Pseudomyrmicinae), Gnamptogenys (Ponerinae = P), Solenopsis (M), Crematogaster (M), Pachycondyla (P), Azteca (Dolichoderinae), Strumigenys (M) and Zacryptocerus (M) they make up more than 50 % of all species collected.

Frequent, with more than 30 species, are also the Attines (leaf cutter ants), and, with at least 25 species, the army ants of the subfamily Ecitoninae. Both groups play important ecological roles, the first by using fungi as food which are cultivated on fresh plant material or on animal or plant detritus, thus being indirectly part of the decomposing system, the second by hunting to a high proportion other ant species, thereby maintaining high fluctuation of ant colonies. About a dozen ant species in Panguana are obligatory living on ant-plants.

Most species in Panguana seem to be typical forest species which are only or predominantly found in forest habitats. So far known, about 50% of the species are at least partly arboreal with regard to their nest or activity habits. Geographical distribution of a number of species is widespread within the Amazonian rain forest, e.g. from Peru to the Guyanas.

The most important parameter for the ant community structure seem to be nest habits, but, activity time and space, food spectrum and recruitment modes, colony size and social organization, aggressive as defensive behaviour play important roles, too.

References

1. Ellenberg, H. 1959. Schweiz. Zeitschrift Forstwesen, 3: 169-187.
2. Kempf, W.W. 1972. Studia Entomologica (N.S.), 15: 3-344.
3. Collingwood, C.A. 1978. EOS, 52: 65-95.
4. Agosti, D. and C.A. Collingwood. 1987. Mitteilungen der Schweizerischen Entomologischen Gesellschaft, 60: 261-293.
5. Collingwood, C.A. 1979. The Formicidae (Hymenoptera) of Fennoscandia and Denmark, Scandinavian Science Press, Klampenborg.
6. Wilson, E.O. 1987. Biotropica, 19: 245-251.

TETRAMORIUM CAESPITUM IN AN IRISH DUNE : SPATIAL DISTRIBUTION AND ORDINATION OF SITES

JOHN BREEN and ANNE O'BRIEN
Thomond College of Education, Plassey, Limerick, Ireland

INTRODUCTION

The study site is a dune system, the Ballyteigue Burrows located in the south-eastern corner of Ireland. *Tetramorium caespitum* is the dominant ant species in the dunes and the other ants present are, in order of importance, *Myrmica scabrinodis*, *Lasius niger*, *Formica lemani, and M. rubra.* Although common nearby, *L. flavus* does not occur on the dunes but it occurs at an adjacent foreshore included in the ordination study below.

NEST DENSITY

A flat stony area 30 x 30 m (about 2300 stones) was checked for ant colonies. The site contained 24 established *T. caespitum* colonies, equal to 37.5 m^2 per colony, 7 incipient *T. caespitum* colonies (eggs plus deälate queens; in one case two queens were co-operating) and 7 established *Myrmica scabrinodis* colonies.

Nearest neighbour analysis [1] showed significant regularity (χ^2 = 99.67; d.f. = 76; P = 0.036) and implies inter-colony competition. The observed colony area can be compared with 43 m^2 in Britain [2] and 49.5 m^2 in Denmark [3].

ORDINATION

The following variables were measured in ten 0.5 x 0.5 m quadrats at each of 12 sites: vegetation (percent cover, bare surface and height), percent sand (> ϕ 500 μm), percent organic matter (400°C, 6 h), integrated temperature by sucrose inversion [4] and number of ants found in the top 10 cm of the quadrat - nests were scored as 10.

The results show a trend of the ants being concentrated in warm sites with short or no vegetation. Kendall's τ rank correlation was highly significant for: *M. ruginodis* negatively with *Thymus praecox* and *Carex* sp., both *T. caespitum* and *L. flavus* positively with *Lotus corniculatus*, and *M. scabrinodis* positively for *Carex* sp. and negatively for *Ammophila arenaria*.

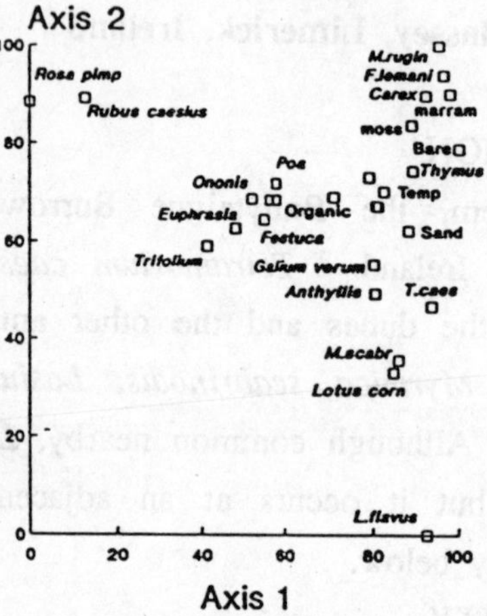

Fig.1 Ordination of variables.

An ordination using recriprocal averaging [5], (fig. 1) showed that all the ant species had similar positions on axis 1 but were spread out on axis 2. Axis 1 seems to represent a gradient from shrubby species through grassy species to bare and high sand content. Axis 2 might represent less established dune (top) to more established.

ACKNOWLEDGEMENTS

Thanks to J. Kennedy and J. Bennett (T.C.E.), P. Thornton and M. Munroe (University of Limerick) for technical assistance. The work was supported by the British Ecological Society.

REFERENCES

1. Thompson, H.R. 1956. Ecology, 37: 391 - 394.
2. Brian, M.V., J. Hibble and D.J. Stradling. 1965. Journal of Animal Ecology, 34: 545 - 555.
3. Nielsen, M.G. 1974. Natura Jutlandica, 17: 92 - 95.
4. Berthet, P. 1960. Vegetatio, 9: 197 - 207.
5. Hill, M.O. 1973. Journal of Ecology, 61: 237 - 249.

PERFORMANCE OF *APIS CERANA* IN RAICHUR, KARNATAKA STATE, INDIA

Shashidhar Viraktamath, S. Lingappa*, Somasekhar.
Department of Agril. Entomology, University of Agricultural Sciences, Raichur Campus, 584 101, India.

Beekeeping is mostly confined to the southern and coastal districts of Karnataka. Hence, studies were made to explore the potentiality and to evaluate the performance of Apis cerana F. colonies in Raichur which is located at 16° 12' N and 77° 21' E and experiences average maximum and minimum temperatures of 33.6° C and 22.1°C, respectively and annual rainfall of 661.3 mm.

Observations on number of empty cells, those with brood, pollen and honey per 6.25 sq.cm. at 10 different places on the central comb in 3 (1987-88) to 7 (1988-89) colonies were recorded at fortnightly interval throughout the study period. All the colonies under observations were kept at an equal strength and free from pests and diseases.

Brood and pollen were found throughout the year (Table 1). However, the major honey flow season occurred during March-April. A maximum of 1.25 kg. honey yield per colony was recorded which was lower than the national average of 4-5 kg./colony[1] and this lower yield may be due to lack of sufficient nectar yielding plants. A severe dearth period occurred during June-September and the colonies had to be fed with sugar solution. Though the variations in brood rearing, storage of honey and pollen followed a similar seasonal pattern reported from southern Karnataka[2], the climatic factors had no significant influence on these variations.

Severe incidence of Galleria mellonella L. was recorded from May to August which resulted in desertation of 4 colonies in the apiary.

---

* Present address: Univ.of Agril. Sciences, Dharwad Campus, 585 005, India.

It is hoped that these studies form a basis for evolving future strategies for promoting beekeeping in this area.

REFERENCES:

1. Phadke, R.P. 1986. Beekeeping in India: present status, constraints and plan of action for further development. Indian Bee Journal, 48 : 29-33.

2. Reddy, C.C. 1980. Observations on the annual cycle of foraging and brood rearing by Apis cerana indica colonies. Journal of Apicultural Research, 19 : 17-20.

Table 1. Variations in brood, pollen, honey and empty cells in Apis cerana colonies in Raichur.

Months[1]	Mean[2] no. of cells per 6.25 square cm. with			
	Brood	Honey	Pollen	Empty
June	17.36	0.0[3]	5.18	16.05
July	18.47	0.0[3]	6.06	14.86
August	20.59	0.0[3]	4.21	11.41
September	24.74	0.0[3]	3.64	9.48
October	28.10	5.97	4.45	5.67
November	25.74	7.86	4.82	2.77
December	25.92	8.70	4.32	2.36
January	24.60	11.54	4.15	1.46
February	24.08	11.66	3.55	2.07
March	22.17	14.84	2.93	1.75
April	22.34	16.60	2.51	0.97
May	20.43	10.39	1.99	8.06

1. Mean of two years; 2. Mean of 3-7 colonies;
3. Fed with 50% sugar solution.

Symposium 9

EVOLUTION AND SIGNIFICANCE OF POLYGYNY IN SOCIAL INSECTS

Organizers : **LAURENT KELLER, Switzerland**
D. CHERIX, Switzerland
ROBERT L. JEANNE, U.S.A.
STEVEN W. RISSING, U.S.A.

PERMANENT AND TEMPORAL POLYGYNY IN SOCIAL INSECTS : A NOTE

Yosiaki Itô

Laboratory of Applied Entomology and Nematology, Nagoya University, Chikusa, Nagoya 464-01 Japan

Both of the two influential hypotheses of the evolution of eusociality, kin-selection theory and parental manipulation theory require high relatedness between workers and reproductives. Colonies of some primitively eusocial wasps and bees are established by association of females, but in many species a dominant female monopolizes oviposition on a nest or eats eggs laid by subordinates, resulting in high intracolony relatedness.

Based on cladistic analysis and data on social structures of the Vespidae, Carpenter (1989) concluded that long-term polygyny was always derived from monogyny, and there is no evidence for evolution of long-term polygyny from a 'rudimentary-caste-containing' stage. This opinion is consistent with a view that, although reproductive division of labour evolves most easily when relatedness is high, it can be maintained under regime of low relatedness after morphologically distinct castes have evolved (Strassmann et al., 1989).

Recent data on intracolony relatedness of eusocial Hymenoptera may support this view. Mean relatedness values of Polistes and Mischocyttarus, which have no distinct castes, are ranging from 0.31 to 0.80 (Strassmann et al., 1989), but values for 3 swarm-founding Polistinae having morphologically distinct castes are 0.11, 0.26 and 0.34 (Queller et al., 1988). These results are consistent with high relatedness values of rudimentary-caste-containing species, Microstigmus comes (Ross and Matthews, 1989) and eusocial bees (e. g. Kukuk, 1989), and low relatedness of some polygynous ants (e. g. Pearson, 1983; Ross and Fletcher, 1985).

Two questions arise. (1) Why could multiqueen social systems evolve in some ants and wasps despite their low intra-colony relatedness? (2) Was evolution of permanent polygyny from facultatively polygynous, rudimentary-caste-containing species impossible?

In eusocial species with morphologically distinct castes, workers never produce female progeny, and therefore cost of altruism is negligible. But there is no explanation on the coexistence of many reproducing females despite conflict among them.

In wasps, possibly most of neotropical polistine genera except Mischocyttarus and Polistes, some species of the subgenus Icarielia (genus Ropalidia), Vespa affinis and Vespula germanica are long-term polygyny. All of them are swarm-founding species constructing enveloped nests and their colony cycle is perennial. An extreme is a huge colony of Stelopolybia vicina, containing 3000 - 4000 queens and more than one million workers (Sakagami, perso. com.), comparable to large polygynous colonies of some ants. Polygyny of V. affinis may have evolved from monogynous, typical vespid ancestor, as that of V. germanica which have evolved from monogynous European ancestor, but there is no strong evidence of evolution of long-term polygyny in swarm-founding wasps from monogynous relatives. There is no report of monogynous species having morphologically distinct castes in the Polistinae.

In the genus Ropalidia, there are 3 social systems. Type 1: Long-term polygyny in Icarielia which construct enveloped nests. Type 2: Facultative polygyny in many species of the subgenus Icariola with non-enveloped nests and no distinct castes, and Type 3: Possible long-term polygyny in species which make non-enveloped nests in cavities and have no distinct castes (e. g. R.(Icariola) socialistica and R. (Anthreneida) sumatorae). The possibility of evolution of Type 1 from Type 2 through Type 3 still remains. I suppose there might be Type 3 species in neotropical wasps, and, if so, this can be a link between swarm-founding polygynous species and Mischocyttarus, which makes non-enveloped nests and have no distinct castes. If we suppose evolution of long-term polygyny from facultative polygyny of rudimentary-caste-containing species, we must consider joint action of kin-selection, parental manipulation, mutualism and colony level selection. Evaluation of the relative roles of these processes may be one of important subjects of study in the 1990's.

References

Carpenter, J.M. 1989. Cladistics, 5:131-144.
Pearson, B. 1983. Behav. Ecol. Sociobiol., 12:1-4.
Queller, D.C., J.E.Strassmann and C.R.Hughes 1988. Science, 242: 1155-1157.
Ross, K.G. and D.J.C.Fletcher 1985. Behav. Ecol. Sociobiol., 17: 349-356.
Ross, K.G. and R.W.Matthews 1989. Anim. Behav., 38:613-619.
Strassmann, J.E., C.R.Hughes, D.C.Queller, S.Tueillazzi, R. Cervo, S.K.Davis and K.F.Goodnight 1989. Nature, 342:268-270.

SERIAL POLYGYNY IN ROPALIDIA MARGINATA: IMPLICATIONS FOR THE EVOLUTION OF EUSOCIALITY

RAGHAVENDRA GADAGKAR, K.CHANDRASHEKARA, SWARNALATHA CHANDRAN, AND SEETHA BHAGAVAN
Centre for Ecological Sciences, Indian Institute of Science, Bangalore 560 012, INDIA.

Ropalidia marginata (Lep.) (Hymenoptera : Vespidae) is a primitively eusocial wasp in which the queens and workers are not morphologically differentiated (1). In peninsular India the species exhibits a perennial indeterminate colony cycle which may include multiple repeats of the typical annual colony cycle shown by temperate polistine wasps (2). Although there is some pre-imaginal caste bias, the roles of queen and worker are largely decided in the adult stage (3). Female wasps eclosing on a nest have several options available to them. These include leaving their natal nest to start their own solitary foundress nests, staying on at their natal nests and assuming the role of a worker, staying on at their natal nest and eventually taking over the role of the queen and leaving their natal nest along with a few workers to start their own multiple foundress nests (2). From several hundred hours of observation of several colonies it appears that only one individual ever lays eggs at any given time in a colony (4, this study). The impression of monogyny created by this observation is misleading because queen replacements are quite common so that workers often rear brood which are the offspring of two or more queens. Moreover, the workers are also offspring of two or more individuals. This phenomenon may thus be called serial polygyny.

In this brief report, we describe some features of serial polygyny in *R. marginata* on the basis of our study of four laboratory and three free foraging colonies. Our results show that queen turnover is quite common and that queen replacements seem to take place at all times of the year. Queen replacements are not accompanied by any significant desertion of the colony by workers. The same workers who were rearing offspring of the previous queen therefore, begin to rear offspring of the new queen. The tenure of queens is highly variable and ranges from 7 to 299 days with a mean of 124 days. The age of the queens at the time of taking over the role of egg laying is also highly variable and ranges from 7 to 78 days. The productivities of the queens can vary enormously and may be measured either as the total number of eggs laid during their tenure which ranges from 19 to 2207, the number of eggs laid per day of their tenure which ranges from 0.61 to 10.2 or the total number of adults produced which ranges from 0 to 394. The proportion of eggs laid by a queen which successfully become adults is rather low, yet highly variable, ranging from 0 to 0.49.

New queens were daughters, sisters or nieces of previous queens and workers were daughters, sisters or nieces of the queens of their colony. Queens of *R. marginata* are known to mate multiply, and use sperm simultaneously from two or more males (5). Consequently, workers rear brood which are complex mixtures of their full- sisters, half-sisters, brothers, nieces, nephews and cousins. Thus, workers sometimes reared brood to whom they had coefficients of genetic relatedness as low as 0.0625 (mother's, sisters' offspring). Our previous work has shown that workers are unlikely to be able to discriminate between different levels of genetic relatedness within the colony (6). These observations make it very unlikely that individuals stay behind on their natal nests and take on worker roles because of the possibility of being able to rear brood which are more closely related to them than their own offspring would be. Instead our observations strongly support the hypothesis that female wasps stay on their natal nests because of the hope of becoming queens in future (2). Even if the probability of becoming a queen is quite small for a given individual, the fitness gained by those that succeed can be so great as to offset the cost incurred by the remaining bearers of the hypothetical "gambling" allele which has been postulated to program its bearers to take the risk of being a part of a social group and await their chances of becoming queens (2). The mean number of adults successfully produced by a queen is 78, a number which is likely to be substantially larger than the number of offspring that a solitary wasp can produce.

TABLE 1: Serial Polygyny in *R. marginata*. Data from four laboratory colonies and three free foraging colonies.

	Min.	Max.	Mean	S.D.
Duration of study per colony (days)	199	606	387	146
Number of queens per colony	2	5	3	1.2
Tenure of queens (days)	7	299	124	89
Age of queens at the beginning of tenure (days)	7	78	38	23
Age of queens at the end of tenure (days)	19	300	139	88
Number of eggs laid	19	2207	331	500
Number of larvae produced	12	1625	258	394
Number of pupae produced	0	679	107	178
Number of adult female offspring produced	0	360	63	99
Number of adult male offspring produced	0	136	15	36
Total number of adult offspring produced	0	394	78	124
Number of eggs laid per day	0.61	10.2	2.5	2.2
Proportion of eggs laid that become adults	0	0.49	0.18	0.16

1. Gadagkar, R. 1985. Proc. Indian Acad. Sci. (Anim. Sci.), 94:309-324.
2. Gadagkar, R. In: Social Biology of Wasps, K.G. Ross and R.W. Matthews, Eds., Cornell University Press, New York, *in press*.
3. Gadagkar, R., Vinutha, C., Shanubhogue, A. and A.P.Gore. 1988. Proc. R. Soc. Lond. B., 233:175-189.
4. Chandrashekara, K. and R. Gadagkar. *this volume*.
5. Muralidharan, K., Shaila, M.S., and R. Gadagkar. 1986. J.Genet., 65:153-158.
6. Venkataraman, A.B., Swarnalatha, V.B., Nair, P. and R. Gadagkar. 1988. Behav. Ecol. Sociobiol., 23:271-279.

LIFE HISTORY STUDIES OF A COMMUNAL HALICTINE BEE, *LASIOGLOSSUM (CHILALICTUS) ERYTHRURUM*

Kukuk, P. F.*, M. W. Blows# and M. P. Schwarz#

*Department of Biological Science, University of Montana, Missoula MT 59812, USA.
#Departments of Zoology and Genetics, La Trobe University, Bundoora Victoria 3083, Australia.

INTRODUCTION

In halictine bees, communal species are found in different lineages than eusocial species, suggesting that communal associations may not be merely transition states in an evolutionary pathway between solitary life and eusociality but are an evolutionarily stable pattern of sociality (1). At present, there are very few detailed studies of communal bees or wasps even though such associations are both numerous (2) and theoretically interesting (3).

In Australia, most species in the genera *Homalictus* and *Lasioglossum* are group-living. All of these appear to be communal even though many species produce up to three generations per season (4,5,6), suggesting that there is selection favoring group life but not a reproductive division of labor. This was thought to be due to a short life span of females so that an overlap of generations, necessary for the evolution of eusociality, was lacking (4). Our data indicate a slight overlap of generations in the field but more importantly, a complete lack could occur if females continue to forage until they die, rather than extending their lifetimes by rearing a limited number of female offspring and then remaining in the nest. Long lifetimes have been observed in the laboratory (6, pers. obs.)

The presence in Australia of numerous communal halictine bees in the genera *Lasioglossum* and *Homalictus* clearly suggests that group life confers some selective advantage but the lack of reproductive castes remains problematical.

MATERIALS AND METHODS

We have undertaken studies of the *L. erythrurum* in southwest Victoria, Australia and here summarize results obtained by triweekly nest excavations. These continued for two active seasons and at less frequent intervals during the intervening inactive season at one nest aggregation plus supplementary excavations from two additional aggregation sites in the Cabboboonee Forest (Sept. l988 - March l990).

RESULTS

During the two seasons at the primary site, *L. erythrurum* appears to be univoltine, in contrast to previous reports (4, 6). Moreover, three nest aggregations within a radius of 10 km were not synchronous in the timing of brood production for either of the two active seasons. Dissections confirmed that virtually all females are mated and have active ovaries in nests where provisioning is taking place with the exception of newly eclosed, young females.

The species is largely protandrous with normal males produced early and large, macrocephalic males (7) eclosing somewhat later, but both on average prior to females. The sex allocation is slightly male biased, with males comprising 57% of the total pupal dry weight in l988-89 and 55% in l989-90. Large colonies appear to occupy reused nests, while new nests are begun by one or a small number of females. In an undisturbed site, most colonies were old and large while in a disturbed site, the colonies were young and smaller.

CONCLUSIONS

The apparent univoltism at this location in contrast to previous observations, plus the lack of synchrony in brood rearing among nearby nest aggregations suggests that females of *L. erythrurum* are very sensitive to local conditions and that they rear brood when conditions permit, i.e. they are opportunistic. This suggests a working hypothesis for explaining the lack of reproductive altruism in Australian halictines based on environmental unpredictability.

If group life is highly advantageous, but the probability of rearing more than one generation of brood per active season fluctuates so that in a considerable proportion of years only one brood can be reared, then this may provide a barrier selecting against parental manipulation of the first brood. From the point of view of the spring reproductives, a strategy of producing a small brood consisting primarily of females workers would be strongly selected against during years when a second, reproductive brood could not be reared due to unpredictable, disadvantageous local conditions.

REFERENCES

1) Kukuk. P. F. and G. C. Eickwort. 1987. In, J. Eder and J. Rembold, eds., Chemistry and Biology of Social Insects. Verlag J. Peperny, Munich.
2) Eickwort, G. C. 1981. In, R. Herman, ed., Social Insects, Vol 2. Academic Press, NY.
3) Trivers, R. H. 1971. Quart. Rev. Biol. 46:35-57.
4) Michener, C. D. 1960. J. Kans. Ent. Soc. 33, 85-95.
5) Knerer, G. and M. P. Schwarz. 1976. Science 194, 445-448.
6) Knerer, G. an M. P. Schwarz. 1977. Zool. Anz. 200, 321-333.
7) Kukuk, P. F. and M. P. Schwarz. 1988. Pan-Pac Ent. 131-137.

11th International Congress—IUSSI 1990, India

VARIABLE COFOUNDING RATES IN ALLODAPINE BEES OF THE GENUS *EXONEURA*

Michael P. Schwarz
Department of Zoology,
La Trobe University,
Bundoora, Vic. 3083 AUSTRALIA

Exoneura bicolor is a univoltine allodapine bee common in south-eastern Australia. In populations occuring in montane, wet sclerophyll forests most nests are built in fern fronds which form spatially discrete clumps around individual tree-ferns. Resulting nest aggregations range in size from a few nests to as many as eighty. Two types of brood rearing colonies occur, those comprised of females re-using their natal nests, and those occuring in newly established nests. Founding of new nests is largely restricted to a short period of time in late spring. The majority of both re-used and newly founded nests are occupied by two or more closely related cofoundresses (r = ca. 0.5), indicating that pleometrotic colony founding effectively involves kin recognition. Hence, it appears that benefits from cooperative nesting are large, but that kin association is favoured when possible. Such kin association may be enabled by the highly clumped distribution of nest substrates in montane forests.

In contrast, a closely allied and probably con-specific population of Exoneura occuring in low-lying open heathlands exhibits very low levels of nest cofounding, although re-used nests are frequently occupied by groups of related females. Relatedness between adult nestmates in re-used nests is not different from populations in montane forests.

In both montane and heathland populations, reproductive efficiency per female greatly increases as a function of the number of cooperating females. Sex ratios in newly founded nests in both populations are strongly female biased. This is probably an adaptation to exploit benefits of cooperation in situations where groups of highly related females can be readily assembled due to their emergence in a common natal nest. However, sex ratios of brood raised in re-used nests are more strongly female biased in montane populations than in heathland areas. This may reflect the high cofounding rates in the former populations compared to haplometrotic founding in the latter.

The low level of nest cofounding in the heathland population may be linked to both the spatial distribution of nesting substrates and temporal variation in substrate availability. Heathland populations of Exoneura nest in Xanthorrhoea flower stalks, which are spatially dispersed rather than clumped. Furthermore, Xanthorrhoea flowers primarily after fires or disturbance, and flower stalks remain structurally useable for nesting for only a few years.

Rates of cofounding in heathland areas may be influenced by substrate distribution in two ways: (1) Low density of existing nest substrates and appearence of new stalks in recently burnt areas may preclude cofounding because of the difficulty of locating kin over large distances; (2) Low density of nests may reduce parasite pressure and lead to lower benefits from cooperative nest guarding than in montane habitats. However, because colony productivity in both areas increases with the number of adult nestmates, these factors can only explain patterns of nest cofounding if costs are incurred by nesting with unrelated females. It is suggested that the allodapine trait of communal progressive rearing allows significant opportunities for nestmate parasitism, and that this in turn limits cooperation when kin groups cannot be readily assembled.

SOCIAL BEHAVIOUR OF COFOUNDRESSES IN THE MULTIPLE-FEMALE NEST IN *POLISTES* STIGMA (HYMENOPTERA : VESPIDAE)IN SOUTH INDIA

Tadashi Suzuki

Department of Biology, Faculty of Science, Tokyo Metropolitan University, Fukazawa, Tokyo 158, Japan

The study of social biology of tropical Polistes wasps is yet virtually lack in Asia. P. stigma is a speices widely found in tropical Asia and probably a commonest Polistes species in woody areas in south India. In this paper I show the social relations among cofoundresses and their behaviour in the pre-emergence multiple-female nest of P. stigma. The proportion of multiple-female nest fouding is low in this species in south India.

Observation was conducted in Coimbatore (11°N,77°E), Tamil Nadu, south India. Behaviour of ten cofoundresses on a new nest were observed. They originated from one colony and associated in the new nest one month after the abandonment of the natal nest. Since all but two cofoundesses oviposited some eggs in the natal nest, most, if not all, cofoundresses were potential egg-layers.

Dominant/aggressive interactions were very rare throughout the pre-emergence period. But dominance hierarchy was found and three or four classes of social-ranking were recognized. Three foundresses were observed to oviposit. Among them one foundress which was the top-ranking individual oviposited more frequently and ate all of other's (subordinates') eggs, resulting in monopolisation of egg production (functional monogyny).

Cofoundresses were divided into four groups according to their foraging work: 1) a principal egg-layer (queen) which laid most eggs and foraged pulp but not other materials, 2) six flesh-foragers which foraged flesh at least once in addition to liquid and/or pulp (two of them oviposited), 3) two non-flesh foragers which foraged liquid but not flesh, 4) a non-forager which did not show any foraging work.

Time distribution among on-nest, on-nest-proximiy (on the substrata to which nests are attached), and off-nest (outside the on-nest and on-nest-proximity) were distinctive among foundress groups. For example, the queen spent most time on the nest though she often moved to the nest-proximity when the nest was small as though she avoided contact with other females on it or let them work, and non-flesh foragers at times spent most or all of the time under observation (5 h/day) off the nest probably without making noticeable foraging activity.

MONOGYNY AND POLYGYNY IN PONERINE ANTS WITHOUT QUEENS

Christian Peeters

School of Biological Science, University of New South Wales, Sydney, **Australia**

The queen caste has disappeared in a small number of ants in the subfamily Ponerinae, and gamergates (mated workers) perform the reproductive role. Recent studies indicate that queenless societies can have either one or several gamergates per colony, which reveals that reproductive inhibition can occur despite the absence of the queen caste. I will review the mechanisms involved in this reproductive regulation, and compare the social characteristics of monogynous and polygynous species without queens. In the queenless Ponerinae, all workers have a spermatheca and are capable of functioning as reproductives, but only those which can mate during the limited period of male activity differentiate as gamergates.

Monogynous Queenless Ponerine Ants

In *Pachycondyla krugeri* [1], *P. sublaevis* [2], *Diacamma australe* [3], *D. rugosum* [4], *Platythyrea lamellosa* [5] and *Streblognathus aethiopicus* [6], only one worker mates and lays eggs in each colony. Social regulation can occur at two physiological levels: (i) sexual attractiveness of workers, and (ii) oogenesis in their ovaries. In *D. australe* and *D. rugosum*, the retention or mutilation of bladder-like appendages on the thorax determines the ability to mate. *Diacamma* is exceptional, however, because in other species little physical interaction occurs between gamergates and workers. Thus inhibitory pheromones must be involved, and these are likely to be volatile since trophallaxis does not occur in ponerine societies. In many species the single gamergate inhibits unmated workers from laying haploid eggs.

Polygynous Queenless Ponerine Ants

In *Ophthalmopone berthoudi* [7], *O. hottentota* [8], *Dinoponera quadriceps* [9], and all the known queenless species of *Rhytidoponera* (e.g. *violacea* [10], sp.12 [11], *metallica* [12], *confusa* and *chalybaea* [13]), more than one gamergate occur in each colony. There appear to be no species-specific ratios of gamergates to virgin workers in a colony. Indeed, in *O. berthoudi*, this ratio fluctuates during the year, as it reaches a maximum following the mating period, and then decreases with the emergence of new workers [7]. Further evidence of a lack of social regulation are the marked variations in the numbers and proportions of gamergates which can exist between colonies at one time of the year.

However, in *D. quadriceps,* young colonies may only have one gamergate, while bigger colonies have several [9]. Fewer inseminated workers are found in *Rhytidoponera* colonies (less than 10%) than in *O. berthoudi*, where up to 108 gamergates have been found in a single nest (consisting of 261 adults). It remains to be determined whether all the gamergates in a colony are equally fecund. The pattern of ovarian activity in unmated workers varies considerably between species.

Social and Ecological Correlates of Gamergate Number

Interspecific comparison of monogynous and polygynous colonies reveal the following: (i) there will be differences in the genetic relatedness of nest inhabitants; (ii) ability to recognize nestmates is unchanged; (iii) the characteristics of colony budding are expected to differ; (iv) gamergate fecundity varies considerably between species, and there is no interspecific relationship between colony size and gamergate number.

The existence of both monogynous and polygynous queenless species affords a useful insight into the evolutionary tendencies determining queen number throughout the ants. The presence of multiple queens in a colony has generally been assumed to be adaptive, either with reference to particular ecological circumstances (e.g. short-lived nest sites) [14], or to kin-selection considerations [15]. Following the repeated loss of the queen caste in different unrelated ponerine genera, workers do not always possess physiological mechanisms which can be improved by selection in order to regulate mating or ovarian activity in others. Thus whether either one or several gamergates reproduce is not solely determined by selective pressures, but may also depend on phylogenetic or mechanistic considerations. Available data support the contention that monogyny or polygyny in queenless species are not associated with distinct ecological profiles (e.g. habitat or microhabitat preferences).

REFERENCES

1. Wildman, M.H. and Crewe, R.M., 1988. Insectes Sociaux, 35: 217-225.
2. Peeters, C., Higashi, S. and Ito, F. Submitted
3. Peeters, C. and Higashi, S., 1989. Naturwissenschaften, 76: 177-180.
4. Fukumoto, Y., Abe, T. and Taki, A., 1989. Physiol. Ecol. Japan, 26: 55-61.
5. Villet, M., Hart, A.P. and Crewe, R.M. In preparation
6. Ware, A., Compton, S. and Robertson, H., 1990. Insectes Sociaux, in press
7. Peeters, C. and Crewe, R., 1985. Behav. Ecol. Sociobiol., 18: 29-37.
8. Peeters, C. and Crewe, R., 1985. South African J. of Zoology, 20: 268.
9. Araujo, C., Fresneau, D. and Lachaud, J.P., 1990. Actes Coll. Insectes Soc., 6:
10. Whelden, R.M., 1957. Annals Entomol. Soc. of America, 50: 271-282.
11. Peeters, C., 1987. Insectes Sociaux, 34: 75-86.
12. Haskins, C.P. and Whelden, R.M., 1965. Psyche, 72: 87-112.
13. Ward, P.S., 1983. Behav. Ecol. Sociobiol., 12: 285-299.
14. Hölldobler, B. and Wilson, E.O., 1977. Naturwissenschaften, 64: 8-15.
15. Nonacs, P., 1988. Evolution, 42: 566-580.

QUEEN NUMBER, SOCIAL STRUCTURE, REPRODUCTIVE STRATEGIES AND THEIR CORRELATES IN ANTS

Laurent KELLER(1) and Luc PASSERA(2)

(1) Musée Zoologique, Palais de Rumine, CP 448, 1000 Lausanne 17, Switzerland
(2) Laboratoire d'Entomologie, Université Paul-Sabatier, 118 route de Narbonne, F-31062 Toulouse Cedex, UA CNRS 333, France

A colony of social insects may have a single functional (i.e. egg laying) queen (monogyny) or more than one (polygyny). Recently, a considerable interest has grown in polygyn colonies of highly eusocial insects. One of the primary aims of these studies is to try to resolve the paradox of why queens associate since they usually have lower fecundity in polygyne colonies compared to monogyne colonies (e.g., 1,2,3,4). Several hypotheses based on kin selection and mutualism have been proposed to resolve this question but generally reproductive success of queens has been asessed through the number of offsprings they (or their relatives) produce. In this paper we wish to stress that monogyne and polygyne species and/or forms of the same species often differ in many other physiological, morphological and behavioural aspects which not only influence the number of offsprings produced and their cost of production, but also the probability these offsprings have to survive and reproduce themselves. Therefore, these shifts in reproductive strategies should be taken in account when assessing the reproductive success of ant queens.

To determine the relation between queen number, mode of colony founding, physiology and morphology of queens, we collected workers and mature winged queens of 24 ant species. Of these species, 13 were monogynous and 11 polygynous. The mode of colony founding was found to be related (X^2 test; P<0.001) to the social structure (monogyny/polygyny) of the species; 12 (92%) of the monogynous species were independent-founding species (i.e., young queens start colony founding without the help of workers), whereas only 1 (8%) exhibited dependent founding (i.e., young queens need the help of workers to start new colonies). In contrast, only 3 (27%) of the polygynous species were independent founding species, and 8 (73%) of them employed dependent founding.

Mature winged queens of species utilizing independent colony founding had a far higher relative fat content than those of species employing dependent colony founding (5). These fat reserves are stored during the period of maturation, i.e. between the time of emergence and mating, and are used to nurture the queen and the brood during the time of colony founding. Mature winged queens of species founding independently but non claustrally were found to have a relative fat content intermediate between the values found for those founding either independently or dependently. This suggests that such young queens rely partially on their fat reserves and partially on the energy provided by the prey they collect to nurture themselves and the first brood during the time of colony founding. Study of the fat content of mature gynes of all species has shown that it gives a good indication of the mode of colony founding.

Queen/worker size dimorphism was higher in species founding independently than in those founding dependently. In species founding independently, queen length and dry weight were on average 2.1 and 16.9 times higher than that of workers, whereas these values were only 1.5 and 4.2 times higher in species founding dependently. Difference in energy content between gynes and workers was nearly six times higher in species founding independently compared to those founding dependently. The production of new gynes therefore requires a much greater energetic cost in species founding independently than in those founding dependently.

Changes over time in the fecundity of queens was investigated in three monogynous, independent colony founding species and two polygynous dependent colony founding species. Fecundity of queens founding independently increased slowly with time, whereas fecundity of queens founding dependently reached the maximum level some weeks after the beginning of the first reproductive season (6).

Finally, considering the reproductive potentiality of the queens must take into account their life-span. A review of the available data in ants indicated that life-span of monogyne independent founding species is far higher than that of polygyne dependent founding species.

From this it is clear that monogyne and polygyne species differ in many life history traits that influence not only the number of offspring they produce, but the probability they have to survive and disperse and we argue that these differences should be considered in further works trying to understand the factors underlying changes in queen number in ants.

REFERENCES

1. Brian, M. V. (1969). Male production in the ant *Myrmica rubra* L. *Insectes Soc.*,16,177-190.
2. Fletcher, D. J. C., Blum, M. S., Whitt, T. V. & Tempel, N. 1980. Monogyny and polygyny in the fire ant *Solenopsis invicta* Buren. *Ann. Entomol. Soc. Am.* 73, 658-661.
3. Keller, L. 1988. Evolutionary implications of polygyny in the Argentine ant, *Iridomyrmex humilis* (Mayr) (Hymenoptera: Formicidae): an experimental study. *Anim. Behav.* 36, 159-165.
4. Vargo, E. L. & Fletcher, D. J. C. 1989. On the relationship between queen number and fecundity in polygynous colonies of the fire ant, *Solenopsis invicta. Physiol. Entomol.* 14, 223-232.
5. Keller, L. & Passera, L. 1989b. Size and fat content of gynes in relation to the mode of colony founding in ants (Hymenoptera; Formicidae). *Oecologia*, 80, 236-240.
6. Keller, L. & Passera, L. In press. Fecundity of ant queens in relation to their age and the mode of colony founding. *Insectes Soc.*

COOPERATIVE COLONY FOUNDING IN SOME DESERT ANTS

Steven W. Rissing, Department of Zoology, Arizona State University, Tempe, AZ 85287-1501 and Gregory B. Pollock, Department of Political Science, Northwestern University, Evanston, IL, 60208, USA

Messor pergandei and Acromyrmex versicolor, are abundant ant species in the desert Southwest United States. Unlike most other ant species, colonies of both are often started cooperatively [1, 2] by unrelated foundresses [3]. While M. pergandei raises its first brood in isolation on stored energy reserves (claustrality), A. versicolor forages prior to first worker eclosion (non-claustrality). We have performed comparative observations and experiments on these species, exploiting especially their differences in claustrality, to understand the intra- and inter-colonial selective forces favoring cooperation among non-relatives. Aspects of this work are summarized here.

MATERIALS and METHODS

Field Behavior. Messor pergandei has a prolonged mating season, and while mating sites have not been found, frequent presence of males in starting nests suggests dispersal from such sites is limited [1]. Acromyrmex versicolor has a more coordinated flight season with small mating swarms occurring near the ground [2]. Comparison of nest density with mating swarm size and number in an area as well as marking experiments suggests dispersal is limited in this species as well. Starting nests of both species are clumped: M. pergandei in sandy ravines [4] and A. versicolor under large trees [2]. When given a choice, foundresses of the former species actively prefer to join larger foundress associations (Krebs and Rissing, unpublished data).

Pleometrotic Advantage. Number of initial workers produced is a linear function of foundress number in M. pergandei; further, weight lost by a foundress while producing the first workers declines as foundress number increases [5]. In both species the first workers produced engage in reciprocal brood raids with neighboring conspecific colonies; in M. pergandei the colony with the most workers (= the colony with the most foundresses) usually wins at such raids to the demise of the other colony [5, 6,7].

Foundress Behavior. While foundresses of both species cooperate in raising a first communal brood [7, 8], A. versicolor foundresses asymmetrically share the relatively dangerous task of foraging: in 44 of 45 laboratory colonies containing 3 foundresses, a single foundress became a foraging specialist [7]. Further, the cooperation displayed among M. pergandei foundresses during claustrality disintegrates as the first workers open the nest and foundresses begin to fight lethally [6]; this does not occur in A. versicolor where laboratory colonies remain polygynous even after 2.5 years [7].

DISCUSSION

Significant inter- and intra-colony selective forces are likely at play in M. pergandei and A. versicolor. Abiotic forces resulting in the clumping of starting nests of both species [2, 4] combined with strong adult territoriality in both [9, 10] result in intraspecific brood raiding among starting colonies. We have shown that worker number (a direct function of foundress number) determines the outcome of such brood raids in M. pergandei [5, 6] and suspect the same for A. versicolor. Such intrademic selective forces appear sufficient to favor cooperative colony founding in both species. Recent discovery of a non-cooperative founding M. pergandei population [11] will permit further examination of the role of inter-colony interactions in favoring intra-colony cooperation in this species.

Messor pergandei and A. versicolor differ with respect to claustrality during colony founding (the former is claustral, the latter not) and fate of co-foundresses following eclosion of first workers (colonies of the former usually reduce to monogyny, the latter not). Foundress condition when the first workers eclose differs between the two: M. pergandei foundresses have lost almost half their mass [5] while A. versicolor foundresses have not (unpublished data). Thus, it may be in the interest of an M. pergandei foundress to reduce the energetic demands on her colony at the risk of her own demise by engaging in lethal queen flights at this time while this is not the case for relatively well-fed A. versicolor foundresses.

REFRENCES

1. Pollock, G.B. and S.W. Rissing. 1985. Psyche 92: 125-134.
2. Rissing, S.W., R.A. Johnson and G.B. Pollock. 1986. Psyche 93:177-186.
3. Hagen, R.H., D.R. Smith and S.W. Rissing. 1988. Psyche 95: 191-201.
4. Rissing, S.W. and G.B. Pollock. 1989. Journal of Arid Environments 17: 167-173.
5. Rissing, S.W. and G.B. Pollock. Insects Sociaux, in press.
6. Rissing, S.W. and G.B. Pollock. 1987. Animal Behaviour 35: 975-981.
7. Rissing, S.W., G.B. Pollock, M.R. Higgins, R.H. Hagen and D.R. Smith. 1989. Nature 338: 420-422.
8. Rissing, S.W. and G.B. Pollock. 1986. Animal Behaviour 34: 226-233.
9. Wheeler, J. and S.W. Rissing. 1975. Pan-Pacific Entomologist 51: 303-314.
10. Gamboa, G.J. 1974. Masters Thesis; Arizona State Univ., Tempe, Arizona.
11. Ryti, R.T. 1988. Pan-Pacific Entomologist 64: 255-257.

SIZE PREFERENCES IN THE CHOICE OF PLEOMETROTIC PARTNERS : COMPETITION IN THE PEACEABLE KINGDOM OF FOUNDRESS ANT QUEENS?

Peter Nonacs
Museum of Comparative Zoology
Harvard University
Cambridge, MA 02138 USA

In many species of ants, new colonies are founded by queens who disperse from their natal nest, find a suitable nest site, and raise the first worker brood primarily from their own metabolic reserves. They do so either entirely alone or as a cooperative venture with one or several other queens (pleometrosis). There are many possible advantages to cooperating including: larger and more rapidly developing first broods, less weight loss per queen, and a headstart in intercolonial competition between young colonies (1). However, in the majority of pleometrotic species, only one queen from the initial group will survive to colony maturity. Therefore, potential for both intragroup cooperation and competition is present.

Because ant queens are so dependent on their own food stores, size may be an important character trait in potential partners. Large size could indicate a desirable mutualist to be joined: a queen with large reserves to put into brood care. However, large size could also mean a formidable competitor to be avoided: a queen that is likely to be more productive or to be able to win a war of attrition. Therefore, I tested if newly-mated, Lasius pallitarsis queens exhibited any size preferences in choosing pleometrotic partners.

A queen was introduced at the base of a small T-maze apparatus in one of two protocols. In one, there was a queen at the end of one of the arms of the T, in a vial containing soil. The other arm contained a similar vial but without a queen. The resident queen was either larger or smaller (as measured by weighing before the experiments) than the introduced queen. In the second protocol both arms of the T-maze contained a queen: one relatively heavy and the other light. One day later the positions of all queens were noted. Immediately thereafter, all introduced queens were removed and given access to a rich, liquid diet for 24 h and then re-introduced to the same choice again. Feeding queens should improve their condition and outlook on life.

The results for the experiments with only one resident queen showed that introduced queens with a large size advantage (>9%) showed no preference between empty and occupied vials. All other queens showed significant avoidance (Table 1). Feeding did not markedly change the behavior of larger introduced queens, but did significantly increase the likelihood of joining in initially smaller queens. In the experiment with two resident queens, the light resident was significantly preferred (joined in 24 of 34 trials) before feeding but not after feeding (light joined in 17 of 35 trials).

Table 1: The one-resident experiments before and after feeding. The results are given as the number of queens joining per all trials. Changes refers to the number of changes per all changes that were from being alone prior to feeding to joining after feeding. Q is the weight differential of the introduced to resident queens.

Queen Size	Unfed	Fed	Changes
$Q > 9\%$	6/13	5/13	1/3
$9\% \geq Q > 0\%$	2/14	3/14	2/3
$0\% > Q \geq -9\%$	4/22	8/21	6/7
$Q < -9\%$	3/13	6/12	3/3

L. pallitarsis queens appear to evaluate at least their own size and condition as regards joining others. All queens prefer to join lighter ones and queens without a large size advantage avoid others. Fed queens, however, act more like large queens. These results are consistent with joining behavior being more affected by competitive rather than mutualistic considerations. The results are also consistent with a theoretical model that predicts discrimination of competitive ability in potential partners, to be an important attribute for pleometrotic species (2).

REFERENCES

1. Rissing, S.W. and G.B. Pollock. 1988. In: Interindividual Behavior Variability in Social Insects, R.L. Jeanne, Ed., Westview Press, Boulder, CO, pp. 179-222.
2. Nonacs, P. 1989. Evolutionary Ecology, 3:221-235.

EXPERIMENTS ON COLONY FOUNDATION IN THE POLYGYNOUS ANT *CARDIOCONDYLA WROUGHTONII*

Robin J. Stuart, Department of Entomology, University of California, Davis CA 95616 USA; and Department of Zoology, University of Vermont, Burlington VT 05405 USA.

Cardiocondyla wroughtonii (Forel) is a small (worker= 1.5-2 mm) tropical tramp ant species. Colonies nest in plant cavities, are often polygynous, and produce winged gynes, winged males, and wingless worker-like (ergatoid) males. Winged sexuals appear to conduct mating flights but all sexuals will mate in the nest. Ergatoid males fight and kill one another so that there is typically only one per nest but there may be many winged gynes and winged males present simultaneously (1, 2, and contained references).

Laboratory experiments were conducted on the colony foundation ability of queens cultured individually (n= 40), in groups of four (n= 20), and with workers (1, 2, or 4 queens, each with 10 workers; n= 4, 2, and 2 respectively). Plastic petri dishes (dia= 7 cm) with tissue culture tubes as nests (40 X 4 mm, half filled with water and plugged with cotton) were used to culture colonies. Food was an artificial diet (1) supplemented with frozen Drosophila.

Queens with workers were highly successful at founding colonies and all produced new workers. No solitary queens produced any new workers but eggs were produced in 10 replicates, larvae in 4, and pupae in 2. Groups of queens produced eggs in all 20 replicates, larvae in 18, pupae in 10, and workers in 4. Mites appeared to be a major factor in the death of brood.

These results suggest that haplometrotic and to a lesser extent pleometrotic colony foumdation are risky ventures for this species. Colony budding as a major means of colony foundation is consistent with observations of the biology of this species both in the laboratory and in the field (1, 2, and refs). Other mechanisms not evident in this study might function to promote colony foundation by unaccompanied queens under field conditions but such mechanisms remain to be demonstrated.

1. Stuart, R. J., A. Francoeur and R. Loiselle. 1987. Naturwissenschaften 74: 548-549.
2. Kinomura, K. and K. Yamauchi. 1987. J. Ethology. 5: 75-81.

QUEEN RECRUITMENT IN POLYGYNOUS AND POLYDOMOUS *FORMICA* POPULATIONS

W. Fortelius, D. J. C. Fletcher* &. D. Cherix**

Department of Zoology, University of Helsinki, SF-00100 Helsinki, Finland
* Program in animal behaviour, Bucknell University, Lewisburg, PA 17837, U.S.A.
** Museum of Zoology, Palais de Rumine, C.P.448, CH-1000 Lausanne 17 Switzerland

Cooperative nestclusters comprising tens or even hundreds of nests, with hundreds of mated queens with developed ovaries in each, regularly occur among species of red wood ants (*Formica rufa* group). The few existing data on relatedness among these coexisting queens (4 and Pamilo et al. in this volyme), together with some behavioural observations (3) indicate that own daughters are recruited back into the natal nest . The present study is concerned with mechanisms regulating queen recruitment in polydomous colonies i.e. how and when virgin females can get mated and join the queen population of polygynous nests. The work is part of an international research project on different aspects of reproduction in a highly polydomous and polygynous "super-colony" of *Formica lugubris* (Zett.) in the Swiss Jura Mountains and is presently continuing (1,2).

EXPERIMENTS AND RESULTS

To test if the workers of a particular nest respond differently to mated and unmated females we did the following tests. Alate females were put in plastic boxes (11 cm^2 x 6 cm deep) which were perforated by holes (diameter=2.5 mm) allowing free passage of workers but not of the bigger sexuals. In each experiment one box with nest material and 10 mated alate females (mated in the laboratory) and another box with nest material and 10 virgin alate females were dug about 10 cm into the tested nest. The testboxes thus coincided in nests producing sexuals with the natal group of alates. After 2 hours the boxes were checked for the number of females alive. Three different nest categories were tested: nests producing mainly female brood (female nests), nests producing mainly male brood (male nests) and nests not producing sexual brood (worker nests).

In all cases unmated females were better accepted by the workers and the survival rate was highest for unmated females in the natal nest (Table 1.). The other main result seen from Table 1. is that virtually all introduced females (irrespective of whether mated or not) were killed in both male and worker nests.

We thus continued by regularly testing unmated females in their natal nest for more than three weeks. A dramatic change in the worker response towards the introduced females coincided with the end of the nuptial flight period for the particular nest (all alates had left). When the natal alate population was present in the nest 84±31% of the introduced nestmates were alive after two hours while the corresponding number was 35±23% when the natal alate population was absent. (Mean of 8 successive tests for both periods).

Recipient nest	Female category Mated	Unmated	Replicates
Female nest -natal	49±39	84±31	8
Female nest -alien	19±20	74±33	14
Male nest	0	15±17	4
Worker nest	12±13	24±23	5

Table 1. Percentage of females alive (mean±sd) in the test box, after 2 hours in the nests.

No effect of the distance to the natal nest on the worker response was detected when unmated females were tested in 6 female nests along a 2 km distance gradient.

In a different set of experiments we simultaneously studied worker response towards alates of both sexes in two nest categories, female nests and nests producing both sexes in about the same numbers (mixed nests), and the occurrence of intranidal matings. In these experiments we used boxes through which neither workers nor sexuals could pass. Instead we added about 100 workers and nestmaterial from the recipient nest together with 20 sexual pairs into each box, before putting it in place well inside the nest (10-20 cm). In this case the boxes were checked after 20 hours.

Virtually all the tested females were inseminated when taken out from the mounds. 99% of the females and 92% of the males were alive after 20 hours in a female nest while the corresponding numbers for females and males in a mixed nest are 71% and 76% respectively. (Means for 4 replicates of each category).

The results indicate that presence of female brood in high numbers affects the worker response towards sexuals. In addition the possibility of intranidal matings clearly exists. The conclusion at this moment is that the best strategy for a female to join the "queen-pool" of a conspecific nest is to mate and stay in the natal nest.

REFERENCES

1. Cherix, D., Chautems, D., Fletcher, D. J. C., Fortelius, W., Gris, G., Keller, L.,Passera, L., Rosengren, R. & Vargo, E. L. 1989. Actes Colloques Insectes Sociaux, 5:45-53.
2. Cherix, D., Chautems, D., Fletcher, D. J. C., Fortelius, W., Gris, G., Keller, L.,Passera, L., Rosengren, R., Vargo, E. L. &. Walter, F. 1990. Ethology Ecology and Evolution (in press).
3. Fortelius, W. 1987. In: Chemistry and Biology of Social Insects, J. Eder & H. Rembold, Eds., Verlag J. Peperny, München, pp. 293-294.
4. Pamilo, P. 1990. Behavioral Ecology and Sociobiology (in press).

MONOGYNY OR POLYGYNY-A RESULT OF DIFFERENT DISPERSAL TACTICS IN RED WOOD ANTS (*FORMICA*; HYMENOPTERA)

Sundström, L.
Dep. of Zoology, Helsinki University
P.Rautatiekatu 13 00100 Helsinki
Finland

INTRODUCTION

Several hierarchical levels can be distinguished in ant populations: population - subpopulation - colony - individual. Formica truncorum (Formica rufa s.l.) have in addition two different colony types: monodomous, monogynous and polydomous, polygynous (1). These colony types occur in different areas of the Finnish archipelago, forming separate populations. Each one of these populations is divided into several subpopulations, represented by separate islands. The two colony types also have different population structures; the monodomous populations show little or no genetical differentiation between the islands (subpopulations), whereas the polydomous ones show considerable allele frequency differences between the islands (1). This pattern indicate strong dispersal in the former group, but weak dispersal in the latter.

The main point of this study is to trace these observed variations in population and colony structure back to individual dispersal and mating behaviour among sexuals.

MATERIAL AND METHODS

Males and females from these two colony types were tested for specific behaviours connected to dispersal and mating, such as precopulatory dealation, median mating date after eclosion, and frequency of mating without prior flight. Males were also tested for choice between mating and flying. Morphological characters, head width, thorax width and wing length were also measured.

RESULTS

There were no differences in dealation behaviour between mono- and polydomous females, when no males were present. However, clear quantitative differences were observed both in male and female mating behaviour indicating different inherent tendencies to mate before dispersal. Monodomous females mated in significantly lower numbers and later than polydomous ones. Males also had a similar effect on both mating frequency and median mating date. In the choice tests monodomous males also flew rather than mated more often than did polydomous males. The results with the observed population structure, with low genetic differentiation indicating high dispersal rates in monodomous populations and the opposite in polydomous ones. In males the behavioural differences were also connected to morphological characters so that males with proportionally longer wings were the ones who flew. These morphological characters also correlated with colony type, the variation being much larger in polydomous colonies.

DISCUSSION

Similar intraspecific variation in colony structure has earlier been observed in Formica exsecta (2). In this species there is also a significant morphological and behavioural dimorphism in máles, but not in females (3). The present results indicate that the colony structure may be a result of two different behavioural phenotypes, dispersers and nondispersers, and that the frequencies of these phenotypes may dictate the colony type (1). The results show differences in mating tactics rather than in dealation tactics, as has been observed in F. rufa s.str. (4)(1). So far, however, nothing is known about worker behaviour in mono- vs polydomous colonies of this species. The fact that there was a clearly higher morphological variation in polydomous than in monodomous males may also indicate an ongoing process of sexual selection (male-male competition) or frequency dependent selection.

LITERATURE

1. Sundström, L. 1989. Actes coll. Insectes Sociaux, 5:93-100
2. Pamilo, P. 1984. Biological Journal of the Linnean Soc. 21:331-348
3. Fortelius, W. Pamilo, P. Rosengren, R. Sundström, L. 1987. Ann. Zool. Fennici 24:45-54
4. Gösswald, K. Schmidt, G.H. 1960. Ins. Soc. 7:279-321

SHIFT IN REPRODUCTIVE STRATEGIES AND ITS CONSEQUENCES IN SEXUALS OF A POLYGYNOUS ANT *IRIDOMYRMEX HUMILIS* (MAYR)

Luc PASSERA(1) and Laurent KELLER(2)

(1) Laboratoire d'Entomologie, Université Paul-Sabatier, 118 route de Narbonne, F-31062 Toulouse Cedex, UA CNRS 333, France
(2) Musée Zoologique, Palais de Rumine, CP 448, 1000 Lausanne 17, Switzerland

The polygynous status of a large number of ant species is often associated with a shift in the reproductive strategies. In many polygyne species, queens need the help of workers to start new colonies (dependent mode of colony founding) and new colonies are established through nest-budding, a process in which inseminated queens move out of the nest accompanied by workers. In a substantial number of species employing nest-budding, one (generally the females) or, both of the two sexes has lost the ability to participate in a mating flight and the gynes are inseminated within the nest. this is the case in the Argentine ant *Iridomyrmex humilis*. In this species males fly, whereas mating flights by gynes is a very rare event. Mating generally occurs within the nest.

To investigate whether shifts in reproductive strategies are associated with physiological and morphological changes, the maturation process during the time between emergence and mating was studied in *I. humilis* and compared with other species exhibiting contrasting reproductive strategies. This comparison suggests that the time of maturation and the amount of carbohydrate and fat stored during this period are associated with the mode of colony founding as well as the presence versus lack of nuptial flight.

RESULTS AND DISCUSSION

Maturation period: Both in the field and in the laboratory, production of *I. humilis* males started before that of winged queen. These males are long lived, so that by the time the winged queens emerged, there were nearly always males ready to copulate. Laboratory experiments showed that females were sexually mature soon after emergence. As early as 24 hours after emergence 18% of the gynes were already mated and by the time they were 3-4 days old, 83% were inseminated. In other ant species and particularly in those participating in a nuptial flight before starting new colonies independently, time of maturation is generally far longer. This difference is probably associated with the need for independent founding species to accumulate large quantities of energy for nuptial flight and the time of colony founding (see below).

Carbohydrate storage: In ants, energy for flying is mainly derived from glycogen (1). To investigate whether the loss of mating flight by *I. humilis* females is associated with a lower glycogen content, changes in the amount of glycogen were determined for males and females during the time of maturation. At the time of emergence, the relative amount of glycogen (g glycogen/g dry weight) was almost identical in both sexes. During the time of maturation the amount of glycogen increased at a strikingly faster rate in males than in gynes. When males reached the age at which gynes mated, they had more than four times as much glycogen compared to gynes. At the time of mating, males had a glycogen content similar to sexuals of other species participating to nuptial flight (2), whereas females had a far lower content thus suggesting that gynes have not accumulated enough glycogen to undergo a nuptial flight. Whether this is a cause or an effect of the virtual deletion of mating flights in this species cannot be determined with certainty, but the available evidence suggest that it is an effect. When the maturation period was extended experimentally by preventing gynes from mating, the relative amount of glycogen greatly increased reaching values similar to those found in females sexuals of species which regularly undergo nuptial flights.

Fat storage: In ants, female sexuals starting new colonies independently, store large reserves of fat between emergence and mating (3). These fat reserves, which may account for more than 70% of the dry weight of mature winged queens, are used to nurture the brood during the time of colony founding. In contrast, the ratio fat/dry weight reached only 25% at the time of mating in *I. humilis*, although there was a slight increase in weight and fat between emergence and mating. This value is similar to that found in other species founding dependently (see Keller and Passera 1989).

Overall the data indicate that lack of nuptial flight and dependent founding in *I. humilis* are associated with a far lower accumulation of energy i.e., carbohydrate and fat and a short duration of the maturation period. We therefore hypothesis that duration of the maturation period of female sexual is related to the amount of energy required for colony founding.

REFERENCES

1. Passera, L., Keller., L., Grimal, A., Cherix, D., Chautems, D., Fortelius, W., Rosengren, R. & Vargo, E. L. In press. Carbohydrates as energy source during the flight of sexuals forms of the ant *Formica lugubris* (Hymenoptera, Formicidae). *Entomol. Gener.*
2. Passera, L. & Keller, L. In press. Loss of mating flight and shift in the pattern of carbohydrate storage in sexuals of ants (Hymenoptera, Formicidae). *J. Comp. Physiol. A.*
3. Keller, L. & Passera, L. 1989. Size and fat content of gynes in relation with the mode of colony founding in ants (Hymenoptera; formicidae). *Oecologia*, 80, 236-240.

11th International Congress—IUSSI 1990, India

THE REGULATION OF POLYGYNY AND QUEEN CYCLES IN RED ANTS (*MYRMICA*)

G.W. ELMES

NERC, Institute of Terrestrial Ecology,
Furzebrook Research Station
Wareham, Dorset. BH20 5AS UK

1. What is a polygynous species?

The regal status of a species or population is the integral of all its colonies'. In practise, populations are called monogynous or polygynous depending upon the most common condition, a dangerous simplification! The frequency distributions for queens was determined for 10 *Myrmica* and, although these had a wide range of averages and proportions of queenless colonies, all fit a Pearson Type III curve. This suggested that all were generated biologically in the same way and that all *Myrmica* species should be considered polygynous. It was shown that the number of queens in a *Myrmica* colony depends upon its worker-number, the exact relationship varying between species. Also, queens in large groups were smaller (not just lighter) than those comprising small groups; this relationship extended to the species level, another indication for a common mechanism of regulation.

Individual colonies can be called polygynous, monogynous or queenless; despite being a common condition the last is not generally accorded the same status as the two types of regality. The populations of *Myrmica* species studied always comprised a mixture of all 3 types and a variety of empirical evidence showed that individual colonies can move between the three reginal states during the course of a year. Therefore care should be taken when interpreting field data in terms of polygyny, unless the history of individual colonies are known.

2. Determination of queen number at the species level

Normally the growth of *Myrmica* colonies is limited by the characteristics of their nest-sites and I believe that nest-site availability limits most *Myrmica* populations. Egg production depends upon "ovary-mass" which depends on queen-size. Therefore in any given nest-site there should be an ideal queen-size to maintain the optimum colony-size. If nest-sites are fairly similar then selection should produce a single, suitably sized queen. However, when nest-site characteristics are ephemeral and variable there is no optimum size for queens, rather a variable-sized queen is required. This can only be provided by a a variable number of smaller queens. This argument resembles the "difficult habitat" hypothesis of Holdobler & Wilson.

The selection for number and size of queens cannot be disentangled. It seems that the advantage of small inexpensive" queens will correlate with nest-site variability. Small queens should have difficulty in claustral colony foundation and gain more by joining existing colonies, their loss of individual fitness should be compensated by increased fitness due to the wider range of potential nest sites. Loss of fitness should be mitigated by joining relatives (as reasoned by Nonacs) but this does not seem to be general in *Myrmica*. Small queens gain relatively more than large queens by joining a group and whereas established groups should prefer a small joiners, larger queens might be permitted to join only queenless colonies. This would produce the assortment of queens observed in wild colonies.

3. Variation within species

Given that *Myrmica* speciates by physiological and social adaptation to the nest-sites available in a particular habitat, then a new habitat should be colonized by the nearest species able to use the new nest sites. Starting from an initial level of polygyny and queen-size the adaptive process would continue. In isolated and unusual local circumstances it becomes a new race, sub-species or species with a modified physiology and possibly a new level of polygyny and queen-size. All these taxonomic units are particularly abundant in *Myrmica*.

Therefore populations should show variation in queen-numbers and, perhaps, queen-size. So far it has been shown that: (1) The number of queens in an average sized colony (Q) varies between populations. (2) Q varies within a population over time. (3) Q changes during the course of a single season reflecting the actual mechanisms of Q adjustment. (4) The size of queen can vary between populations.

4. Cycles and Trends in queen-number

Q fluctuated in the same way in 4 adjacent *M. rubra* populations, indicating an extrogenous cause such as weather. An apparent 10 year cycle could be explained by summer temperatures which happened to cycle during the study. Despite huge variations during the course of each season, Q declined steadily in a Polish population of *M.limanica*; probably in response to a regenerating marshland habitat.

In two proximate populations of *M.sulcinodis*, Q followed a 4-5 year cycle. No extrogenous correlations could be found suggesting that it was endogenous. A simple recruitment model with a density dependant element, can generate such a cycle within a colony with periodicity depending on queen- longevity. However, the phasing of individual colonies, necessary to produce a cycle at the population level, must be extrogenous. This problem needs more theoretical study.

4. Microgynes and Parasites

In some species selection might produce queens that are no bigger than their workers (microgynes). Several studies of the only free-living European microgynous species (*M.ruginodis microgyna*) have not resolved whether it is a polymorphism of *.M.ruginodis* or a separate species. There are several microgynous species in North America and study of these might help to understand this problem.

A disadvantage of polygyny and reduced queen size is that populations with fewer queens might be prone to parasitism from more polygynous ones, especially if they have evolved a microgyne form. This might be held in check when there is a niche for free-living, highly polygnous colonies but might evolve into a permanent parasitic relationship if this niche is lost: eg. the microgyne parasite of *M.rubra*.

FUNCTIONAL MONOGYNY OF *LEPTOTHORAX ACERVORUM* IN JAPAN

ITO, F.
Graduate School of Environmental Science,
Hokkaido University, Sapporo 060, Japan

A small myrmicinae ant, *Leptothorax acervorum* is widespread over Eurasia and north America. According to Buschinger (1968), this species is facultatively polygynous in Europe, many colonies containing multiple queens laying eggs. The author investigated the social organization of *L. acervorum* in Furano, northern Japan.

There were multiple dealates in 54 of 102 colonies collected. In 50 of the multi-dealate colonies, the number of egg layers was only one, even in the colonies which contained many inseminated dealates; in one colony no egg layer was involved; and in three colonies, two or three dealates were laying eggs. Even in the last three colonies, however, only one dealate was a prime egg-layer which had long ovarioles and dense accumulation of yellow bodies; and other layers had short ovarioles and tiny yellow bodies, suggesting that they laid few eggs. Thus, Japanese *L. acervorum* shows functional monogyny which has ever been described in few species of tribe Leptothoracini.

RELATEDNESS IN NEO-TROPICAL POLYGYNOUS WASPS

Colin R. Hughes, David C. Queller and Joan E. Strassmann
Department of Ecology and Evolutionary Biology, Rice University,
P.O. Box 1892, Houston, Texas, 77251, U.S.A.

Many neo-tropical wasp colonies are founded by large swarms which contain many queens. Mature colonies contain hundreds or thousands of individuals, often with little or no morphological caste differences. In these colonies, there are often as many as 10 queens and sometimes even hundreds.

High queen number makes the maintenance of eusociality in this group difficult to explain because it is expected to lead to lowered relatedness between workers and the brood they rear. With just ten singly mated, outbred queens contributing eggs equally to the brood, relatedness is expected to fall below 0.1, requiring a benefit to cost ratio greater than five to explain continued helping by the workers. Hamilton himself recognized this difficulty and suggested that it may not exist in nature due to the asynchronous production of sexuals within a population and limited dispersal of swarms[1]. This, he reasoned, could lead to significant inbreeding and thereby, higher intracolony relatedness. West Eberhard suggested a second resolution after observing a *Metapolybia* nest[2]. She documented a decline in queen number starting soon after swarming that ended when just one queen was left; at this point a reproductive brood was produced and queen number rose. This "cyclical monogyny" would be reflected in cyclical variation in relatedness with reproductives being produced when intra-colony relatedness peaked.

We measured levels of intra-colony relatedness in 4 species of this enigmatic group of wasps (*Polybia occidentalis*, *Polybia sericiea*, *Polybia emaciata*, and *Parachartergus colobopterus*) in order to detail just how difficult it is to explain their continued sociality and to evaluate the above hypotheses. Females were collected from their nests at two field sites in Venezuela then kept frozen until analyzed using standard starch gel electrophoresis techniques. Relatedness values were calculated from individual phenotypes at polymorphic loci using the method of Queller and Goodnight[3].

Relatedness (±S.E.) varied from a high of 0.34 ± 0.05 in *Polybia occidentalis* (21 colonies, 197 individuals) to a low of 0.10 ± 0.07 in one of two populations of

Parachartergus colobopterus (15 colonies, 161 individuals) that were sampled[4]. In no population of any species studied was there any indication of inbreeding.

Relatedness within colonies of polygynous tropical wasps is not as low as had been expected from knowledge of queen number alone. This makes the problem of explaining the maintenance of sociality in these wasps less difficult. However, in at least some species, most notably here in *Parachartergus colobopterus,* relatedness is quite low. Four explanations seem possible to explain why workers in such colonies do not reproduce directly. First, the benefit to cost ratio could be high enough to explain worker behavior in the absence of any other special factors. Second, workers may actually be a sub-fertile caste, less able than queens to lay eggs. Third, it is possible that the relatedness value we measured is not the true value of importance to the workers; it is conceivable that workers could aid more closely related brood preferentially, as has been suggested recently in honey bees[5]. Fourth, relatedness among colony mates may not always be as low as we measured; a situation which would arise if West Eberhard's cyclical monogyny hypothesis[2] is applicable.

We investigated these questions in more detail in one population of *Parachartergus colobopterus*. Morphometric analyses of queens (females whose ovaries contained at least one mature oocyte) and workers revealed that there were no differences between these two groups in head width, length of scape, or number of hamuli. Further, there was no correlation between the head width and the number of mature oocytes in the ovary. This strongly suggests that this species has no morphological castes and that workers are not physically unable to perform reproductive functions.

Relatedness among a random sample of females from this population was higher than that found at our other field site; $r \pm S.E. = 0.32 \pm 0.11$ (24 groups, 375 individuals). The most intriguing finding was that relatedness among queens was significantly higher than relatedness among the random sample[6]. This higher relatedness largely resolves the disparity between the relatedness value calculated from queen number and that calculated from allozyme variation. High relatedness among queens is also compatible with West Eberhard's hypothesis that reproductive individuals are produced when only one queen is left in the colony[2]. However, analysis of the number of queens in new, young, mature and declining colonies failed to provide any support for the cyclical monogyny hypothesis. There was no decrease in queen number as colonies aged[6].

1. Hamilton, W. D. 1972. Ann. Rev. Ecol. Syst. 3: 193-232.
2. West Eberhard, M. J. 1972. Science 200: 441-443.
3. Queller, D. C. and K. F. Goodnight. Evolution 43: 258-275.
4. Queller, D. C., J. E. Strassmann and C. R. Hughes. 1988. Science 242: 1155-1157.
5. Page, R. E., G. E.Robinson, and M. K. Fondrk. 1989. Nature 338: 576-579.
6. Strassmann, J. E., D. C. Queller, C. R. Solís and C. R. Hughes. submitted.

THE GENETIC AND SOCIAL STRUCTURE OF POLYGYNOUS SOCIAL WASP COLONIES (VESPIDAE : POLISTINAE)

M. J. West-Eberhard

Smithsonian Tropical Research Institute
Correspondence: Escuela de Biologia, Universidad de Costa Rica, Ciudad Universitaria, Costa Rica, Centroamerica

CORRELATES OF POLYGYNY IN SOCIAL WASPS

Swarm-founding neotropical social wasps seem to fall into two categories [1, 2]: "periodically monogynous" species, in which queen number sometimes falls to one; and "permanently polygynous" species, in which monogyny rarely, if ever, occurs. As a rule, monogyny occurs in relatively small colonies, and there are indications that a single queen is unable to reproductively dominate large colonies [3, 4]. Thus, polygyny is generally associated with large colony size in social wasps [2, 3, 5].

Marked caste dimorphism (size and morphological differentiation between queens and workers, determined prior to adulthood) also correlates with large colony size in the swarm-founding wasps, at least in species with queens larger than workers [6]. This may be due to the greater long-term "security" of these large colonies; in contrast, the caste-flexibility of adults in small colonies allows potential queens to serve as a reserve worker force which may sometimes save them from extinction [7]. Thus, in social wasps the most specialized worker-queen dimorphisms occur in the most highly polygynous species, due to the common correlation of both polygyny and dimorphism with large colony size.

EVOLUTIONARY HYPOTHESES, AND A TEST: THE GENETIC STRUCTURE OF COLONIES

Evolutionary explanations of worker behavior have depended on either family structure (kin selection, or maternal manipulation in matrifilial groups) or mutualism (with some probability of reproduction by workers). Both are inapplicable to permanently polygynous societies with pre-adult caste determination, since (with outbreeding) kinship would be low [1] and workers are incapable of oviposition. Three hypotheses might explain the persistence of sterile workers in such species: (a) Clonal-group Hypothesis. Extreme inbreeding maintains a high degree of genetic relatedness among groups members; (b) Purely Mutualistic Queen Hypothesis. Queens are outbreeding, unrelated mutualists, each contributing a quota of workers that maintains a group essential for reproduction; (c) Lineal Fission Hypothesis. Kinship of colony members is high due to colony fissioning (swarming) along kinship lines.

These hypotheses were examined in a genetic study of Agelaia (formerly Stelopolybia) multipicta in Costa Rica. A. multipicta has pre-adult caste determination: queens and workers are discrete morphs differing in size and color [8]; and no monogynous colony has been recorded. Starch-gel electrophoresis of 1076 females (289 queens and 787 workers) from 11 colonies inhabiting an area of 2 square kms., using an esterase polymorphism, revealed a high degree of within colony relatedness (0.265 + 0.081), with low inbreeding (-0.80 + 0.099) [8]. Queens had a higher mean intracolony relatedness and intercolony variance (0.685 + 0.345) than did workers (0.265 + 0.067).

These results (moderately high relatedness with low indices of inbreeding) contradict the Clonal Group and the Pure Mutualism Hypotheses. The intracolony relatedness observed is similar to that recently recorded for other swarm-founding species [9], and resembles the worker-brood relatedness for half sibs (0.250) characterizing multiply mating monogynous species such as honeybees. This could support kin-selected beneficence on the part of workers and facilitate the evolution of cooperation among queens. The higher intracolony relatedness (and intercolony variance in relatedness) of queens may indicate that lineal fissioning of queens (but not workers) occurs and/or that reproductives are produced by a reduced number of queens; one very large colony contained only two queens after emitting daughter swarms [2].

1. West-Eberhard, M. J. 1973. Proc. VII Congress IUSSI, London, 396-403.
2. Jeanne, R. L. In press. In: The Social Biology of Wasps, K. G. Ross and R. W. Matthews, Eds., Cornell.
3. West-Eberhard, M. J. 1977. Proc. VII Congress IUSSI, Wageningen, pp. 223-227.
4. West-Eberhard, M. J. 1978a. Science 200: 441-443.
5. West-Eberhard, M. J. 1978b. J. Kansas Ent. Soc. 51(4): 832-856.
6. Jeanne, R. L. 1980. Ann. Rev. Entomol. 25:371-396.
7. West-Eberhard, M. J. 1981. In Natural Selection and Social Behavior. R. D. Alexander and D. W. Tinkle, Eds., Chiron Press, N.Y. pp. 3-17.
8. West-Eberhard, M. J., J. A. Lobo, and J. Azofeifa. In press. Proc. Nat. Acad. Sci. U.S.A.
9. Queller, D. C., J. E. Strassmann and C. R. Hughes 1988. Science 242: 1155-1157.

EFFECTS OF GROUP MEMBERSHIP ON INDIVIDUAL REPRODUCTIVE SUCCESS: MEASUREMENT OF MULTILEVEL SELECTION IN A WORKER PARTHENOGENETIC ANT, *PRISTOMYRMEX PUNGENS*

Kazuki TSUJI

Laboratory of Applied Entomology and Nematology, Faculty of Agriculture, Nagoya University, Nagoya 464-01, Japan

Association of multiple reproductives in eusocial insects has been a subject in the evolutionary ecology. Hamilton's rule (br-c>0 or <0) based on genetic theory has offered a straightforward analytical method for empirical studies on the adaptive values of such social behavior. His original model, however, assumes qualitative variation in a particular character, thus can be applied to only situations where animals have discrete alternatives such as helping others or not. This inclusive fitness approach is not appropriate to know the effect of quantitative characters on the fitness of participants in a social interaction. Recently quantitative genetics methods have been developing to the extent that they enable us to measure selection acting on characters of organisms, such as social insects, living in structured populations; that is "measurement of multilevel selection". In spite of great potential of these methods to empirical studies of social insects, few studies have used them (Queeler & Strassmann 1988). Two methods (mainly) have been proposed to measure multilevel-selection: one is Price's (1972) partitioning of covariance (selection differential), and the other is Heisler & Damuth's (1987) contextual analysis. The contextual analysis is useful to know whether or not group membership affects individual fitness, while the covariance partitioning is applicable to detect difference in the direction and strength of selection acting on the same character between within-colony level and between-colony level (see Wade 1985; Heisler & Damuth 1987). I tested in Pristomyrmex pungens what level of characters affects individual fitness of reproductive workers using these methods.

P. pungens is a queenless and obligate worker parthenogenetic species. All young workers are intranidal workers and reproductive. But old workers stop oviposition and begin to forage. Each worker emerges in summer and dies by the end of autumn of the next year. The mean individual reproductive success in a colony thus can be directly estimated from the colonial reproductivity, i.e. the number of newly emerged workers divided by that of old workers emerged last year. I collected 40 colonies in a field near Gifu, Japan on August 5 1988, and estimated their reproductivity. Furthermore, to know individual relative fitness within each colony, I picked 20 intranidal workers randomly from each colony and measured their ovarian volumes whose relative values were used as values of relative individual fitness within each colony. Absolute individual fitness (to the total population) was calculated as the reproductivity of the colony multiplied by the within-colony relative ovarian volume of the individual. These values of fitness in different levels were used for correlation (and multiple regression) analyses to the several measurements of characters: colony size, extranidal worker ratio, mean head width, coefficient of variation (CV) of head width and larger worker ratio at the colony level, and the head width at the individual level.

The covariance components at the between-colony level revealed significant negative directional and disruptive selections on colony size (this was similar to the reproductivity effect) and negative directional selection on within-colony CV of head width. On the other hand, significant positive directional and disruptive selections on individual head width were detected at within-colony level. This means that selection on the deviation of body size from the mean in each colony is opposite in direction between within- and between-colony levels; that is average sized workers were more "altruistic".

In the contextual analysis, the large sample size (N=800) seemed to make selection on many characters statistically significant. A noteworthy result is that significant stabilizing selection was detected in the extranidal worker (post-reproductive forager, old worker) ratio. This result is consistent with the hypothesis that cooperative colonies (all workers reproduce first and forage later) are maintained by the colony-level selection for the best intranidal/extarnidal ratio. (Supplementary laboratory experiments have also shown that colonies with extranidal worker ratio 5-10%, common in the field, were most productive.)

Although individual body size under natural selection at the within-colony level was shown by the covariancepartitioning method, the contextual analysis found only aggregative group characters (the mean and the CV of head width), rather than individual head width itself, significantly related to individual worker's relative fitness (to the total population). This result seems to support Oster & Wilson's (1978) argument that once eusociality established natural selection operates at colony level (at least in the narrow meaning of "selection" process, which excludes intergeneration transfer of the character, heritability). The fact that this species seems most distant from eusocialty among ants (Tsuji 1988) may suggest that colony level selection may be important not only for eusocial insects but non-eusocial organisms living in groups whose members are depending on each other.

So far the quantitative genetic models of multilevel selection have no straightforward presentation of the relationship of within- and between-group heritability while Hamilton has shown it as relatedness. With solving this problem, the quantitative genetic methods can be strong tools for empirical studies of social insects.

REFERENCES

1.Heisler L, Damuth J 1987. Am. Nat. 130: 582-602.
2.Price G R 1972. Ann. Hum. Genet. 35: 485-490.
3.Queeler D C, Strassmann J E 1988. In Clutton-Brock T H (ed) Reproductive Success, The Univ. of Chicago Press, pp.76-96.
4.Oster G, Wilson E O 1978. Caste and Ecology in Social Insects. Princeton.
5.Tsuji K 1988. Behav. Ecol. Sociobiol. 23: 247-255.
6.Wade M J 1985. Am. Nat. 125: 61-73

RELATEDNESS AND QUEEN NUMBER IN *LEPTOTHORAX LONGISPINOSUS*

JOAN M. HERBERS & ROBIN J. STUART
Department of Zoology
University of Vermont
Burlington VT 05405

The tiny forest ant *L. longispinosus* is facultatively polygynous, and the proportions of queenless, monogynous, and polygynous nests can vary among geographic populations (Herbers 1986, 1989). Such variation in queen number can affect the relatedness structure among workers within the nest and thus have a strong impact on kin selection dynamics.

METHODS

Nests of *L. longispinosus* were excavated in both Vermont and New York at each of four seasons. Nests were taken to the laboratory and scored for electrophoretic variation at six enzyme loci. Electrophoretic data were analyzed by the methods of Queller and Goodnight (1989) to estimate coefficients of relatedness among workers for each nest, as well as populational weighted averages.

RESULTS

Totals of 184 (Vermont) and 316 (New York) nests were assayed. In these samples, from zero to 12 (Vermont) or 28 (New York) queens per nest were found. Relatedness estimates, averaged over all nests in a sample, are shown in Figure 1. There was seasonal variation, but only the spring Vermont sample showed deviation from a general pattern of relatedness hovering around 0.5. Individual estimates of relatedness for each nest were examined further. There was no significant difference between sites, among seasons, or among replicate plots (Kruskal-Wallis tests, P>.05).

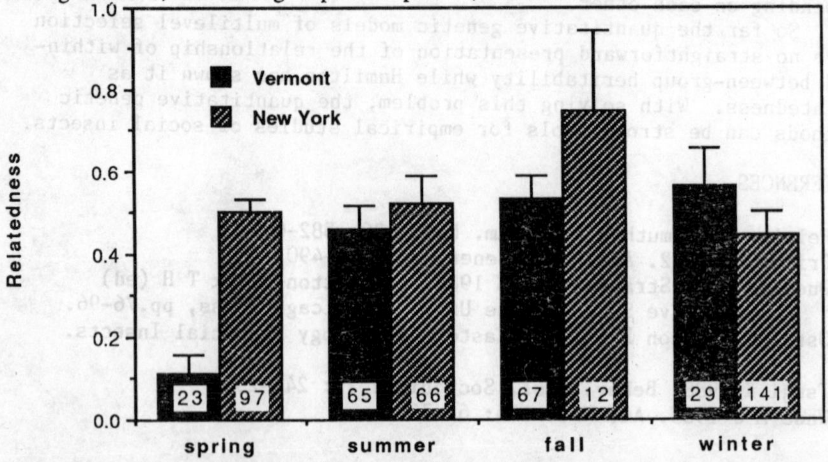

Figure 1. Relatedness among workers within nests of *L. longispinosus*. Means, standard errors, and sample sizes are indicated.

Relatedness estimates were examined for differences among nests based on the number of resident queens. When data from all seasons and both sites were pooled, there was a strong negative correlation among queen number and relatedness (Spearman's $\rho = -0.117$, $P<.01$). That is, relatedness among workers declined as queen number increased. When the data were examined separately for each site, however, this relationship was significant only in New York ($P<.001$), not in Vermont. Furthermore,the pattern of differential relatedness based on queen number in New York was particularly strong in the winter collections.

DISCUSSION

In general, workers were closely-related to their nestmates; with one exception, relatedness fell between 0.4 and 0.6. Thus, facultative polygyny docs not necessarily imply weakened genetic structure within the social unit. We do not know why the spring sample from Vermont showed low relatedness values; we note that it is a small sample, however. Whether this particular result is artifactual or represents an important source of seasonal variation awaits further studies.

The retention of high relatedness among worker nestmates even under polygyny suggests that the multiple queens are themselves relatives. This inference is borne out by two lines of evidence. First, incipient colonies are almost exclusively monogynous, suggesting that polygynous associations arise secondarily rather than through cooperative colony foundation. Second, newly-mated queens can be re-accepted by their natal nests but are rarely allowed into others (R. Stuart, pers. obs.). Thus the effect of polygyny on worker relatedness in *L. longispinosus* appears to mirror its congener in Sweden (Douwes et al. 1987), where resident queens of *L. acervorum* are closely related as well.

There was a curious site-dependency in the relation between queen number and relatedness among worker nestmates. In New York, workers in polygynous nests shared fewer genes with their nestmates than did workers in monogynous nests than did workers in queenless nests. No such pattern was observed in Vermont, however. There are other ecological differences that suggest kin selection dynamics operate differentially between the sites (Herbers, 1990). We also must interpret our results in terms of seasonal polydomy: a given colony fractionates into several discrete nesting units in spring, then re-coalesces for overwintering. The negative correlation between queen number and relatedness was strongest in winter-collected plots, when presumably entire colonies were together. In summer, however, colony boundaries are diffuse, making predictions about queen number and relatedness somewhat specious.

ACKNOWLEDGMENTS

This study was supported by grants from the National Science Foundation and Vermont EPSCoR.

REFERENCES

Douwes, P, L. Sivusaari, M. Niklasson , and B. Stille. 1987. Relatedness among queens in polygynous nests of the ant *Leptothorax acervorum*. Genetica 75:23-29.

Herbers, J.M. 1986. Nest site limitation and facultative polygyny in the ant *Leptothorax longispinosus*. Behav. Ecol. Sociobiol. 19:115-122.

Herbers, J.M. 1989. Community structure in north temperate ants: temporal and spatial variation. Oecologia 81:201-211.

Herbers, J.M. 1990. Reproductive investment and allocation ratios for the ant *Leptothorax longispinosus*. Amer. Nat. (in press)

Queller, D.C. and K. F. Goodnight. 1989. Estimating relatedness using genetic markers. Evolution 43:258-275.

GENETIC STRUCTURE OF *MYRMICA RUGINODIS* POPULATIONS

Seppä, P.
Department of Genetics
University of Helsinki
Arkadiankatu 7
SF - 00100 HELSINKI

Genetic structure of populations of a common palearctic ant, Myrmica ruginodis (Nylander) was studied by an electrophoretic survey. It was determined as genetic relatedness (genotypic correlation) within colonies in a population. Total of 9 populations from southern Finland were studied and 3-5 loci were used in the analysis.

Genetic relatedness was studied between both worker and queen nest mates. Worker nest mates were studied in 9 populations, one of which consisted of 7 subpopulations in closely located islands. The number of colonies collected / population was 17 - 64, and the number of workers analysed was 6 - 8 / colony. In one of these populations colonies were also excavated in order to collect all the queens. 25 % of the colonies turned out to contain more than one queen and the genetic relatedness between these nest mates was determined. The estimates of genetic relatedness as well as distribution of genotypes in these colonies were compared between queens and corresponding worker force.

According to this study M. ruginodis is a slightly polygynous species determined as relatedness between worker nest mates within a colony. However, the actual number of queens present in a colony varies a lot (0 - 10) mean being 1.4. Especially the frequency of colonies with no queens is surprisingly high, about 25 %. These results are pretty well consistent with those previously presented and the reason for such a high variability in the social structure is probably due to the ability of worker caste to recruit new queens to the colony. Significance of these results in the evolution of social organization and especially in the evolution of polygyny is discussed.

Symposium 10

PEST ANTS : PRESENT AND FUTURE

Organizer : **ROBERT K. VANDER MEER, U.S.A.**

SEED-HARVESTING BY ANTS IN AUSTRALIAN AGRO-ECOSYSTEMS

A.N. ANDERSEN, C.S.I.R.O. Division of Wildlife & Ecology, Tropical Ecosystems Research Centre, PMB 44 Winnellie, NT 0821, Australia.

Seed harvesting by ants is virtually ubiquitous in Australia, occurring in all major vegetation types in all climatic zones. The harvester ants include specialist granivores (e.g. certain species of *Monomorium*) that are comparable to those occurring in deserts elsewhere in the world, but most have broadly omnivorous diets. These seed-eating omnivores belong to a wide variety of genera, including *Monomorium*, *Pheidole*, *Meranoplus* and *Tetramorium* (Myrmicinae), *Melophorus* and *Prolasius* (Formicinae) and *Rhytidoponera* (Ponerinae). They feed on a broad range of seeds, selecting them opportunistly according to seed size, morphology and availability.

The Problem

The impact of harvester ants on natural vegetation is poorly understood, but they are known to be a serious impediment to plant establishment in anthropogenic habitats. Harvester ants can be pests wherever seeds are sown, such as in lawn plantings, revegetation programs, and forestry operations. However, although the extent of the problem is unclear, they appear to cause most problems in agricultural situations, where land clearing often results in a proliferation of seed-eating *Pheidole* species. Densities of *Pheidole* nests can reach $8m^{-2}$, and seeds can be completely removed within a few days. Harvester ants can influence the success of pasture grasses (both native and introduced), pasture forbs, and a variety of crops. They can reduce total plant density, influence species composition, and affect the persistence of desired species.

Past Solutions

The control of harvester ants has traditionally involved the use of insecticides, which have been applied by spraying the ground, using poison baits, or soaking or coating seeds. Although pesticides can markedly improve rates of plant establishment, there are several reasons to be concerned about their widespread use. First, the general environmental toxicity of many pesticides is now considered to be unacceptably high. Second, the public is becoming increasingly concerned about the possibility of pesticide residuals on their foodstuffs. Third, some insecticides are known to be phytotoxic, reducing germination and growth rates. Fourth, pesticides often kill a wide range of soil organisms, therefore having a widespread effect on soil ecology. Fifth, the indiscriminate use of insecticides might unnecessarily promote the evolution of resistance in pest species. Alternative methods of harvester ant control are therefore becoming increasingly attractive. These include coating or covering seeds with non-toxic substances, sowing at higher rates to compensate for losses, sowing the seeds during periods of low ant activity, and compacting the soil immediately after sowing.

The Future

Seed-harvesting by ants in Australian agro-ecosystems is a poorly defined problem. Assessment and control is *ad hoc* - State agricultural departments are aware that it occurs, but have no systematic management programs dealing with it. Harvester ants are commonly lumped together with other insect pests, so that insecticidal treatment is usually not targeted specifically for them. The extent of the problem, and consequently its cost, is therefore not clear. There is, however, a perception among management authorities that it is probably greater than is currently recognized. For example, there is a widespread problem of lack of sowing success, although of course many factors potentially contribute to this.

One priority area for the future is therefore to define the extent of the seed-harvesting problem. Attention should focus on (i) the distribution and abundance of harvester species in agro-ecosystems; (ii) rates of seed-removal by the ants; (iii) the impact of these losses on different plant species.

If it is established that harvester ants are a widespread and serious problem, then attention should focus on developing integrated management techniques that do not rely so heavily on insecticidal control. In addition to the alternative methods mentioned above, several other options are worth pursuing. These include: (i) using specially-bred strains or alternative plant species that are less susceptible to harvester ants; (ii) bio-control of harvester ant populations; and (iii) coating seeds with repellents rather than with toxins.

PEST ANTS OF INDIA AND THEIR MANAGEMENT

G.K.VEERESH
Senior Professor and Head,
Division of Plant Soil Science,
UAS G K V K, BANGALORE 560 065, India

A detailed account of the Pest Ants of India' has been given by Veeresh (1989). A breif review on some of the pest ants and their significance is given hereunder.

A. Ants as Direct Pests to Crops:

Solenopsis geminata F.: The red ant Solenopsis geminata F. attacks potato, brinjal, lablab; leaves of Cajanus, ailanthus, cucumber, tomato, ripe fruits of mango, papaya, sapota, flowers and flower buds of brinjal and bhendi, citrus stem, banana pseudostem etc. It was also employed in the biological control of Lac pests and cut worms in lawns (Veeresh and Rajanna, 1981).

Dorylus orientalis West Wood : It is a pest in gardens on vegetables, bulbs and underground tubers, on larvae and pupae of bees in the hives. In recent years, it has become a serious pest of groundnut (Arachis hypogea) in the coastal belt of Karnataka. Affected plants bear apparently, good pods, with a small black hole, contain nothing but black powder when opened.

Myrmecine Ants as Pests: The three myrmecine ants, Myrmicaria brunnea Saunders, Pheidolqiton diversus (Jerdon) and Tetramorium smithi (Mayr) are serious pests of the bhendi crop (Abelmoschus esculentus Moench) (Abraham and Remamony, 1978).

B. Ants as Indirect Pest to Crops:

The agricultural importance of ants lies in the injury they cause, in an indirect manner, on a variety of cultivated crops, ornamental plants, trees and shrubs by their symbiotic association with a host of mealy bugs, scale insects, aphids, fulgorids, psyllids, membracids and lycaenid caterpillars.

Ant species associated with some of the serious sucking insect pests comprise mainly from the subfamilies of Formicine (Camponotus, Oecophylla, Anoplolepis), Dolichorinae (Dolichodorus and tapinoma) and Myrmicinae (Solenopsis, Crematoqaster, Lapomyrmex, Meranoplus,

Monomorium, Pheidole, Myrmicaria etc.).

C. Ants Harmful to Man's Interests:

Ants carry away the seeds sown in the fields and nursery beds, from godowns and dwellings, attack woodwork, masonry, loosen the foundation, and root system by excavating soil at the base and disrupt irrigation (Veeresh and Gubbaiah, 1984; Veeresh, 1987); invade houses for flour, fats and sugars; attack beehives causing desertion of bee colonies.

D. Management of Ant pests:

Several workers have attempted to check the menace, by banding with sticky substance, treating breeding places with insecticides, keeping baits containing poison, application of insecticide at the time of sowing etc.

E. References:

Abraham, C.C. and Remamony, K.S., 1978. J. Bombay nat. Hist. Soc., 75 : 242-243.

Veeresh, G.K. and C. Rajanna, 1981, UAS Tech. Series No. 37 :218:222.

Veeresh, G.K. and Gubbaiah, 1984. J. Soil Biol. Ecol., 4: 65-73.

Veeresh, G.K., 1987. in J. Eder, and H. Rembold, Verlag J Peperny Munchen, P. 667-68.

Veeresh, G.K., 1989. pest Ants of India in Applied Myrmecology : A World Perspective.

THE PEST ANTS OF SOUTH AMERICA

H.G. FOWLER, Departamento de Ecologia & Centro Para o Estudo de Insetos Sociais, Instituto de Biociências, UNESP, 13500 Rio Claro, São Paulo, Brazil

THE PRESENT SITUATION

Without a doubt, the leaf-cutting ants are the most recognized pest ants of the Neotropics. Not all species of *Atta* and *Acromyrmex* are of economic importance, but all are democratically treated as pests by growers.

A second group of species of the genera *Paratrechina, Azteca, Wasmannia, Iridomyrmex, Solenopsis, Tapinoma, Camponotus* and *Acropyga* have been reported as occassional agricultural pests in certain areas, principally due to their protection of pest Homoptera, and secundarily, in interferring with crop maintenance and harvest. No systematic control is practiced against these ants presently, although in the past species of *Azteca* were heavily treated in some cocoa plantations.

The third group of pest ants includes species of *Tapinoma, Monomorium, Iridomyrmex, Pheidole, Solenopsis* and *Camponotus*, and is the least known. This group is large and very heterogeneous and includes structural and public health pests, as well as exotic species which greatly modify existing faunas. Most of these have been shown to be potential vectors of hospital infections, while others have been demonstrated to decimate local ant faunas, and even in some cases, the vertebrate fauna. Currently, except for efforts in controlling *Wasmannia auropunctata* in the Galapagos Islands, none are subject to control measures, and their impact and risk factors are largely ignored.

CURRENT GAPS IN OUR KNOWLEDGE

Leaf-cutters: We lack precise information of interspecific relations, nest densities, and which species are truely involved as pests under what densities and under which cropping systems. Nothing is known of the symbiotic fungus, or its role in detoxifying specific insecticides, nor do we have reliable data on forage quantities used.

Other agricultural species: No information is available mapping any species to a particular species of Homoptera, and we are unable to predict which associations are relativamente constant and which are opportunistic, with the exception of *Acropyga*. Data relating Homoptera species and densities to crop productivity or pathogen epidemology are lacking, and thus the true role of ants remains unknown. Some tramp species of *Tapinoma* and *Wasmannia* have obvious associations with Homoptera in certain agricultural settings, such as banana plantations, but their role in controlling other pests is unknown. Mosaic information is needed to manipulate ant communities to achieve major pest control under plantation settings.

Structural, Public Health and Ecological Pests: For nearly all types, preliminary data are not available. For example, in Brazilian hospitals, the number of resident ant species varies from 8 to 18, and generally all species are native and all are potential pathogen vectors, and may contribute to the 2X rate of hospital infections compared with the developed world. Due to diversity, control will be difficult, and public education services need to be developed. The ecological pests have hardly been studied, but all known cases involve exotic species. The lack of sufficient and efficient quarantine services is striking in all of South America. Strategies must be developed to minimize the effects of these ants in unique habitats, such as the Galapagos Islands, before their biodiversity is lost.

THE FUTURE OF PEST ANT CONTROL

For leaf-cutting ants, we are now witnessing a shift toward management instead of control, and this will accelerate as more data becomes available. Possibilities include the development of more ecologically sound toxicants and even a renewed frontal attack against the symbiotic fungus. However, for other pest ants, classical control techniques must be developed, coupled with perhaps management in some cropping systems.

LASIUS NEGLECTUS, A NEW POLYGYNOUS PEST ANT IN EUROPE?

J.J. Boomsma[1], A.J. van Loon[1], A.H. Brouwer[1] and A. Andrásfalvy[2]

[1]Department of Population Biology and Evolution, University of Utrecht, The Netherlands. [2]Vegetable Crops Research Institute and Company, Budapest, Hungary.
Present address J.J.B.: Institute of Ecology and Genetics, University of Aarhus, DK-8000 Aarhus C, Denmark.

A new polygynous ant, *Lasius neglectus*, was recently described from Budapest, Hungary [1]. The workers of this new species are very similar to workers of the common European species *Lasius alienus*, but the sexuals are clearly smaller than *alienus* queens and males. These ants were only recognized as distinctly different (by A. Andrásfalvy some 15 years ago), after it had become clear that their nests were polygynous and that their foraging behaviour (mass-tending of aphids on trees and readily entering buildings in search for food and nesting space) was quite unlike that of *L. alienus*.

There are several reasons to classify this species as a pest ant. Firstly, *Lasius neglectus* has almost certainly been introduced to its type-locality in Budapest. It did not occur there about 20 years ago and its allozyme characteristics are so distinct from those of sympatric *L. alienus* [2] that the possibility of recent evolution of the polygyny-(pest ant)-syndrome can be safely excluded. Secondly, *L. neglectus* has so far only been found in non-natural suburban areas in Budapest, where it reaches very high densities and where it effectively excludes the other ants which normally occur at such sites [1]. Thirdly, it is apparently expanding its distribution area, although its rate of dispersal seems low compared to other pest ant species.

The significantly smaller body size of queens and males of *Lasius neglectus* is likely to be associated with the polygynous colony structure and the inferred high probability that the males mate within their natal nests [1,2]. Provisional estimates based on one polymorphic marker-locus indicate a very low relatedness and significant inbreeding among worker-nestmates. A limited amount of mother-offspring analyses for this marker-locus also showed that single mating is the rule [2].

Lasius neglectus is the first polygynous species of its subgenus in Europe. The only other known polygynous *Lasius* s.s., *L. sakagamii* from Japan [3], is more similar to *L. niger*, and thus unlikely to be the closest relative of *L. neglectus*. There are, however, some striking parallelisms in ecology and behaviour between these two species, such as intranidal mating, retention of daughter-queens in the nests, reduced territoriality and nearly complete domination of the habitat [4]. The important difference in the context of insect-pests is, however, that *L. sakagamii* seems to prefer stable undisturbed habitats [3,4].

Even though the currently known distribution area of *L. neglectus* is limited to about two square km plus a few recent infestations in the Budapest area, we stress that this ant is a real pest in that area, as its workers penetrate buildings in massive numbers. It is also important to note that further accidental dispersal seems extremely easy. The ants readily make bud-nests in the soil of indoor flower-pots and one such pot probably suffices to start an infestation elsewhere [1].

At present it is impossible to evaluate the future status of *L. neglectus* as a pest ant. Its current distribution suggests that the problem is yet confined to the capital of Hungary, but that should not make us over-optimistic about the possibilities to prevent its further spread in the longer run. Much more information is required, both on its natural dispersal behaviour, and on possible other infestations in central Europe, and on the possible source populations from which this ant may have been introduced. We hope that our communications will encourage myrmecologists to search deliberately for multiple queens and small sexuals in any aggregation of tree-dwelling, *alienus*-like *Lasius* ants that they might encounter.

REFERENCES

1. Loon, A.J. van, J.J. Boomsma and A. Andrásfalvy. 1990. Insectes Sociaux, in press.
2. Boomsma, J.J., A.H. Brouwer and A.J. van Loon. 1990. Insectes Sociaux, in press.
3. Yamauchi, K. and K. Hayashida. 1970. Journ. Fac. Sci., Hokkaido Univ., Ser. VI, Zool., 17: 510-519.
4. Yamauchi, K., K. Kinomura and S. Miyake. 1981. Insectes Sociaux, 28: 279-296.

THE ECONOMIC IMPORTANCE OF THE IMPORTED FIRE ANT IN NORTH AND SOUTH AMERICA

PATTERSON, RICHARD S.
USDA-ARS Insects Affecting Man and Animals Research Laboratory, P.O. Box 14565, Gainesville, Florida 32604, USA.

The origin of fire ants is South America where 17 different species exist. Solenopsis geminata may have been introduced into the Northern Hemisphere 200-300 years ago as the new world was settled. The introduction had little impact on the native ants. Solenopsis richteri, a native of Argentina and Uruguay was introduced into Mobile, Alabama where it was identified in the early 1930's as a serious problem because of its aggressive nature. Then approximately a decade later a second aggressive species, S. invicta, the red imported fire ant, a native of northern Argentina, Brazil and Paraguay was found inhabiting the mobile area. Within a few years, fire ants, especially S. invicta,, began spreading across southeastern United States. The red imported fire ant quickly became the dominant ant species displacing the other three Solenopsis species as well as many other native ants found in suitable fire ant habitats. S. richteri which is slightly more cold tolerant is now confined to the northern border areas where S. invicta is less successful. S. xyloni and S. geminata are found mainly in areas where S. invicta has not established itself or in drier habitats.

At present S. invicta have become established in 11 states and Puerto Rico and infests ca 100 million hectares. It has been introduced in plant material (infestations in soil of nursery stock) into eight additional states but has so far been prevented from permanent establishment. S. invicta also has the potential to become established in indoor locations throughout the United States. With the transportation of plant material and other goods from country to country, it is not unthinkable to have S. invicta become a cosmopolitan species. In the United States, fire ants can cause a loss of 50-100 million dollars (U.S.) to our economy through direct damage and control costs. If the ants spread to their full potential in the USA, this figure could double or triple. In the USA, fire ants are becoming a more important urban pest invading homes, restaurants, hospitals, etc. inflicting painful stings on the inhabitants. It is also an important agricultural pest destroying crops in the field and in storage.

Fire ants can inflect a painful sting. Human reactions to stings range from local pruritus and erythema to anaphylactic shock and death. Since S. invicta is very aggressive and builds tremendous colony numbering millions of workers all of which are capable of inflicting stings. It is no wonder that fire ants are ranked as a major urban arthropod pest.

TIME'S ARROW AND THE PEST STATUS AND MANAGEMENT OF LEAF-CUTTING ANTS

L.C. FORTI[1] and H.G. FOWLER[2], (1) Departamento de Defesa Vegetal, Faculdade de Agronomia, 18600 Botucatu, São Paulo, and (2) Departamento de Ecologia, Instituto de Biociências, UNESP, 13500 Rio Claro, São Paulo, Brasil

Even before the Europeans colonized Latin America, leaf-cutting ants were the prominent pest of indigeneous agriculture, and became the scourge of colonial agriculture (1). Although before the advent of toxic baits, many countries legislated mandatory ant control (2,3), such measures faded in the 1950's as control measures became relatively simple and more inexpensive. However, even though we now have the technology in hand to effectively control species of *Atta*, and to a lesser extent species of *Acromyrmex*, the war has not been won. We breifly review the present situation and then we look into the crystal ball to predict what is in store for the future.

THE PRESENT

In the not too distant past, the war was a losing battle. Growers learned what they could and could not grow to escape the ravages of ant harvest. Many myths evolved as to the use of trap plants, such as jack beans or sesame, or even the use of the crazy ant *Paratrechina fulva* to reduce populations of *Atta*. A number of chemical substitutes then supplanted these measures, principally sodium cyanide and carbon disulfide, which was used until 1968. Methyl bromide appeared around 1950 and quickly dominated the market, but was highly toxic to human applicators. The big breakthrough occurred however with the incorporation of chloronated hydrocarbons in baits, following fire ant development Of these, Mirex is still the most widespread in usage. Pest status has been attributed automatically to any leaf cutter species, and often our knowledge of biology, ecology and damages is insufficient, and even baits fail to control certain species in certain situations.

THE FUTURE

Based upon our experience, the documented problems, and the current socio-economic trends in Latin America, we can make the following unerring predictions. First, due to rapid clearing of large portions of the Amazon basin, large continuous pastures are being formed in previously low-density leaf-cuttin ant areas, leading to eventual invasion of grass-cutting or disturbed habitat species, such as *Atta capiguara, Atta laevigata, Atta bisphaerica* and members of the *Acromyrmex landolti* complex, which is now occurring in Peru and central Brazil. This pattern is also repeated south of the Amazon as forested areas which served as dispersion barriers are cleared. Plantation crops and wood or pulp plantations are being over-run by species of *Acromyrmex* following control of *Atta*, and this pattern should continue. Second, the promising develop ment of management instead of control is now begining to take shape. By obtaining density to production costs, rational economic decisions can be made, and as chloronated hydrocarbons are being phased out by environmental concerns, slower acting, but more expensive, substitutes will dominate. Included in management are cultural techniques, especially the identification of risk areas, and even the use of colonies of *Atta* to regulate colony densities. We additionally do not see a viable future for biological control or natural plant products in the near future, because it is with these that these ants have coevolved. However, the change in mentality from control to management should ensure economic considerations, which will also be environmentally sound.

References

1. Cherrett, J.M. 1986. In: Economic Impact and Control of Social Insects, S.B. Vinson, Ed, Praeger Press, New York. pp. 165-192.
2. Oliveira-Filho, M.L. de. 1935. Boletim Agricola, 35:541-610.
3. Jacoby, M. 1944. Boletim da Sociedade Brasileira de Agronomia, 7:85-94.

ANT PEST MANAGEMENT- WHAT ARE THE TECHNIQUES AND THE POTENTIAL OF PHYSIOLOGY, BEHAVIOR AND GENETICS FOR FIRE ANT CONTROL ?

S. B. Vinson
Department of Entomology
Texas A&M University
College Station, TX 77843-2475, USA

The imported fire ant, *Solenopsis invicta* Buren, invaded the United States over 50 years ago and has become the major ant pest in the U.S.A. This ant has infested over 971 thousand sq. kilometers in densities ranging from 50 to 1,000 mounds per hectare, despite major control efforts. Although social insects in general are difficult to control [1] the imported fire ant (IFA) appears to be particularly resilient. The high worker (over 200 per day) and gyne production, colonies (mounds) with several hundred thousand ants each and, in the polygyne form, with several hundred fertile queens per mound and with 50 to 1,000 interconnected mounds per hectare, make control a challenge.

PRESENT TECHNIQUES

Management of ants can presently be divided into three basic options: 1) Nest removal, which is not practical with the IFA as mounds consist of hundreds of thousands of workers and may have more than one fertile gyne, 2) Sanitation, but due to the omnivorous feeding habits of the IFA and large foraging area, this form of ant management is extremely difficult, and 3) Control through the use of pesticides. The use of pesticides can be further divided into three approaches.

a) Surface Treatment. Use of a contact pesticide applied to surfaces can be effective in temporary suppression of foraging workers. However, surface treatments are rarely effective in eliminating nests, unless a long-lasting, residual compound is used. Elimination of foragers can reduce the vigor of the colony, but the colony recovers. Often colonies in treated areas move. Repeated surface treatment can be effective, but can also be expensive and may have environmental consequences.

b) Drenching or Treating Individual Nests (Mounds). Individual nest treatment with insecticide is generally very effective. There are a number of variations of this technique ranging from injected products, applications of dusts or granules, drenching of nests with the product in water, or application of moderately volatile products in liquid form [2]. The efficacy of these techniques varies, but in each case there is a percentage of nests where the product fails to reach the queen and the nest survives. Often nest movement occurs and workers may adopt a new queen [3].

Several problems occur with the use of individual nest treatments. Small nests are not discovered and treated, and some nests are located in protected locations where they are not accessible (under cement or rocks, at the base of large trees or foundation of a house or the ant nests occur on inaccessible property). In the polydomas, multiple queen form of the IFA [4], it is difficult to treat all of the mounds that make up a colony thus, parts of the colony and the queens within escape treatment. In each case these nests serve as a source of reinfestation. In the multiple queen fire ant with mound densities of over 1,000 per ha the use of individual mound treatment is too expensive and environmentally unsafe to be a viable control approach [2].

Baits. The use of baits for IFA management consists of a food attractant (any highly polyunsaturated vegetable oil) in which the control agent is dissolved. The mixture is applied to a carrier for application convenience. While baits are effective in ant control [1], there are problems. Many toxicants act too rapidly or have a limited range of concentrations over which they act. Foragers consuming fast acting toxicants fail to transport material back to the colony before they become affected or killed. If the dose is reduced so foragers are not affected, then the colony threshold is not reached. Thus, the baits are limited to a select group of toxicants [5] or control agents that have a slow or non-toxic mode of action (examples are the insect growth regulators).

Ants socially feed, thus allowing compounds to move through the colony, but this results in dilution of material. Compounds that act on small ant colonies of 100s of workers fail when provided to colonies of 100s of thousands of workers due to the complex and alternate food sources of such large colonies. The nature of the bait can influence food flow patterns within a colony and the ultimate target and dilution factors that any bait additive may obtain [6]. Another problem is that some colonies are satiated while others prefer to feed on items other than the bait.

Problems are inherent in the oil bait used for IFA control. The bait has a short environmental half-like due to oxidation of the polyunsaturated feeding stimulants. Further, sun light, high temperature and humidity all reduce bait half-life so that timing of application when ants are foraging is important. As a result of these problems many colonies escape bait control and serve as a source of reinfestation.

POTENTIAL OF PHYSIOLOGY, BEHAVIOR AND GENETICS FOR CONTROL

Advances in these fields can either be used to improve older management techniques or provide new approaches. Additional basic studies of the IFA is not likely to provide new ideas for mound removal or sanitation. However, basic studies may improve delivery systems for other management approaches. Baits can benefit from two areas of research. One is through studies of food attractants, preference, and food flow. In addition to oil, the IFA is stimulated to feed by certain carbohydrates and amino acids (proteins) [7], but developing formulations for liquid and solid water based baits, their application, and environmental stability are unsolved problems. However, such baits may become important in conjunction with biological controls agents or some physiologically active compounds.

The second area for bait improvement is to determine the source, function and identity of semiochemicals involved in ant communication. Once identified, some of these behavioral chemicals may be used to increase bait attraction, mask or overpower the repellency of some bait products, increase bait specificity to reduce their impact on nontarget organisms, or to influence the movement of products into the colony. However, potential problems exist with the use of semiochemicals, including different responses to pheromones with different ages of ants, the need for a proper sequence of pheromones to elicit the proper response, and habituation or even resistance to the use of these compounds.

Studies in physiology and behavior may provide not only products that act in new ways to effect control, but may offer new management approaches. It may be best to develop concepts for some physiological systems using insects other than the fire ant with which such studies may be difficult. Such areas may include the neuroendocrine system, digestion, nutrient absorption, excretion, cuticle synthesis, molting, circulatory system, immunity, general metabolism, and fat body physiology. One such success was the identification of juvenile hormones which were subsequently shown to have potential for IFA management [8] and have resulted in products [2]. In most cases new products would be placed in a bait or applied to a nest and subjected to the same problems in reaching the target as described for toxicants. There are some physiology research areas in which ant studies may yield specific opportunities. The isolation and identification of semiochemicals that influence reproduction, growth, development, and social organization, or influence behavioral aspects of food flow, kin recognition, brood tending, mating behavior and aggression have control potential. Once identified these compounds may have direct effects useful in management, but may require application as a bait or drench with problems alluded to earlier. Other areas of IFA physiology that have management potential have been discussed several times [9].

One important area that remains poorly understood concerns sex allocation and reproduction interactions leading to colony foundation. What are the physiological changes that occur? What is the role of weather and semiochemicals in flight initiation, sex location, mate selection, mating, landing site selection, dealation, muscle degeneration, and initial colony foundation? Understanding the physiological basis, could offer new management approaches.

Life tables for the IFA are particularly lacking in regard to growth characteristics in the field, influence of food availability, and mortality factors on both worker ant abundance and colony density, and colony longevity. Sources of food and energy flow into IFA colonies and differences between monogynous and polygynous forms are unknown. Such ecological data would be of value in developing management strategies.

Effective ant parasites exist, a large number of species have been isolated from fire ants [10]. How parasites locate and enter ant colonies, how colonies become exposed to such agents, routes of invasion, barriers to invasion and effects of parasites on an ant colony are poorly understood. Studies of the behavior, physiology and chemical ecology of ant parasites could lead to approaches that could improve the effects and success of such agents.

One additional approach to ant management is through the use of genetics. Increasing the levels of sterility or the introduction of lethal genes may have promise, but these techniques require the ability to produce large numbers of males and understand the mating system. More importantly, the study of genetics of the IFA is fundamental to understanding the population structure of the fire ant complex essential in developing biological control strategies and in providing the most effective management tactics.

REFERENCES CITED

1. Vinson, S. B. 1986. Economic Impact and Control of Social Insects, Pergamon Press, New York. p. 422.
2. Drees, B. M. and S. B. Vinson. 1989. Fire Ants and their Control, Texas Agri. Ext. Ser. Bryan, TX. B-1536, p. 12.
3. Tschinkel, W. R. and D. F. Howard. 1978. Behav. Ecol. Sociobiol, 3:297-310.

4. Bhatkar, A. P. and S. B. Vinson. 1987. In: Chemistry and Biology of Social Insects, J. Eder and H. Rembold, Eds., Verlag J. Peperny, Munchen, pp. 599-600.
5. Williams, D. F. 1983. Fla. Entomol, 66:162-172.
6. Sorensen, A. A., T. M. Busch and S. B. Vinson. 1983. Physiol. Entomol., 8:83-92.
7. Ricks, B. L. and S. B. Vinson. 1970. J. Econ. Entomol., 63:145-148.
8. Vinson, S. B. and R. Robeau. 1974. J. Econ. Entomol., 67:584-587.
9. Vinson, S. B. 1983. Fla. Entomol., 66:126-139.
10. Jouvenaz, D. P. 1983. Fla. Entomol., 66:111-121.

FUTURE CONTROL STRATEGIES : FIRE ANT PHEROMONES

VANDER MEER, R.K.

USDA-ARS Insects Affecting Man and Animals Research Laboratory, P.O. Box 14565, Gainesville, Florida 32604, USA.

INTRODUCTION

The fire ant is an agricultural pest, but has received most notoriety because of its aggressive behavior and potent venom. About 1% of the population develop allergic reactions and it is estimated that one-third of the population in an infested area is stung at least once each year. The situation is compounded by its large numbers and the fact that the ant thrives wherever man disturbs the environment.

Baits were developed for fire ant control because they offered large scale treatment possibilities, as well as a method that required much less insecticide. Our laboratory has been instrumental in the discovery and development of several effective bait formulations. All of these baits have some affect on non-target organisms, especially other ants. One goal of our laboratory is to determine how pheromones produced by the fire ant can be used to develop more species-specific control methods.

FIRE ANT PHEROMONES

The following social insect behavioral responses are pheromone mediated (1) Alarm, (2) Attraction, (3) Recruitment to a new food source or nest site, (4) Trail following, (5) Brood care, (6) Assistance at molting, (7) Recognition of nestmates, (8) Caste regulation, (9) Control of competing reproductives, (10) Worker and sexual excitant during mating flights, and (11) The rendezvous of male and female sexual during mating flights [1]. We know very little about mating flight pheromones (sex pheromones, etc.) or caste regulation in fire ants. These are areas that have potential in control strategies but require basic research. Assistance at molting and brood care have been thought to be pheromonally controlled and a fire ant "brood pheromone" has been reported; however, the actual existence of brood pheromones is in doubt [2]. Fire ant nestmate recognition research has provided many benefits [3]; however, the associated chemical cues are not chemically discrete (derived from heritable and environmental sources) [3]. The possibility of disrupting the recognition process would be most favorable in polygyne populations, but research is lacking.

Much behavioral information has been accumulated about the control of competing reproductives in fire ant colonies. The fire ant queen produces a dealation inhibitory primer pheromone that prevents dealation and ovariole development in coexisting female sexuals. Intriguingly, workers respond to greater than usual amounts of this pheromone by executing supernumery queens, until the pheromone level is sufficiently reduced [4]. Introduction of this pheromone into a colony could induce workers to execute their own queen. Evaluation of this hypothesis awaits the isolation, identification and synthesis of the active compound(s). Of similar interest is queen control over male and female alate production by influencing the behavior of worker toward male and female brood [5]. Chemical studies have not been done on this system.

Recruitment pheromones produced by the Dufour's gland elicit, depending on the situation, orientation, alarm, migration to a new nest site, or recruitment to a food source [6]. Artificial induction of these behaviors would result in disruption of the ant's social organization and perhaps aid in control. But the most obvious way to use pheromones is through the incorporation of worker attractant pheromones into baits. This would make the baits more efficient and species-specific. Two attractants have been chemically defined. (A) The attractant associated with recruitment has two components, but preliminary studies indicate that at greater than physiological levels a single component significantly attracts workers. We are currently attempting to make this compound available via synthesis. The queen also produces a worker attractant, which has been isolated and identified as a three component mixture (A, +B, -B, +C, and -C). Two compounds (B and C) are optically active; however, through comprehensive olfactometer bioassays we determined that significant attraction could be obtained with A in combination with ±B or even with ±B itself. Evaluation of the two attractant pheromones incorporated in bait particles is in progress.

There are many possible ways to use pheromones as adjuvants to fire ant control. It remains for us to keep our minds open and continue the basic research that will lay the ground work for the innovative pest ant control of the future.

REFERENCES

1. Vander Meer, R. K. 1982. Semiochemicals and the red imported fire ant (*Solenopsis invicta* Buren) (Hymenoptera: Formicidae). Florida Entomol., 66: 139-161.
2. Morel, L., Vander Meer, R.K. 1988. Do ant brood pheromones exist? Ann. Entomol. Soc. Am. 81: 808-815.
3. Vander Meer, R.K., Obin, M.S. and Morel, L. 1990. In: Applied Myrmecology: A World Perspective. R.K. Vander Meer, K. Jaffe, and A. Cedeno, Eds., Westview Press, Boulder CO., In press.
4. Fletcher, D.J.C. 1986. In: Fire Ants and Leaf-Cutting Ants: Biology and Management. C.S. Lofgren and R.K. Vander Meer, Eds., Westview Press, Boulder, CO., pp. 184-191.
5. Vargo, E.L. 1990. In: Applied Myrmecology: A World Perspective. R.K. Vander Meer, K. Jaffe, and A. Cedeno, Eds., Westview Press, Boulder CO., In press.
6. Vander Meer, R.K. 1986. In: Fire Ants and Leaf-Cutting Ants: Biology and Management. C.S. Lofgren and R.K. Vander Meer, Eds., Westview Press, Boulder, CO., pp. 201-210.

EFFECTS OF THE IGR'S PYRIPROXYFEN AND FENOXYCARB ON COLONIES OF *PHEIDOLE MEGACEPHALA* (FABRICIUS)

NEIL J. REIMER[1], B. MICHAEL GLANCEY[2], AND J. W. BEARDSLEY[1].

1. DEPARTMENT OF ENTOMOLOGY, UNIV. OF HAWAII, HONOLULU, HI, USA
2. USDA-ARS, 1600 SW 23RD DR., GAINESVILLE, FL, USA.

The big-headed ant, Pheidole megacephala (Fabricius), is a seriuos pest in pineapple, coffee, taro, and other crops in Hawaii due to their behavior of tending various homopterans [1]. No materials are currently registered for use against ants in Hawaiian crops, therefore, new toxicants are being sought for use in insecticidal baits. This paper reports on the effects of two IGR's, fenoxycarb and pyriproxyfen, on colonies of P. megacephala.

Materials and Methods.

Thirty-six big-headed ant colonies were collected from pineapple stumps in Hawaii and brought back to the laboratory. Each colony was set up in plastic shoe boxes with food (modified Banks diet [2]), water, and a nest chamber (módified Bishop cell [3]). Each colony was reduced to comparable numbers of workers, brood, and soldiers and allowed to stabilize for 2 weeks then starved for 4 days. After 4 days of starvation, each of 12 colonies were fed 0.5 ml of 2% fenoxycarb in soybean oil, 12 colonies fed 0.5 ml of 2% pyriproxyfen, and 12 colonies fed neat soybean oil as controls. All colonies were returned to their regular diet after 24 hr and the soybean oil removed. The number of eggs, volume of brood, and number of workers and soldiers were counted for each colony at weekly intervals. Queens from 3 replications were removed at 5, 6, and 7 weeks for histological examination.

Results and Discussion.

Queens in colonies treated with pyriproxyfen laid significantly fewer eggs than controls 2 wks after treatment. Oviposition ceased in all pyriproxyfen treated queens at 6 wks after treatment with no recovery during the 15 wks of the study. Brood volumes in pyriproxyfen treated colonies declined after 2 weeks due to pupal death. No brood were found after 4 weeks. Fenoxycarb treated queens laid significantly fewer eggs than controls after 2 wks. Fenoxycarb treated colonies shifted caste differentiation from workers to males at 3 wks after treatment. At 5 wks, 90% of larvae were male, the remaining 10% developed into intercastes.

Histological examination revealed that both materials had similar effects (reduced egg size, vacuolated ovarioles, and undeveloped nurse cells). However, the primary effect for pyriproxyfen was egg resorption; and for fenoxycarb, retrogression of ovarian tissue.

Field applications in pineapple with both materials, formulated as baits, resulted in reductions in ant densities within 11-12 weeks after treatment.

References.

1. Reimer, N. J., J. W. Beardsley, and G. C. Jahn. 1990. Pest ants int the Hawaiian Islands. In R. K. Vander Meer, K. Jaffe, and A. Cedena [eds.], "Applied Myrmecology: A World Perspective". Westview Press.

2. Banks, W. A., C. S. Lofgren, D. P. Jouvenez, C. E. Stringer, P. M. Bishop, D. F. Williams, D. P. Wojcik, and B. M. Glancey. 1981. Techniques for collecting, rearing, and handling imported fire ants. USDA, AAT-S-21. 9pp.

3. Bishop, P. M., W. A. Banks, D. F. Williams, C. E. Stringer, J. A. Mitchell, and C. S. Lofgren. 1980. Simple nests for culturing imported fire ants. J. Ga. Entomol. Soc. 15: 300-304.

BIOLOGICAL CONTROL OF FIRE ANTS : A LOOK INTO THE FUTURE

JOUVENAZ, DONALD P.

USDA-ARS Insects Affecting Man and Animals Research Laboratory, P.O. Box 14565, Gainesville, Florida 32604, USA.

The red and the black imported fire ants (IFA), Solenopsis invicta Buren and Solenopsis richteri Forel, currently infest ca 10^8 ha in the southeastern United States and Puerto Rico. Recently, they have been accidentally transported across the arid Southwest to Arizona and California. Should they become established in the West, their range will increase substantially. In addition, a polygynous form having denser populations and which is sometimes more difficult to control with chemicals is spreading within the population.

As medical and agricultural pests, IFA have been quarantined and assaulted with a succession of pesticides at a cost of at least $200,000,000 (1990 dollars). Despite these efforts, IFA are thriving in the United States. Pesticides are useful for temporary, local suppression, but cannot provide a permanent solution. Biological control, however, may provide at least a permanent amelioration of the problem.

The primary goal of USDA research is the establishment of a complex of specific natural enemies of IFA in the United States, where they are now essentially absent. In South America, IFA are beset by a complex of pathogens, parasites and social parasites, and symbiotic predators. Two of these, a nematode and a socially parasitic ant, may be able to destroy established colonies (Jouvenaz, elsewhere in this proceedings); a virus of unknown virulence also occurs. The remaining organisms appear to be debilitating agents well adapted to their hosts. They should not be neglected, however, for a complex of these organisms may exert sufficient stress on IFA to alter the competitive balance in favor of our native ant fauna.

Polygynous IFA may prove especially vulnerable to natural enemies, for their intercolonial barriers are diminished (at least within nest complexes) and, due to their high population densities, they may be under greater food stress as well. In addition, the large and numerous corpse piles associated with polygynous populations present a problem of hygiene. Could the apparent absence of polygynous IFA in Mato Grosso be due to natural enemies?

Parasites and their hosts tend to coevolve towards mutual tolerance. Hokkanen and Pimentel (1984) contend that introductions for biological control involving new parasite-host associations were successful 75% more often than those employing evolutionarily old associations. Since many of the natural enemies of IFA seem to be debilitating rather than directly lethal to colonies, the probability or degree of successful biological control might be increased by utilizing new associations and/or increasing the virulence of old pathogens through biotechnology.

New parasites and pathogens of IFA might be sought from other species of ants, especially other Solenopsis spp., but the probability of success does not appear high due to a prevailing pattern of host-specificity. Competitors, however, may hold brighter prospects.

Buren (1983) proposed the introduction of a complex of 20-30 carefully selected exotic species of ants to competitively displace IFA. Despite obvious technical and regulatory difficulties and the potential for environmental harm, this approach may have merit, especially in concert with a complex of specific natural enemies.

Biotechnology presents the possibility of transforming old, attenuated pathogens into new, virulent ones. The endoparasitic yeasts of IFA are prime candidates for genetic engineering. Since they produce no toxins or histopathology, it may be possible to transform them to produce toxins of our choice, insect hormones, or even semiochemicals to disrupt colony organization. They may well prove host-specific for Solenopsis spp., and can be mass-produced and transmitted per os. Their invasive ability declines during in vitro culture, but this common problem with entomopathogenic fungi might be overcome through biotechnology and/or improved culture techniques.

In view of the behavioral complexity, ecological versatility, and high reproductive potential of IFA, it is easy to be pessimistic about biological control. Yet, Sr. Antonio C. C. Pereira, EMBRAPA entomologist and USDA collaborator, showed me a farm in the State of Mato Grosso, Brazil on which an IFA population crash appeared to have occurred. We found the site almost devoid of IFA, whereas less than a year before it had been heavily infested with IFA having a Thelohania solenopsae (Microsporida) infection rate of at least 70%, and other pathogens were also present. Porter et al. (elsewhere in this proceedings) found fire ant populations to be significantly lower in Brazil than in the United States, but the causes are unknown.

Literature Cited

Buren, W. F. Buren, W. F. 1983. Artificial faunal replacement for imported fire ant control. Fla. Entomol. 66: 93-100.

Hokkanen, H. and D. Pimentel. 1984. New approach for selecting biological control agents. Can. Ent. 116: 1109-1121.

Symposium 11

EVOLUTION OF EUSOCIALITY IN ARTHROPODS OTHER THAN THE HYMENOPTERA AND ISOPTERA

Organizers : **YOSIAKI ITO, Japan**
WILLIAM A. FOSTER, United Kingdom

CASTE AND DIVISION OF LABOR IN THE NAKED MOLE-RAT, AN EUSOCIAL MAMMAL

Stanton Braude
Insect Division, Museum of Zoology
University of Michigan
AnnArbor, MI 48109-1079 USA

The naked mole-rat, Heterocephalus glaber, is the only vertebrate known to be eusocial. Like many Isopteran species, naked mole-rats live in colonies consisting of a queen, her mate or mates and a large number of sterile workers. The natural history of this species and the caste structure among the workers will be discussed.

H. glaber is completely fossorial; burrows can extend over an area as great as 200 X 300 meters with tunnels totaling over 3 kilometers in length. Some of these tunnels are only a few centimeters below the surface, but many are as deep as a meter. Naked mole-rats feed on the roots, tubers and bulbs of various plant species. Their only known predators are snakes which enter their burrows when soil from fresh digging is being thrown out.

Although eusocial insects have been studied for hundreds of years, the eusociality of the naked mole-rat was only recently discovered (Jarvis, 1981). I have investigated the division of labor and caste structure of this species both in the laboratory and in the field.

Jarvis (1981) first described the caste structure of naked mole-rats as consisting of reproductives, large non-workers, intermediate sized infrequent workers, and small frequent workers. Early studies of captive naked mole-rats at the University of Michigan also found a negative correlation between worker size and the frequency of various cleaning and foraging tasks in the colony (Isil, 1983; see also Lacey and Sheman, 1990). Digging, however, was not possible for these captive mole-rats because they were housed in a tunnel system of glass tubing and plastic boxes.

In 1983 I provided the captive colonies of naked mole-rats at the University of Michigan with cork in dead-end tunnels as a digging substrate. I found that the largest animals in the colony, previously described as non-workers, were in fact the most frequent diggers.

In 1986 I began a field study of naked mole-rats in Meru National Park, Kenya. My behavioral studies have focused on the one observable naked mole-rat behavior, the kicking of soil out of the burrow. This behavior is called volcanoing because the spurts of soil ejected from the hole, and the conical mound which forms, resemble a miniature volcano erupting. This also appears to be a particularly dangerous task since the volcanoer's posterior is directly exposed to predators, and volcanoers are the first individuals that a predator would encounter. Brett (1990) observed snakes entering volcanos on five occasions.

In 1986, 1987 and 1988 I identified the individuals which were involved in volcanoing in 3 colonies. The entire colonies were then trapped and the relative size of the volcanoers was determined. In two colonies the volcanoers were clearly among the largest animals in their colonies. In the third colony animals of all sizes were found to be volcanoing. In that colony, however, the queen gave birth to a litter only a few days later. It is not surprising that parturition would interfere with the normal division of labor in the colony. This is in fact seen in captive colonies.

References

Braude, S. 1990. Which naked mole-rats volcano? In: The Biology of the naked mole-rat (P. Sherman, J. Jarvis and R.D. Alexander eds.). Princeton University Press, Princeton.

Brett, R. 1990. The population structure of naked mole-rat colonies. In: The Biology of the naked mole-rat (P. Sherman, J. Jarvis and R. D. Alexander eds.). Princeton University Press, Princeton.

Isil, S. 1983. A behavioral study of the naked mole-rat, *Heterocephalus glaber*. Unpublished Masters thesis, University of Michigan, Ann Arbor.

Jarvis, J. 1981. Eusociality in a mammal: cooperative breeding in naked mole-rat colonies. Science 212:571-573.

Lacey, E. and P. Sherman 1990. Social organization of naked mole-rat colonies: evidence for division of labor. In: The Biology of the naked mole-rat (P. Sherman, J. Jarvis and R. D. Alexander eds.). Princeton University Press, Princeton.

BROOD CARE IN THE WHIP SCORPION *THELYPHONUS INDICUS* S.

Y.L. Ramachandra & Geetha Bali
Department of Zoology, Bangalore University,
Bangalore 560 056

The pregnant female occupied a burrow where the eggs were laid. The young ones after emerging were found to climb on the back of the mother clinging with the help of tarsal adhesive discs to the abdominal segments all round forming a ring (fig.1). The mother with the young ones was found to assume the aggressive posture described earlier(1). Any mechanical stimulation led to an aggressive response characterized by further extension of the pedipalps, elevation of the cephalothorax further above the ground arching the abdomen and further stiffening of the body. This is interesting because, the animal under ordinary circumstances is fairly timid in its disposition especially when there is any unusual extraneous stimulus. If such a stimulus is of low magnitude, the animal orients its appendage as well as the body towards the source of the stimulus while it runs away rapidly constantly rotating the whip until it reaches a hiding place if the stimulus is strong. But the female carrying young ones showed none of this behaviour and exhibited aggressive responses to minute mechanical stimuli. Under laboratory conditions such aggressive responses were seen only when the adults encountered each other under conditions of overcrowding and starvation. The mother remained sedentary in this posture for many days without feeding.

Interestingly, when the brood was separated from the mother, the behaviour of the mother instantaneously changed. It was no more found to be aggressive. It assumed the normal posture with the ventral surface lying close to the ground, body being more flexible and pedipals not widely extended. Mechanical stimuli given at this juncture led to escape response and the animal ran away timidly. The mother which made no attempts to eat its brood and prevented other members in the group from eating them by its aggressive disposition was now found to voraciously devour the young ones the moment they were separated. These studies indicate that the presence of the brood profoundly modifies the behaviour of the mother inhibiting feeding and inducing an aggressive behaviour which are of immense importance in the survival of the species.

(1) Geetha Bali and S.D.Moro. 1988. Revue Arachnologique,7(5) : 189-196.

REPRODUCTIVE CONSEQUENCES OF PERIODIC SOCIALITY IN SPIDERS

Helga Sittertz-Bhatkar, Electron Microscopy Center, Texas A&M University, College Station, Texas, USA 77843

Periodic sociality or quasisociality [1] in spiders involves living together, showing tolerance against aggression, interact, and cooperate during a certain period of the life cycle. Mother spider provisions food for the spiderlings with trophallaxis [2] through their first molts outside the cocoon till dispersal. The progression of nursing behavior in the cosmopolitan family Theridiidae was compared in two European (Theridion impressum and T. sisyphium) and two neotropical species (Theridula faceta and T. gonygaster) through ethological and micromorphological studies. The state of predigested fluid and midintestinal digestive gland of the mother during nursing were studied histologically. In T. impressum, mother perceives the vibrational signals from her spiderlings along the fine threads on which they rest. A group of about 100 spiderlings develop at a time. The mother passes partly digested regurgitate to the aggressive spiderlings crowding near her mouth, while wiping away others crawling on to her body.

Differential regurgitate formation in A) T. impressum and B) T. sisyphium. Ex-excretion vacuole (V), K-nucleus, N-food vacuole, R-regurgitate vacuole, Rz-absorptive cell: a-released, b-basal cell; S-secretion vacuole: 1-empty, 2-filled; Zt-released apical cell, Zw-lumen cell).

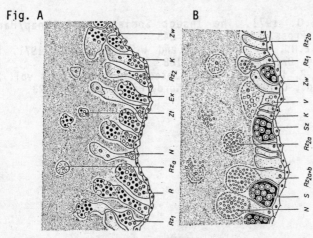

Fig. A B

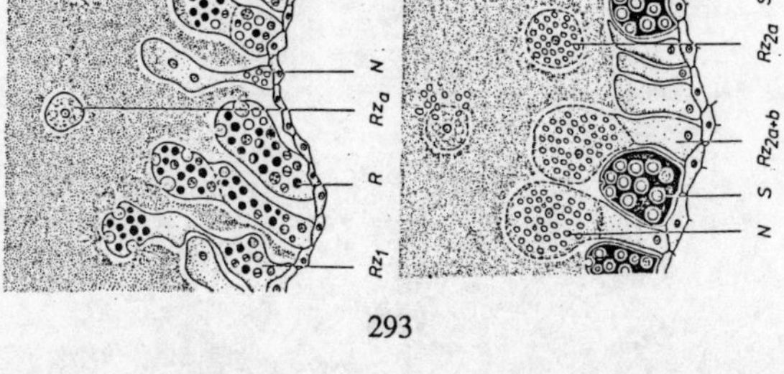

The nursing is in a passive defense of the mother against her own youngsters, similar to intercolonial trophallaxis in ants [3] or appeasement behavior in higher animals. In T. sisyphium, a progeny of 50 spiderlings is fed on completely predigested regurgitate. They approach in a file towards the mother and are fed singly without an aggressive encounter. The regurgitate from T. impressum mother remains partly digested (Fig. A) as a result of the nursing pressure; that in T. sisyphium is completely digested prior to feeding the singly approaching spiderlings (Fig. B). T. impressum mother does not feed herself for 3 days. She kills her prey and predigests it with the midintestinal enzymatic fluid. The spiderlings suck the digested fluid from the prey. Both, the absorptive and secretory cells, of the mother degenerate and the midintestinal lumen stays empty. The growing spiderlings aggressively attack the mother, that is unable to feed them on solicitation, and as a result suck her body contents digested with her own enzymes. The spiderlings molt and disperse and the social behavior ends. In contrast, T. sisyphium mother survives, captures prey together with the youngsters and feeds them till their 3rd or 4th molt and then dies of a natural death. Neotropical Theridiidae shows a further progression in periodic sociality. The mother dissolves a spot on the egg sac with her regurgitate and 26-45 spiderlings emerge. She secrets a snare of threads on which they disperse and feeds them individually with her regurgitate. They develop uniformly, unlike those of T. impressum. The mother does not eat during nursing. All the spiderlings molt in concert and both mother and the spiderlings disperse leading a solitary life. Similar brood provisioning occurs in T. gonygaster. Cross-fostering of young between the two Theridula species is successful. Allospecific spiderlings initiate nursing in the mothers through solicitation. As the periodic sociality in Theridiidae progresses, there seems to be decrease in progeny size: about 100 in T. impressum, 50 in T. sisyphium and 25 in the Theridula species. All these spiders develop larger densities in a short time period than their solitary counterparts. The youngsters of Theridion hibernate through the winter and reproduce in numbers in the spring in the central European conifer plantations. T. faceta occupies up to 80% of Theridiidae in Phaseolus vulgaris agroecosystem in Tabasco, Mexico; each bean leaf containing a female. The species develops 3 generations in a single 3-month cropping period and proves to be a major biotic suppression agent of the pestiferous Homoptera.

References

1. Wilson, E.O. 1971. The Insect Societies. Belknap/Harvard University Press, Cambridge. p. 548.
2. Kullmann, E.H., H. Sittertz, and W. Zimmermann. 1971. Bonner Zoologische Beiträge, 22: 175-188.
3. Bhatkar, A.P. 1983. In: Social Insects in the Tropics vol. 2, P. Jaisson, Ed., Université Paris-Nord, Paris. pp. 105-123.

THE EVOLUTION OF SUBSOCIALITY AND MATING SYSTEMS IN THYSANOPTERA

Bernard J. Crespi
Animal Behavior Research Group
Department of Zoology, South Parks Road
University of Oxford
Oxford 0X1 3PS, U. K.

Social and sexual systems among thrips species studied thus far may be classified into three main groups: (1) species showing maternal defense of eggs (Elaphrothrips, Bactridothrips, and Gigantothrips), (2) species in which females oviposit communally but do not defend their eggs (Hoplothrips, Sporothrips, Idolothrips, and Anactinothrips), and (3) species that form galls, in which females or both sexes fight to the death during gall initiation (Oncothrips, Onychothrips, and Kladothrips).

Maternal care of eggs appears to be common among spore-feeding thrips, some of which also reproduce by viviparity. In Elaphrothrips tuberculatus, the primary selective cause of maternal care is cannibalism by viviparous females and oviparous females that have not yet started their clutches of eggs. Localization of ovipositing females by maternal care allows males to defend individual females. In E. tuberculatus, and probably many other thrips species with maternal care, males fight with their forelegs and abdomens and engage in post-copulatory mate guarding. Such mate-guarding, and the production of only daughters from eggs in facultatively-viviparous species, creates conditions where broods probably often comprise a group of all females related by 3/4.

Communal oviposition, the selective cause of which is unknown, localizes groups of breeding females and makes them defensible by males. In Hoplothrips spp., Sporothrips amplus, and Idolothrips spectrum, males fight in defense of oviposition areas. Hoplothrips spp. exhibit male "fighter-flier" polymorphisms, and large, wingless males tend to win fights. Colonies of Hoplothrips persist on shelf fungi for many generations, and their extremely polygynous mating system may engender high relatedness among males and females in colonies. In the spore-feeding S. amplus and I. spectrum, colonies apparently persist for a single generation; in I. spectrum, but probably not in S. amplus, a male guards an egg mass against predators until the eggs have hatched. Anactinothrips gustaviae, studied by Kiester and Strates (1984) in

Panama, also oviposits communally; in this species, adults and larvae forage in coordinated groups on patches of lichen on tree trunks, and males probably fight each other in defense of oviposition sites.

Australian gall-forming thrips engage in lethal fighting among foundress females (Oncothrips and Onychothrips) or within each sex (Kladothrips). Adults of the inquiline species Csirothrips watsoni and wing-reduced second-generation females of Oncothrips tepperi may defend galls against predators and other inavders. Onychothrips arotrum and Oncothrips tepperi produce highly female-biased first-generation broods, which suggests that mating usually occurs prior to dispersal; however, males are winged in both species.

Traits favoring or indicative of sociality in thrips include haplodiploidy, generation overlap, the presence of maternal care in some species, female-biased sex ratios, wing polymorphism, mobile juveniles, and, in gall species, female foreleg polymorphism and a defensible, multigenerational habitat. Factors mediating against thrips sociality include a general lack of selective opportunities for helping behavior, the "unimprovable" nature of most thrips habitats, and the absence of a sting or equivalent.

Kiester, A. R. and E. Strates. 1984. Social behaviour in a thrips from Panama. J. Nat. Hist. 18:303-314.

THE SOCIAL AND SEXUAL BEHAVIOR OF AUSTRALIAN GALL THRIPS

Bernard J. Crespi
Animal Behavior Research Group
Department of Zoology, South Parks Road
University of Oxford
Oxford 0X1 3PS, U. K.

Collections, observations and experiments were used to investigate the social and sexual behavior to six gall-forming and four inquiline thrips species (Thysanoptera: Phlaeothripidae) in the semi-arid zone of New South Wales and Queensland, Australia. Kladothrips rugosus females form galls on the phyllodes of Acacia pendula. The galls persist for one generation. During gall initiation, females fight to the death, males fight to the death, and one female or one adult of both sexes remains alive within the gall to breed. Oncothrips tepperi females form galls on A. oswaldi that persist for two or more generation. In this species, only the females engage in gall initiation and lethal fighting. Female Koptothrips flavicornis invade young galls of O. tepperi, kill the foundress, and rear a brood of their own. First-generation offspring of O. tepperi comprise 18-337 winged females (mean 70.0), 3-46 winged males (mean 11.4), and 2-10 wing-reduced females (mean 4.5) with greatly enlarged forelegs. Most young second-generation galls of O. tepperi contain one breeding winged females and several dead wing-reduced females. These data suggest that wing-reduced female O. tepperi are either sterile soldier morphs that protect the first brood, or that they engage in fighting among themselves in circumstances as yet unknown.

Onychothrips arotrum, Onychothrips tepperi, and Oncothrips antennatus form galls on Acacia aneura. In O. arotrum and O. tepperi, but not in O. antennatus, females engage in lethal fighting during gall initiation. Adult offspring sex ratios of O. arotrum range from 2-14% male (mean 5.5%), which suggests that mating occurs within galls prior to dispersal; however, males of this species are winged. Foundress female O. arotrum are often still breeding after their offspring have become adult within the same gall. Csirothrips watsoni invades galls of O. arotrum, apparently after most or all of the adult O. arotrum have dispersed. Laboratory observations indicate that C. watsoni adults kill adult O. arotrum, using their enlarged forelegs. C. watsoni adults plug the gall entrance using the cast exuviae of O. arotrum, and they sometimes exhibit striking defensive behavior, either maternal or among siblings.

An undescribed species of Hoplothripini forms large, woody, multi-generational stem galls on Casuarina trees. Adult sex ratios in colonies vary from 7-33% male (mean 19%), and the eggs are sufficiently large that females develop only one or perhaps two at a time. Hoplothripini sp. galls are invaded by two species, Thaumatothrips froggatti and an undescribed species, which both systematically kill the Hoplothripini sp. present, using their enlarged forelegs. Galls containing both Hoplothripini sp. and these inquilines sometimes contain partitions between the species, which may have been built by the Hoplothripini sp. adults.

Gall thrips, which exhibit haplodiploidy, female foreleg and wing polymorphism, female-biased sex ratios, extreme forms of physogastry, overlap of generations, and mulitgenerational habitation of a defensible resource, offer unique opportunities to assess the roles of relatedness, ecology, and demography in the evolution of social behavior.

COLONY DEFENCE BY FIRST INSTAR NYMPHS AND DUAL FUNCTION OF ALARM PHEROMONE IN THE SUGAR CANE WOOLLY APHID, *CERATOVACUNA LANIGERA*

Norio ARAKAKI

Okinawa Prefectural Agricultural Experiment Station,
4-222 Sakiyama-cho, Naha, Okinawa, 903 Japan

All nymphal stages and apterous adults of Ceratovacuna lanigera secreted droplets containing an alarm pheromone from their abdorminal cornicles when stimulated with a pin. When fresh droplets on a piece of filter paper were put near the aphids, 1st-instar nymphs originating from apterous adults attacked it (Fig. 1 A), but the advanced instar nymphs and adults escaped from the spot. Nymphs and adults originating from alate adults showed escaping behavior only. These different responses strongly correlated with morphological differences. The frontal horns of 1st-instar nymphs produced by apterous adults were very long and pointed sharply at the tips. However, in the advanced instars, the horns were very short. The ratio of hind leg length to the body length of the 1st-instar nymph was large, but decreased in advanced instars. The horns of the 1st-instar nymphs produced by alate adults were quite short, and there were no differences in their length among the 1st and advanced instars. Ratio of the hind leg length to body length of the 1st-instar nymph was small.

When a syrphid larva, Allograpta javana, was placed on a colony of aphids, many 1st-instar nymphs attacked it using their frontal horns (Fig. 1 B). During the attack, these nymphs spontaneously secreted droplets and adhered them to the syrphid body. When these droplts were placed on the body, the syrphid larva was agressively attacked by additional 1st-instar nymphs.

Fig. 1.

A: First-instar nymphs attacking a tip of the piece of paper, where droplets of cornicle secretion were daubed.

B: First-instar nymphs attacking a syrphid larva. Some aphids crawled under the larva, secreting droplets, and tried to adhere them to the syrphid's body.

EXPERIMENTAL EVIDENCE FOR EFFECTIVE AND ALTRUISTIC COLONY DEFENCE AGAINST NATURAL PREDATORS BY SOLDIERS OF THE GALL-FORMING APHID *PEMPHIGUS SPYROTHECAE*

WILLIAM A. FOSTER

DEPARTMENT OF ZOOLOGY, UNIVERSITY OF CAMBRIDGE

DOWNING STREET,

CAMBRIDGE CB2 3EJ, U.K.

INTRODUCTION

Aoki's [1] report that the woolly aphid *Colophina clematis* has a special soldier caste of 1st instars was an exciting discovery because it established that there was a third Order (Hemiptera) of eusocial insects, thus providing a fresh arena in which to test ideas about the evolution of eusociality [2]. Soldier behaviour has been described in a number of other aphid species and it is now thought that eusociality has evolved at least four times independently in the Aphidoidea [3,4]. However, there are remarkably few quantitative studies of the interactions of soldier aphids with predators that are known to be natural enemies of the particular species being studied.

The present observations were carried out on *Pemphigus spyrothecae* which makes spiral galls on the leaf petioles of *Populus nigra* and is the only European aphid in which soldier behaviour has been described [5]. Unlike other European *Pemphigus* species, *P. spyrothecae* does not host-alternate but completes its entire life-cycle on the primary host, poplar. The foundress makes the gall in the spring and parthenogenetically produces thick-legged 1st instars (the "soldiers") that grow up into wingless adults (2nd generation). These produce both thick-legged ("soldier") and normal-legged 1st instars, and these latter grow up into winged sexuparae (3rd generation) that fly from the galls to poplar trees and give birth to the sexuals, which mate and produce eggs that overwinter in the bark of the tree [6].

RESULTS AND DISCUSSION

The thick-legged 1st instar soldiers are able to protect the aphids in the gall from being eaten by a range of aphid predators, including *Anthocoris minki* (Heteroptera:Anthocoridae), which appears to be the major natural predator of the gall generations of *Pemphigus* in the U.K. Most of the aphids that attacked the predators were themselves killed. The soldiers were able to kill predators introduced into natural galls and seemed able to protect the non-soldiers: a mean of between 9 and 20 % of the soldiers died in these experiments but only between 1 and 5 % of the non-soldiers died. Experiments were devised in which individuals of *A. minki* were free to enter and leave the gall (Figure 1). The soldiers were effective in

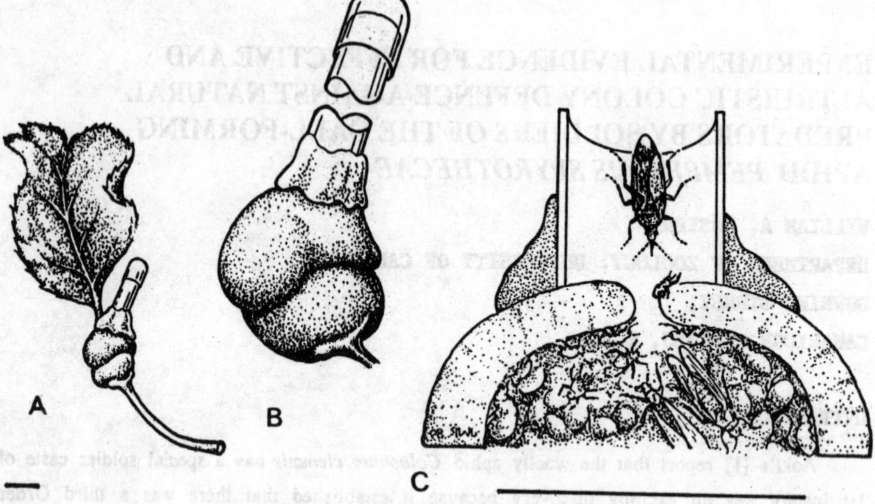

Figure 1. Experimental arrangement of gelatine capsule attached over the exit hole of a gall of *Pemphigus spyrothecae* on a poplar leaf.

A. Intact leaf, gall and capsule. Scale bar 1cm

B. Closer view of gall and capsule

C. Section through base of the capsule and the gall in the region of the exit hole. Adult *A. minki* in the capsule, adult and juvenile aphids in the gall. Scale bar 1 cm.

preventing the predator's access to the gall and in killing those predators that did manage to get in.

Natural galls were set up which had experimentally manipulated numbers of soldiers and non-soldiers within them, and a single predator was introduced into each gall. With 100 soldiers and 30 non-soldiers, the predator (juvenile *A. minki*) died in 18 out of 20 experiments. With no soldiers and 80 non-soldiers in the galls, the predator survived in 10 out of 10 experiments. This clearly demonstrates that it is the soldiers alone that kill the predators and that the primary role of the soldier caste is the defence of the aphid colony.

REFERENCES

1. Aoki, S. 1977. Kontyu, Tokyo, 45:276-282.
2. Aoki, S. and Kurosu, U. 1989. Japanese Journal of Entomology, 57:407-416.
3. Aoki, S. 1987. In: Animal Societies: Theories and Facts, Y. Itô, J.L. Brown, J. Kikkawa, Eds. Japan Sci Soc Press, Tokyo, pp 53-65.
4. Itô, Y. 1989. Trends in Ecology and Evolution, 4:69-73.
5. Aoki, S. and Kurosu, U. 1986. Journal of Ethology, 4:97-104.
6. Lampel, G. 1960. Z. Angew. Ent. 47:334-375.
7. Lampel, G. 1968-9. Bull. Naturforsch Ges Freiburg, 58:56-72.

SUBSOCIALITY OF THE AUTRALIAN GIANT BURROWING COCKROACH, *MACROPANESTHIA RHINOCEROS* (BLATTODEA : BLABERIDAE)

Tadao MATSUMOTO

Department of Bilogy, College of Arts & Sciences,
The University of Tokyo, Megro, Tokyo 153, Japan

The Australian giant burrowing cockroach Macropanesthia rhinoceros is an ovoviviparous blaberid that displays brood care and is the largest and bulkiest Blattarid in Australia (1). The body length of the largest adult male reaches about 8cm and the live weight reaches 30g. Little information has been avalilable to date on the ecology in the field. The cockroaches (Blattaria) are of special interest among the presocial insects because they are closely related to the Isoptera (termites), all of whose members are eusocial (2). Some species dwelling inside wood and feeding of it have a monogamous family life. The link between adults and nymphs in the xylophagous cockroaches Cryptocercus (Cryptocercinae) and Salganea (Panesthiinae) is long lasting, and ca last the whole nymphal life (3, 4, 5). The appearearance of a monogamous family structure in xylophagous cockroaches and termite is a true convergence (6). The Australian giant burrowing cockroach also belongs to the Panesthiinae and make family life as Salganea. The present paper deals with familial associaltion, growth of nymphs, population density and other field observation of the cockroach. The study sites are located in the open eucalypt woodland near Mount Garnet (17°41'S. 145°07'E, alt.680m) in northeastern Queensland. The field studies made five times from October 1987 to October 1989. Sixteen quadrats (each 8m x4m) were investigated. The surface of the sandy soil in quadrats was removed to a depth of about 10cm using a scoop. Then, a trench about 50cm wide and 5cm deep was dug carefully along the nest burrow using shovel for the heavy work, and a towel for the finer work. 301 nest burrows were opened in the quadrats and in burrows around the quadrats, and examined the cockroaches, food strages, feces, predators and sometimes gests in the burrows. The distribution pattern of nest burrows is almost random. The cockroaches prefer litter on the ground surface as food, this litter contains dead leaves, woods, grasses and herbes. 85 adults and 123 single nymphs (old an middle ages) were collected in total. And 32 families consiting of a group of small nymphs together with either an adult pair of a female were found. This evidence shows that the cockroach forms family groups in spring and that intimate adult female-offspring relationships continue for about half a year. The nymphs pass through to be six or seven instar until next spring, then they disperce from their natal burrows to make their own. It is considered that Macropanesthia was derived from the ancestor of a xylophagous cockroach similar tha rain forest Pansthini, and secondary became adapted of savanna life.

References

1. Roth,L.M. 1977. Australian Journal of Zoology, Supplment Ser. 48,1-112.
2. Wilson, E.O. 1971. The Insectes Societies; Belknap Press of Harvard University, Cambridge.
3. Nalepa, C.A. 1984. Behavioral Ecology and Sociobiology, 14;237-279.
4. Seelinger, G. & Seelinger U. 1983. Zeitschrift Tier-Psychologie, 61: 315-333.
5. Matsumoto, T. 1986. In: Chemistry and Bilogy of Social Insects, Eder & Rembold, Ed., Verlag J. Peperny, p.394.
6. Gautier J.P. et. al. 1988. In: The Ecology of Social Behavior, C.N. Slobodchikoff, Ed., Academic Press, pp.335-351.

PRIMITIVE SOCIAL BEHAVIOUR IN SUBSOCIAL STAPHYLINID BEETLES

Wyatt T D

Department for External Studies, University of Oxford,
1 Wellington Square, OXFORD OX1 2JA, UK

For air-breathing terrestrial insects the intertidal saltmarsh, covered by the tide twice a day, is a difficult and harsh habitat. However, for those few species able to exploit the habitat there are rich food sources, such as the thick algal carpet on the mud surface, and few competitors (except grazing fish feeding at high tide). The subsocial staphylinid beetle, *Bledius spectabilis*, which burrows in the *Salicornia* zone of northern European saltmarshes [1], was used by Wilson [2] as his example of a pioneering species able to exploit a difficult habitat largely as a result of its elaborate maternal care.

Following the description of its ecology by Bro Larsen [3,4], I investigated the role of the mother beetle in protecting her larvae from the effects of the tide [5]. The size and shape of the maternal burrow proved to be important with a narrow neck keeping out the initial stages of the tide and giving time for the female to respond by blocking the neck. Equally important may be the unblocking at low tide to allow diffusion of oxygen into the burrow. The maternal care, for the first half of the first instar, also appears to be important in reducing mortality from a predatory beetle *Dichierotrichus gustavi* [6] and a parasitoid wasp *Barycnemis blediator* [7].

However, the pattern of parental care in *Bledius* species and its independent evolution in non-related species in two other families [3,8,9] subjected to the same habitat seems to suggest that maternal care in *B. spectabilis* has been selected for primarily for its protection from the harsh environment [2,8].

Other examples of subsocial behaviour in the Staphylinidae will be reviewed and parental care in different beetle families will be compared in the light of subsocial behaviour in other insect orders.

REFERENCES

1. Wyatt, T.D. & Foster, W.A. 1988. Ecological Entomology 13: 433-464.

2. Wilson, E.O. 1975. Sociobiology. Belknap Press, Mass.

3. Larsen, E.B. 1936. Vidansk Medd Fra Dansk Naturh Foren Kobenhaven, 100, 1-232.

4. Larsen, E.B. 1952. Trans of the 9th Int. Congress of Entomology 1: 502-506.

5. Wyatt, T.D. 1986. Behavioral Ecology and Sociobiology 19: 323-331.

6. Wyatt, T.D. & Foster W.A. 1989. Animal Behaviour, 38: 778-785.

7. Wyatt, T.D. & Foster W.A. 1989. Behavior 110 : 76 -92

8. Wyatt, T.D. 1987. New Scientist 116 (1581 8 October 1987), 50-51.

9. Van Wingerden, W.K.R.E., Littel, A., & Boomsma, J.J. 1981. In: Final report of the section 'Terrestrial Fauna' of the Wadden Sea working group 10 (eds. Smit CJ Den Hollander J van Wingerden WKRE & Wolff WJ).pp 101-125

Symposium 12

PHYLOGENY AND EVOLUTION OF THE FORMICIDAE

Organizer : **J. BILLEN, Belgium**

COMPARING DIFFERENT HYPOTHESES ABOUT THE ORIGINS AND PATTERNS OF ANT DIVERSITY

Cesare Baroni Urbani
Zoologisches Institut der Universität, Rheinsprung 9, CH-4051 Basel, Switzerland

The family Formicidae has been regarded as a solid monophyletic group by most students who dealt with its phylogeny [1]. Baroni Urbani [1] gave the first comprehensive, logical analysis of ant phylogeny, which in this paper will be referred to as the Reference Phylogeny. This phylogeny is based on a certain number of assumptions, like Fitch parsimony of character states [2], character coding, and monophyly of the involved taxa. Modifying one or more of these assumptions will give different results. The possible merits of phylogenies employing alternative hypotheses are discussed here. Search for the shortest tree (i.e. the one with the lowest number of character evolution steps) has been made by McClade [3] and PAUP [4] .

1. Assumptions about Character Evolution
Instead of Fitch parsimony allowing binary characters to be reversible and multistate characters to be unordered, the more conservative (and realistic) Wagner parsimony has been applied to the same data matrix. According to this hypothesis multistate characters are always ordered [5]. This hypothesis can be further extended by considering evolution of binary and multistate characters as irreversible.

2. Assumptions about Character Coding
The Reference Phylogeny accepted polymorphic characters within taxa. The Spartan hypothesis considers that all characters coded as polymorphic for one taxon are wrong or misunderstood. Under this assumption, they were removed, and the reduced data matrix contains 22 characters instead of 27.
Another alternative is the tolerance hypothesis which codes polymorphic characters within a taxon as their most common state only. This hypothesis is based on the assumption that rare character states within a taxon represent documented cases of evolutionary reversion which do not affect the main pattern of character evolution.

3. Assumptions about Monophyly

One of the most critical points in the Reference Phylogeny is the acceptance of weak synapomorphies characterizing the subfamilies Ponerinae and Myrmicinae. The lumper hypothesis considers these subfamilies as synonyms. A set of possible alternative hypotheses may split one or both these subfamilies into more taxa of subfamilial rank in the hope to delimit more solid monophyletic groups. These hypotheses are called the splitter hypotheses. One hypothesis, due to Clark [6], which divides the Ponerinae into 6 subfamilies, is tested here.

Testing of these hypotheses implied examination of 1'942 different cladograms. The Reference Phylogeny shows some advantages over all phylogenies produced using alternative hypotheses either in terms of Consistency Index, or in terms of Treelength, except for that constructed using the Spartan hypothesis. In the latter, as was to be expected, reducing the number of characters by all the equivocal ones, gives a considerable decrease in treelength and a remarkable increase in terms of Consistency Index. This phylogeny, however, is inferior to the Reference Phylogeny in showing a higher number of ambiguous branchings, i. e. of arbitrary dichotomies not supported by synapomorphic characters.

Hence, the Reference Phylogeny appears to represent the best available representation of ants' evolutionary history, at least for the time being, but demonstrates the great need for additional, reliable characters.

References

1. Baroni Urbani, C. 1989. Ethology Ecology & Evolution, 1: 137 - 168.
2. Fitch, W. M. 1971. Systematic Zoology, 20: 406 - 416.
3. Maddison, W. P. and Maddison D. R. 1987. MacClade, version 2.1. A phylogenetic computer program distributed by the authors.
4. Swofford, D. L. 1989. PAUP. Phylogenetic Analysis Using Parsimony, version 3.0b. A computer program distributed by the Center for Biodiversity, Illinois Natural History Survey, Champagne, Illinois.
5. Kluge, A. G. and Farris, J. S. 1969. Systematic Zoology, 18: 1 - 32.
6. Clark, J. 1951. The Formicidae of Australia. Vol. I: Myrmeciinae. C.S.I.R.O., Melbourne, pp. 13 - 16.

PHYLOGENY OF THE FORMICIDAE AND THE BEHAVIOURAL ANALYSIS OF SOME ARCHAIC AUSTRALIAN ANTS

P. Jaisson*, F. Nicolosi*, R. W. Taylor** and D. Fresneau*

** Ethologie et sociobiologie, (CNRS n° 667), Université Paris-Nord, 93430 Villetaneuse, France.*
*** CSIRO, Division of Entomology, Canberra, Australia.*

INTRODUCTION

The rediscovery of the "dinosaur ant" *Nothomyrmecia macrops* Clark in 1977 by Taylor opened a new era in the studies on ants phylogeny. Instead of the two previously described **Poneroid** and **Myrmeciioid** Complexes [6], branching on the common trunk of the phylum, Taylor [4] hypothesized a different branching with two phylads : the **Poneroid** and **Formicoid**. This resulted from the assumption that the key-character separating the two branches was the tubulation (Poneroid) or non-tubulation (Formicoid) of the fourth abdominal segment. Thus, *N. macrops* was located on the last Complex, far from another archaic Australian ant genus, *Myrmecia*, which has a tubulated abdomen. This view reinforced the existence of the autonomous Nothomyrmeciinae subfamily [3]. It also valued the tubulation of the abdomen as a major feature in ants phylogeny. However, as expressed by Taylor, it should be falsified if tubulation occurred more than once.

Recently, morphological and biochemical studies of the exocrine glands in both genera [1][2] showed an undeniable resemblance. This inclined to return back to the previous view which joined them together within the same subfamily, Myrmeciinae. In this context, we have carried out a comparative ethological study in order to know whether or not behavioural traits would support any of the two above mentioned assumptions.

RESULTS

1) The behavioural data recorded from two marked *N. macrops* colonies were compared with colonies of *M. fuscipes* and *M. pilosula*). All three species showed more or less marked polyethism. But only *N. macrops* and *M. pilosula* workers were able to realize distinct specific tasks, which is a clear primitive trait. However, *N. macrops* individuals performed much less interactions.

2) The analysis of the interindividual distances between members of a same colony evidenced a very special feature in *Nothomyrmecia* : each worker or queen is surrounded by a "self-space" and avoid having close contacts with its nestmates, as shown by *figure 1*. In ants, we never observed something similar. In this respect, *N. macrops* societies seem to offer a striking compromise between individual and social tendencies. The same measurements, applied to *Amblyopone australis* (a primitive ponerine genus) provided results close to those of *Myrmecia* and clearly distinct from *Nothomyrmecia*. Moreover, the evolved ponerine *Ectatomma ruidum* revealed still more nearness between individuals, as compared with both *Amblyopone* and *Myrmecia*.

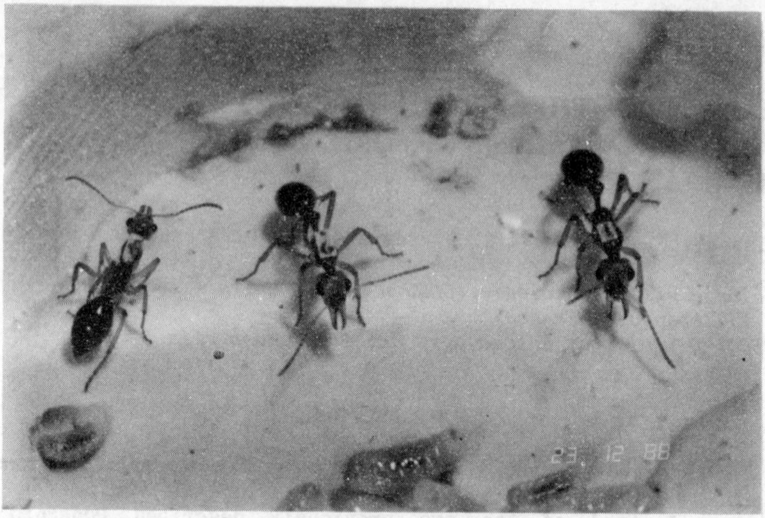

Figure 1 : Three workers of a *N. macrops* colony in culture, surrounded by their typical "self-space".

If the interindividual distances may reflect the level of social interactions, thus the social evolution, the different species studied revealed three possible steps : 1°) the most archaic, with separated individuals surrounded by their own "self-space" (*Nothomyrmecia*); 2°) a more evolved step with closer individuals (*Myrmecia*, *Amblyopone*); 3°) the highly evolved level, where individuals are very close together (*Ectatomma*).

DISCUSSION

If it makes sense to join the behavioural data to the previous morphological and biochemical arguments, we should assume that :

1. If *Nothomyrmecia* has to be well separated from *Myrmecia*, it should be closer to the previous Clark's *Promyrmecia* genus (represented by *M. pilosula*).
2. If the abdominal tubulation occurred only once, the *Nothomyrmeciinae* should be located at the top of the common phylogenetic trunk, before the bifurcation.

An alternative possibility, mentioned by Wilson [5], is that ants would be a diphyletic group, being *Nothomyrmecia* the most primitive living form in one of the phyla.

REFERENCES

1. Billen, J. 1988. Actes Coll. Insectes Sociaux, 4 : 27-33.
2. Billen, J., Jackson, B.D. and E.D. Morgan. 1988. Experientia, 44 : 715-719.
3. Clark, J. 1951. The Formicidae of Australia, vol. I : Myrmeciinae. CSIRO, Melbourne. p. 230.
4. Taylor, R.W. 1978. Science, 201 : 979-985.
5. Wilson, E.O. 1971. The Insect Societies. Harvard University Press, Cambridge, p. 548.
6. Wilson, E.O., Carpenter, F.M. and W.L. Brown. 1967. Science, 157 : 1038-1040.

PHYLOGENETIC, BIOGEOGRAPHIC, AND EVOLUTIONARY INFERENCES FROM THE DESCRIPTION OF AN EARLY CRETACEOUS SOUTH AMERICAN MYRMECIINAE

C.R.F. Brandão
Museu de Zoologia da Universidade de São Paulo
Av. Nazaré, 481, São Paulo, SP, 04263, BRAZIL

INTRODUCTION

Cariridris bipetiolata, recently described (1) from a single fossil found within a fine grained limestone piece from Santana do Cariri, Ceará, Northeast Brazil, in the Araripe Basin (Santana Formation, Crato Member - Upper Aptian - Lower Cretaceous - between 107 and 114 million of years B.P.,), represents the first Southern Hemisphere Mesozoic record and possibly the earliest known formicid. The specimen is considered to be a worker and the genus to belong to the Myrmeciinae. This subfamily, represented today only by the Australian Myrmecia, also includes the Baltic amber (Oligocene) worker Prionomyrmex longiceps (redescription and male in 2), and Ameghinoia piatnizkyi, described from three alate females embedded in limestones pieces from San Carlos de Bariloche, Argentina, probably of Lower Miocene or Upper Oligocene age. In this essay I explore the phylogenetic, biogeographic, and evolutionary inferences provided by the discovery of this new Cretaceous fossil ant.

DISCUSSION

The Formicidae phylogenetic system is undergoing profound changes (3, 4, and this Symposium) and a comparison of fossil and extant Myrmeciinae with other Cretaceous ant fossils might help to clarify the comparable holomorphology and hence to establish the states of key characters in ant phylogenetic reconstruction. The mid-Cretaceous Sphecomyrminae fossils were originally described from New Jersey. It is accepted today that they have occured across a wide portion of present-day North America. Other Sphecomyrminae fossils were recently found in Siberia and Kazakstan (5 and references). All known specimens came from formations dated as Cenomanian to Santonian (100 to 80 m.y.B.P.). Sphecomyrminae may have retained the relatively short scape, and the narrow, short, curvilinear, bidentate "wasp-like" mandible, while the fully constricted petiole is probably derived. Myrmeciinae is defined by the following synapomorphies: multiple serially arranged teeth at the mandibular blade (though not observable in C. bipetiolata, but very much alike in other extinct and Recent Myrmeciinae), the peculiar shape of the petiolar node. and the clearly differentiated postpetiole. The mandibles and petiole attachement of the generalized Ponerinae are therefore to be interpreted as plesiomorphous. Some important features as, for instance, palpal counting, eventual tubulation of the fourth abdominal segment, structure of the mesonotum, presence of the metapleural gland, counting of tibial spurs and tarsal claws, details of wing venation and genitalia are not visible in limestone preserved fossils. If we assume, however, that these characters, or at least some of them, are conservative at the subfamily level, then the study of Myrmecia could indicate their generalized states. Sphecomyrminae and Myrmeciinae could be considered the stem of, respectively, the Formicoid and Poneroid clades (modified from 6). The separation between them must have occured at the Lower Cretaceous.

By the beginning of the Cretaceous, at the time Cariridris appeared, Gondwana was a large land mass, still connected with South Laurasia; at its Eastern portion the Tethys Sea was beginning to close. At the end of the period the South Atlantic has widened into a major ocean, and the Mediterranean was clearly recognizable. Australia remained attached to Antarctica and this to South America, all with tropical to subtropical climates. It seems altogether possible, by the occurence of Prionomyrmex in the Baltic amber, that Myrmeciinae occured once in Africa and Antarctica. The few Myrmeciinae fossil specimens thus far discovered may indicate that either they were not so abundant as Recent ants or that they forage individually. Even a deposit as rich as Santana Formation yielded just one fossil ant.

Accepting the only criterium available for evaluating the probability of eusocial versus solitary levels in fossils ants (7), C. bipetiolata agrees with Recent and fossil ant workers, in opposition to winged Cretaceous formicids and modern wingless aculeate females. Moreover the enlarged prothorax is comparable to modern ant workers prothoraxes. Eusociality in Formicidae will have, then, to be extended to more than 100 m. y. B. P.

REFERENCES

1. Brandão, C.R.F.; Martins-Neto, R.G. and Vulcano, M.A. in press. Psyche.
2. Wheeler, W.M., 1914. Schrift.Physik.-oekon. Ges. Königsberg, 55: 1-142.
3. Hölldobler, B. and Wilson, E.O., in press. The Ants Harvard Univ. Press.
4. Baroni Urbani, C., 1989. Ethol. Ecol. Evol., 1: 137-168.
5. Wilson, E.O., 1985. Psyche, 92: 205-216.
6. Taylor, R.W., 1978. Science, 201: 979-985.
7. Wilson, E.O;, 1987. Paleobiol., 13: 44-53.

THE STING APPARATUS AND PHYLOGENY OF THE ANTS

Charles Kugler
Biology Department
Radford University
Radford, VA 24142 U. S. A.

In the last 36 years there have been essentially two hypotheses of the internal phylogeny of the Formicidae; that of Brown [1] and Wilson et al. [2], and that of Taylor [3]. Both agree on the following: 1) early in ant evolution the ancestors to the Ponerinae (including Cerapachyini), Dorylini, and Myrmicinae separated from the ancestors to Nothomyrmecia, Aneuretinae, Dolichoderinae, and Formicidae, 2) the Dorylinae (including Ecitonini, Cheliomyrmecini, and Aenictini) are derived from primitive Ponerinae, 3) the Myrmicinae are derived from the Ponerinae, 4) the Pseudomyrmecinae share common ancestors with Myrmecia, 5) the Dolichoderinae evolved from the Aneuretinae, 6) the Aneuretinae shared a common ancestor with Nothomyrmecia, and 7) the Formicinae are more closely related to Nothomyrmecia than to Myrmecia. However, Wilson et al. suggested that the Fromicinae evolved directly from Nothomyrmecia-like ancestry, whereas Taylor said that first the Aneuretinae and Nothomyrmecia separated, then the formicines arose from the aneuretines. Taylor also proposed that Myrmecia and the Pseudomyrmecinae, instead of arising from Nothomyrmecia as previously thought, evolved from ponerine ancestors, and thus from the opposite side of the deep split described in 1).

Though stinging ability varies widely in ants, neither phylogeny used sting characteristics. Kugler [4] showed that the sting apparatus has potential taxonomic value in the Formicidae. It is a complex of eight scletites derived from three abdominal segments, and is a rich source of characters that may vary at the species or genus level. In this study I extend these investigations to look for characters that may help us differentiate ant subfamilies and reconstruct their histories.

The sting apparatus is compared in representatives of nine proposed subfamilies of ants: Nothomyrmecinae, Myrmecinae, Cerapachyinae (1 genus), Ponerinae (7 genera), Myrmicinae (72 genera), Ecitoninae (3 genera), Dorylinae (1 genus), Dolichoderinae (3 genera), and Formicinae (5 genera). An hypothesis of the phylogenetic relationships of the subfamilies based on this complex character system is presented and compared with hypotheses based on external anatomy. Preliminary analysis tends to support many traditional views of evolutionary relationships, but provides no particular support for a separating Nothomyrmecia from the Myrmecinae.

REFERENCES

1. Brown, W. L., 1954. Insectes Sociaux, 1 (1) : 21-31.

2. Wilson, E. O., Carpenter, F. M. and Brown, W. L., 1967. Psyche, 74 (1) : 1-19.

3. Taylor, R. W., 1978. Science, 201 : 979-985.

4. Kugler, C., 1978. Studia Entomologica, 20 (1-4) : 413-548.

PHYLOGENETIC ASPECTS OF EXOCRINE GLAND DEVELOPMENT IN THE FORMICIDAE

J. Billen

Zoological Institute, K.U.Leuven, Naamsestraat 59, B-3000 Leuven (Belgium)

All of the approximately 10,000 known ant species are eusocial, and together form the cosmopolitan family Formicidae. According to the author, they are classified in nine [1], ten [2] or eleven [3] extant subfamilies. This simple numerical difference mostly deals with the respective estimation towards the attribution of the subfamily status to a particular phylogenetic group, while the classifications presented generally consider two complexes, albeit with some characteristic differences. The resulting phylogenetic trees are mainly based on morphological criteria, although the most recent cladistic analysis diagram by Baroni Urbani [2] includes behavioural aspects as well. Also other approach methods have recently been found to furnish useful evolutionary information in insects, as is e.g. the case with chemotaxonomy and the analysis of mitochondrial DNA. In our examination on the occurrence and development of the exocrine glands in the Formicidae, we found it too provides additional evolutionary information [4 to 8].

The high number of exocrine glands in social insects is mainly due to their crucial role in the elaboration of pheromonal substances. Although the general appearance of the exocrine system and the development of the ubiquitous major glands is relatively similar in all ant species, there are a number of other glands that are only found in one or a few groups, for which they may form a valuable diagnostic character. The Formicidae as a family are thus characterized by the presence of postpharyngeal and metapleural glands. At the subfamily level, the Dolichoderinae and Aneuretinae share the exclusive occurrence of the Pavan gland and the very much enlarged pygidial gland, which therefore support the generally accepted relationship between both groups. The Formicinae, on the other hand, are unmistakable by their peculiar venom gland, that shows a dorsal, cap-like convoluted gland portion. The venom gland is also used as a subfamily character with regard to the appearance of its two secretory filaments : in the Dolichoderinae, these are lobate, whereas they are long and slender in the other subfamilies. The opposite situation, however, does also exist, as we could ascertain from recent observations in Australian *Leptomyrmex* (a dolichoderine with long filaments) and *Meranoplus* (a myrmicine with lobate filaments), and therefore warrants for cautious use of venom gland morphology in a phylogenetic context.

The Dufour gland, although anatomically probably the most simple of all glands, exhibits different ultrastructural epithelial types in most of the subfamilies [4,5]. These different types support the existence of subfamily groups as such, but do not readily allow to trace a developmental pattern to explain their evolution. It does, however, reveal a similar fine structure in *Myrmecia* and *Nothomyrmecia*, which is not found as such in ants from other subfamilies. This apparent relationship between both Australian subfamilies is further supported by their similar gland chemistry, and particularly by their exclusive sharing of a sting bulb gland, which we very recently discovered inside their sting [8]. For these reasons, we believe that Myrmeciinae and Nothomyrmeciinae are more closely related [1] than is reflected by their actual and rather distant position [2,3].

References

1. Brown, W.L. 1954. Ins. Soc., 1 : 21-31.
2. Baroni Urbani, C. 1989. Ethol. Ecol. Evol., 1 : 137-168.
3. Taylor, R.W. 1978. Science, 201 : 979-985.
4. Billen, J.P.J. 1986. Entomol. Gener., 11 : 165-181.
5. Billen, J.P.J. and Gotwald, W.H. 1988. Zool. Scr., 17 : 293-295.
6. Hölldobler, B. 1982. In : The Biology of Social Insects, M.D. Breed, C.D. Michener and H.E. Evans, Eds., Westview Press, Boulder, Colorado, pp. 312-317.
7. Jessen, K. 1987. In : Chemistry and Biology of Social Insects, J. Eder and H. Rembold, Eds., Peperny Verlag, München, pp. 445-446.
8. Billen, J. 1990. Int. J. Insect Morphol. & Embryol., 18 (in press).

EXOCRINE GLAND CHEMISTRY AND PHYLOGENY OF ANTS

E.D. Morgan, Department of Chemistry, University of Keele, Keele, Staffordshire. ST5 5BG England.

The phylogeny of ants is almost wholly based upon morphological characters. The graphical representation of that phylogeny, from Brown [1], Wilson [2] or as modified by Taylor [3] presents 10 to 12 subfamilies. Baroni Urbani has recently attempted to derive a phylogeny based on behaviour [4], but has pointed out the difficulties in finding suitable characters in a group of entirely eusocial behaviour.

Biochemical features are used at times at systematic indicators, but have not been applied to the Formicidae. Ants are the largest group of social insects and are richly endowed with chemicals for communication. A number of chemical types are therefore available to use for this purpose [5], such as the venon, and the secretions of the postpharyngeal gland [6], as yet little known, and the better known mandibular and Dufour gland secretions. Mandibular glands of ants contain a limited range of volatile compounds and in many species there are none at all, nevertheless, mandibular glands do at times provide information in constructing a subfamily tree. Pavan's gland, filled with monoterpenoid compounds, or iridoids, found only in Dolichoderinae, is both a morphological and chemical characteristic for this subfamily [5, 7]. The Dufour gland, present in females and workers of all aculeate Hymenoptera provides a range of volatile secretions of not too complex, but varied composition and the glands of a relatively large number of species (more than 100) have been examined [7, and not yet reviewed]. The Myrmicinae (~ 25 species examined) all contain sesquiterpenoids, and usually linear alkanes and alkenes as well, except the genus Messor (5 species) which has no sesquiterpenes. The Formicidae (> 60 species) differ in usually having oxygenated compounds (alcohols, ketones, esters etc) in addition to the hydrocarbons, though some species do not. However these latter have undecane as chief hydrocarbon while the myrmicines have pentadecane, heptadecane or longer chains as chief component. The Pseudomyrmecinae (3 species) contain isopropyl esters, rare in insects and not as yet found elsewhere in ants. The Myrmeciinae, on very limited evidence (3 species) divide into two groups, supported by other evidence [8]. The "jumpers" and the "inch" or "bulldog" ants. M. pilosula which belongs to the "jumpers" contains only oxygenated compounds. M. gulosa and M. nigriceps, members of the "inch" group contain a mixture of hydrocarbons and oxygenated compounds [9]. Nothomyrmecia macrops, the only member of the subfamily Nothomyrmicinae, closely resembles the "inch" group of Myrmeciinae from a chemical viewpoint [10]. The Dufour glands of Ponerinae have not, so far, been shown to have any characteristic features, but their mandibular glands all contain alkylpyrazines, a character shared by a few dolichoderines and formicines and a single Myrmicine. The one example of the Dorylinae analysed shows hydrocarbons only, but of longer chain length than found in Myrmicinae and Formicinae. Information on other subfamilies is being collected.

REFERENCES

1. Brown, W.L., 1954. Insectes Sociaux, 1: 21-31.
2. Wilson, E.O., 1971. The Insect Societies, The Belknap Press of Harvard University Press, Cambridge, Mass.
3. Taylor, R.W. 1978. Science, 201: 979-984.
4. Baroni Urbani, C., 1989. Ethology Ecology and Evolution, 1: 137-168.
5. Attygalle, A.B., and Morgan, E.D., 1984. Chemical Society Reviews, 13: 245-278.
6. Bagneres, A.G., and Morgan, E.D., 1990 (submitted).
7. Blum, M.S., and Hermann, H.R., 1978. In: Arthropod Venoms S. Bettini, Ed. Springer Verlag, Berlin p. 801-894.
8. Greenslade, P.J.M., 1979. A guide to Ants of South Australia, South Australian Museum, Adelaide.
9. Jackson, B.D., Billen, J.P.J., and Morgan, E.D., 1989. Journal of Chemical Ecology, 15: 2191-2205.
10. Billen, J.P.J., Jackson, B.D., and Morgan, E.D., 1988. Experientia, 44: 715-719.

CAN WE USE THE DIGESTIVE TRACT FOR PHYLOGENETIC STUDIES IN ANTS?

F.H. Caetano

Department of Biology, Institute of Biociencias, UNESP, 13500 Rio Claro, SP, Brasil

The digestive tract of adult ants in general is quite uniform. On the other hand, some morphological characteristics appear exclusively in some subfamilies, while others, in spite of being rare, occur in some genera of different subfamilies. The dilated oesophagus forming a thoracic crop e.g. occurs in physogastric queens of *Myrmica*, *Lasius*, *Monomorium*, *Stenamma* and *Leptothorax* [1], in workers of almost all Ponerinae [2], in *Sericomyrmex moreirai*, in *Dolichoderus attelaboides*, and in some castes of *Acromyrmex*. The proventriculus bulb was thoroughly studied by Eisner [3], who proposed a phylogeny based on this organ, and suggested that the Myrmicinae (except the tribe Cephalotini) have lost the proventriculus bulb due to their feeding on solid foods. It thus should not have a barrier for the food passage. However, it is known that ants in the tribe Attini essentially have a liquid diet, althoug they lack the bulb. Both characters discussed so far (presence of thoracic crop and absence of proventricular bulb) also occur in the Mutillidae, whose diet only consists of liquid material of animal origin. Another characteristic, present in both Mutillidae and Formicidae, is the external aspect of the ventriculus. In Apidae and Vespidae, the ventriculus presents transverse external undulations, originating from transverse muscle rings, and thus giving it a characteristic appearance. Mutillidae and Formicidae, however, do not show such undulations.

The ileum of these insects apparently is a long and narrow tube. In the Cephalotini, it shows a dilated median portion, filled by about ten types of bacteria and by one filamentous fungus [4]. In some castes of *Acromyrmex* and in *Sericomyrmex moreirai*, this region also presents a dilation in the anterior region, next to the entrance of the malpighian tubules. Such configuration is constant in almost all Ponerinae [2]. In this dilated portion of *Paraponera clavata* and *Dinoponera australis*, there are longitudinally arranged spines. These are different, however, in the first and second portion, the first having only one tooth, the second up to 10 teeth. Both portions are separated by an epithelial fold. Thus, we can state that, at least in the Ponerinae, the ileum comprises two portions. Another evidence for this division is the presence of different cell types in both regions. In *Neoponera vilosa*, the first region has cells which do not exhibit absorption but only secretory features, while in the second region, cells only show absorption characteristics [5]. This subdivision of the ileum in two portions is also known from the Mutillidae, although they here are clearly separated by a transversal ridge. Such a ridge in ants is only found on histological sections [2].

In conclusion, the digestive tract in the Formicidae has more similarities with the Mutillidae than with Apidae and Vespidae. To our opinion, the only region which is similar in Formicidae, Apidae and Vespidae, is the proventriculus, which in some species presents a certain structural similarity, however, without any clear functional significance. In order to have a clear and more realistic idea of this situation as a whole, more studies are to be done on the Formicidae and Mutillidae, as well as a detailed study in the Tiphiidae.

References

1. Petersen-Braun, M. and Buschinger, A. 1975. Ins. Soc., 22 : 51-66.
2. Caetano, F.H. 1988. Naturalia, 13 : 129-174.
3. Eisner, T. 1957. Bull. Mus. Comp. Zool. Harvard College, 116 : 438-490.
4. Caetano, F.H. 1989. In : Insect Endocytobiosis : Morphology, Physiology, Genetics, Evolution, W. Schwemmler and G. Gassner, Eds., CRCPress, Boca Raton, chap. 5.
5. Caetano, F.H. 1989. XII Col. Soc. Bras. Micr. El., Caxambu - MG, 155-156.

PHYLOGENETIC RELATIONSHIPS AND CLASSIFICATION IN THE ECTATOMMINI (HYMENOPTERA : FORMICIDAE)

John E. Lattke, Fundacion Terramar,
Apartado 89.000, Caracas 1080-A, Venezuela

The ponerine ant tribe Ectatommini as presently known is made up of 9 genera with a varied distribution around the globe. The genus Acanthoponera has been pointed out as the most primitive, with the genera Heteroponera, Rhytidoponera, Gnamptogenys, Aulacopone, Proceratium, and Discothyrea more or less group around it. Another group was made up of Paraponera and Ectatomma. The distribution of the genera has been explained as a consequence of competitive displacement, with fossils from Baltic, and Sicilain Amber, plus Florissant Shale provoding evidence of a supposed, previous wider distribution (1).

These genera were examined for the use of new characters in evaluating their relations. Characters of the mesosomal sternum were found very useful. Aulacopone was not examined. The primitive genus Myrmecia was used as an outgroup.

The following synapomorphies were determined for Ectatommini: compound eyes at cephalic midlength or posterad, a sulcus partially or totally impressed around the compound eye, frontal lobes that only partially cover antennal insertions, development of a lamella around anterior clypeal and mesepisternal borders, a median cephalic carina (indistinct or secondarily lost in Gnamptogenys)and absence of empodia. The tribe can be divided into two groups: the Ectatomma group characterized by a bidentate prosternal process, and metacoxal fossae that open directly into the posterior metasternal envagination. Of its 3 genera, Ectatomma is considered the most primitive, with Rhytidoponera, and Gnamptogenys forming sister, derived genera. The other group is the Heteroponera group, made up of this genus and Acanthoponera.

They share a triangular prosternal process and closed metacoxal fossae separated from the metasternal envagination. This leaves Ectatommini with only 5 genera.

Paraponera is placed within its own genus group as the combination of primitive and derived traits make it unlike any of the other groups. Some of its autapomorphies are antennal scrobes, petiolar spiracles at the lateral base of the node, spinulate lateral hypopygidial margins, and an anteroventral postpetiolar process hidden within the petiole.

Discothyrea and Proceratium are considered close to each other on account of the frontal-clypeal configuration, the evenly rounded posterior petiolar border, and larval bosses. Aulacopone is provisionally grouped with the last two genera due to the frontal clypeal configuration.

The reliability of the fossil evidence is seriously questioned as most specimens are in poor conditions, or do not coincide with apomorphies determined for Ectatommini. Ectatommine distribution is explained as a consequence of the separation of South America from Palaeoantarctica, but after the fission of Africa. Proceratium and Discothyrea are considered more ancient due to their almost cosmopolitan distribution.

References

1. Brown, W.L. 1958. Bulletin of the Museum of Comparative Zoology, 118(5):175-362.

ECOLOGY AND PHYLOGENETIC RELATIONSHIPS WITHIN NEOTROPICAL MEMBERS OF THE GENUS *GNAMPTOGENYS* (HYMENOPTERA: FORMICIDAE)

John E. Lattke, Fundacion Terramar
Apartado 89.000, Caracas 1080-A, Venezuela

Neotropical members of the ponerine ant genus Gnampto- were studied in order to elucidate phyletic trends within the group. The study is mostly based on external worker morphology, and the related genus Ectatomma is used as an outgroup. Sixty-one species were determined, and divided into 5 large species groups.

The Holcoponera group has the greatest amount of plesiomorphies: triangular mandibles with dorsal rugulae, convex anterior clypeal margin, head wider posterad than anterad, deeply impressed promesonotal suture, lack of propodeal armature, broadly sepatated teeth of the prosternal process, and high petilar node. Most of its members are epigeaic, generalist predators. Two major subgroups are recognizable. The porcata subgroup has a posteriorly inlined petiolar node with a brief anterior peduncle, and an anteriorly produced subpetiolar process. The striatula subgroup has relativelt erect nodes with no anterior peduncle and a variably shaped subpetiolar process. In the other species groups the promesonotal suture is absent or weakly impressed, the petiolar node becomes lower, and the teeth of the prosternal process are close. Most are considered derived from a Holcoponera ancestor.

The rastrata group has propodeal teeth, tends to a straight anterior clypeal border and well impressed metanotal groove. They apparently are all specialized predators of millipedes.

The mordax group has experienced shortening of the scapes, suntriangular to subfalcate mandibles, elongated head, and straightened second gastric segment. Most are cryptic dwellers of leaflitter and there is a trend towards dietary specialization beetles and other ants.

The tornata group marks a trend towards subtriangular mandibles with a smooth dorsum, subquadrate heads, laterally acutely angular clypeal lamella and no transverse mesosonal impressions. Many species prey upon beetles and one is an exclusive arboreal nester.

The haenschi group is made up of one species with uncertain placement within the genus. Its combination of median cepahlic carina, frequent presence of ocellus, atrophied eyes and anteriorly placed propodeal spiracle make it unique. It is either a very derived Holcoponera product, or represents a pre-Holcoponera lineage. Nothing is known of its biology.

Most species are inhabitants of mesic, forested areas, small nests constructed in decomposing wood, and rarely passing 500 adults. A few species are polygynous, ergatoid queens are unknown.

UNRAVELLING THE *CAMPONOTUS FULVOPILOSUS* SPECIES COMPLEX (HYMENOPTERA : FORMICIDAE)

H.G. Robertson,

Camponotus is one of the most diverse genera of ants in the world, probably containing in excess of 1000 species [1]. The taxonomy of the group is in a chaotic state partly because of the difficulty in distinguishing between inter- and intra-specific variability. The object of the present study has been to use the *Camponotus fulvopilosus* species complex in southern Africa as a model for understanding species relationships in the genus as a whole. We examined differences between species and populations in morphology, habitat, and chemistry of mandibular and Dufour's gland secretions.

The four species making up the group are listed below. They share the common distinguishing feature of thick, blunt hairs on the dorsal surface of the gaster.

Camponotus fulvopilosus (De Geer)
  = *Camponotus fulvopilosus* var. *detritoides* Forel **syn. n.**
  = *Camponotus fulvopilosus* var. *flavopilosus* Emery **syn. n.**
Camponotus brevisetosus Forel **stat. n.**
Camponotus detritus Emery
Camponotus storeatus Forel **stat. n.**

Camponotus fulvopilosus is the most widely distributed of the four species, occurring in savanna and arid regions of southern Africa and extending as far north as Angola and the Congo [2]. The other three species are found on the periphery of the latter distribution: *C. brevisetosus* in Natal and the eastern Transvaal; *C. detritus* in the dunes of the Namib Desert; and *C. storeatus* in the mountains of the southern Cape. Little is known about *C. brevisetosus* - it has been rarely collected and was not re-located during the present study.

The species can be distinguished morphologically from one another mainly on the basis of differences in colour, hair length and hair distribution. There is considerable variation within *C. fulvopilosus* and one cannot exclude the possibility of there being more than one species within its present definition.

Habitat separation between species was found in certain regions. In the Namib Desert *C. fulvopilosus* is found in the gravel plains whereas *C. detritus* occurs only in the sand dunes. At a site near Willowmore in the southern Cape we found *C. storeatus* on a north-facing slope in a rocky, semi-succulent-shrub habitat whereas *C. fulvopilosus* was found only on the sandy flood plain of the river at the base of the slope. At two other sites north of Willowmore we found the two species in the same habitat, even foraging on the same bushes.

In addition to the morphological and habitat differences between species, there were also chemical distinctions. The composition of the male mandibular glands differed in *C. fulvopilosus* and *C. detritus* and there were also differences between *C. fulvopilosus*, *C. detritus*, and *C. storeatus* in the composition of worker Dufour's glands.

The *C. fulvopilosus* complex shows the typical pattern found in many African ant groups of a widely distributed, variable species from which more narrowly distributed species appear to have speciated. We still have a poor understanding of the geographical variation in morphology and chemistry of *C. fulvopilosus* which will only be remedied by further collecting.

1. Brown, W.L. 1973. A comparison of the Hylean and Congo-West African rain forest ant faunas. In: Tropical Forest Ecosystems in Africa and South America: a comparative review, B.J. Meggers, E.S. Ayensu, and W.D. Duckworth, Eds, Smithsonian Institution Press, Washington, D.C., pp. 161-185.
2. Wheeler, W.M. 1922. Ants of the American Museum Congo Expedition. A contribution to the myrmecology of Africa. Bulletin of the American Museum of Natural History 45: 1-1139.

STRUCTURAL FEATURES ON THE RECTAL PADS OF *NEOPONERA VILLOSA* (FORMICIDAE : PONERINAE)

F.H. CAETANO
Department of Biology, Institute of Bioscience, UNESP
13500 - Rio Claro, SP, Brazil

The rectum epithelium shows a great structural variations, from a simple epithelium of the rectum wall it pass to a hight complexity in the rectal pads. In this one we find 3 types of cells. In the Hymenoptera rectal pads, between the basal epithelium and the principal cells of the pads, there is a papillary lumen (Martoja - Ballan - Dufrançais, 1984 and Caetano, 1984 e 1988). The rectal pads and rectum of **N.villosa** are constituted by four types of cells. This cells shows owns characteristic, until the nucleus level, inclusively. These cells are : Principal cell of the pad, Basal epithelium cells of the pad, Junctional cells and "intermediary cells".

The principal cells shows = tight junctions, septate desmosomes, and interdigitations in the lateral membrane with many mitochondria.

The basal epithelium has deep invaginations in both apical and basal membrane. Between then there are few cytoplasmic organelles and the cytoplasm exhibit low electron density.

The junctional cells have hight electron density and they are located between the principal cell and the basal epithelium of the pad, and the rectum epithelium. This cells are thin and they are dilated at the nucleus region. Between the basal epithelium and the junctional cells occur a group of cells which we called "intermediary cells".

These cells are large and exhibit hight electron density. In its cytoplasm we can see many mitochondria and vacuoles. The function of these cells is unknown yet.

A COMPARATIVE STUDY ON THE STING GLANDS OF THE ANTS *DACETON ARMIGERUM* AND *ACANTHOGNATHUS* SP. (HYMENOPTERA: MYRMICINAE)

MATHIAS, M.I.C. , F.H.CAETANO, M.A.L.PIMENTEL and M.E.M. TOMOTAKE - Department of Biology, Institute of Bioscience, UNESP 13500 - Rio Claro - SP, Brazil

The ants **Daceton** and **Acanthognathus** are considered the more generalizad genus of Dacetini tribe. They have distinct foraging behaviour: **Dacenton** is exclusively arboreal and **Acanthognathus** is epigaeic. They have a distinct size too, **Dacenton** is bigest than **Acanthognathus.** This genus retain a large eyes, multi-segmented antenae and palpi. These suggest that this one forage mainly at night, like your related Australian genus **Orectognathus** (BROWN JR and KEMPF,1969). By other hand **Dacenton** forage during the day and the visual stimuli is very developed (WILSON, 1962). The poison gland comprise a two free filaments that arise in a large reservoir (with a thin wall) from this part a long and narrow duct.

The Dufour's gland is a club-shaped sac that open close to the poison gland duct at the sting base by a narrow and short duct.

In **Acanthognathus** the Dufour's gland is bigger than its poison gland, but in **Daceton** the Dufour's gland is very small (smaller than the poison gland duct).

These dates may be suggest that **Daceton** needs a large poison gland to capture the prey and since its use the visual stimuli (WILSON,1962), they don't use the trail pheromone like others Myrmicinae (**Monomorium** and **Solenopsis**) (ALI et alii, 1988) produced by the Dufour's gland . **Acanthognathus** use only Collembola in its nourishment and the long mandibules may be enough to take it and their foraging is during the night or inside the litter, so that trail pheromone must be essencial for its survival.

Symposium 13

SOCIAL POLYMORPHISM: HOW AND WHY?

Organizer : **CHARLES NOIROT, France**

EVOLUTION OF POLYMORPHISM IN ISOPTERA: DEVELOPMENTAL AND BEHAVIOURAL CONSTRAINTS

Charles NOIROT
Laboratoire de Zoologie, Université de Bourgogne
6 Bd. Gabriel, 2100 DIJON, FRANCE

In opposition with social Hymenoptera, two separate sterile castes, soldiers and workers, occur in termites. These two "neuter" castes appeared independently and their evolution was canalized by obvious differences both in postembryonic development and behavioural constraints, which allowed much more diversification in the soldiers than in the workers. Whereas the soldiers differentiated very early and only once (monophyletism), the workers, absent in many primitive species, are polyphyletic (1).

The flexibility of postembryonic development is basic for the polymorphism. The deviations from the normal line (ending in imagoes) follow very different rules in workers and soldiers respectively (1). For the workers, the differentiation of imaginal organs linked with reproduction and dispersal does not occur, thus the morphology remains larval-like. Such a very simplified development does not favour the diversification of the worker caste. By contrast, the soldier development is astonishingly convergent with that of holometabolous insects, with a transitional inactive stage (presoldier) to be compared with a pupa (2). We do not know how this holometaboly was acquired, but as soon as this evolutionary jump was realized, it allowed very varied morphogenesis, true metamorphoses. Additionally, both types of neuter development may occur in succession during the individual life ("temporal polymorphism" : 3), as soldiers frequently differentiate from workers.

As regards the behaviour, workers are generalists, even when a polyethism is observed. This multifunctionality strongly constrained their morphological diversification, as evidenced by the mandibles (only tools of these workers) the structure of which was markedly conserved. In opposition, soldiers are specialists, with a very simplified behaviour. They neither dig, nor build, nor care the brood, nor forage nor even feed by themselves (they must be fed by their nestmates). They are solely specialized in defense (their role in recruitment, in some species, is a derived condition). This drastic simplification seems primitive, as it is observed in all the species. Thus the soldiers are free of the behavioural constraints so obvious in the workers, and again this is well seen in the mandibles, utilized only to fight according to astonishingly diverse adaptations (4) ; they may even regress in the nasute soldiers ! In some species, the soldiers are still more nutritionally dependent on the workers, by loss of their intestinal symbionts : they receive a liquid food, probably saliva (2,5). A strong reduction of the gut volume ensues, which lightens the soldier and increases its ergonomic efficiency. Such an evolution occured independently in about 20 genera, belonging to 2 families and 6 subfamilies. But in many cases, the gut reduction allowed an hypertrophy of defense gland reservoirs (frontal or salivary), another way to produce more successful soldiers.

The differences between the two castes as regards their diversification are good examples of the weight of constraints in evolution. As soon as the types of development and behaviour were acquired, they canalized along divergent pathways the evolution of soldiers and workers respectively.

1. Noirot, Ch. and Pasteels, J.M. 1987. Experientia, 43: 851-860.
2. Deligne, J. 1970. Thesis, Univ. Bruxelles.
3. Noirot, Ch. and Bordereau, C. 1990. In: Morphogenetic Hormones in Arthropods, A.P. Gupta, Ed., Rutgers Univ. Press, New Brunswick (in press).
4. Deligne, J., A. Quennedey and M.S. Blum. 1981. In: Social Insects, H.R. Hermann, Ed. Academic Press, New York, Vol. II, pp. 1-76.
5. Noirot, Ch. 1955. Ann. Sci. Nat. (Zool.), (11) 17: 399-595.

TERMITE WORKERS : A MODEL FOR THE STUDY OF SOCIAL EVOLUTION

Yves Roisin
Laboratoire de Biologie animale et cellulaire, C.P. 160, Université libre de Bruxelles, 50 av. F.D. Roosevelt, 1050 Bruxelles, Belgium

INTRODUCTION

Termites have long been neglected in studies of social behaviour evolution, for two major reasons : first, the popularity of genetic models based on hymenopteran haplodiploidy; second, the fact that all termites are fully eusocial and thus should not provide intermediate situations between solitariness and eusociality [1].

But the relevance of these reasons is now rapidly declining. Genetic models have not proved satisfactory to explain the fixation of a worker caste in insect populations [2,3]. On the contrary, recent behavioural studies have shown that an eusocial worker phenotype could ensue *de facto*, with little genetic change, whenever interactions among members of social groups favour the evolution of the behavioural traits of non-reproductive helpers [4,5]. Second, termite caste systems have recently undergone a thorough reappraisal. Although termite *soldiers*, present throughout the order, are typical eusocial altruists, an eusocial *worker* caste, absent in the lower termites, has arisen separately from soldiers and at least twice during the evolution of the Isoptera [6]. The purpose of this paper is to pinpoint aspects of termite caste systems that raise testable hypotheses as to their evolution, likely to improve our general understanding of the origins of sociality.

PECULIARITIES OF TERMITE SOCIALITY

Two routes to less- or non-reproductive helpers are currently recognized : in the subsocial route, societies originate from familial units comprising parents and their offspring; in the parasocial (communal) route, they result from associations of individuals, related or not, from the same generation [7-9]. Both routes are open to Hymenoptera and Vertebrates; by contrast, since termite workers and soldiers are specialized larvae [6,10], there is no doubt their sociality evolved via the subsocial route. In addition, the genetic structure of termite societies seems totally free from complications arising from exchanges of reproductives or genes between established colonies (e.g., through the readoption, common in hymenopterans, of females mated outside the nest). A likely basis for offspring care by parents in termites is the need for transmittal of intestinal symbionts, as in cryptocercid cockroaches [11]. In familial units, overlap of generations and cooperative brood care arise when young individuals postpone dispersal flight as well as sexual maturation within the nest, and help their parents raise additional offspring. At this stage, the fundamental characteristics of a non-reproductive helper phenotype are present.

HYPOTHESES AND PERSPECTIVES

Delayed dispersal seems favoured by the opportunity of taking advantage of a massive source of food, such as a piece of rotting wood; hence the prediction that pluripotential individuals in societies of primitive termites should be able to appreciate chances of future colony development, and remain as helpers under favourable conditions, but proceed toward the winged adult in unfavourable situations. Although suggested by some authors [12], this relationship still awaits confirmation.

Larvae delaying wing development still retain the alternative of helping their parents or reproducing as neotenics within the mother nest. Neotenics benefit from an increased relatedness with the brood, but their participation in reproduction is genetically neutral for the colony's founding pair. No relatedness argument could thus support the hypothesis of parental manipulation. However, breeding within the mother nest is exceptional in the parents' presence, excepted in species where colonies reproduce by budding. One explanation could be that an increase in colony population is much more efficient than an increase in individual size to enhance the young colony's chances of survival, and thus both components, direct and indirect, of its members' fitness. The young would therefore find an advantage in investing energy in other siblings rather than in their own growth and maturation. Empirical data do indeed indicate that incipient social insect colonies as a rule produce smaller than normal workers [8,13,14].

Although attractive, the two above-mentioned hypotheses are purely intuitive, qualitative, and thus of limited significance. Their further development should be supported by models asserting quantitatively the selective value of delaying dispersal and breeding, where individual and colony fitnesses would be considered simultaneously [e.g. 15]. Termites are especially suitable for such studies, due to the genetic structure of their societies and the various levels of advancement towards eusociality reached by their working stages.

ACKNOWLEDGEMENTS

Supported by a postdoctoral fellowship from the FNRS (Belgium) and a FRFC grant (No. 2.4513.90). Contribution No. 202 from King Léopold III Biological Station, Laing Island, Papua New Guinea.

REFERENCES

1. Starr, C.K. 1979. In: Social Insects, Vol. I, H.R. Hermann, Ed., Academic Press, New York, pp. 35-79.
2. Craig, R. 1979. Evolution, 33: 319-334.
3. Pamilo, P. 1984. Behavioral Ecology and Sociobiology, 15: 241-248.
4. Michener, C.D. 1985. Fortschritte der Zoologie, 31: 293-305.
5. West-Eberhard, M.J. 1987. In: Chemistry and Biology of Social Insects, J. Eder and H. Rembold, Eds, J. Peperny Verlag, München, pp. 369-372.
6. Noirot, C. and J.M. Pasteels. 1987. Experientia, 43: 851-860.
7. Michener, C.D. 1969. Annual Review of Entomology, 14: 299-342.
8. Wilson, E.O. 1971. The Insect Societies. Bellknap Press, Cambridge, Mass. p. 548.
9. Brown, J.L. 1987. Helping and Communal Breeding in Birds. Ecology and Evolution. Princeton University Press, Princeton, N.J., p. 354.
10. Noirot, C. 1989. Ethology, Ecology and Evolution, 1: 1-17.
11. Nalepa, C.A. 1984. Behavioral Ecology and Sociobiology, 14: 273-279.
12. Watson, J.A.L. and J.J. Sewell. 1981. Sociobiology, 6: 101-108.
13. Noirot, C. 1985. In: Caste Differentiation in Social Insects, J.A.L. Watson, B.M. Okot-Kotber and C. Noirot, Eds, Pergamon Press, Oxford, pp. 75-86.
14. Roisin, Y. 1988. Zoomorphology, 107: 339-347.
15. Roisin, Y. 1987. Experientia Supplementum, 54: 379-404.

POLYMORPHISM AND POLYETHISM IN *COPTOTERMES HEIMI* (WASMANN) (RHINOTERMITIDAE : ISOPTERA)

H.R. Pajni and C.B. Arora,
Department of Zoology, Panjab University, Chandigarh-160014.
(India)

Coptotermes heimi (Wasmann) shows the differentiation of two types of workers and neotenic reproductives. Two common larval stages are followed by the divergence of four lines. One of the lines forms the pre-soldier which undergoes another moult to form a soldier. The second line yields the minor workers. The third line forms the major workers which moult into the adult stage. The fourth line signals the beginning of the reproductive caste which moults twice to produce mature males and females. Some of the winged forms during the third, fourth or fifth stage of development develop as functional reproductives (neotenics). The soldiers and major workers are females but minor workers are males.

The reproduction is carried out by the neotenics and the primary reproductives have never been encountered. Both the major and minor workers and their differentiated larval stages take part in all the activities. Major workers have more participation in the outdoor work of foraging and construction of galleries. The minor workers concentrate more on the repair of the nest and the galleries. Whereas foraging is performed mostly by `the adult workers, gallery extension and gallery repair are accomplished by the penultimate larvae and the adults to an equal extent. The construction and repair of nest are attended to by more of penultimate larvae than the adult workers. Out of the minor workers with 14, 15 and 16 antennal segments, the first mentioned show relatively much less contribution as compared to the latter two categories.

Polymorphism in C. heimi projects the family Rhinotermitidae to have a variable type of development. It is slightly advanced than that in the genera Prorhinotermes and Reticulitermes which show longer undifferentiated larval period and retrogression in the reproductive line, but is relatively less developed than that of genus Schedorhinotermes (1). In fact, the genera Coptotermes and Schedorhinotermes and the family Hodotermitidae show the most advanced type of development in lower termites which approaches close to the pattern of polymorphism in the higher termites (1).

The polyethism in C. heimi marks the first study in the family Rhinotermitidae. The pattern is almost similar to what has been described in Nasutitermes dunensis (2) and Microcerotermes beesoni (3), the two carton nest-forming species of family Termitidae. The participation of the two types of workers and their larvae in different activities has also been noted in many ground dwelling Termitidae (1; 3).

References

1. Watson, J.A.L.; B.M. Okot-Kotber and CH. Noirot (1985). Current Themes in Tropical Science, Vol. 3. Caste Differentiation in Social Insects. Pergamon Press, Oxford : 405pp.

2. McMahan, E.A.; P.K. Sen-Sarma and S. Kumar (1983). Biometric, polyethism and sex-ratio studies of Nasutitermes dunensis Chatterjee and Thakur (Isoptera: Termitidae). Ann. Entomol., 1(1) : 15-25.

3. Kumar, S. (1986). Comparative Study of Polymorphism and Polyethism in the Indian Termites Odontotermes obesus and Microcerotermes beesoni (Isoptera:Termitidae). Ph.D. Thesis, Garhwal University, Srinagar, Garhwal : 168pp.

MONOMORPHISM AND POLYMORPHISM IN AFRICAN PONERINAE WITH A SPECIALIZED ALIMENTARY DIET : DOES THIS HAVE ANY BEARING ON THE TYPE OF PREY?

DEJEAN A.* and SUZZONI J.P.**
* Laboratoire d'Ethologie, Université Paris XIII, Villetaneuse, France.
** Laboratoire d'Entomologie, Université Paul-Sabatier, Toulouse, France.

Two species where the worker caste is monomorphic (*Leptogenys bubastes* and *Plectrotena gabonensis*) are compared to other species where the worker caste is polymorphic (*Megaponera fœtens* and *Centromyrmex bequaerti*).

Leptogenys bubastes is a predator of Isopods. Their mandibles are arched with a very lateral insertion which enables a wide opening. The large prey are seized by the edge of the carapace, turned over and stung on the ventral side. Workers hunt alone but recruitment by chemical trail is possible for the transport of a large prey. In *Plectrotena* the mandibles are slightly arched and set very laterally on the cephalic capsule, thus enabling a large opening and seizing Iulidæ, their main prey. Stinging is obligatory. We again find a recruitment through a chemical trail for the transport of large prey.

The *Megaponera fœtens* are predators of Macrotermitinæ. Majors forage alone. After discovering Termites, they return empty handed to the nest to recruit nestmates. There is formation of a column with the 3 subcastes. The minors, with short legs enter the Termite galleries with ease whereas the majors are more efficient in the prey transport. Societies of *Centromyrmex bequaerti* are found in the wall of *Macrotermes* or *Cubitermes* termitaries or in banks close to such termitaries. The workers go through narrow galleries in order to hunt.

It appears that in these Ponerinæ, there are two ways in specialization: to the hunting of prey supplied with a carapace corresponds modified mandibles and to the Termites (with soldiers) corresponds polymorphism. However, contradictory arguments may be put toward the genus *Leptogenys* which includes species specialized in Termites predation. Likewise, it is difficult to prove a correlation between one of the worker subcaste and the elimination of Termite soldiers. In *Megaponera* all the subcastes enter the galleries and in the narrow hunting galleries of *Centromyrmex bequaerti*, we find majors and minors in an indistinct order.

COLONY DEMOGRAPHY AND FITNESS IN *TRACHYMYRMEX SEPTENTRIONALIS*

S.N. Beshers and J.F.A. Traniello
Boston University, Boston, MA USA

INTRODUCTION

The theory of caste evolution put forward by Oster and Wilson predicts that the caste distribution function (CDF) should be directly related to colony fitness, as measured by alate production [1]. This prediction has been tested in two ant species with completely dimorphic workers, Pheidole dentata and Colobopsis impressus, with mixed results. In P. dentata the CDF did not predict fitness at all [2], while in C. impressus the proportion of soldiers was significantly related to alate production [3]. We have performed an additional test with the attine ant Trachymyrmex septentrionalis, a weakly polymorphic species.

METHODS

A comparative study was done with populations in Florida (FL) and Long Island (LI). Entire colonies were excavated and censused shortly before the mating flights, when mature alates were present. Workers were sorted into size classes by head width measurement, and into age classes by color.

RESULTS AND DISCUSSION

Demographic analysis of the two populations reveals a constellation of differences, which may be related to the length of the growing season: about eight months in FL in half that in LI. In both populations, mature colonies produce only alates in the spring, and rear workers after the mating flights. FL colonies attain larger size and produce significantly more alates than LI, both in absolute numbers and in relation to colony size, while LI alates are larger. LI workers are slightly more polymorphic: head width ranges from 0.7-1.3 mm in LI and 0.8-1.2 mm in FL workers, but the mean head widths in the two populations are not significantly different.

Caste profiles change with colony development: small colonies have workers in a narrow size range centered on a head width of .95 mm, and the full range of caste diversity appears as colonies mature.

Fitness was measured as both total alate biomass and alate biomass per worker. Surprisingly, in mature colonies (400 to over 1000 workers), alate production was not correlated with colony size. Neither measure of fitness was significantly related to the mean, minimum, or maximum head width, the head width range, or the frequency of any size class in the colony.

Comparison of the two populations supports the hypothesis that the CDF has been shaped in relation to ecology, since colonies at the far ends of the species' geographical range have consistently different demographies, with the LI population being more polymorphic. However, no relationship between the CDF and fitness was found. This could be because low variance in demography obscures the correlation, behavioral flexibility in workers dampens the effects of suboptimal demography, or high variance in alate production results from microhabitat differences between colonies. An interesting possibility is that colonies might be investing less than their maximum available energy in reproduction as a consequence of life history strategy decisions. Colonies of many species are known to follow characteristic schedules of growth, caste proliferation, and reproduction [1]. We suggest that caste differentiation could be usefully studied in relation to life history decisions such as the number of new workers to rear, the caste distribution of these workers, and the investment tradeoff between growth and reproduction.

REFERENCES

1. Oster, G.F. and E.O. Wilson. 1978. Caste and Ecology in the Social Insects, Princeton University Press, Princeton.

2. Calabi, P. and J.F.A. Traniello. 1989. Behavioral Ecology and Sociobiology 24:69-78.

3. Walker, J. and J. Stamps. 1986. Ecology 67(4):1052-1062.

FOOD AND THE EVOLUTION OF SOCIALITY IN BEES

Hayo H.W.Velthuis

Laboratory of Comparative Physiology, Univ. Utrecht, The Netherlands.

At first sight it appears as if bees are very uniform in their food requirements. All bees visit flowers to collect pollen, from which the protein is derived, and to collect nectar or oil, the energy source. However, there has been a very important speciation, leading to about 20.000 species of bees (Michener, 1972), and it is supposed that the competition for these food components has been an important factor leading to specialization for certain flower types (Baker & Hurd, 1968).

The protein source, the pollen, varies tremendously in its properties according to the floral origin. There are important differences in size, in the heaviness of the pollen wall, the presence or absence of air chambers etc., leading to variation in the ratio protein content/volume. In addition, there is variation in amino acid composition, in the mineral content etc. (Stanley & Linskens, 1974).

Before the protein can be utilised it has to be removed from the interior of the pollen grain. Not much is known about the mechanism of digestion. Probably, the germinating pores of the pollen grain allow enzymes of the bee to attack it. In the digestive process the pollen wall is not dissolved, and therefore the ratio protein content/volume could have great impact for the net gain of the digestion, and thus for the growth of the larva and the rate of oogenesis in the adult female. Especially in bee larvae, that, until their final instar, store the waste products in their intestine, the size of the resulting adult is expected to be correlated with that ratio.

Female bees invest most of their energy in brood care, that is nest construction and foraging. Probably, their reproduction is limited by the investment in brood care rather than by their capacity to produce eggs. As a consequence the number of offspring of a solitary female is generally in the order of 6-15. This low number allows the female of many species to produce relatively large eggs (Iwata & Sakagami, 1964) for which she has to digest the pollen. We could interpret this as another form of parental care, for the large egg results in a larger first instar larva, and the larger larva probably has a better start in digesting the bee bread in its cell.

In relatively simple forms of sociality the females in the nest are morphologically the same. However, they develop a dominance hierarchy that is characterized by one female producing the majority of the offspring. Often this means that, in comparison to the model solitary female, such a queen should have an increased rate of egg production. It is conceivable that this rate of egg production soon will become limited by her maximum capacity to digest pollen. A first adaptation to reach a still higher egg laying rate would be to increase the net rate of protein delivery by the intestine. One way in obtaining this is to partly change the diet, namely in consuming the eggs laid by the workers in her colony. The queen is, so to say, parasitizing on the digestive capacity of her nestmates. Such egg eating occurs in the Halictidae as well as in the stingless bees, and also among laying worker bees of the bumblebees and the honeybees. From my reasoning it is clear that a queen should not inhibit her nestmates to produce eggs, instead she should stimulate them. Absence of inhibition is not an indicator of incomplete dominance, as has been said in the literature so often.

A second adaptation to obtain a further increase in the rate of egg laying would be to reduce the size of the egg. This is very prominent in the honeybee. However, if a larger first instar larva would be an advantage, given the problems for the larva related to digesting pollen grains, then the development of smaller eggs in the course of social evolution would constitute a constraint to that development. It is interesting to note that this seems to be compensated for by the development of the hypopharyn- geal gland. In solitary bees, such as Xylocopa and Centris, the gland is relatively small, but in the social bees, the Meliponini, the Bombini and especially in the honeybees, Apis, it is very large (C. da Costa Cruz, 1960). In the honeybee the product constitutes the major part of the royal jelly, that is fed to the queen larvae, and of the worker jelly, fed to the worker and drone larvae. In the stingless bees, however, the amount of free protein and amino acids in the larval food is small (Hartfelder, 1986). This led us (Velthuis & Sommeijer, in press) to suppose that the original function of the gland, present in the solitary bees, is the production of an enzyme, used by the producer itself. In derived systems the enzyme could also be secreted into the brood cell and be an aid for the larva in digesting the bee bread. In the stingless bees the secretion could have mainly this enzymatic function, while in the honeybee apparently the proteinaceous character is prevalent.

In Apis this enabled the development of a third type of adaptation to achieve a still higher egg production by a single individual, the use of the hypopharyngeal gland secretion for the transfer of protein from worker to the queen. Apis appears to be the only taxon where the queen does no longer digest the pollen herself, but parasitizes fully on the workers. The shift to the hypopharyngeal gland secretion means a frequent trophallactic contact between queen and workers. This does not occur in any other bee group, and where it occurs it has probably a different, possibly a more original function. In stingless bees frequencies of trophallactic contacts between the queen and the workers are related to the phase of cell construction, thus indicating to the queen when egg laying is becoming possible (Sommeijer & de Bruijn, 1984).

In conclusion, the evolution of the bees not only resulted from a competition for the food available in flowers, leading to speciation, but also for the food inside the nest, leading to several forms of sociality.

References

Baker, H.G. & Hurd, P.D. 1968. Intrafloral ecology. Ann. Rev. Entomol. 13:385-414

Cruz, C. da Costa 1960. Contribuição ao estudo da evolução das abelhas. Thesis, Rio Claro.

Hartfelder, K. Trophogene Basis und endokrine Reaktion in der Kastenentwicklung bei Stachellosen Bienen. Thesis, Tübingen.

Iwata, K. & Sakagami, S.F. 1966. Gigantism and dwarfism in bee eggs in relation to the modes of life, with notes on the number of ovarioles. Japan. J. Ecol. 16:4-16.

Michener, C.D. 1972. The Social Behavior of the Bees. Harvard Univ. Press, Cambridge, Mass.

Sommeijer, M.J. & de Bruijn, L.L.M. 1984. Social behaviour of stingless bees: "bee-dances" by workers of the royal court and the rhythmicity of brood cell provisioning and oviposition behaviour. Behaviour 89:292-315.

Stanley, R.G. & Linskens, H.F. 1974. Pollen. Springer Verlag, Berlin, Heidelberg, New York.

Velthuis, H.H.W. & Sommeijer, M.J. 1990. Social and hormonal regulation of caste dimorphism in stingless bees. In: A.P.Gupta, Ed., Morphogenetic Hormones of Arthropods, Vol. I, part III. Rutgers Univ. Press, New Brunswick, NJ. in press.

Symposium 14

REPRODUCTIVE FITNESS AND EUSOCIAL ORGANISATION

Organizers : **J.J. BOOMSMA, The Netherlands**
G.W. ELMES, United Kingdom

BREEDING SYSTEMS AND KIN SELECTION IN SOCIAL HYMENOPTERA

Kenneth G. Ross
Department of Entomology
University of Georgia
Athens, GA 30602 U.S.A.

The theory of kin selection has had a profound impact on how we think about the evolution of insect societies and thus on the questions asked in empirical studies. Kin selection can be approached from one of two complementary theoretical perspectives. In the *inclusive fitness perspective*, Hamilton's Rule shows that an altruistic trait can spread via kin selection if the ratio of an actor's fitness losses to the beneficiaries' fitness gains exceeds the average relatedness (r_{xy}) between interactants [*1*]. In social insects such interactants will usually be nestmates. In the *population genetics perspective*, kin selection is divided into a within-group component, involving selection among individuals within societies, and a between-group component, involving competition among societies [*2*]. Altruistic traits selected against within societies can nonetheless be favored in the population if between-group selection is sufficiently strong. The relative strengths of the two levels of selection are influenced by the distribution of genetic variability within and between colonies, a measure of which is average nestmate relatedness (r_{xy}). Thus, from either theoretical perspective, r_{xy} is an important parameter for studies of kin selection in social insects.

Over the past 15 years the usefulness of measuring r_{xy} has sometimes been questioned. Two reasons are: 1) values of r_{xy} cannot be used to evaluate Hamilton's Rule in the absence of values for the fitness parameters, 2) natural history data suggest the frequent occurrence of multiple matrilines and/or patrilines in colonies, the presumed effect of which is to depress r_{xy} to 'unacceptably low' levels. Despite these objections relatedness estimation has enjoyed a recent renaissance due to both technical and conceptual advances. On the technical side, the rapid spread of molecular and population genetics into insect sociobiology has made robust estimates of relatedness readily obtainable, whereas estimation of fitness parameters in natural populations remains a difficult task. On the conceptual side, it has become clear that assessing the effects of multiple mating and polygyny on r_{xy} is more complicated than simply tallying up numbers of mates or queens.

The factors influencing r_{xy} in a colony are components of the 'breeding system' [*2*] and can be grouped into two general categories (Table 1). The first includes factors acting through the number of matrilines present and, if more than one, their relationships to one another and temporal stability. The second includes factors acting through the number and nature of the patrilines. The first factor in each category is the actual number of queens or matings, which if greater than one has often been assumed to lead to values of r_{xy} incompatible with a prominent role for kin selection. However, any variance in maternity or paternity apportionment, or relatedness between colony matriarchs and patriarchs, will lead to levels of r_{xy} in excess of those expected by considering only the number of matrilines or patrilines. This said, there is now increasing interest in teasing apart the components of the breeding system to determine the role of each in generating r_{xy} in particular model systems.

In the fire ant *Solenopsis invicta* we have information on all components of the breeding system. Fire ant queens invariably mate once so the second group of factors is not of concern. In monogyne *S. invicta* the first group of factors is also irrelevant since there is but a single queen and r_{xy} takes the highest possible value for outbred female Hymenoptera, 0.750 [*3*]. In polygyne *S. invicta*, however, the first group of components of the breeding system is of great interest. Average relatedness for nestmate workers (r_w) can be estimated as: $r_w = (1/H_n)(r_s) + ((H_n-1)/H_n)(r_q/4)$, where r_s is relatedness of sisters, r_q is relatedness of nestmate queens, and H_n is the effective harmonic mean number of queens [4,2]. Genetic studies have shown that r_w=0.059, r_s=0.750, and r_q=0.011 [3]; thus H_n=12.7. Because this estimate of 12.7 queens assumes that all reproduce equally, it is a minimum estimate. If there is any significant point or temporal variance in maternity apportionment, as has been shown in the laboratory [5], the true harmonic mean number of queens in the population will be considerably greater than 12.7. Note also that the arithetic mean will be higher yet than the true harmonic mean – thus it is of interest that the arithmetic mean number of mated queens per nest from this population has been estimated at ca. 35 [6]. From the above it is seen that, in polygyne *S. invicta*, neither the effects of variance in maternity apportionment nor consanguinity among mated queens compensates for the high number of matrilines present in a nest to rescue r_w from very low values.

Table 1. Components of the 'breeding system' influencing relatedness in social groups.

1. *Factors acting through the number and nature of matrilines.*
   a. Actual number of reproducing females.
   b. Apportionment of maternity among females (point variance).
   c. Temporal changes in apportionment of maternity (temporal variance).
   d. Relatedness of reproducing females in group.
2. *Factors acting through the number and nature of patrilines.*
   a. Actual number of matings by each reproducing female.
   b. Apportionment of paternity for offspring of each female (point variance).
   c. Temporal changes in apportionment of paternity (temporal variance).
   d. Relatedness of mates (e.g. inbreeding, fraternal polyandry).

In the yellowjacket wasp *Vespula squamosa*, monogyny is the rule, so the first group of breeding system components affecting r_{xy} can be ignored. Relatedness of female nestmates was estimated from genetic markers as r_{xy}=0.403 [7], from which the effective harmonic mean number of matings per queen (H_m) can be estimated from: $r_{xy} = 0.25 + 1/2H_m$ [8,2]. The value of H_m estimated, 3.3, is a minimum because it assumes equal contribution to paternity by a queen's mates. Overall variance in paternity apportionment through colony development was determined with genetic markers for several natural colonies, leading to the conclusion that the actual harmonic mean number of matings is 5-6 [incorrectly reported in 7]. Again, the arithmetic mean number of matings would be even higher. Clearly, variance in paternity apportionment among the multiple mates of a queen boosts relatedness above what one would assume it to be on the basis of mating frequency alone.

In conclusion, the tools are available to probe in detail the importance of breeding system components in determining average nestmate relatedness. Several considerations indicate that r_{xy} will often be higher than might be predicted from superficial knowledge of colony queen number and mating frequency. First, the appropriate summary measure for inferring population-wide effects of polygyny and polyandry on average r_{xy} is the harmonic mean number of queens and male mates [2], which returns a higher value of r_{xy} than the arithmetic mean. Further, any departures from equal contributions to maternity or paternity will lead to elevated r_{xy}, as will consanguinity among reproductives. Continuing study of breeding system components and their effects on r_{xy} should tell us much about how kin selection influences social organization.

1. Hamilton, W. D. 1964. Journal of Theoretical Biology, 7: 1-16.
2. Wade, M. J. 1985. Journal of Theoretical Biology, 112: 109-112.
3. Ross, K. G. and D. J. C. Fletcher. 1985. Behavioral Ecology and Sociobiology, 17: 349-356.
4. Queller, D. C., J. E. Strassmann, and C. R. Hughes. 1988. Science, 242: 1155-1157.
5. Ross, K. G. 1988. Behavioral Ecology and Sociobiology, 23: 341-355.
6. Vargo, E. L. and D. J. C. Fletcher. 1987. Physiological Entomology, 12: 109-116.
7. Ross, K. G. 1986. Nature, 323: 798-800.
8. Page, R. E. 1986. Annual Review of Entomology, 31: 297-320.

HOW TO CAPITALIZE ON RELATIVE RELATEDNESS ASYMMETRY ?

J. J. Boomsma

Department of Population Biology and Evolution, University of Utrecht, The Netherlands
Present address: Institute of Ecology and Genetics, University of Aarhus, 8000 Aarhus - C, Denmark

The question, "How adaptive is the sex ratio for workers of eusocial Hymenoptera?", has been the subject of debate ever since the appearance of Trivers and Hare's seminal paper in Science in 1976 [1]. Their claim that the overall female bias in the sex-investments of ants is likely to be related to a continuing process of kin-selection on workers, was challenged by Alexander and Sherman [2], both for practical reasons (such as biases in the estimation of investment), and on theoretical grounds (arguing that the alternative hypothesis of Local Mate Competition (LMC) was at least as likely). Since then, only the kin-selection or relatedness asymmetry (RA) hypothesis has gained some further support [3, 4, 5, 6], but it also became clear that most of the variance in the colony-level sex ratio remained unexplained.

Recent theoretical studies [7, 8] have emphasized the possible role of relative LMC and relative RA for the adaptive adjustment of the colony sex ratio. The central notion in these papers is that apart from mean (population-level) degrees of LMC and RA (which may both be close to their minimal values), there are often also variable colony-level degrees of LMC and RA, which can be expressed as values relative to those means. Frank's [7] argument is that males in large colonies may suffer more from competition with related males than those in small colonies and that we should therefore expect a positive correlation between the degree of female bias and colony size. Boomsma and Grafen [8] argued, on the other hand, that in populations where RA is variable due to multiple mating, polygyny or worker-reproduction, colonies with relatively high RA (e.g. full-sib-worker colonies) should specialize on producing females, whereas colonies with relatively low RA (e.g. half-sib-worker colonies) should make mostly males. Both effects (alone or in combination) are thus expected to induce a considerable amount of extra variation in the colony-level sex ratio. The two predicted patterns of variation are to some extent confirmed by the available comparative data on ant sex ratios [8, 9]. So, the fact that these covariances of the sex ratio with colony-characteristics were not included in the original RA- and LMC-hypotheses might explain why further evidence from more detailed population studies has been so limited.

The relative RA hypothesis is somewhat vulnerable because of its explicit assumption that workers should somehow be able to assess their colony-level RA [but see 10]. The good thing is, however, that this extended model enables us to test the operation of kin-selection by comparing numerical sex ratios of different RA-classes of colony, without having to worry primarily about a proper estimate of investment in terms of energy [11]. Also, polygynous ants, which were previously considered as much less suitable for testing the population-level RA-hypothesis, have now become the more ideal test-cases, because many such species are likely to have a high among-colony variance in RA.

Two further consequences of the relative RA hypothesis turn out to be interesting. The first concerns the worker-queen conflict over the sex ratio. If workers should appear to be able to capitalize on their relative (colony-level) RA by producing a more extreme sex ratio than expected on the basis of the mean RA in the population or even the mean RA in their own class of colony, the worker-queen conflict has also become a variable. Assuming that the overall population sex ratio is female biased, because of at least partial worker-control in all colonies, it follows that any mother queen would gain fitness if her workers raised only males. As in the relatively low-RA colonies also the workers should favor males, this implies that there will be no worker-queen conflict left in these colonies. In the relatively high-RA colonies, however, the worker-queen conflict can reach an absolute maximum when the (full-sib) workers are selected to raise only females, whereas their mother would prefer all males. If there is a certain cost to having a worker-queen conflict (e.g. oophagy by workers to remove excess haploid eggs), this may actually give a selective advantage to lower-RA (male producing) colonies, which in turn (but under quite restrictive conditions) might have helped to promote evolutionary developments towards multiple mating (Boomsma, in prep).

The second consequence refers to primitively eusocial species such as halictid bees. If part of the summer-generation colonies are eusocial (with helping workers) and another part are either parasocial (with replacement queens) or solitary, we also end up with a case of variable RA, i.e. relatednesses are symmetrical in the solitary and parasocial colonies, but asymmetrical in the eusocial colonies. Even if worker-control in the eusocial colonies is very limited and ineffective, we would still have some selection for eusocial colonies to specialize (relatively) on females and for parasocial colonies to make (relatively) more males. Recent studies [12, 13] have shown that such adaptive colony sex ratios do indeed occur in halictid bees.

REFERENCES

1. Trivers, R.L. and H. Hare. 1976. Science, 191: 249-263.
2. Alexander, R.D. and P.W. Sherman. 1977. Science, 196: 494-500.
3. Ward, P.S. 1983. Behavioral Ecology and Sociobiology, 12: 301-307.
4. Herbers, J.M. 1984. Evolution, 38: 631-643.
5. Have, T.M. van der, J.J. Boomsma and S.B.J. Menken. 1988. Evolution, 42: 160-172.
6. Bourke, A.F.G., T.M. van der Have and N.R. Franks. 1988. Behavioral Ecology and Sociobiology, 23: 233-245.
7. Frank, S.A. 1987. Behavioral Ecology and Sociobiology, 20: 195-202.
8. Boomsma, J.J. and A. Grafen. 1990. Evolution, in press.
9. Nonacs, P. 1986. Quarterly Review of Biology, 61: 1-21.
10. Ratnieks, F.L.W. 1990. Evolution, in press.
11. Boomsma, J.J. 1989. American Naturalist, 133: 517-532.
12. Yanega, D. 1989. Behavioral Ecology and Sociobiology, 24: 97-107.
13. Mueller, U.G. 1990. Proceedings 11th International Congress of IUSSI, Bangalore, India, this volume.

MALE PARENTAGE AND SEXUAL DECEPTION IN SOCIAL INSECTS

Peter Nonacs
Museum of Comparative Zoology
Harvard University
Cambridge, MA 02138 USA

INTRODUCTION

One of the defining characters of eusociality among Hymenoptera is reproductive division of labor. The production of female offspring is commonly limited to one or a few mated individuals. However, because of haplodiploidy, any unmated female (e.g., a worker) with ovaries can potentially lay male eggs. Indeed, Bourke (1) has shown worker-laying to be quite common, although occurring mostly in situations where queens are absent or infertile. The latter point appears paradoxical because workers are more closely related to their sons than their brothers. Therefore, kin selection theory would predict that they, as caretakers of the brood, should replace queens' male eggs with their own in order to maximize their inclusive fitness. Ratnieks (2) proposed an evolutionary scenario which could prevent such worker behavior. If queens are multiply mated or colonies are polygynous, then there is a the following hierarchy of worker preferences in males: their own > full sisters' > mother's > all others. If the "all other" category is large relative to the first two, then a strategy of worker-policing is favored. Workers can maximize inclusive fitness by keeping everyone except their mother from laying male eggs. However, this model cannot work in colonies with one, singly-mated queen because there is no "all other" category and all worker eggs are preferable to the queen's. However, even in these species queens appear to often monopolize male production. I propose they do so by a process of sexual deception (3).

MODEL AND DISCUSSION

Workers could have little trouble in distinguishing worker from queen eggs because they come from physically distinct individuals. However, queens produce both male and female eggs. It would be selectively disadvantageous for workers to randomly destroy and replace queen eggs, because the gain in relatedness of replacing brothers would be more than offset by the loss of closely related sisters. Therefore workers would need to be able to discriminate between brothers and sisters at some point in early larval development.

To calculate the cutoff point, assume that X is the total amount of energy invested in male production by a given colony and n is the cost of a single male. Brothers and nephews are assumed equally costly and not to vary in qualitative ways. Therefore, the inclusive fitness for raising brothers is $X/n \cdot G$ and for worker-produced males (mostly nephews) it approximates $(X - x)/n \cdot G$, where x equals what must be put into the queen's sons before they can be recognized as male and therefore is not available for investment in worker-produced males and G is the appropriate degree of relatedness. The inclusive fitnesses of the two options are equal at $x = X/3$. If discrimination occurs before this point in development, worker males should be raised: after this point - queen males. Moreover, if workers make mistakes (i.e., identifying females as males and destroying them), then the cutoff point occurs even earlier in development. With a 10% error rate, widespread male replacement is not favored past 1/5 of the total development cost.

There is basically only anecdotal evidence as to when workers begin to be able to discriminate brothers from sisters. Recently, this has been explicitly tested in <u>Camponotus</u> <u>floridanus</u>. Workers were presented pairs of larvae or pupae (one either sex) to retrieve back to the colony. They showed no significant sex bias in retrieval at any stage of larval development, but were significantly faster in retrieving females at the pupal stage (Nonacs & Carlin, unpub. data). This may indicate that <u>C. floridanus</u> workers cannot tell the sex of their brood until late in development. Of course, it is possible that workers could tell, but did not care with larvae. This cannot be ruled out until the treatment of larvae is observed in stressed and normal colonies.

REFERENCES

1. Bourke, A.G.F. 1988. Quarterly Review of Biology, 63: 291-311.
2. Ratnieks, F.L.W. 1988. American Naturalist, 132: 217-236.
3. Nonacs, P.N. 1990. In: Evolution of Sex Ratio in Insects and Mites, D.L. Wrensch and D.A. Krainacker, Eds., Chapman Hall, In press.

11th International Congress — IUSSI 1990, India

SEX-RATIOS AND POLYGYNY IN *MYRMICA* ANTS

G.W. ELMES

NERC, Institute of Terrestrial Ecology,
Furzebrook Research Station
Wareham, Dorset. BH20 5AS UK

1. Introduction

Sex-ratio analysis is useful for investigating the evolutionary strategy of monogynous Hymenoptera. But, is it any use for polygynous species such as *Myrmica* ? Trivers & Hare showed that polygynous species (where queens are related) should have male biased sex-ratios and, using data for *Myrmica*, they demonstrated that species having more queens produced more males. *Myrmica* is a popular genus for investigating polygyny because it is common, has tractable colony sizes and, although its species are ecologically similar, it has a wide range of queen-numbers. However, theoreticians who use the genus to test models of polygyny, often over-simplify assumptions for the relatedness of queens, their longevity and reproductive strategy and often wrongly attempt to discriminate between monogynous and polygynous species.

The team at Furzebrook Research Station has made many laboratory studies of the social physiology of *Myrmica* and this should be accomodated when interpreting any field data, such as sex-ratios. In brief summary the main results are: 1. All species of *Myrmica* are polygynous and recruit new, usually unrelated queens during autumn and the following spring. Queens are short-lived, averaging 2 seasons. 2. All workers can lay eggs although some species lay more than others. Young, spring-workers, reared from over-wintered brood, lay most eggs; usually they are larger than summer-workers. 3. Most males are produced by workers, the larvae over-winter and generally they are treated like worker-larvae. 4. Gynes mostly develop from large larvae that undergo a proper diapause during winter. They can be switched to workers during spring, becoming the particularly large workers that contribute most to the following year's males. Suppressed gynes can in some ways be thought of as vicarious male production. 5. The presence of queens causes workers to lay fewer male eggs, favour small larvae and discriminate against large gyne-biased larvae (queen-effect).

2. Predictions for sex-ratios

Polygyny by unrelated queens does not alter the overall symmetry between the value of males and gynes, therefore, if queens control the sex-ratio, they should prefer a 1:1 ratio (Trivers & Hare).n exception is during their first spring in an already established colony, when they have no interest in the sex-ratio but should prefer workers to gynes in order to maximise their own future production.

If laying-workers control the sexuals, they also prefer a 1:1 ratio if the brood are sons/nephews and sisters. This applies during the initial regnum of a foreign queen but in the next year the female progeny are from the new queen while the males belong to the old workers. Therefore the majority of workers should prefer a very male-biased sex ratio. This bias should reduce in the 3rd year and revert to 1:1 by the 4th year, by which time the queen is senescent.

3. Data for *Myrmica sulcinodis*

I measured queen (Q) and worker (W) numbers and the spring production (P) of *M.sulcinodis* for 11 years at Stoborough and Winfrith heaths. Q cycled with a 4-5 year periodicity at both sites. The cycle might have been extrogenous, ie. weather prevented recruitment of new queens or caused excessive mortality of old queens, or it might have been endogenous, in which case the natural cycle of most of the colonies must have been synchronized by one or more extrinsic events. Fortunately, regardless of its cause, the cycle for Q enabled the processes that occurred in individual colonies to be estimated by the destructive sampling of a subset of the population. Overall, male production (M) was a constant proportion of P. Gyne production (G) was very variable but never exceeded 50% of P. This result might derive from the 1:1 preference of both males and queens, given that the worker production can in some senses be considered as delayed male production.

For detailed analyses only the 75 colonies from Stoborough that produced >10 sexuals were used. G/P decreased steadily over the 11 years but, despite this trend, G/P clearly cycled and was negatively correlated with the Q cycle. This result was expected from the empirical understanding of queen-effect. M/P showed no trend but also cycled, lagging a year behind that for G/P. It showed the sudden rise in M followed by a three year decline that is expected when workers were controling the investment in males. When combined as sex-ratio (M/G+M) a third cycle was produced that lagged behind that for Q and had an overall trend for increasing male investment. Similar analyses at Winfrith showed very male-biased sex-ratios; gynes were too few to detect any tendency for a cycle in G/P although there was an indication that M/P cycled in relation to Q.

4. Conclusions

Provided that the cycle in queen numbers, based upon a sample of colonies, mirrors the events that would take place in an individual colony over the same period, then the sex-ratio follows a pattern expected if the queens mostly control gyne production and the workers male production. A conflict of interest between the queens and workers, after new unrelated queens are recruited, is prolonged over several seasons when the preferred sex-ratio ranges from male-biased to 1:1.

In addition the sex-ratio varies independently of the social structure of the colony. At Stoborough the investment in M and new workers both increased over 11 years while W remained more or less static. Exact interpretation of this is difficult without data for summer-worker production, which was not measured. However, I believe that changes caused by a regenerating heathland habitat, must have reduced summer-worker production and caused more larvae to overwinter. This put more pressure on resources in spring, reducing G and resulting in increased male-biased ratios. The very strongly male-biased population on the more barren and sparsely vegetated Winfrith Heath, probably always has had fewer summer resources compared to Stoborough.

POLYGYNY AND REPRODUCTIVE ALLOCATION IN ANTS

Pamilo, P., Seppä, P. and Sundström, L.
Department of Genetics and Department of Zoology
University of Helsinki, Finland

The relationships among coexisting queens in a secondarily polygynous colony can be described as behavioural (dominance) or genetic (relatedness); the inclusive fitness of a queen depends on the proportion of offspring produced by each queen and on the relatedness among the queens. Even though relatedness among queens is such an important variable, few studies have actually estimated it in nature. A general impression from these studies is that the queens are as related to each others than are the worker nest mates. This indicates that nests recruit own daughters as new reproductives. When no reproductive dominance exists, the expected relatedness r depends on the queen number n as $r = 3/(3n + 1)$.

Dominance hierarchies among queens result in different representation of their genes in the offspring generation. In extreme form, this leads to functional monogyny. An alternative is that the queens contribute unequally in sexual brood and worker brood (6). We have collected data from Formica sanguinea and Myrmica ruginodis to test this possibility. The same data sets are used to test the hypothesis that the observed bimodal distributions of colony sex ratio (4) result from worker control. Namely, workers should bias colony sex ratios on the basis of relatedness asymmetry (1). The extent of this asymmetry depends on the genetic heterogeneity of the colony, e.g. on the level of polygyny.

When colonies are polygynous, population sex ratios depend on the relative success of the dispersing queens and the queens recruited back to the natal colony. If all daughters disperse, the inclusive fitness of the old colony members is proportional to the sum $g_f v_f x/X + g_m v_m y/Y$, where g_f and g_m are the relatednesses of female and male offspring to the individual concerned, the v's are sex-specific reproductive values, x and y are the proportional investments in females and males by a colony, and X and Y are the respective proportions in the whole population (3). The prediction is that, under worker control, the optimal sex allocation approaches the 1:1 ratio with increasing polygyny, unless the queens are unrelated. If polygynous colonies recruit daughters back as new reproductives, part of the worker force represents an investment supporting the females. The sex ratio, calculated from sexual individuals only, should then be $Y = 1/(1 + kq)$ where q is the contribution in future generations by the dispersing daughters (relative to all daughters) and the factor k varies from 1 to 3 depending on who controls sex allocation (queens or workers) and on the relatedness of the coexisting queens (and assuming monandry). The published sex ratio data from 22 polygynous ants, when fitted to the above model, indicate that q is likely to be close to 0.5, i.e. there will be equal numbers of grandchildren through dispersing and non-dispersing daughters.

In many situations the threshold for recruiting new queens is lower for workers than for old queens (2,5). However, polygyny (as well as polyandry) can reduce the queen-worker conflicts over male production and sex allocation and might be regarded as queens' way of manipulating the workers. This is true if the colony sex ratio is simultaneously biased to take into account the increased genetic heterogeneity. Although the queen-worker conflicts may play a role in the evolution of polygyny, the main determinants are likely to be ecological (5), particularly when we note the large intraspecific variation of the queen number in some species (7).

1. Boomsma, J.J. and A. Grafen. 1990. Evolution (in press).
2. Nonacs, P. 1988. Evolution, 42:566-580.
3. Pamilo, P. 1990. The American Naturalist (in press).
4. Pamilo, P. and R. Rosengren. 1983. Oikos, 40:24-35.
5. Rosengren, R. and P. Pamilo. 1983. Acta Entomologica Fennica, 42:65-77.
6. Ross, K.G. 1988. Behavioral Ecology and Sociobiology, 23:341-355.
7. Sundström, L. 1989. Actes des Colloques Insectes Sociaux, 5:93-100

REPRODUCTIVE ALLOCATION IN PONERINE ANTS WITH OR WITHOUT QUEENS

Christian Peeters

School of Biological Science, University of New South Wales, Sydney, Australia

The morphological dimorphism between ant queens and workers represents an adaptation for the performance of reproductive and helper roles respectively. In general, queens have enlarged wing muscles and fat reserves which can be broken down to feed the first generation of workers, thus increasing queen success during independent colony foundation. However in many species of ants, colony division occurs instead of foundation by individual queens. This is an exceptional strategy of reproductive allocation since its major component is the entourage of workers which leave the parental colony [1].

Colony division (either fission or budding, [2]) is particularly widespread in the Ponerinae, where it is usually associated with two significant evolutionary modifications: (a) permanently-wingless ("ergatoid") queens - these have a reduced thorax and occur in various species belonging to 15 genera [3]; (b) elimination of the queen caste (winged or ergatoid) and reproduction by gamergates (mated workers) [4]. Selection for these alternative strategies may be related to the limitations of a simple social system in harsh environments, e.g. dry habitats with a seasonal effect.

Low success rate during independent colony foundation

The frequent occurrence of colony division in the Ponerinae appears to be an evolutionary response to the problems associated with independent foundation by dealate queens. In this phylogenetically-primitive subfamily, queens are restricted to partially-claustral foundation. The newly-mated queens are unable to rear the first worker brood without hunting outside the nest and retrieving insect prey, because there is no trophallaxis among nestmates (including larvae), and there is only a small difference in size between queens and workers (thus wing muscles are small). This can lead to high mortality in particular environments, and the successful foundation of new colonies may then depend on the annual production of a large number of female sexuals. However, an increased investment in new queens may not be possible when there is an interaction between small colony size, solitary foraging, and occurrence in habitats where annual unpredictability in the availability of food coincides with the period of production of new sexuals [5].

<u>Colony division and the occurrence of ergatoid queens or gamergates</u>

Selection may favour processes of colony division in species characterized by both reproductive investment which is limited, and a low success rate during independent foundation. This then eliminates the need for the costly production of winged queens. One evolutionary trend was for queens to become wingless (aerial dispersion by queens does not occur with fission), with an associated progressive reduction in the size of their thorax. In other species, the replacement of queens by gamergates was possible because workers in most ponerine species have a spermatheca. An understanding of the transition from queens to gamergates awaits more details of the process of larval differentiation.

<u>Comparing patterns of reproductive investment</u>

Data on reproductive allocation are available for very few ponerine species, and seldom exist in the form of dry weight ratios. It is necessary to investigate how patterns of reproductive investment vary according to the types of reproductive individuals. In species with winged queens the investment ratios are female-biassed (e.g. *Rhytidoponera confusa* [6]), but the effects of small colony size and limited caste dimorphism (both characteristic of many of the Ponerinae) on investment ratios remain to be determined. It is not known how reproductive allocation (especially the number of males produced) differs in species with ergatoid queens or gamergates. Although both exhibit processes of colony division, the size of their colonies can differ dramatically. Queenless species have between 10 and a few hundred workers [5], while species with ergatoid queens can have dramatically larger colonies [3]. The available data indicate that very few new ergatoids are produced annually, which may be the antecedent of the pattern found in army ants [7]. This strategy is possible because ergatoids have a low mortality rate during colony division (they are protected by workers). Thus selection on the parental colony may have led to a reduction in the number of new ergatoid queens produced, since only two or three can be successful during colony fission [2].

REFERENCES

1. Macevicz, S., 1979. American Naturalist, 113: 363-371.
2. Franks, N.R. and Hölldobler, B., 1987. Biol. J. Linnean Soc., 30: 229-243.
3. Peeters, C., 1990. Insectes Sociaux (in press).
4. Peeters, C., 1987. In: Chemistry and biology of social insects, J. Eder and H. Rembold (eds), Verlag J. Peperny, Munich, pp. 253-254.
5. Peeters, C. and Crewe, R.M., submitted
6. Ward, P.S., 1983. Behav. Ecol. Sociobiol., 12: 285-299.
7. Schneirla, T.C., 1956. Insectes Sociaux, 3: 49-69.

SEXUAL PRODUCTION IN *PARATRECHINA FLAVIPES* (SMITH) (HYMENOPTERA : FORMICIDAE)

K. Ichinose

c/o Prof. C.R. Baltazar, Dep. Entomol., UPLB, College, Laguna 4031, Philippines

Collection Of Nests

Since *P. flavipes* completes its sexual production by September and makes nuptial flights in next late June to early July in the Tomakomai Experimental Forest, southern Hokkaido (2), I collected nests of the ant species there, totally 495, during September to early June over three years, 1983 to 1985. Of the collected nests, sexual(s) were produced in 95 nests, of which 19 were monogynously queenright and the rest queenless. I used the 95 nests for the present study. Some of the nests formed one colony togehter with each other (polydomy), although any queenright colony had one or no queenright nest. The numbers of queenright and queenless colonies were 19 and 52, respectively.

I examined dry weight of 20 individuals of each sex and obtained 1.682 ± 0.164 mg for females and 0.092 ± 0.019 mg for males. The values give the relative weight of a male to a female, 18.3, which I adopt for calculating sex ratios, $SR = M/(18.3F + M)$, where SR is the sex ratio and F and M are the individual numbers of females and males, respectively.

Sex Ratio

The mean sex ratios of queenright and queenless nests were 0.600 and 0.563, respectively, while those of colonies were 0.562 and 0.532. The mean numbers of females and males produced in queenright and queenless nests were 2.00 and 29.8, respectively, and in colonies

4.43 and 25.7. The mean numbers of individuals produced in a nest or colony should be employed instead of the mean ratios to avoid great contribution of nests or colonies which produced not more than a few individuals of either sex. The mean numbers of individuals give the ratios of 0.448 and 0.447 to queenright nest and colony, respectively. Neither of the ratios significantly differs from the 0.5 ratio, which is expectedly controled by queens ($\underline{X}^2$-test, $\underline{P} > 0.05$). On the other hand, the sex ratios of queenless nest and colony were 0.240 and 0.223, respectively, insignificantly different from 0.25 ($\underline{X}^2$-test, $\underline{P} > 0.05$) expected by the worker control.

The above results support the prediction by Trivers and Hare (3); queens prefer the sex ratio of 0.5, while workers 0.25. Boomsma (1) claims that sex ratios would be overestimated if the ratios were obtained in dry weight. He suggests that we should make some adjustments for the artifact to obtain more realistic sex ratios. However, the difference in the sex ratio between queenright and queenless conditions was so evident in the present study that I could conclude that there is queen-dominant conflict over the sex-ratio control between the queens and workers in _P. flavipes_ colonies. Because of the strong dominance of the queen, workers must leave the queenright mother nest and make independent queenless colonies to attain their preferable sex ratio.

REFERENCES

1. Boomsma, J.J. 1989. The American Naturalist, 133(4): 517-532.
2. Ichinose, K. 1987. Kontyu, 55(1): 9-20.
3. Trivers, R.L. and H. Hare. 1976. Science, 191(4224): 249-263.

A PATH ANALYSIS APPROACH TO EMPIRICAL STUDY OF ALLOCATION RATIOS IN SOCIAL INSECTS

JOAN M. HERBERS
Department of Zoology
University of Vermont
Burlington VT 05405

Theoretical explorations of allocation ratios in eusocial hymenopterans have implicated many factors that must be evaluated by the empiricist. Unfortunately, many such factors are either not directly measurable (e.g. local resource competition) or may be confounded (e.g. polydomy and polygyny). The empiricist is left with the very difficult task of partitioning the characteristically large variance in allocation ratios among many putative causal factors.

The methods of path analysis, developed by Sewall Wright (1921, 1934), provide a singularly powerful analytical tool within which to frame hypotheses about allocation ratios, and by which to test those hypotheses with data (Herbers 1990). The path model (Fig. 1) describes the effects of colony demographics (queen number [Q] and worker number [W]) on total investment in new queens [F] and males [M]. Together, investment in the sexes sums to total reproductive investment [I]. Finally, the derived parameter of allocation ratio [R], or proportional investment in males, can be examined with respect to the direct and indirect effects of all other variables in the network.

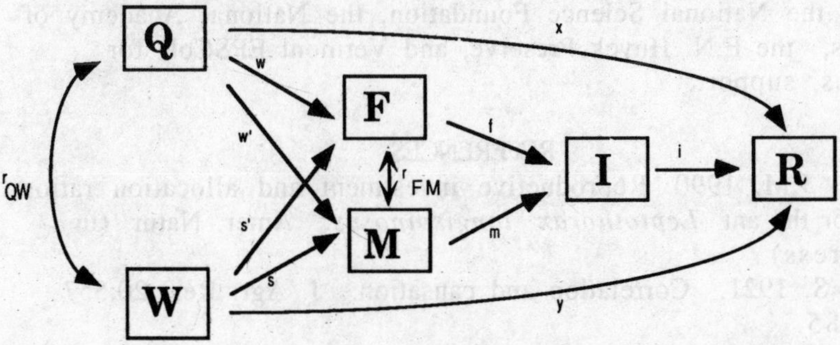

Figure 1. A path model for studying allocation ratios in social hymenopterans

The path model (Fig. 1) separates out the direct effect of one variable on another from indirect effects operating through correlation with other causal variables. Thus, queen number may have a direct effect on investment in males (q') as well as indirect effects operating through its correlation with worker number ($w'*r_{QW}$). Furthermore, the entire path diagram can be decomposed to test particular hypotheses of causality. For example, the magnitude and direction of the effect of variable queen number on investment in males or females can be compared to the corresponding effects of variable worker number to examine putative queen-worker conflict. Similarly, the consequences of varying male investment on total reproductive investment serves as a measure of the strength of local mate competition.

I applied the path model to sex allocation data from the tiny forest ant *Leptothorax longispinosus*. Two populations of this species differ consistently in important parameters of their reproductive biology: in New York, both males and females are heavier than their counterparts in Vermont; furthermore, the New York population typically has an allocation ratio more female-biased. In both sites, considerable annual variability for population allocation ratio occurs, requiring separate analyses to be run for each year of data collection. Results show that I can exclude differential rearing costs, local resource competition, and local mate competition as important factors. Most interestingly, there was consistent evidence of conflict between queens and workers over allocation ratios, which was principally mediated by conflict over investment in new queens. Even so, a large portion of the variance in investment parameters remained unexplained; thus unmeasured ecological parameters, probably resource acquisition ability, must be estimated for future work.

ACKNOWLEDGMENTS

I thank the National Science Foundation, the National Academy of Sciences, the E.N. Huyck Preserve, and Vermont EPSCoR for generous support.

REFERENCES

Herbers, J.M. 1990. Reproductive investment and allocation ratios for the ant *Leptothorax longispinosus*. Amer. Natur. (in press)

Wright, S. 1921. Correlation and causation. J. Agr. Res. 20:557-585.

Wright, S. 1934. The method of path coefficients. Ann. Math. Stat. 5:161-215.

FACULTATIVE SEX-RATIO ADJUSTMENT IN *AUGOCHLORELLA STRIATA* (HALICTIDAE: HYMENOPTERA) : A TEST OF KIN-SELECTION AND SEX-RATIO THEORY

Ulrich G. Mueller
Section of Neurobiology and Behavior
Cornell University, USA

Hamilton (1964, 1972) hypothesized that asymmetries in relatedness among siblings of eusocial Hymenoptera should selectively favor workers biasing the ratio of investment toward females, thus creating conflict over the colony's sex-ratio between workers and their mother. In monogynous colonies, reproductive females are symmetrically related to their male and female offspring (both r=1/2), while workers are asymmetrically related to their brothers (r=1/4) and sisters (r=1/4+1/2n; n=number of matings by the mother), but symmetrically related to their nieces and nephews (both r=1/8+1/4n). Previous studies attempting to test Hamilton's hypothesis verified the predicted female bias, but failed to exclude alternative hypotheses, such as local mate competition, which could also explain sex-ratio biases.

In an attempt to test Hamilton's hypothesis while controlling for local mate competition, I studied a population of the primitively eusocial bee *Augochlorella striata* (Halictidae), testing the prediction that eusocial colonies (mother-daughter associations) should produce a more female biased ratio of investment compared to parasocial colonies (sister-sister associations resulting from the loss of the mother). Only workers in eusocial colonies can capitalize on the relatedness asymmetry by biasing the sex-ratio toward females. No such asymmetry exists in parasocial colonies, regardless of whether a female herself reproduces, or one of her sisters.

A. striata is a ground-nesting sweat bee with colony size of 2-8 females which cooperatively construct and provision earthen combs of 5-30 cells. In a single nest aggregation of *A. striata,* 40 monogynous nests with individually marked females were randomly assigned at the beginning of the worker phase to two experimental conditions, parasocial (foundress removed) and eusocial (foundress remained). Foundresses were then removed during the first week of the worker phase. 30 days later (the developmental time from egg to adult), all nests were excavated and intact combs reared out in the laboratory. Naturally occurring nest failure had reduced sample sizes to 13 and 15 nests in the parasocial and eusocial conditions, respectively. Nine of the 15 eusocial nests had lost the foundress by the time of excavation, subdividing this sample into a supersedure (n=9) and a truly eusocial (n=6) condition.

Both the numerical sex-ratio (SR_n=% males) and the investment sex-ratio (SR_i=% male dry weight) were significantly more male biased in the parasocial (SR_n=67.1%; SR_i=51.6%) and in the supersedure nests (SR_n=57.7%; SR_i=44.8%) than in the eusocial nests (SR_n=44.7%; SR_i=26.4%). This difference partly resulted from the fact that parasocial and supersedure nests ceased provisioning earlier in the season than did eusocial nests, which, in conjunction with the protandrous production of offspring in *A. striata*, caused a more female biased sex-ratio in the eusocial nests. An analysis controlling for this seasonal effect by considering only those offspring produced during the early part of the season when all nests were still actively provisioning, however, still confirmed the predicted difference in numerical and investment sex-ratios between parasocial and eusocial nests.

Social behavior in *A. striata* may therefore have evolved not only in response to ecological constraints limiting reproductive options, but also in relation to genetic variables intrinsic to the haplodiploid system of sex-determination.

ALLOCATION TO WORKERS IN THE ANT *LEPTOTHORAX LONGISPINOSUS*

Vickie L. Backus, Department of Zoology, University of Vermont, Burlington VT U.S.A. 05405-0086

There has been much attention paid to reproductive allocation in ants, (c.f. [1]) but very little is known about how nests allocate resources to workers. In this study I have examined allocation to workers in two populations of the temperate forest ant *L. longispinosus*.

Nests were collected during the summer of 1988 from two sites; Mallett's Bay State Park in Vermont (n=30 nests) and the E. N. Niles Huyck Preserve in New York (n=49 nests). Nests were returned to the laboratory, censused and maintained until November when nest contents were dried and weighed.

The mean number of workers in each nest was significantly higher in NY. Mean worker weight and total investment in workers were also significantly higher in NY.

A model was proposed predicting that queen number, worker number and mean queen weight determine both mean worker weight and total worker investment. Each predictor has a direct effect on the dependent variable. In addition there are indirect effects due to correlations between the predictor variables. Path analysis [2] is a technique that allows the direct and indirect effects of a predictor variable to be quantified.

Table 1 shows the path coefficients calculated for these data. The similarity of the path coefficients in Table 1A shows that the two populations are have a similar goal for total worker investment. Table 1B shows that the two populations are determining mean worker weight differently, for example in VT queen number is positively related to mean worker weight while in NY the relationship is negative.

Table 1 Path Analysis of Models for or Total Worker Investment (A) or Investment in Mean Worker Weight (B).

		VT	NY
A	Queen Number to Total Worker Investment	-.008	.112
	Worker Number to Total Worker Investment	.757	.788
	Mean Queen Weight to Total Worker Investment	.137	.035
B	Queen Number to Mean Worker Weight	.347	-.632
	Worker Number to Mean Worker Weight	.216	.608
	Mean Queen Weight to Mean Worker Weight	-.413	.014

1. Nonacs, P. 1986 Ant reproductive strategies and sex allocation theory. The Quarterly Review of Biology. 6:1-21.
2. Li, C.C. 1975 Path Analysis--A Primer. Boxwood Press, Pacific Grove CA.

WORKER POLICING IN SOCIAL INSECTS

Francis L. W. Ratnieks

Department of Entomology, University of California, Berkeley, CA 94720, USA

Insect workers are rarely sterile. Except a few ant genera, hymenopteran workers have ovaries and can produce offspring [1]. In species without a morphologically distinct worker caste, workers can mate and produce offspring of both sexes. In dimorphic species workers typically cannot mate, but can produce haploid males. In some termites, workers may be able to metamorphose into reproductives [2]. However, social insect workers frequently do not reproduce directly. This paper discusses the possibility that direct worker reproduction may be prevented by worker-worker inhibition--that is by worker policing [3].

Worker policing can be defined by its effect as any action by one worker which reduces direct reproduction by other workers [3]. In contrast to queen policing [1], which is *one against many*, and should be most effective in colonies with few workers, worker policing is *many against one*, and could be effective in all but very small colonies, see Fig. 1.

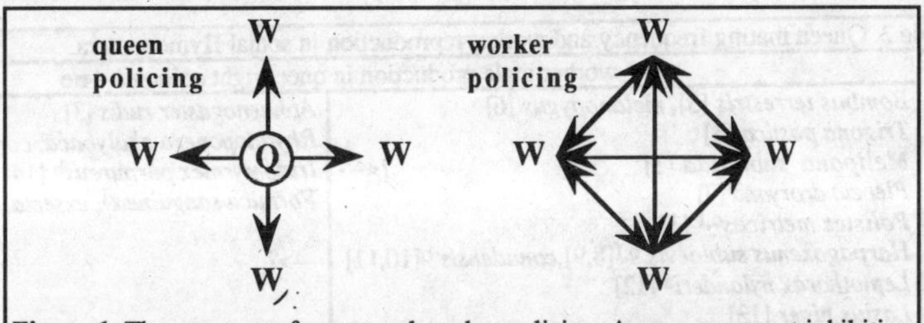

Figure 1. The structure of queen and worker policing. Arrows represent inhibition.

To evolve, both worker and queen policing must have a mechanism, and must increase the inclusive fitness of the actor. Possible mechanisms include physical domination of reproductive workers and/or eating of worker-laid eggs. Queen policing results in increased production of offspring versus grandoffspring, and gives a fitness benefit to queens for male and female production in both diploids and haplodiploids. In worker policing the fitness benefits to workers depend on the sex and ploidy of the offspring, queen mating frequency, and the identity of the mates of sibling reproductives, Table 1.

Table 1. Relatedness inequalities between a worker and the reproductive offspring of various colony members, according to offspring sex, ploidy, identity of sibling mates, *n* (queen mating frequency).

conditions		*n*	male	female
haplodiploid	reproductive workers	< 2	self > worker > queen	queen > self > worker
	outmated or unmated	> 2	self > queen > worker	self > queen > worker
diploid	sib. repros outmated	= 1	self = queen > sibling reproductive	
	sib. repros mate to sibling	= 1	self > queen = sibling reproductive	
	sib. repros outmated	> 1	self > queen > sibling reproductive	
	sib. repros mate to sibling	> 1	self > queen = sibling reproductive	

In the production of males, hymenopteran workers in matrifilial monogyne colonies are more related to workers' sons (0.375) than queen's sons (0.25) when $n = 1$ (*n* is the short term effective queen mating frequency [4]). However, for $n > 2$, workers are more related to queen's sons than workers' sons $(0.125 + 0.25/n)$ [3]. This suggests that workers will gain from policing other workers if mating frequency is > 2. Note that workers are most related to their own male offspring (0.5), and should continue to attempt to reproduce if policing is not fully effective, unless this lowers colony productivity or survival [3].

The relationship between queen mating frequency and worker male production in queenright colonies gives indirect support for worker policing in Hymenoptera, Table 2. Species with $n > 2$ do not have worker-derived males, whereas worker-derived males occur in over half the species with $n < 2$. Absence of worker male production the latter group could be due to queen policing, or self policing if worker reproduction lowers colony productivity [3, 5]. Preferential removal of worker-laid versus queen-laid male eggs by honey bees [16], provides direct support. Over 24h workers removed 99% of worker-laid but only 55% of queen-laid male eggs. Figure 2 shows additional data showing egg removal probabilities.

Worker policing may also be important in other situations. Inhibition of reproduction by potential workers may aid the evolution of eusociality [17]. In termites, possible worker policing has been noted [2, 18], in which workers prevent each other from becoming reproductivess by destroying developing wing buds. Table 1 also shows that in termites, worker policing is expected for all *n*, unless supplemental reproductives mate with sibs. The latter is a possible explanation for inbreeding cycles in termites [18].

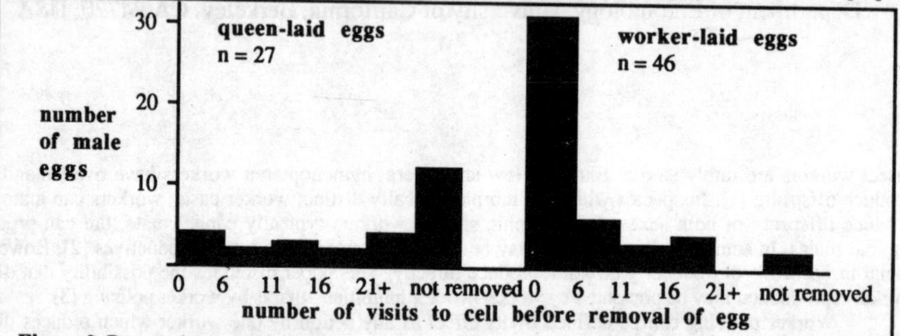

Figure 2. Removal rates of worker- and queen-laid male eggs transferred into drone cells. Data was collected using an observation hive not rearing drones in November in Berkeley.

Table 2. Queen mating frequency and worker reproduction in social Hymenoptera

n	yes---worker male production in queenright colonies---no	
< 2	*Bombus terrestris* [3], *melanopygus* [6]	*Aphaenogaster rudis* [3]
	Trigona postica [3]	*Rhytidoponera chalybaea, confusa* typeA [5]
	Melipona subnitida [3] ?<---	*Iridomyrmex purpureus*[b] [14]
	Plebeia droryana [7]	*Formica sanguinea*[d], *exsecta, pressilabris* [3]
	Polistes metricus[c,f] [3]	
	Harpagoxenus sublaevis [b,f][8,9],*canadensis* [b][10,11]	--->?
	Leptothorax nylanderi[a] [12]	
	Lasius niger [13]	
≈ 2		*Vespula squamosa* [3] ($n = 1.96$)
> 2		*Apis mellifera*[e], *cerana* [3, 4]
		Vespula maculifrons [3]
		Atta sexdens [g][4]

data insufficient for firm conclusion of *n* (a) or worker male production (b); occasional worker-derived male (c); worker-derived males in lab (d); 1/1000 males worker-derived (e) [15]; effective queen policing likely (f); worker male production not observed in queenright or orphaned colonies (g), E. O. Wilson, pers. comm.

1. Oster, G. F. and E. O. Wilson. 1978. Caste and Ecology in Social Insects. Princeton University Press, Princeton, New Jersey.
2. Zimmerman, R. B. 1983. Behavioral Ecology and Sociobiology 12: 143-145.
3. Ratnieks, F. L. W. 1988. American Naturalist 132: 217-236.
4. Page, R. E. 1986. Annual Review of Entomology 31: 297-320.
5. Cole, B. J. 1986. Behavioural Ecology and Sociobiology 18: 165-173.
6. Owen, R. E. and R. C. Plowright. 1980. Journal of Heredity 71: 241-247.
7. Machado, M. F. P. S., E. P. B. Contel and W. E. Kerr. 1984. Genetica 65: 193-198.
8. Buschinger, A. and U. Winter. 1978. Insectes Sociaux 25: 63-78.
9. Bourke, A. F. G. 1988. Behavioral Ecology and Sociobiology 23: 233-245 & 323-333.
10. Buschinger, A. and T. M. Alloway. 1979. Zeitschrift für Tierpsychologie 49: 113-119.
11. Buschinger, A. and T. M. Alloway. 1978. Insectes Sociaux 25: 339-350.
12. Plateaux, L. 1981. In: Biosystematics of Social Insects, P. E. Howse and J.-L. Clément, Eds, Academic Press, London and New York.
13. van der Have, T. M., J. J. Boomsma and S. B. J. Menken. Evolution 42: 160-172.
14. Halliday, R. B. 1983. Insectes Sociaux 30: 45-56.
15. Visscher, P. K. 1989. Behavioural Ecology and Sociobiolgy 25: 247-254.
16. Ratnieks, F. L. W. and P. K. Visscher. 1989. Nature (Lond.) 342: 796-797.
17. Stubblefield, J. W. and E. L. Charnov. 1986. Heredity 57: 181-187.
18. Ratnieks, F. L. W. 1989. Conflict and Cooperation in Insect Societies. Thesis. Cornell University.

REPRODUCTIVE CONFLICT, FITNESS, AND LIFE HISTORY STRATEGY IN SLAVE-MAKING ANTS

A.F.G. Bourke*, N.R. Franks**, B. Ireland***

* Department of Zoology, University of Cambridge, CB2 3EJ, U.K.
** School of Biological Sciences, University of Bath, BA2 7AY, U.K.
*** School of Mathematical Sciences, University of Bath, BA2 7AY, U.K.

Kin selection theory predicts the occurrence of reproductive conflicts within Hymenopteran societies (1). In particular, it predicts queen-worker conflict over sex investment ratios, and queen-worker and worker-worker conflict over male production. This presentation summarizes our studies to date of such conflicts in the monogynous slave-making ant Harpagoxenus sublaevis.

Our work suggests that in H. sublaevis queens achieve their preferred 1:1 sex investment ratio contrary to the workers' interests (2). This conclusion holds even when we allow for the effects on the queen-produced sex ratio of male production by orphaned colonies (3). Queens pheromonally prevent male production by workers in queenright conditions, so that most worker reproduction occurs in orphaned colonies. In both queenright and queenless colonies, workers form dominance orders as a result of worker-worker competition for egg-laying rights. High-ranking workers inhibit egg-laying in subordinates with physical aggression. Compared to dominant workers, subordinates appear to participate more in scouting (searching for host species colonies) and slave-raiding (stealing host worker pupae) (4).

Recently, we constructed a life history model for H. sublaevis which explores the consequences for queen and worker fitness of (a) male production by orphaned workers, and (b) the division of labour in orphaned colonies between dominant egg-laying workers and subordinate slave-raiding workers (5). The model incorporated parameters derived from field and laboratory studies of H. sublaevis (2,4,6). It used numerical optimization to find the optimum life history strategy for a colony queen. This was defined as that policy of allocating resources between slave-maker worker, queen and male production that yielded the greatest rate of spread of copies of the queen's genome.

The results of the model accorded well with independent field data on H. sublaevis life history strategy. We also ran variants of the model which assumed either (a) that workers are totally sterile, or (b) that queenless workers do not slave-raid. These showed that queen fitness fell both when workers are sterile and when orphaned workers fail to raid.

Orphaned colonies produced most males when subordinate workers slave-raided. Such workers incurred a cost of raiding of almost zero, since both their chances of becoming dominants and of dying on slave-raids were small. Thus, by Hamilton's rule, it pays reproductively-inhibited subordinate workers to raid for slaves which will rear the subordinates' nephews.

These findings confirm previous suggestions that queens posthumously favour worker reproduction (7), and that the division of labour among workers is linked to reproductive conflict among them (8). They therefore underline the importance of queenless worker reproduction in retarding the evolution of total worker sterility (9), and of within-colony reproductive conflict as an influence on eusocial organisation.

REFERENCES

1. Trivers, R.L. and H. Hare. 1976. Science, 191: 249 - 263.
2. Bourke, A.F.G., T.M. van der Have and N.R. Franks. 1988. Behavioral Ecology and Sociobiology, 23: 233 - 245.
3. Bourke, A.F.G. 1989. Evolution, 43: 913 - 918.
4. Bourke, A.F.G. 1988. Behavioral Ecology and Sociobiology, 23: 323 - 333.
5. Franks, N.R., B. Ireland and A.F.G. Bourke. 1990. Behavioral Ecology and Sociobiology (In press).
6. Bourke, A.F.G. 1987. Ph.D. thesis, University of Bath.
7. Alexander, R.D. 1974. Annual Review of Ecology and Systematics, 5: 325 - 383.
8. West-Eberhard, M.J. 1981. In: Natural Selection and Social Behavior, R.D. Alexander and D.W. Tinkle, Eds., Chiron, New York, pp. 3 - 17.
9. Bourke, A.F.G. 1988. Quarterly Review of Biology, 63: 291 - 311.

Symposium 15

BEHAVIOURAL ONTOGENY OF INDIVIDUALS AND COLONIES

Organizers : **PRASSEDE CALABI, U.S.A.**
NORMAN F. CARLIN, U.S.A.

REGULATION OF COLONY DEVELOPMENTAL PLASTICITY IN *APIS MELIFERA*

Gene E. Robinson*, Robert E. Page, Jr.+, Colette Strambi†, Alain Strambi†

*Department of Entomology, University of Illinois, Urbana, IL 61801; +Department of Entomology, University of California, Davis, CA 95616; †Centre National Recherche Scientifique, Laboratoire de Neurobiologie, Marseille, France

INTRODUCTION

Colony developmental plasticity, a response to ongoing variation in colony age demography and resource availability, involves adjustments in the proportions of individual workers engaged in various tasks. The developmental plasticity of insect colonies has been well documented, but underlying mechanisms are poorly understood. We summarize here results (1, 2) that suggest that two processes are involved in colony-level regulation of division of labor: 1) plasticity in age-dependent behavior is a consequence of modulation of juvenile hormone (JH) titers by extrinsic factors, and 2) stimuli that can affect JH titers and age polyethism elicit variable responses among genetically distinct workers within a colony.

HORMONAL CONTROL OF DEVELOPMENTAL PLASTICITY

JH is involved in the control of age polyethism in adult worker honey bees. Low titers are associated with nest tasks such as brood care ('nursing') during the first 1-3 weeks of the bee's ca. 6-week life, whereas a higher titer at about 3 weeks of age induces foraging. In two experiments, we tested the hypothesis that environmental cues modulate the intrinsic rise in JH, providing a mechanism for plasticity in age polyethism.

1. 'Single-cohort colonies' (N=2), each composed of 2000, 1-3-day-old bees, were established to uncouple the usually tightly linked factors of worker age and behavioral status for two tasks: nursing and foraging. We collected groups of precocious foragers and normally aged nurses when they were 7-10 days old, and normally aged foragers and overaged nurses when they were 21-24 days old. Radioimmunoassays revealed an association between behavioral status and JH titer, independent of age. In each single-cohort colony, pooled samples of foragers had significantly higher hemolymph levels of JH than pooled samples of nurse bees, at both ages. Similar results were obtained for changes in age demography that occur naturally during colony fission.

2. Observations were made on individually labelled bees in a colony to determine which ones were acting as nurses or foragers. After removing most young, nursing-age, bees, we identified a group of foragers that reverted to nursing activities. Hemolymph levels of JH in reverted nurses were significantly lower than in foragers. The lower titer in nurse bees probably represents a drop in JH that is

associated with reversion, because the reverted nurses were once foragers themselves in the same colony.

These results, coupled with those demonstrating that treatment with JH, JH mimic, or JH analog induces changes in age polyethism, support the hypothesis that JH acts as a colony integrating mechanism.

GENETIC CONTROL OF DEVELOPMENTAL PLASTICITY

Because of polyandry and sperm mixing, honey bee colonies are composed of numerous subfamilies of workers, each subfamily descended from the colony's queen and one of her mates. We collected precocious foragers, normally aged nurses, normally aged foragers, and overaged nurses, as in Experiment 1. Using allozymes as paternity markers (in single-cohort colonies derived from instrumentally inseminated queens), we detected significant genotypic differences between workers in the likelihood of becoming a precocious forager or a normally aged nurse in 7 out of 9 colonies, and becoming an overaged nurse or a normally aged forager in 4 out of 8 colonies. These results demonstrate that genotypic differences in sensitivity to changing conditions can also affect colony integration.

DISCUSSION

Some environmental and colony stimuli, unknown at this time, apparently can modulate the intrinsic rise in JH to accelerate, retard, or reverse worker behavioral development. Factors in addition to genotype that may influence sensitivity to changing colony conditions include worker age and experience. A major gap in our knowledge of colony operation is how the needs of a colony are perceived by its workers. Individual workers probably do not possess 'global awareness' of the colony's state because it is hard to imagine that they have the capacity to integrate sensory information on all activities in their complex society, in which tens of thousands of individuals perform many different tasks in parallel. Therefore, plasticity in worker differentiation is probably based on responses to a subset of stimuli in the colony. The discovery of hormonal and genetic mechanisms of behavioral integration may lead to a greater understanding of the perception of colony needs by workers.

REFERENCES

1. Robinson, G.E., Page, R.E., Strambi, C., and A. Strambi. 1989. Science 246: 109-112.
2. Robinson, G.E., Page, R.E., Strambi, C., and A. Strambi. Submitted.

We thank M.K. Frondrk for expert technical assistance. Supported by an Ohio State University Postdoctoral Fellowship (G.E.R.), NSF grants BSR-8800227 (G.E.R.) and BNS-8719283 (R.E.P.), and the CNRS (C.S. and A.S.).

LABOR DIVISION IN HONEYBEE COLONIES IS INFLUENCED GENETICALLY BY THE LEARNING ABILITY OF HONEYBEES

Ch. Brandes*
Institut fuer Neurobiologie der FU Berlin, Koenigin-Luise Str.28-30, Berlin and Institut fuer Bienenkunde, Karl-von-Frisch Weg 2, Oberursel, Fed.Rep.Germany

Labor division in honeybees is influenced to a high degree by age. Particular tasks are performed by specific sub-groups in an age-dependent manner. Environmental and genetic factors, however, also affect labor division. I tested the hypothesis that differences in learning ability would affect labor division in a colony composed of bees from bidirectionally selected learning strains.

To produce lines with high and low learning worker bees of the race Apis mellifera capensis were selected and used to rear queens. Selection was done over several generations and lines were obtained that showed different ranges of learning abilities.

Worker bees from the selected lines differed in proboscis extension reflex conditioning (PER). Conditioning consisted of odors paired (CS+) or unpaired (CS-) with an unconditioned stimulus (US). The US was a drop of sugar water presented to the bees antennae and proboscis. Bees from the high line responded to CS+ much more frequently than did bees from the low line.

Five different lines, two with high learning scores, two with intermediate scores and one with low scores, were used to examine the relation between learning and labor division. A colony was composed with 4000 age-marked bees from these 5 lines. Social behavior of individual worker bees in this colony was observed for 4 weeks. Eight different tasks were assayed: Four active tasks (feeding/observing larvae, feeding nestmates, constructing combs, grooming), 2 foraging-related tasks (dancing, following dancers) and 2 passive tasks (receiving food and be groomed). All lines participated in labor division in the same age dependent manner. An influence of learning was found that resulted in a correlation between learning and the magnitude of participation in feeding or in constructing combs. There also was a correlation between learning and all active inside tasks taken together. High learners participated more in active tasks than low learners. Although no significant correlation was detected between learning and passive inside tasks low learners showed a higher percentage of receiving food and were groomed more often than high or intermediate learners. No correlation was detected between learning and outside tasks. But here again, low learners differed from high learners. The former showed higher activity in following dancers but lower activity in active dancing.

Taken together these results show a strong influence of learning on labor division in 2 out of 8 tasks. In the remaining 6 tasks a non linear relationship between learning and labor division may exists. Differences exist between low and high or low and intermediate learners but not between intermediate and high learners.

* Present adress: Brandeis University, Department of Biology, Bassine 235, Waltham, MA 02254-9110, USA

LIFE HISTORY OF COLONY AND INDIVIDUAL BEHAVIOUR

Paul Schmid-Hempel

Zoologisches Institut der Universität

Rheinsprung 9, CH-4051 Basel, Switzerland.

Workers of highly eusocial hymenoptera gain fitness by helping their mother to raise reproductive offspring. Up to a point, colonies that attain a alrge size at the time when reproduction takes place can produce more sexuals and hence gain larger genetic representation in the next generation. As a consequence, colonies are expected to grow at maximum possible rate during the ergonomic stage and individual worker behaviour should be selected within this context of life history (Houston, Schmid-Hempel & Kacelnik, Am.Nat. 131: 107, 1988). Among other things, the rate at which a colony develops depends on the rate at which resources are delivered to the hive through the foraging effort of its members and how resources are converted into increments of colony state. Individual foraging strategies should thus not only depend on resource availability, but also on the state of the colony at different stages of the life cycle.

As part of this problem, we have now tested whether nectar-collecting behaviour of individual worker of the honeybee depends on colony size, a variable that is continously changing over the development of the colony. Experimental colonies were set up to contain either with 10'000 adult bees (SMALL) or 35'000 bees (LARGE), while worker : brood ratios were kept constant in the two treatments. Individual workers from these two types of colonies could collect nectar from artificial flowers, similar to the paradigm used by Schmid-Hempel (J.Anim.Ecol. 56:209, 1987). On average, workers from SMALL colonies visited fewer flowers per mean foraging trip than those from LARGE (40.1 + 1.1 vs. 44.8 + 1.1, F = 11.05, P = 0.01), but spend longer time handling each flower (7.3 + 0.4 s vs. 5.8 + 0.4 s, F = 5.64, P = 0.02). Other components of foraging behaviour did not differ among tretaments. When compared with predictions from a simple model, bees from LARGE behaved somewhat closer to maximisation of energetic efficiency than those from SMALL. The observed behavioural differences between the two treatments were most pronounced early in the season. Moreover, behaviours paralleled the relative development of the colonies over the duration of the experiment (Wolf & Schmid-Hempel, Behav.Ecol. Sociobiol, in press). It is concluded that individual short-term behaviour is affected by colony size. The exact relationship and its consequences for fitness are intricate. A fuller picture will be gained by also considering the level of individual activity over longer time scales.

COLLECTIVE DECISION-MAKING IN HONEY BEES : HOW COLONIES CHOOSE AMONG NECTAR SOURCES

Thomas D. Seeley[1], Scott Camazine[1], and James Sneyd[2*]

[1]Section of Neurobiology and Behavior, Mudd Hall, Cornell University, Ithaca, NY 14853, USA

[2]Department of Biomathematics, UCLA School of Medicine, 10833 La Conte Ave., Los Angeles, CA 90024 USA

[*]Present address: Centre for Mathematical Biology, Mathematical Institute, 24-29 St. Giles', Oxford OX1 3LB, UK

SUMMARY

A honey bee colony can skillfully choose among different nectar sources. It will selectively exploit the most profitable sources in an experimental array and can rapidly shift its foraging efforts following changes in the nectar source array (Figure 1). How does this colony-level ability emerge from the behavior of the individual bees within the colony? The answer comes from understanding how bees modulate their colony's rates of recruitment and abandonment for different nectar sources in accordance with the different level of profitability of each source. Each forager performs a multifactorial modulation of her foraging behavior in relation to nectar source profitability according to the following pattern. As profitability increases, the probability of abandoning the nectar source decreases, the tempo of foraging increases, and the intensity of dancing increases. This raises the question: How does a forager assess the profitability of her particular nectar source? A series of experiments reveals that this is done without comparison among the available nectar sources in the field. The foragers do not directly compare the relative profitability of different nectar sources, nor do the food storer bees compare the profitabilities of different nectar loads and transmit that information to the foragers. Rather each forager assesses the profitability of her own nectar source, and evaluates its absolute profitability based upon an innate scale of nectar source quality. Hence each forager makes an independent evaluation of her nectar source based upon the limited information that she has gathered herself. Such a system can operate effectively if all the foragers have a common, shared scale for evaluating the profitability of a nectar source, and have a shared set of rules for modulating their behavior in relation to this variable. Even though each forager operates independently, together they will generate a coherent colony-level response to different food sources in which better nectar sources are heavily exploited and poorer ones abandoned. Based upon these rules for recruitment and abandonment, we present a scheme in the form of a flow diagram, by which the colony-level pattern of foraging emerges from the behavioral rules of the individuals (Figure 2). The diagram describes the movement of a group of nectar foragers having a choice between two nectar sources, such as those in the experimental setup of Figure 1. At any moment a bee can either be at one of the following 7 compartments: 1) at the hive unloading nectar from nectar source A, 2) at the hive unloading nectar from nectar source B, 3) at nectar source A, 4) at nectar source B, 5) dancing for food source A, 6) dancing for food source B, or 7) following a dancer. The arrows represent the possible movements of bees from one location to the next. This scheme has been incorporated into a differential equation model simulating the buildup of bees at the two feeders. The simulation successfully describes what colonies actually do. Nectar-source selection by honey bee colonies can be thought of as a process of natural selection among alternative nectar sources as foragers from more profitable sources "survive" (continue visiting their source) and "reproduce" (recruit other foragers) better than do foragers from less profitable sources. In this form of decentralized decision-making the pattern of selective food source utilization emerges as a self-organizing process based on the dynamic interactions of the foragers, without any need for a higher-level decision-maker to oversee the process.

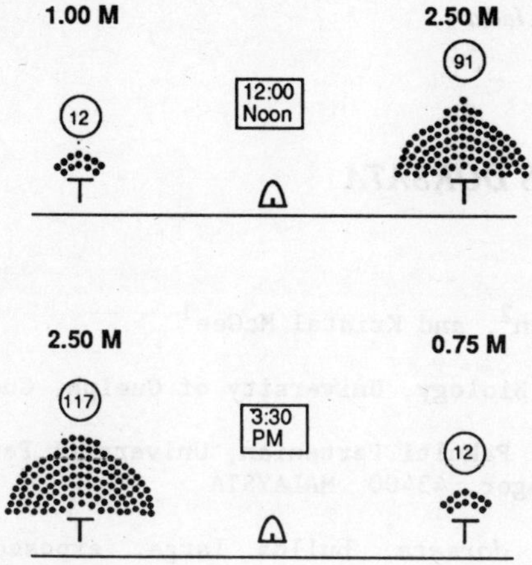

Figure 1. Selective utilization of nectar sources by honey bee colonies. When given a choice between two nectar sources which differed in profitability (molarity of sucrose), the colony consistently directed the majority of its foraging effort onto the richer source of nectar. The number of dots above each feeder denotes the forager group size for that feeder, i.e., the number of different foraging bees that visited that feeder at least once in the half hour preceding the time indicated. For several days prior to the start of the observations, a small group of bees was trained to each feeder. Thus on the morning of the experiment both feeders had equivalent histories of low-level foraging. At noon, the profitabilities of the two feeders were reversed.

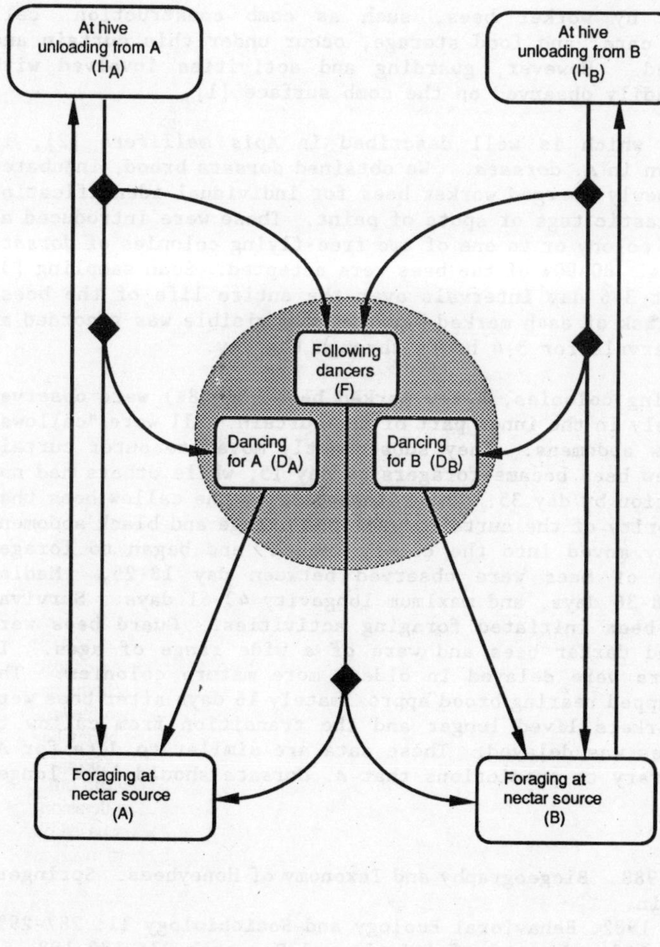

Figure 2. Flow diagram describing how bees select among two nectar sources differing in profitability. At any given moment each nectar forager in the colony is in one of the seven compartments shown (H_A, H_B, F, etc.) Each black diamond represents a decision branch point, representing the behavioral rules of recruitment and abandonment. For example, after unloading her nectar a bee must decide whether or not to abandon her food source. If she chooses to continue foraging at the food source, she must then decide whether to dance or simply return to the nectar source without recruiting.

AGE POLYETHISM IN *APIS DORSATA*

Gard W. Otis[1], Makhdzir Mardan[2], and Kristal McGee[1]

[1]Department of Environmental Biology, University of Guelph, Guelph, Ontario, CANADA N1G 2W1
[2]Plant Protection Department, Fakulti Pertanian, Universiti Pertanian Malaysia, Serdang, Selangor, 43400 MALAYSIA

The giant honey bee, *Apis dorsata*, builds large, exposed nests throughout much of tropical Southeast Asia. The comb is protected by a protective curtain of three to six layers of worker bees. Many of the tasks performed by worker bees, such as comb construction, cell cleaning, brood care, and food storage, occur under this curtain and cannot be viewed. However, guarding and activities involved with foraging are readily observed on the comb surface [1].

Age polyethism, which is well described in *Apis mellifera* [2], is presently unknown in *A. dorsata*. We obtained *dorsata* brood, incubated it, and marked newly emerged worker bees for individual identification with numbered plastic tags or spots of paint. These were introduced at night to a caged colony or to one of two free-flying colonies of *dorsata* at the UPM campus. 80-90% of the bees were accepted. Scan sampling [3] was performed at 3-6 day intervals over the entire life of the bees. The behavioral task of each marked bee that was visible was recorded at five-minute intervals for 3-4 hours through the day.

In the free-flying colonies, a few marked bees (38-58%) were observed almost immediately in the inner part of the curtain. All were "callows" with pale yellow abdomens. They subsequently moved to outer curtain positions. A few bees became foragers by day 15, while others had not made the transition by day 35. Regardless of age, the callow bees that made up the majority of the curtain developed orange and black abdomens at the time they moved into the colony "mouth" and began to forage. Maximum numbers of bees were observed between day 18-25. Median longevity was 28-36 days, and maximum longevity 43-61 days. Survival was high until bees initiated foraging activities. Guard bees were both callows and darker bees and were of a wide range of ages. In general behaviors were delayed in older, more mature colonies. The caged colony stopped rearing brood approximately 16 days after bees were introduced. Workers lived longer and the transition from callow to orange/black bees was delayed. These data are similar to data for *A. mellifera*, contrary to predictions that *A. dorsata* should live longer [4].

REFERENCES

1. Ruttner, F. 1988. Biogeography and Taxonomy of Honeybees. Springer-Verlag, Berlin.
2. Seeley, T.D. 1982. Behavioral Ecology and Sociobiology 11: 287-293.
3. Kolmes, S.A. 1984. Journal of Apicultural Research 23: 189-198.
4. Dyer, F.C. and T.D. Seeley. 1987. Journal of Experimental Biology 127: 1-26.

REPRODUCTIVE DOMINANCE BETWEEN BUMBLE BEE WORKERS

N. Pomeroy[1] and R.C. Plowright

Department of Zoology, University of Toronto, Toronto, Ontario, Canada M5S 1A1

[1](Present Address: Department of Botany and Zoology, Massey University, Palmerston North, New Zealand)

INTRODUCTION

The ontogeny of bumble bee colonies is of special interest both in terms of the shift in reproductive dominance from the foundress to (some of) the workers, and in maximising colony size for pollination by avoiding early production of queens and males. The suggestion by Röseler [1] that the foundress produces male eggs in response to the stress of increasing worker numbers on the comb has met with both support [2,3,4] and dissent [5]. Most studies seem to have concentrated on dominance behaviour, even if followed by dissection of workers. We wanted to more directly quantify the oogenesis-suppressing strength of individual workers of *Bombus (Pyrobombus) impatiens* Cresson. Our results have implications for understanding the dominance interactions between the foundress and her workers, and hence colony ontogeny.

METHODS

"Test" workers (whose dominance was being measured) were confined with callow (less than 12 h after emergence) "target" workers in small wooden boxes (25x25x77mm unless otherwise stated) with 50% sucrose solution and daily replaced pollen at 30°C and 60% relative humidity for 6 days, after which all workers were frozen. Ovaries were removed, fixed in Bouin's fluid and weighed after surface blotting. Further details appear elsewhere [6].

Size Fourteen various-sized workers were confined together until 3 days old, then confined individually with a younger worker until the older one laid eggs. Thus it was intended to have all "test" workers in an equivalent reproductive state. Then the "test" worker was confined with a pair of "target" workers as above. Worker size was measured as the length of the marginal cell of the wing.

Age Pairs of approximately equal-sized callow workers were aged for 1, 2, 3, 5, 7, 9, 12, 16 or 20 days before placing each worker individually with a pair of "target" workers as above.

Group Size Expt. 1 (performed early in the research programme with the aim of determining an adequate group size for dominance experiments): Eight solitary callows, four groups of 2 and three groups of 3 were confined for 6 days as above. Expt. 2: Three pairs, three groups of 4, two groups of 6 and two groups of 8 workers were confined in larger boxes (25x38x77mm) but otherwise as above.

All ovarian weights were $\log_{10}$-transformed and the two values for the pair of target workers in each size and age boxe were averaged and treated as a single datum.

RESULTS

Worker Size and Age There was no significant effect of size when workers were tested for dominance strength immediately after laying their first eggs, but the smaller workers were older when tested and this may have counteracted any size effect, as the age trial did show an decrease in ovarian weight of target worker with increasing age of test worker ($r = -0.489$, $P < .05$). We pooled the size and age data and obtained the following double regression: $\mathbf{\log (wt) = 3.301 - 0.27(size) - 0.02(age)}$ where: *wt* = ovarian weight (x 10^{-5}g) of target workers, *size* = marginal cell length (mm) of test worker, and *age* = number of days test worker is older than target workers. Student's *t* for *size* = 1.85 ($P < .05$) and for *age* = 3.03 ($P < .001$, both one tailed). Aggressive interactions were seldom seen.

Group Size Figure 1 shows the combined results of the two trials.

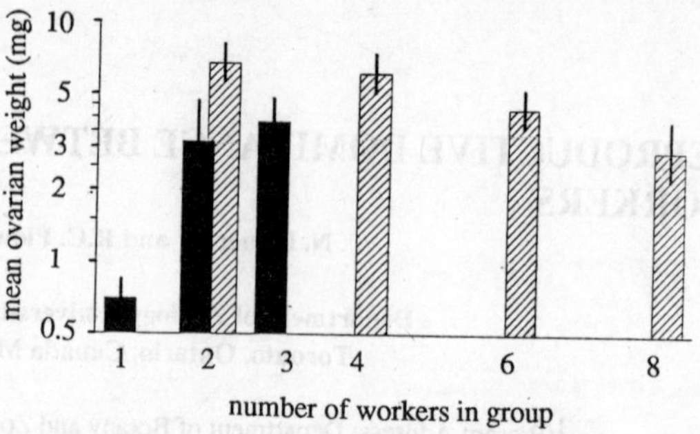

Figure 1. Ovarian weights of workers confined in different group sizes. Black columns: expt. 1, floor 77x25 mm; shaded columns: expt. 2, floor 77x38 mm. Vertical lines indicate Standard Error of mean.

Highest mean ovarian development ocurred in groups of two to four workers and declined in single workers and in larger group sizes. The decline was significant as group size increased from 4 to 8 workers ($r = -0.353$, $P < .05$).

DISCUSSION

The slight effect of worker size and the stronger effect of age is consistent with results for *B. terrestris* [3,7]. Substitution of values for the mean size of queens (4.09 mm marginal cell) and mean ovarian weight (0.54 mg) of target workers confined with them under the same conditions as above in the regression equation yields an "age" of 23 days, i.e. queens are, for their size, no more dominant than 23-day old workers. So there is no need (at least in *B. impatiens*) to postulate *any* qualitative difference in dominance mechanism between queens and workers.

Although at low densities (< 2 bees/10 cm2) workers have a mutually stimulating effect on oogenesis as reported by Free [8], at higher and more natural densities there seems to be mutual inhibition even though the individuals would not be considered "dominant". The decrease (of 0.32 logarithm units) in mean ovarian size as group size rose from 4 to 8 workers was equivalent to the increased inhibitory effect of a dominating worker being 16 days older (estimated from the partial regression for age). This leads to the idea of an additive "dominance titre" which may be received either from many subordinate bees or a few dominant bees. In this way the increasing density of workers on the comb could begin to affect the queen as postulated by Roseler [1] long before the appearance of overt dominance behaviour (such as used to define the "competition point" by Duchateau and Velthuis [5]). A mechanistic notion of domination by frequency x intensity of contacts initially seems contradicted by the fact that dominant workers seek rather than avoid the queen [3]. But perhaps they are able to saturate their receptors for the low volatility [6,9] pheromone and thus block the dominance signal which may normally need "on/off" contacts for reception.

REFERENCES

1. Röseler, P.-F. 1967. Zoologische Jahrbücher: Abteilung für Allgemeine Zoologie und Physiologie der Tiere, 74: 178-197.
2. Honk, C.G.J. van, P.-F. Röseler, H.H.W. Velthuis, and J.C. Hoogeveen. 1981. Behavioral Ecology and Sociobiology, 9: 9-14.
3. Doorn, A. van, and J. Heringa. 1986. Insectes Sociaux, 33: 3-25.
4. Tod, C.B. 1986. Socio-economic effects on colony size in the bumble bee *Bombus terrestris*. M.Sc. thesis, Department of Botany and Zoology, Massey University, Palmerston North, new Zealand.
5. Duchateau, M.J. and H.H.W. Velthuis. 1988. Behaviour, 107: 186-207.
6. Pomeroy, N. 1981. Reproductive dominance interactions and colony development in bumble bees, Ph.D. thesis, Department of Zoology, University of Toronto.
7. Röseler, P.-F. and I. Röseler. 1977. Proceedings of the 8th International Congress of the IUSSI, Wageningen, pp. 232-235.
8. Free, J. B. 1957. Proceedings of the Royal Entomological Society of London, A, 32: 182-184.
9. Röseler, P.-F., I. Röseler and C.G.J. van Honk. 1981. Experientia, 37: 348-351.

INCUBATION OF BROOD BY BUMBLE BEES : ISOLATION OF A PHEROMONE INITIATING THIS BEHAVIOUR IN *BOMBUS RUDERATUS* (FABR.)

D.R. Greenwood*, and R.P. Macfarlane

DSIR Biotechnology and DSIR Plant Protection, Department of Scientific and Industrial Research, Palmerston North, New Zealand.

Bumble bees maintain their brood above ambient temperatures by incubation. All castes perform this task which dissipates body heat through the ventral surface of the abdomen to the underlying wax covered brood. Heinrich [1] showed that incubation activity was triggered by a pheromone of low volatility which was deposited on the brood and detected by the antennae.

A bioassay was developed to aid in the purification of the incubation pheromone from *Bombus ruderatus*. Paired glass fibre discs, one impregnated with a test fraction and the other with an appropriate solvent as a blank, were air-dried and mounted onto plaster moulds resembling brood clumps. Captive *B. ruderatus* queens (ca 20), screened for their ability to respond to the pheromone, were introduced into individual observation boxes containing the disks and their behaviour recorded 4x over 15 min under red light.

Clumps of fresh brood wax freed of eggs, larvae and pupae were extracted with dichloromethane, and the resulting extract was subjected to flash chromatography on silicic acid followed by preparative TLC. Reversed-phase HPLC, ion exchange chromatography and solvent partitioning were used for further purification. Examination of the resulting active material by capillary GC-MS indicated that the pheromone was associated with the unesterified long-chain fatty acid fraction and was readily oxidisable.

The incubation pheromone of bumble bees is species-specific. It is different in nature and origin to brood pheromones from other bees, ants and wasps which are produced by the developing larvae. Brood-derived pheromones may also exist in bumble bees. The bumble bee incubation pheromone is deposited intentionally, presumably since honey pot wax is not active, and therefore is involved in brood marking and recognition. It has a calming influence on nearby queens and workers. The incubation pheromone of bumble bees may aid rearing by screening queens for broodiness and encouraging nest initiation by the secretion of wax.

[1] Heinrich, B. 1974. Pheromone induced brooding behaviour in *Bombus vosnesenskii* and *B. edwardsii* (Hymenoptera: Bombidae). *J. Kansas Ent. Soc.* 47(3): 396-404.

PRESENCE OF BROOD INFLUENCES CASTE IN THE SOCIAL WASP *POLISTES EXCLAMANS* (HYM. : VESPIDAE)

Carlos R. Solís. Dept. of Ecology and Evolutionary Biology. Rice University. P.O. Box 1892. Houston, TX. 77251. U.S.A.

INTRODUCTION

According to theoretical ergonomic models, caste should be fixed at early stages during development and, in seasonal colonies, sexual forms should be produced in discrete batches, one generation before the end of the growing season [1]. This kind of optimization strategy requires that workers renounce direct reproduction, and rely solely on queen reproduction as means of expressing their fitness. Furthermore, it requires that the colony be able to predict the end of the growing season in order to be able to make the switch from worker to reproductive manufacture at the appropiate time. While this is the case in many ant and bee species, ecological constraints such as unpredictable climate fluctuations, high levels of parasitation and predation, and conflicts of reproductive interest [2] can keep some species from adopting the same strategies.

The social wasp *Polistes exclamans* builds annual colonies which are started in the spring by overwintered foundresses. It suffers high but unpredictable levels of predation and parasitation [3], as well as frequent queen supersedure [4]. This two factors alone can select against early individual commitment to a particular caste since nest or brood loss can lead to loss of all reproductive opportunities for individuals committed to the worker caste, and queen supersedure can cause workers to raise brood towards which they are more distantly related than to their own progeny.

I investigated the effect of the loss of brood, the nest and the worker force on caste differentiation into workers and future foundresses in mid and late summer colonies of *P. exclamans*. These events would cause, at that point in the season, almost certain colony failure.

I expected that females that experienced brood loss this late in the season or emerged in colonies with no brood, would become future foundresses, while females in colonies with brood would become workers.

Workers and future foundresses of *P. exclamans* can not be distinguished by their external morphology. They can, however, be told apart by the relative development of the fat body [5]; differential survival at cold temperatures [6; 7], amount of body water [7], and differences in metabolic rate associated with changing temperatures [7].

METHODS

Mid and late summer colonies of *P. exclamans* were collected in 1985 and 1986. In the laboratory, the colonies were subjected to the following treatments that would, simulate, respectively brood loss due to parasitation, nest loss due to vertebrate predation, worker force reduction due to predation away from the nest, and natural colonies: 1) removal of all eggs and larvae, leaving pupae and adults on the nest; 2) removal of the whole nest, leaving just adults; 3) removal of all adults, leaving the nest with all of its brood intact; and 4) control colonies. Colonies were reared in the laboratory under prevailing photoperiod and fed ad libitum on an artificial diet for 16 days, allowing the pupae, where present, to hatch. These treatments yielded 6 kinds of females: i) field eclosed females in broodles colonies, ii) laboratory eclosed females in broodless colonies, iii) field eclosed females without a nest or brood, iv) laboratory eclosed females with a nest and brood but without field eclosed females, v) field eclosed females with brood and nest, and vi) laboratory eclosed females with brood and nest. Caste was assayed by subjecting females to 5° C temperatures, and determining survival every three days. Levels of reserve lipid, triacylglycerols, were assayed by thin layer chromatography before subjecting wasps to cold temperatures. Females were also compared in terms of fresh weight.

RESULTS AND DISCUSSION

In all cases, wasps that were present on the nest or emerged in colonies where the brood had been removed (i.e. broodless colonies and colonies with no nest) showed significantly higher survivor at cold temperatures than those where brood was present (i.e. colonies with field adults removed but with brood present or controls) ($p<0.005$ for all comparissons). A seasonal effect was also observed, as wasps from late summer experiments showed significantly higher survival than wasps from mid summer experiments ($p<0.005$). Age at which exposure to broodless colonies ocurred also influenced cold survivor, since wasps emerged in the laboratory showed better survivor than those emerged in the field ($p<0.005$). The results show that *P. exclamans* females can respond to brood loss and acquire characteristics of future foundresses in its absence. The data also show that caste differentiation in this species is mediated not only by physical factors such as time of the year, but also by age, and, more importantly, by social factors such as presence of brood and opportunities for indirect reproduction.

The analysis of reserve lipids showed that wasps from broodless colonies did not accumulate significantly higher levels of triacylglycerols than wasps in colonies with brood ($p>0.6$). No differences in fresh weight were detected either ($p>0.7$). This suggests that for wasps from broodless colonies to be able to survive that cold exposure period better than wasps with brood, metabolic differences between the two kinds of females would be present. The absence of differences in food reserves is of special importance in this case, because it shows that differential survival was not observed only because wasps from broodless colonies had been able to store food that they would otherwise pass to larvae, but rather that different metabolic processes were taking place in wasps from broodless and broodright colonies.

The results show that *Polistes exclamans* does not show the kind of production schedules predicted by theoretical models, neither does it produce individuals destined to be members of a particular caste. It rather produces a kind of female that makes it possible to respond to environmental contingencies which might otherwise cause the individual or the colony to loose all its reproductive potential. This is in contrast with ant, bee and termite species in which caste is determined very early in development, sometimes as early as the egg stage [8;9;10;11].

REFERENCES

1. Oster, G.F. and E.O. Wilson. 1978. Caste and Ecology in the Social Insects, Princeton University Press, Princeton, NJ.
2. Jeanne, R.L. 1986. Monitore zoologico italiano (N.S.), 20: 119-133.
3. Strassmann, J.E. 1981. Ecology, 62: 1225-1233.
4. Strassmann, J.E. 1985. Zeitschrift zur Tierpsychologie, 69: 141-148.
5. Eickwort, K. 1969. Insectes Sociaux, 16: 67-72.
6. Strassmann, J.E., R.E. Lee Jr, R.R. Rojas and J.G. Baust. 1984. Insectes Sociaux, 21: 291-301.
7. Solís, C.R. in preparation
8. Wilde, J. de and J. Beetsma. 1982. Advances in Insect Physiology, 16: 167-246.
9. Wheeler, D.E. 1986. American Naturalist, 128: 13-34.
10. Suzzoni, J.P., and L. Passera. 1984. Insectes Sociaux, 31: 155-170.
11. Watson, J.A.L. and H.M. Abbey. 1987. Insects Sociaux, 34: 291-297.

COMPARISON OF COLONY STRUCTURE AND BEHAVIORAL ONTOGENY IN THREE EXTRAORDINARY ANT SPECIES OF JAPAN

Kazuki TSUJI[1)], **Katsusuke YAMAUCHI**[2)].

1) Laboratory of Applied Entomology and Nematology, Faculty of Agriculture, Nagoya University, Nagoya 464-01, Japan. 2) Department of Biology, Faculty of Education, Gifu University, Gifu 501-11, Japan.

We demonstrate social organization and colonial life history of three Asian ant species of different subfamilies and will offer generalized questions about extraordinry reproduction in ants.

1. Pristomyrmex pungens (Myrmicinae)

As the social organization of this ant was already reported (Tsuji 1988, 1990), we will show new data clarifing the degree of reproductive specialization among workers. This species is queenless, and workers reproduce parthenogenetically (thelytoky). All young workers stay in the nest and reproduce but old workers stop oviposition and forage outside the nest. Most common criticism for Tsuji (1988) is as follows: As not all ants collected inside the nest had fully mature oocyte, some intranidal workers may not oviposit. We respond to this as follows. The worker of this ant normally has only two ovarioles (each ovary contains a single ovariole), and the maturation rate of oocyte is not so high, thus even reproductively active individuals may have no mature oocyte especially when they are just after an oviposition. In the field colonies and a laboratory colony, mature oocytes distributed randomly or evenly among workers. Individually isolated workers also have shown random variation in the number of eggs laid per day. This suggest there is little specialization in oviposition activity among intranidal workers.

2. Cerapachys biroi (Cerapachynae)

Colonies of this species found in Okinawa islands, southern Japan so far contained no gyne and no males, though the ergatoid queen of this ant was described in other countries. Workers isolated from their colonies before their adult emergence laid eggs which all grew into workers (thelytoky). This species is an opportunistic nester and group predator of other ant species. Observations of a few colony suggests that there seems to be a stronger specialization of oviposition activity among workers related to body size and age than that of P. pungens.

3.Technomyrmex albipes (Dolicoderinae)
This is one of the most common ant species in the Japanese subtropic islands, Okinawa. Field collection and laboratory observation revealed that no dealeted queen (except one) was found in many field colony samples, though colony produced winged males and females in large number in May and June. Winged sexals perform nuptial flight and can do independent founding. In the mature colonies, however, reproduction is performed by wingless ergatoid females inseminated by wingless ergatoid males after within colony mating. Up to half of the adult population of each colony is inseminated ergatoid females, the others being non-reproductive true workers (ergatoid males occupied only small fraction of the adult population). Only workers perform extranidal activities. A morphological distinction between workers and ergatoid females is the possession of spermatheca in the latters. We inferred that dispersion and foundation of new colonies were made by outbred winged sexuals, while maintenance and enlargement (by budding) of the colony is performed by inbred wingless sexuals (ergatoid males and females).

Characteristics shared by these three species are opportunistic nesting, budding, world wide distribution, being in island populations distant from continents (in our studies), inbreeding (or parthenogenesis that may be similar to inbreeding in genetic consequence), association of multiple reproductives and worker (or ergatoid reproduction). These empirical data may help us to understand such a extraordinal reproduction syndrome as well as to solve the problems of maintenance and evolution of sexual reproduction.

Refernces

1.Tsuji, K. 1988. Behav. Ecol. Sociobiol. 23: 247-255.
2.Tsuji, K. 1990. Anim. Behav. (in press).

DIVISION OF LABOUR AND REPRODUCTION IN THREE AFRICAN SPECIES OF PONERINE ANTS IN THE GENUS *PLATYTHYREA* ROGER

M.H. Villet

Platythyrea schultzei, P. sp. A and P. lamellosa lack a morphological queen caste. Dissections and intensive behavioural observations showed that in each colony a single ant had mated and monopolised diploid reproduction. Such mated laying workers of P. schultzei also inhibited oogenesis in their nestmates, while the presence of mated workers of P. lamellosa appear to prevent oviposition but not oogenesis in other colony members. Behavioural evidence suggests that mated workers of P. sp. A might maintain their status by inhibiting mating in other workers. A nest of P. arnoldi was found to contain a functional queen and several laying workers, both virgin and mated [1]. This implies that the type of reproductives found in the other three species were derived from the worker caste, and were not morphologically cryptic queens [2].

Cluster analysis of behavioural profiles of all members of a colony revealed clusters of workers filling different social functions or roles. Age was clearly implicated in producing this division of labour in P. schultzei and P. sp. A, presumably through the effects of learning and changing physiology. Statistical evidence of a link between size and role was found in P. lamellosa only. It may have been due to a correlation between body size, seasonal changes in larval nutrition and age polyethism and not to a direct causal relationship.

REFERENCES

1. Villet, M.H. 1989. Ph.D. thesis, University of the Witwatersrand.

2. Ward, P.S. 1983. Behavioural Ecology and Sociobiology, 12: 285-299.

SPATIAL ORGANIZATION IN THE NEST DURING COLONY ONTOGENY IN THE PONERINE ANT *PACHYCONDYLA (NEOPONERA) APICALIS*

Fresneau D. & Corbara B.

Laboratoire d'Ethologie et Sociobiologie, URA CNRS 667, Université Paris XIII, F-93430 Villetaneuse, FRANCE.

The spatial organization of *Pachycondyla apicalis* colonies has been studied, by using an automated photographic technique (Corbara et al., 1986), during the successive demographical stages of their development. It appears that spatial and social organization are generally highly correlated, especially when the colony comprise more than 30 individuals (see Figure).

The consequences concerning the functioning of the societies as concern division of labour and social interactions between individuals are studied and discussed. According to spatial data (absolute localisations in the linear nest, relative localizations, individual distances to the different categories of brood) it is possible to characterize two critical points during the genesis of a *P. apicalis* society. The first is linked with the differenciation between external and internal tasks. The second is a consequence of the segregation of the different categories of brood.

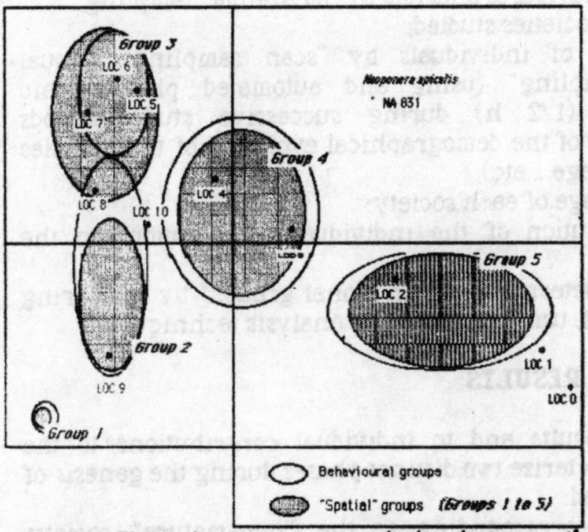

Figure : There is a strong correlation between the tasks performed by individuals and their localisation in the nest. Groups are obtained by mean of a clustering analysis of spatial data and behavioural data. The society consists of one queen and 33 workers.)

Corbara, B., Fresneau, D., Lachaud, J.-P., Leclerc, Y., Goodall, G. An automated photographic tecnique for behavioural investigations of social insects. Behavioural Processes, 13, 237-249.

EVOLUTION OF THE DIVISION OF LABOUR DURING SOCIETY ONTOGENY IN PONERINE ANTS (HYMENOPTERA, FORMICIDAE)

Corbara B.*, Fresneau D.*, Lachaud J.-P.** & Dejean A.* °

*Laboratoire d'Ethologie et Sociobiologie, URA CNRS 667, Université Paris XIII, F-93430 Villetaneuse, FRANCE.
**Centre de Recherches en Biologie du Comportement, Université Paul-Sabatier, F-31062 Toulouse, FRANCE.
°Laboratoire de Zoologie, Faculté des Sciences, Yaoundé, CAMEROON.

Individual analysis of the behaviour of each member of an insect society makes possible an in-depth study of the mechanisms of its social organization. The evolution of this social organization, in terms of division of labour, has been studied from the foundation stage, in four species of Ponerinae

MATERIAL AND METHODS

Four species of neotropical Ponerinae have been studied : *Ectatomma ruidum, Ectatomma tuberculatum, Pachycondyla villosa* and *Pachycondyla apicalis*. The procedure consists of four successive steps :

- determining a grid for a behavioural analysis by "ad libitum" sampling;
- marking individuals of the societies studied;
- recording the behaviour of individuals by "scan sampling" (visual recording) or "instantaneous sampling" (using and automated photographic technique) at regular intervals (1/2 h) during successive study periods corresponding to the different stages of the demographical evolution of the societies (first worker stage, second worker stage ...etc);
- data analysis, i.e. for each stage of each society:
  - describing the evolution of the individual contributions to the different social tasks,
  - when it is possible, determining "functional groups" by comparing and aggregating behavioural profiles, using multivariate analysis techniques.

RESULTS

According to group-level results and to individual contributions to the different tasks , it is possible to characterize two distinct phases during the genesis of the societies in the four species studied:

- during the first phase, corresponding to the "non-mature" society, individual behavioural profiles are essentially determined by the social environnement (essentially the presence of brood),
- during the second phase, corresponding the the "mature" society, age polyethism seems to be the main factor determining behavioural profiles. Nevertheless, the ants show a very high degree of individual behavioural flexibility which has already been demonstrated by the mean of social regulation experiments.

THE FORMATION OF POLYETHIC STRUCTURES IN THE TERMITE *MACROTERMES BELLICOSUS*

R.H.Leuthold*, L.Barella*, J.L.Deneubourg° and S.Goss°

*University of Berne, Division of Animal Physiology, Erlachstrasse 9a, 3012 Berne, Switzerland

°University of Brussels, Unit of Behavioural Ecology, BP 231, 1050 Brussels, Belgium

Macrotermes bellicosus, as all Macrotermes species, has two worker castes: major workers (20%) and minor workers (80%), the former being male, the latter female. Such a caste ratio is unique in this genus (Gerber et al., 1988) and may well be the expression of highly developed polyethic specialisation in this caste.
In contrast to other species, extranidal exploration and building activities on virgin ground are performed almost exclusively by minor workers and only 1 to 3% of strolling majors. Investigations made in experimental arenas have shown that only after food is discovered does the extranidal population undergo drastic transformation in caste ratio, major workers comprising up to 75% (Lys and Leuthold, 1987). Recent discoveries, however, have revealed that the development of castes involved in foraging depends on the consistency of food, the predominance of hard food leading to a significantly higher ratio of major workers. The major workers are indeed the specialists capable of biting off bits of hard food. How does the society regulate caste involvement according to this functional need?
The consecutive steps of this integrative regulation process were analysed in a flat transparent chamber where the same food substance (lime wood) was offered either in a solid piece or in a piece of pressed sawdust. No soil was added to the arena. From a series of analyses of observations we propose a hypothetical model to explain social structuring in terms of a self-regulating mechanism. We are fully aware that our simple statements are valid only for this simplified experimental arrangement with no soil in the arena.
Our 5 main statements are:

1) Both worker castes are recruited to food only by their own caste;
2) Only carriers of food are able to instigate recruitment in the nest;
3) Only 20% of the food collected is carried directly to the nest, the remaining food being transferred one or several times from one worker to another;
4) The minor workers accept food transfer better than the major workers;

5) Food is transferred more successfully to workers who are already carrying one or two food particles than to those carrying nothing.

From this a hierarchical order of probability (P) of food transfer between the possible combinations of workers meeting is concluded:

P (**A**:**a**) > P (**A**:**A**) > P (**A**:a) > P (**A**:A) and

P (**a**:**a**) > P (**A**.**A**) > P (**A**:a) > P (**a**:A)

where **A** and **a** are major and minor foodcarrying workers, A and a are major and minor non-foodcarrying workers.

The model of caste regulation in simple words works as follows: with hard food only the major workers (with more powerful mandibles) are able to bite off pieces. Therefore the majority of food transfers will take place within this same caste. Recruitment will result in an equilibrium of a majority of major workers according to the hierarchy of food transfer probability.

In the case of soft food the minor workers are also able to take pieces. Because of the higher probability of food transfer from **A** to **a** the major workers lose the majority of pieces by transferring them to minor workers which then carry them to the nest and recruit their own caste. In this situation an equilibrium with a majority of minor workers participating in foraging will be reached.

This model is surprising in its simplicity and represents a basis for explaining the social process of integration as seen in this particular situation.

A mathematical formulation of this hypothetical model will be simulated on a computer. The value of such procedure is to provide a method guide for specific experiments in the future which will be aimed at quantifying missing parameters in order to understand the real biological process. The parameters used in the model are the number of major or minor workers able to be recruited, the rate of recruitment, which is partly controlled by the quality of food, and the different probability of food exchange between the termites (minor-major, loaded-unloaded, etc.).

REFERENCES

C.Gerber, S.Badertscher and R.H.Leuthold (1988)
Polyethism in Macrotermes bellicosus (Isoptera).
Insectes Soc. 35, 226-240.

J.-A.Lys and R.H.Leuthold (1987)
Caste polyethism during the process of food acquisition in the termite Macrotermes bellicosus.
In: Chemistry and Biology of Social Insects (eds. Eder and Rembold), Peperny, Munich, p.116.

Symposium 16

CHEMICAL SIGNATURE IN SOCIAL INSECTS

Organizer : **J.L. CLEMENT, France**

CHEMOTAXONOMIC FINGERPRINTING WITH ANT NATURAL PRODUCTS

M. S. Blum, T. H. Jones, and H. M. Fales
Laboratory of Chemical Ecology, University of Georgia, Athens, Georgia 30602 USA

Many of the allomones and pheromones produced by ants have a relatively restricted distribution [1] and these compounds can thus be of considerable chemotaxonomic signifigance. Studies of the chemistry of the venomous alkaloids produced by myrmicine species in the genera Solenopsis and Monomorium have demonstrated that a host of these nitrogen heterocycles are characteristic products of relatively few species [2]. These alkaloids, which include indolizidines, piperidines, pyrrolidines, pyrrolizidines, piperideines, and pyrrolines [3], are present in distinctive mixtures in the venoms of many species, and can constitute diagnostic fingerprints for members of these taxa. Novel alkaloids fortify the venoms of various species [1] and these compounds can provide a useful character for taxonomic analyses. Recent studies have shown that a major species of Monomorium, restricted to New Zealand, may actually consist of four "populations" based on each form possessing a highly distinctive venom fingerprint in terms of alkaloidal constituents [4]. Similarly, an Indian species of Monomorium generates a venom that is fortified with a complex mixture of trans-2,5-dialkylpyrrolidines, cis isomers of these alkaloids, and their N-methyl analogues [5], thus providing a novel venomous fingerprint. Other examples of qualitatively distinctive venoms being produced by these myrmicine species have been recently described [6].

Males of a variety of formicine species produce sex-specific compounds in their mandibular glands that are often quite species specific. Males of many Camponotus species produce sex pheromones that are characterized by the presence of idiosyncratic compounds that may be of considerable chemotaxonomic utility [7,8]. In some cases, all castes of a species may biosynthesize the same mandibular gland constituents [9], a development that may also be useful as a chemotaxonomic character. The synthesis of qualitatively distinctive blends of pheromones in the mandibular glands of males in the formicine genus Myrmecocystus [10] serves to emphasize that sex-specific natural products may have a widespread distribution in the Formicinae.

REFERENCES

1. Jones, T.H., M.S. Blum, and H.M. Fales. 1982. Tetrahedron, 38: 1949-1958.

2. Jones, T.H. and M.S. Blum. 1983. In: Alkaloids: Chemical and Biological Perspectives, S.W. Pelletier, Ed., John Wiley & Sons, New York, Vol. 1, pp. 33-84.

3. Blum, M.S. 1985. In: Bioregulators for Pest Control, P.A. Hedin, Ed., ACS Symposium Series 276, American Chemical Society, Washington, DC, pp. 393-408.

4. Jones, T.H., S.M. Stahly, A.W. Don, and M.S. Blum. 1988. Journal of Chemical Ecology, 14: 2197-2212.

5. Jones, T.H., M.S. Blum, P. Escoubas, and T.M. Musthak Ali. 1989. Journal of Natural Products, 52: 779-784.

6. Jones, T.H., M.S. Blum, A.N. Anderson, H.M. Fales, and P. Escoubas. Journal of Chemical Ecology, 14: 35-45.

7. Brand, J.M., R.M. Duffield, J.G. MacConnell, M.S. Blum, and H.M. Fales. 1973. Science, 179: 388-389.

8. Brand, J.M., H.M. Fales, E.A. Sokoloski, J.G. MacConnell, M.S. Blum, and R.M. Duffield. 1973. Life Sciences, 13: 201-211.

9. Blum M.S., R.R. Snelling, R.M. Duffield, H.R. Hermann, and H.A. Lloyd. 1988. In: Advances in Myrmecology, J.C. Trager, Ed., E.J. Brill Publishing Company, New York, pp. 481-490.

10. Lloyd, H.A., M.S. Blum, R.R. Snelling, and S.L. Evans, 1989. Journal of Chemical Ecology, 15: 2589-2599.

CHEMICAL RECOGNITION OF NESTMATES IN ANTS — AN EVOLUTIONARY APPROACH

Klaus Jaffe

Departamento de Biologia de Organismos
Universidad Simon Bolivar
Apartado 89000, Caracas 1080, Venezuela

The capability of discerning nestmates from other conspecifics seems essential for kin selection theory but not necessarily for colony functioning. Nestmate recognition can be detected experimentally only through differential behavior of workers toward conspecifics. Normally aggressive behavior is thought to be involved, but this is not always evident from experiments. Primitive ants or small colonies lack aggressive behavior but not nestmate recognition systems. Our results suggest that aggressive exclusion of non nestmates is a feature of complex ant societies. Colony mixing in young colonies or in small colonies from socially primitive species is mainly avoided thanks to the orientation mechanisms of workers which make colony mixing an improbable situation. Thus, efficient nestmate recognition systems are evolved only as a complement to aggressive intercolony competition mechanisms. In this sense we might classify the known nestmate recognition systems among ants on a continuum created by two extreme situations: no nestmate recognition and no aggression, and nestmate recognition with aggression toward non-nestmates. As intercolony aggression is regulated by other pheromones as well, such as alarm pheromones and territorial markers, nestmate recognition in aggressive species is linked to the agonistic pheromones.

Much research has been focused on whether nestmate recognition signals are genetically fixed or environmentally determined. The general conclusion is that both factors determine the recognition signals. If this is generally true, than genetically determined signals are of primary importance because they will always be present and will always differ among colonies; thus they will be used as reference during imprinting. Recognition signals of environmental origin will only amplify those which are genetically determined, among closely related colonies. In the absence of genetically determined nestmate recognition signals, environmentally determined signals will be of little use among neighboring colonies subjected to similar environmental stimuli.

Our observations with mature *Atta* colonies in the field revealed that intercolony aggression are very rare events, triggered by competition for food or space, which normally is partitioned thanks to a chemical territorial mark secreted by workers around the nests and on trails leading to very attractive food sources. On the other hand, incipient *Atta* colonies are more tolerant toward workers from other incipient nests and show less aggression toward conspecifics, although killing of incipient *Atta* nests by mature neighbor colonies is very common. Thus, the ontogeny of agonistic behavior of colonies from socially complex species parallels the evolution nestmate recognition in ants.

The signals involved in nestmate recognition systems vary from cuticular hydrocarbons or chemicals absorbed on the cuticular hydrocarbons to very volatile chemicals in the alarm pheromone. Nestmate recognition signals can be any chemical produced by ants. It is known that most pheromones are colony specific, thanks to the different blends of the same chemicals constituting the pheromone of the species. Evolution tends to simplify the signal used for nestmate recognition, probably in order to make the recognition fast and from a distance. Thus, alarm pheromones are the first to be detected in intra-specific encounters and are used as primary nestmate recognition signals. Other chemicals are also recognized colony-specifically but are detected by workers only in specific circumstances and can thus be considered of secondary importance.

Ants seem to learn the familiar odors shortly before or after eclosion. Any strange odors will then be recognized as foreign and will eventually provoke aggression. This causes that intruder ants are normally the first to detect non-familiar odors, triggering alarm behavior, and thus making themselves evident as intruders.

Thus, nestmate recognition can be achieved by different ways, but if considered as part of agonistic behavior, it has to be seen as the mechanism by which ants orient their aggression in intraspecific encounters. In this respect, evolution tends to simplify the recognition signal, and ants from species with complex societies use alarm pheromone components to achieve nestmate recognition and to identify conspecific intruders.

DISCRIMINABILITY OF FAMILIAR KIN AND NON-KIN WITHIN CARPENTER ANT COLONIES

Norman F. Carlin and Stefan P. Cover
Museum of Comparative Zoology Laboratories, Harvard University,
Cambridge, Mass. 02138 U.S.A.

Workers of many social insect species learn to recognize familiar nestmates as a class distinct from all other conspecifics, and usually exclude the latter from their colony. In genetically heterogeneous colonies, containing more than one queen or queen(s) that mated more than once, workers would also improve their inclusive fitness by preferentially helping members of their own matriline or patriline. Studies of honey bees (1) and carpenter ants (2) indicate that workers are capable of distinguishing among subclasses of nestmates on the basis of kinship. However, the biases reported are quite weak, and may be experimental artifacts in colonies of artificial genetic heterogeneity. In colonies of a polygynous carpenter ant, Camponotus planatus, we found no evidence of kin discrimination among naturally cohabiting matrilines.

Queens were separated from their original colony to obtain brood, which was reared to adulthood by nurse workers from the same colony. Workers and virgin queens that eclosed in these rearing groups were individually marked and returned to their group of mother queens. Five C. planatus colonies with four queens each, and one with two queens, containing an average of 69±20 workers and 8±5 virgin queens, were observed for a total of 16 hours each. Grooming, food exchange, antennation and jerking (a behavior which may represent recruitment or aggression) were recorded among workers (N=10,999 acts), between workers and virgin queens (N=850) and between workers and mother queens (N=1822). No consistent kinship bias occurred in any of these interactions. We are repeating this experiment with artificial colonies, combining worker lineages originating from different populations, to determine whether within-colony discrimination may appear among genotypes that do not naturally coexist. We conclude that social insect workers may be unable to discriminate familiar kin from familiar non-kin, preventing them from obtaining the inclusive fitness benefit of within-colony kin recognition in natural colonies.

REFERENCES

1. Page,R.E. and M.D. Breed. 1987. Kin recognition in social bees. Trends Ecol. Evol., 2: 272 - 275.
2. Carlin,N.F., B. Holldobler and D.S. Gladstein. 1987. The kin recognition system of carpenter ants (Camponotus) III: Within - colony discrimination. Behav. Ecol. Sociobiol., 20: 219 - 227.

EFFECT OF KAIROMONES AND ALLOMONES DURING PREDATION BY ANTS

Dejean A. and Corbara B. Laboratoire d'Ethologie et Sociobiologie, URA CNRS 667, Université Paris XIII, F-93430 Villetaneuse, FRANCE.

Kairomones and allomones are interspecifical chemical messages. They are discernable through the adaptative advantage which is in the receiver organism for kairomones (this organism thus receipt informations from substances which are often pheromones for the emetting organism). Adaptative advantage is in the producing organism for the allomones.

Material and methods

This work was achieved in Quintana-Roo, a state of Mexico. It consists in seeking cases where, in the predator-prey relations, the action of a kairomone or an allomone may be observed or suspected and then demonstrated. The showing up of the effect of these two kinds of products may be carried out by choice tests between two types of refuges, by using an olfactometer or a technique of washing and soaking in water of washing off of other ones's.

Results

Leptogenys mexicana are specific predators of Isopoda. In the laboratory broods, Isopoda introduced in the hunting area of *Leptogenys* have a tendancy to take refuge in the ant nest, even if there is other possibilities. Two test were carried out to establish whether Isopoda are attractive to *Leptogenys*. In one case, Isopoda introduced in a hunting area are forced with a choice between three identical refuges, one containing *Leptogenys* workers; in the other, an olfactometer is used. In the two cases, attraction is evident with P>0.001.

Faced with Rhinotermitidae soldiers, *Pachycondyla villosa* hunting workers present a posture of "prudence" whereby they throw back their antennae and lift their hind legs. Thus, they avoid contact of these appendages with the mandibles of the prey. The Rhinotermitidae workers on the other hand are antennated during capture. On washing soldiers and then soaking them in water that was used to wash workers, the posture of "prudence", may be removed. Reciprocally, by washing the workers and soaking them in water that was used to wash soldiers, the posture of "prudence" may be induced. There is a cuticular product emanating from the soldiers which acts as a kairomone on the *P. villosa* workers.

The *Eciton burchelli* are general predators that often attack societies of a number of species of ants. Their approach triggers escape reactions and abandon of the nest in different species. This escape may be induced by blowing into a tube containing *Eciton* workers. There is thus a substance emanating from the *Eciton*, which provoques the reaction of the prey. This escape is more or less well organized. *Camponotus abdominalis* workers, for instance, escape individually in a confused manner, some workers carrying brood comb.

When a column of *Eciton* invests a *Pheidole megacephala* nest, the latter react by conter-attacking. The *Eciton* , however, remain dominant and a descending column is formed in which a number of workers transports *Pheidole* broodcomb. All the *Eciton* workers of the descending column (whether they transport prey or not) are attacked by their nesmates of the ascending column. The aggressed individuals remain motionless and left themselves be massacred. In one or two hours, tens of thousands of *Eciton* workers lie helpless on the ground. Later, they are brought back to the nest along with their prey.

By soaking *Eciton* workers in water used for washing *Pheidole* and then replacing them in their column, we established that these workers are aggressed and then massacred. Workers, soaked in pure water and reintroduced into their column are not attacked. There is thus a product emanating from the *Pheidole* which is captured by the cuticle of the *Eciton*. This product acts as a "belated allomone" and provoques a massacre amongst the *Eciton*.

FIDELITY OF THE GYNES OF *MONOMORIUM PHARAONIS* TO WORKER TRAILS

Awinash P. Bhatkar, Department of Entomology, Texas A&M University, TAES, College Station, Texas, USA 77843-2475

Monomorium pharaonis is a polygynous, polydomous species. The fidelity of gynes of M. pharaonis to the trails of their own workers was compared. The scout workers were allowed to discover a food source and lay trails up to the nests containing a number of gynes. The paths of the recruited, trail-following workers and gynes were traced adapting the technique described earlier [1, 2]. Both laboratory and natural colonies were used; tests on natural colonies were conducted on the home kitchen counter top. First, a scout was allowed to find a bit of an insect part placed 20 cm distant from the nest on a cover slip. Next, the trail of the homing scout and fidelity of a recruited forager or dealated gyne following the trail was traced with a fine felt-tipped pen on a transparent sheet held 5 mm over the trail. About 20-30 trails from 4 laboratory (1-4 in Fig.) and 2 natural colonies (5 and 6) were analyzed.

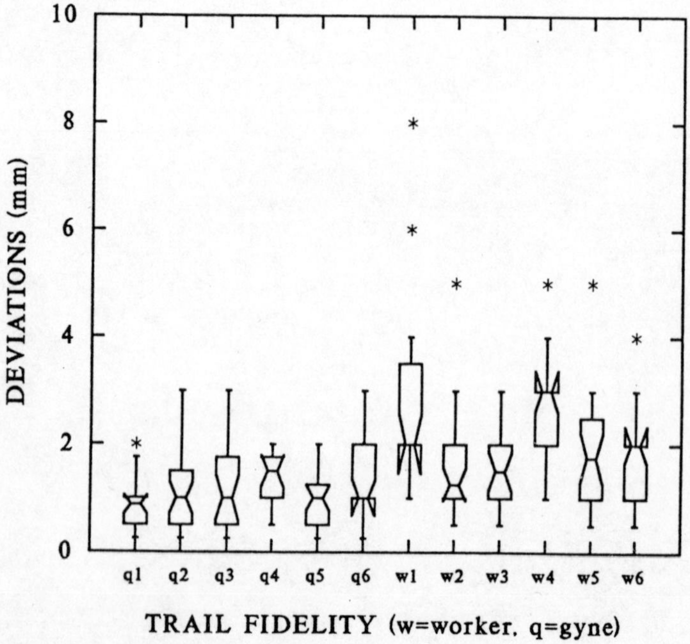

Analysis of the data showed that the recruited gynes of these species made less directional and distance deviations from trails than the workers. The graphical analysis [3, 4] of the directional deviation indicates that the gynes deviated about 0.8-1.8 mm from the worker trails. There was no significant difference between the colonies (except 4), as indicated by the overlapping 95% confidence intervals (notches) around medians in the box plots. Foraging workers from the same colonies (with similar numbers in Fig.) following the trails made significant directional deviations (about 1.5-3.5 mm, and outlier asterisks), an exploratory behavior adapted to finding an additional food resource. Thus, gynes showed more fidelity to the worker trails than the foragers. Further testing indicated that the gynes did not lay trails. These findings are consistent with the behavior of _M. pharaonis_ gynes during sociotomy together with a group of workers. New colonies are generally established near a constant resource find. The contents of worker venom glands resulted in recruitment of gynes to artificial trails, which may be suggestive to the population manipulation utility of the trail substance.

References

1. Bhatkar, A. P. 1982. Folia Entomologica Mexicana, 53: 75-85.
2. Wilson, E. O. 1962. Animal Behaviour, 10: 148-158.
3. McGill, R., Tukey, J.W., and W.A. Larsen. 1978. The American Statistician, 32: 12-16.
4. Wilkinson, L. 1988. SYGRAPH, SYSTAT, Inc., Evanston, Illinois. p. 922.

CHEMICAL SECRETIONS AND SPECIES DISCRIMINATION IN *CATAGLYPHIS* ANTS (HYMENOPTERA; FORMICIDAE)

E.D. Morgan [1], D. Agosti [2] and Sarah J. Keegans [1]

1. Department of Chemistry, University of Keele, Keele, Staffordshire. ST5 5BG. England.
2. C/O British Museum (Natural History), London. SW7 5BD. England

The development of species concepts and proper species description require both experimental and descriptive studies, and are not based on morphological characters alone. Any convenient approach based on such characters as morphological, behavioural or chemical, may be used alone or in combination to discriminate a species, there is no *a priori* best method. Chemical information may be helpful, but has as yet been used for ants on very few occasions, c.f. for example, [1].

An attempt to use chemical information to help distinguish species is illustrated by a study of the group of species of *Cataglyphis* (Formicidae:Formicinae) spread across North Africa and Asia Minor. Morphologically some of these species are easy to differentiate, others are very difficult [2].

The Formicidae use chemical communication extensively and have a number of exocrine glands that can provide characteristic chemical secretions [3]. From evidence so far available, these exocrine glands produce secretions that are relatively constant in composition, intermediate in complexity, characteristic of the species and therefore potentially useful in taxonomy. Where the secretion is volatile, it is suitable for analysis by gas chromatography and identification by mass spectrometry. Single glands from individual ants are examined by the method used. Components comprising as little at 5×10^{-9} g in the secretion can be identified and quantified.

Samples of glands were collected in the field and sealed in glass capillaries for laboratory analysis. Samples were taken from individuals from the same nest, from adjoining nests of the same species, from different localities and from other species, closely or distantly related. The metapleural glands and antennal glands did not provide sources of volatile compounds. The postpharyngeal glands can contain species-characteristic mixtures of hydrocarbons [4], but these have been little studied. *Cataglyphis nodus* had pentacosane as the major component in this gland, *C niger* contained 11- and 13-methylheptacosane (unresolved mixture) as the major component, as did other less closely related species. The mandibular glands of *Cataglyphis* frequently contain only one or no volatile compounds (e.g. *C. savignyi* [5]). Its usefulness is limited, but occasionally helpful. Both *C. nodus* and *C. niger* contain C_7 alcohol, two other species from the *bicolor* group contain nonanal and decanal, some species from other groups contain citronellol and citronellal.

The Dufour gland, present in all Formicidae and in almost all cases filled with volatile substances, contains a particularly rich mixture of hydrocarbons and oxygenated compounds in most Formicinae, including *Cataglyphis*. While the quantity varies (1 to 10µg) from one individual to another, the percentage composition remains relatively constant. Up to 40 compounds have been identified in this group including minor components present at 10^{-8} g or less. *C. niger* and *C. nodus* are very similar with pentadecane the major component, they differ in the proportion of undecane present and in the amount of minor oxygenated components, whereas *C. frigidus* has tridecane and undecane as their major components.

Our study is incomplete, but does show differences, even between closely similar species as *C. nodus* and *C. niger*, both in the *bicolar* group. *C. ruber* of the *albicans* group is quite different. Samples from the *aenescens* group of species are different chemically, as is one species from the *altisquamus* groups. Thus far, chemical similarity appears to run parallel to morphological similarity.

Analyses of some other *Cataglyphis* species have already been described by Hefetz and Orion [6].

REFERENCES

1. Ali, M.F., Attygalle, A.B., Morgan, E.D., and Billen, J.P.J., 1987, Comparative Biochemistry and Physiology, 88B: 59-63.
2. Agosti, D., 1990 (submitted).
3. Billen, J., 1987 In Chemistry and Biology of Social Insects. J. Eder and H. Reinbold, Eds. Verlag Peperny, Munich.
4. Bagneres, A.G., and Morgan, E.D., 1990 (submitted).
5. Ali, M.F., Billen, J.P.J., Jackson, B.D., and Morgan, E.D., 1989, Biochemical Systematics and Ecology, 15: 647-654.
6. Hefetz, A., and Orion, T., 1982, Israel of Journal of Entomology, 16: 87-

NESTMATE RECOGNITION IN MONOGYNE AND POLYGYNE POPULATIONS OF THE FIRE ANT *SOLENOPSIS INVICTA* BUREN

VANDER MEER, R.K.*, OBIN, M.S.# AND MOREL, L.*

*USDA-ARS Insects Affecting Man and Animals Research Laboratory, P.O. Box 14565, Gainesville, Florida 32604, USA.

#Department of Zoology, university of Florida, Gainesville, Florida 32611, USA

INTRODUCTION

It is generally accepted that nestmate recognition (worker discrimination of colony members from non-colony members among conspecifics) is based on learned "chemical cues," a subset of overall colony odor [1,2]. These cues can be derived from individuals, queens and/or workers ("discriminators") or from the environment [3]. The recognition process has two elements: chemical cues on the intruding ant's surface and the resident worker's "neural imprint" or sensory template. If the intruder's cues match the template of the resident no aggression occurs; however, if they do not match, agonistic behaviors may ensue. An aggression bioassay was used to compare nestmate recognition in polygne and monogyne *S. invicta* populations.

Polygyne and monogyne fire ant colony workers use a combination of environmental and heritable cues. All of these represent a sub-set of colony odor [2]. Environmental cues are in a continuous state of flux. We have demonstrated, through the analysis of cuticular hydrocarbons, that heritable cues are also dynamically changing. Although cuticular hydrocarbons have not been directly linked to nestmate recognition, they serve as a very good general model for the fate of heritable cues on the ant's surface. Heritable cues produced by workers or queen(s) are rapidly distributed to other colony members through grooming, trophallaxis, etc. [4,5]. Since both heritable and environmental cues are dynamic rather than static the neural template must also be dynamic. Thus a worker must, during its lifetime, continually update its perception (template) of colony odor and nestmate recognition cues [6].

POLYGYNE VS MONOGYNE NESTMATE RECOGNITION

Within a polygyne population no aggressive behavior was observed for intercolonial introductions, although there was heightened awareness (rapid antennation) compared to intracolonial controls. This supports the contention that territoriality within polygyne populations is almost nonexistent. Therefore, polygyne *S. invicta* populations fall under definition of a super- or unicolony. In contrast, in monogyne populations intercolonial introductions resulted in high levels of aggression.

Polygyne workers were not aggressive toward monogyne workers whether they were intruder or resident. But monogyne workers were highly aggressive toward polygyne workers. Heterospecific intruders *(S. richteri)* were vigorously attacked by polygyne workers, which negates the argument that polygyne workers' overall aggressivity may be diminished.

Environmental cues can be dampened through uniform rearing conditions. Based on our results it appears that in natural *S. invicta* polygyne populations, both environmental and heritable cues are dampened.

Nestmate recognition cues should be spread throughout a polygyne population by social interactions and unchallenged movement of workers. But, a uniform cue pattern throughout a population is not expected. We propose that since polygyne workers contact a large variety of cues, environmental and heritable, the observed lack of aggression is due to a less discriminating template rather than uniform cue profile. The asymmetric aggressive behavior of monogyne and polygyne *S. invicta* workers supports this hypothesis. At any point in time each polygyne worker has a distinct cue pattern, as for monogyne workers; however, the polygyne worker has a broad template compared to that of monogyne workers.

POLYGYNE POPULATION STRUCTURE?

No evidence was found for structure within a polygyne population based on aggression bioassays, although there was a heightened awareness for introductions of workers from close but different polygyne populations (about 30 km apart) in Florida. The same was true for introductions of polygyne workers from central Georgia (ca. 300 km apart) - they were not recognized as different by Florida polygyne workers. In contrast, workers from two Texas polygyne populations (ca. 1,000 km from Florida) were vigorously attacked by Florida polygyne workers. At the present time it is not possible to determine if the heightened aggression is due to heritable or environmental differences.

REFERENCES

1. Breed, M., and Bennett, B., (1987). In: Kin Recognition in Animals. D.J.C. Fletcher and C.D. Michener, Eds., John Wiley & Sons, New York, pp.243-286.
2. Vander Meer, R. K. 1988. In: Interindividual Behavioral Variation in Social Insects. R.L. Jeanne, Ed., Westview Press, Boulder, CO, pp. 223-255.
3. Hölldobler, B. and C.D. Michener. 1980. In: Evolution of Social Behavior: Hypotheses and Empirical Tests. H. Markl, Ed., Verlag Chemie, Weinheim, pp. 433-439.
4. Sorensen, A. A., Fletcher, D. J. C., and Vinson, S. B. 1985. Distribution of inhibitory queen pheromone among virgin queens of the ant, *Solenopsis invicta*. Psyche 92: 57-69.
5. Vander Meer, R. K., and Wojcik, D. P. 1982. Chemical mimicry in the myrmecophilous beetle *Myrmecaphodius excavaticolis*. Science, 218: 806-808.
6. Vander Meer, R.K., Saliwanchik, D. and Lavine, B. 1989. Temporal changes in colony cuticular hydrocarbon patterns of *Solenopsis invicta*: Implications for nestmate recognition. J. Chem. Ecol. 15: 2115-2125.

IDENTITY OF CUTICULAR HYDROCARBON PROFILE AMONG WORKERS OF THE ANT WHICH IS MAINTAINED BY THE PRESENCE OF THE QUEEN WOULD BE THE NESTMATE RECOGNITION CHEMICAL CUE

by R. YAMAOKA and H. KUBO

Department of Applied Biology, Kyoto Institute of Technology, Matsugasaki Sakyoku Kyoto 606 JAPAN

Among social insect, presence of colony odor which was thought to be the most important nestmate recognition chemical cue had been pointed out but the chemical nature is still unknown. Here we present the clear result that the cuticular hydrocarbon profile of ant might be a origin of colony odor and that would be controled by the queen.

Difference of cuticular hydrocarbon composition among 45 species of ants in Japan was examined briefly using capillary GC and GC-MS. All of the ant examined had a completely differrent hydrocarbon composition. This result indicate that the hydrocarbon composition might be species recognition cue of the ant. On same species, in the same colony no difference on the composition was observed between queen, males and workers. In the same species, *Formica japonica*, *Camponotus japonicus*, *Formica sp. 5* and *Tetramorium caespitum* the different colony member had the same hydrocarbons but the relative intensity between them (cuticular hydrocarbon profile; CHP) was appropriately different on all of the ant species examined. On the contrary in the same colony, CHP were examined indivisually and all of the members had completely identical profile. These results suggested that the CHP could be the colony recognition cue if the ant can recognize even the small difference of the relative contents of the hydrocarbon by their anteni.

Mechanism of the uniformity of the CHP was examined by capillary GC. Just after digging of a field colony of *Formica sp. 5* consisted of about 100 workers and one original queen was devided into two. Group A. was formed by a queen and a half of the workers. Group B. was only formed by remainder half of the workers. Both groups reared in the plastic box(5cm x 15cm x 7cm h.) on the same condition. After the separation, in group A all of the members made a mass around the queen while group B gradually lost their aggrigation behavior. Twenty days after separation, the queen in group A was moved into group B. The workers of group B aggregated around the queen gradually and made a mass, on the contrary the workers of group A lost their aggregation gradually. Before the separation the CHP of three individuals was exactly the same and that was also identical with the profile obtained by 5 workers at one time. Ten days after separation the profile of the queenless colony lost the uniformity, another word, relative amount of the each hydrocarbons to most abandant one hentriacontadiene changed. But the total amount of the hydrocarbons did not change and about three microgram per individual. While on queenright colony shape of the profile little bit changed compared with that of ten days before but the identity of the profile of the indivisuals maintained tightly. All of the change of CHP were shown on Fig.1 by circular diagram. All of the results suggested that the queen performs very important role to the identity of CHP in all the colony members.

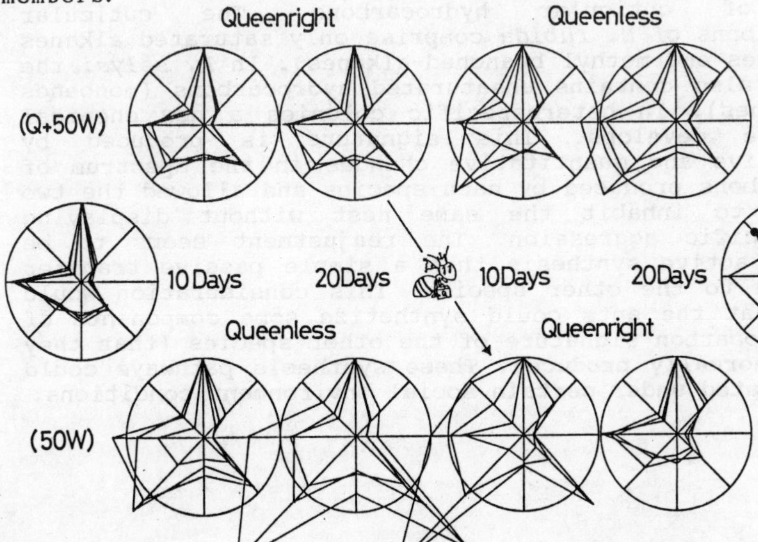

CHANGE OF THE SPECIFIC CHEMICAL SIGNATURES OF TWO ANTS SPECIES REARED IN MIXED COLONIES (*FORMICA SELYSI*, FORMICINAE AND *MANICA RUBIDA*, MYRMICINAE)

C. Errard*, A.G. Bagnères and C. Lange*****

*Laboratoire d'Ethologie et Sociobiologie, U.R.A.667, Université Paris XIII, 93430 Villetaneuse, France. **CNRS-UPR 38, 31 Chemin Joseph Aiguier, 13402 Marseille, France. ***Laboratoire de Chimie Organique Structurale, Université P.M. Curie, 4 place Jussieu, 75005 Paris, France.

In this investigation we studied *Formica selysi* (Formicinae) and *Manica rubida* (Myrmicinae) that live in the same habitat in the French Alps.

We used the experimental model of artificial mixed societies as an effective technique for testing the influence of early conditioning callow workers to a mixed colony odour upon interspecific recognition and modification of the species-specific cuticular hydrocarbons.

It was shown that there is a sensitive period for establishing preferential relations between individuals of these species. Thus, the association of *F. selysi* and *M. rubida* must take place within 24h. after emergence so that heterospecific interactions do not induce aggression.

The cuticular hydrocarbons of *F. selysi* and *M. rubida* reared in single species and in mixed species colonies were determined using gas chromatography (GC) and GC-mass spectrometry. In colonies containing both species, each species modified its species-specific recognition odour. This odour is composed, at least in part, of cuticular hydrocarbons. The cuticular hydrocarbons of *M. rubida* comprise only saturated alkanes (n-alkanes and methyl branched-alkanes). In *F. selysi* the mixture also contains unsaturated hydrocarbons (monoenes and dienes). In heterospecific colonies a new chemical signature develops. This signature is produced by qualitative and quantitative changes in the spectrum of hydrocarbons produced by each species and allowed the two species to inhabit the same nest without displaying interspecific aggression. The reajustment seems to be more an active synthesis than a simple passive transfer from one to the other species. This consideration would imply that the ants could synthetize some components of the hydrocarbon signature of the other species (that they do not normally produce). These synthesis pathways could be activated under certain social environment conditions.

COMMUNICATION BETWEEN THE PARASITIC ANT *FORMICOXENUS PROVANCHERI* AND ITS HOST *MYRMICA INCOMPLETA* (HYM. FORMICIDAE)

by A. LENOIR*, N. BARBAZANGES*, R. YAMAOKA**
C. ERRARD* and A. FRANCOEUR***

* Laboratoire d'Ethologie et Sociobiologie, URA CNRS 667, Université Paris Nord, F-93430 Villetaneuse; ** Kyoto Institute of Technology, Faculty of Textile Science, Kyoto 606 Japan; *** Laboratoire de Biosystématique, Université du Québec à Chicoutimi, Chicoutimi, Québec, Canada.

Formicoxenus provancheri is an ant species living in close relation to its host *Myrmica incompleta*. It constructs small nests in the earthen walls of the host colony or in the immediate vicinity, where its brood is apart from the *Myrmica* chambers. It is strictly dependant of its host and cannot live apart: this is called *xenobiosis* . *Formicoxenus* workers are very attracted by *Myrmica* adults and brood which are abundantly licked, probably to find food particles and cuticular secretions. *Formicoxenus* workers generally forage in the host colony and are rarely aggressed. When introduced into an alien unknown host colony of *Myrmica* they are strongly aggressed. The aggressed *Formicoxenus* present appeasement postures, similar to the sexual calling of other *Formicoxenus*, which inhibit the host aggressiveness.

Formicoxenus brood is isolated and protected from the host because *Myrmica* ergates eat *Formicoxenus* larvae when they can find them.

How can the *Formicoxenus* ergates introduce into the *Myrmica* colony where they are tolerated? Cuticular hydrocarbons were analyzed and it appeared that cuticular hydrocarbons profiles are very similar between the two species. It is also noteworthy that intercolonial recognition is weak in *M. incompleta*. In this highly polygynous species the societies are not closed and individuals from any other colony are integrated easily. This phenomenon is a predisposing factor for parasitism and tolerance towards intruders with a similar chemical signature.

THE POISON APPARATUS OF *MESSOR BOUVIERI*: CHEMICAL AND BEHAVIOURAL STUDIES

Philip J. Wright, Brian D. Jackson and E. David Morgan, Departments of Biological Sciences and Chemistry, University of Keele, Keele, Staffordshire, ST5 5BG, England.

Species of Messor, a genus of Old World harvester ants, have a reduced and probably non-functional sting, unlike most other myrmicine ants. Workers of M. bouvieri Bondroit were collected from Tunisia and Spain. Chemical analysis on single glands was carried out by gas chromatography-mass spectrometry using a solid sampling technique [1].

The poison gland has two major components, the alkaloids anabasine (mean 3μg/gland) and anabaseine (40ng/gland). Anabasine is also the major component of the poison gland of M. ebeninus [2] and anabaseine has been found in species of Aphaenogaster [3], a genus taxonomically close to Messor. In contrast, other members of the genus examined, M. minor, M. capitatus and M. mediorubra, do not contain alkaloids. The venom of M. bouvieri also contains a trace (9ng) of 3-ethyl-2,5-dimethylpyrazine. The synthetic compound was found to induce trail following [4].

Using a circular trail following biassay, extracts of the poison and Dufour glands showed equal activity, but there was no additive or synergistic effect when combined [4]. However, workers showed a clear preference for the Dufour gland extract in a Y test biassay. The Dufour gland contains mainly unsaturated linear hydrocarbons, the most abundant being 6,9-heptadecadiene (50%) and 3,6,9-heptadecatriene (10%). Double bond positions were determined by 3 microchemical methods. M. minor and M. capitatus contain mainly saturated hydrocarbons [5].

References

1. Morgan, E.D. and Wadhams, L.J. 1972. J. Chromat. Sci., 10:528-529.

2. Coll, M., Hefetz, A. and Lloyd, H.A. 1987. Z. Naturforsch., 42C:1027-1029.

3. Wheeler, J., Olubago, O., Storm, C.B. and Duffield, R.M. 1981. Science, 211:1051-1052.

4. Jackson, B.D., Wright, P.J. and Morgan, E.D. 1989. Experientia, 45:487-489.

5. Ali, M.F., Billen, J.P.J., Jackson, B.D. and Morgan, E.D. 1989. Biochem. Syst. Ecol. 17:469-477.

RUBBING MOVEMENTS AT PERCHES AND TEGUMENTAL GLANDS IN THE LEGS OF *POLISTES DOMINULUS* MALES (HYMENOPTERA, VESPIDAE)

Laura Beani and Carlo Calloni

Dipartimento di Biologia Animale e Genetica. Università di Firenze. Via Romana 17. 50125 Firenze, Italy.

Males of *Polistes dominulus* (Christ) (= *P. gallicus* (L.) of the previous authors) return day after day to sunlit prominent landmarks, where they defend tiny adjacent territories, such as few branches of a high tree or a particular segment of a pole. Patrol flights are combined with stops on selected perches and a conspicuous abdomen dragging behaviour, significantly ($P< 0.01$) more frequent in high density of intruders. When they drag the 5th, 6th, and 7th gastral sternites against the substrate, they simultaneously stretch out their legs and drag them for few centimetres over the surface.

This peculiar behaviour, significantly ($P< 0.001$) related with prolonged grooming sequences involving legs, prompted us to verify the occurrence of any tegumental glands in the legs. In fact abdomen rubbing can be interpreted as a pheromonal release, and it is associated, in many polistine wasps, with well developed class 3 cell glands (according to Noirot and Quennedy). Semi-thin sections show distinct secretory units in all the legs, in the femur, tibia and especially the tarsus, resembling sternal glands (peripheral ellipsoidal nucleus, branched cuticular structure in the cell core explainable as the end apparatus, long excretory ducts). Scanning electron micrographs show helical basiconica sensilla with pores, near the base, which are probably duct openings. Tegumental glands in the legs were never previously described in polistinae wasps (and in wasps at all), except as adhesive secretory organs in the last tarsal segment.

OLFACTORY CUES IN NEST RECOGNITION BY SOLITARY BEES (*LASIOGLOSSUM FIGUERESI*; HALICTIDAE) AS A PREADAPTATION FOR THE EVOLUTION OF KIN ASSOCIATIONS

W.T. Wcislo
Department of Entomology, Snow Hall
University of Kansas
Lawrence, Kansas 66045-2119, U.S.A.

"Homing" ability is characteristic of aculeate Hymenoptera [e.g., refs. in 1,2], and is assumed to be an important "preadaptation" for multiple evolutions of insect sociality [e.g., 3,4]. This assumption is imprecise. An ability to recognize one's nest, and distinguish it from another, is more important than homing ability per se because integrative social interactions are modulated by chemical cues [6], while homing primarily relies on visual cues [refs. above]. This report concerns manipulations of recognition cues at nests of a solitary bee, Lasioglossum figueresi [2]. Secondly, it develops the hypothesis that the use of olfactory cues for nest recognition helps explain broad patterns of social evolution within Hymenoptera.

METHODS AND RESULTS

Nests were in vertical earthen banks, and had turrets at the entrances. Prior to the first foraging trip on any day, each female made a brief "locality study" to learn cues for topographic orientation. After a training period, bees returned to their nests and entered them without hesitation if local visual landmarks were displaced.

In choice experiments returning bees were presented with their own and an alien turret at their nest entrances. Following antennal contact, females always entered their turret, and rarely entered foreign turrets. Yet if turrets were switched between nests, then bees entered the correct nests without hesitation.

When both turrets and the inner nest entrances were swabbed with hexane, then returning bees delayed entering nests. Hexane applied to the outside of the entrance had no effect. If the turret was removed prior to applying hexane to the burrow, and the turret then replaced, bees entered without hesitation. Following the application of whole-body extracts of female bees to inner nest entrances, bees entered nests after a longer elapsed time and required a greater number of approach flights.

DISCUSSION AND CONCLUSIONS

Bees and sphecid wasps form a monophyletic group (Apoidea) within which social behavior has evolved more frequently in the former than the latter [4]. The use of chemicals in nest construction, and hence their availability as cues for individual nest recognition, may be preadaptations for the evolution of group-associations along kinship lines [also 6]. The phyletic distribution of sociality is consistent with this hypothesis: most sphecid wasps are solitary or communal [3,7], and apparently do not use chemicals in nest construction [refs. in 2]. The only known eusocial sphecids [8; possibly 9,10] are all in the Pemphredoninae, which use chemical secretions in nest construction [11] and one solitary pemphredonine uses chemical cues in nest recognition [12].

REFERENCES

1. Turner, C.H. 1923. Transactions of the Academy of Sciences, St. Louis, 24 : 27-45.
2. Wcislo, W.T. manuscript. Nest recognition in a solitary sweat bee, Lasioglossum figueresi (Hymenoptera: Halictidae).
3. Wilson, E.O. 1971. The Insect Societies, Harvard Univ. Press.
4. Michener, C.D. 1974. The Social Behavior of the Bees, Harvard University Press, Cambridge.
5. Fletcher, D.J.C. and Michener, C.D. 1987. Kin Recognition in Animals, John Wiley & Sons, Chichester.
6. Hefetz, A. 1987. Physiological Entomology, 12 : 243-253.
7. Evans, H.E. and Hook, A.W. 1986. Sociobiology, 11:275-302.
8. Matthews, R.W. 1968. Science, 160 : 787-788.
9. McCorquodale, D.B. and Naumann, I.D. 1988. Journal of the Australian Entomological Society, 27 : 221-231.
10. Matthews, R.W. and Naumann, I.D. 1988. Australian Journal of Zoology, 36 : 585-597.
11. Malyshev, S.I. 1968. Genesis of the Hymenoptera, Methuen, London
12. Steinmann, E. 1976. Bulletin de la Sociétie entomologique Suisse, 49 : 253-258.

FANNING WORKERS DURING THE LAYING OF THE QUEEN : A FORM OF CHEMICAL COMMUNICATION IN THE STINGLESS BEE *MELIPONA FAVOSA*?

Sommeijer, M.J. and W. Minke
Utrecht University, Bee Research Dept., P.O.Box 80.086, 3508 TB Utrecht, Netherlands.

INTRODUCTION

During former studies of the oviposition behaviour of stingless bees, we observed a typical form of fanning behaviour of workers during the final phase of the "Provisioning and Oviposition Process (POP)". This was especially pronounced in certain *Melipona* species. The fanning behaviour of *Melipona* workers during POP appears very similar to the scent-marking posture of honeybees at the moment of dispersing either secretions that are released from the exposed Nasanow gland or from the base of the sting apparatus by intensive fanning by the wings and holding the abdomen raised [1]. In *Melipona* this typical behaviour only occurs during the final phase of the POP.

The possible chemical communicative nature of this behaviour was also proposed by Zucchi [2] after observing the same phenomenon in *M. subnitida*. The objective of our behavioural study is to collect information concerning the function of this behaviour.

MATERIALS AND METHODS

This behaviour was carefully described in nests of *Melipona favosa* and *M. beecheii* with quantitative measurements of its temporal occurrence during POP.

The occurence of other forms of fanning behaviour in the nest of *M. favosa* was recorded. The occurrence of fannning behaviour was also studied in nests of *M. favosa* where workers were laying reproductive eggs.

RESULTS

Various forms of fanning behaviour can be distinguished. The typical form of fanning takes exclussively place during the last phase of POP. This behaviour is characterized by the standing still of the worker. The abdomen is raised with the body axis in a position of 45 degrees and the wings are vibrating constantly. Comparing the occurrence of this form of "POP-fanning" of a normal queen-right colony, to that occurring in a colony with workers producing reproductive eggs, it appeared that in the latter case the fanning was not different of form, but occurred at a different moment. In a normal colony its starts when the queen takes place on the cell to oviposit. In the colony with reproductive egg-laying workers, it started not before the broodcell was being sealed by the laying worker. The fequency of this type of fanning behaviour varied according to the condition of the colony: it is relatively less frequent in weak colonies.

The analysis of the spatial occurrence of the various types of fanning behaviour at different locations of the nest, indicated clearly that specific forms of fanning behaviour are carried out in the broodnest area, and on storage pots where food is being discharged. Little fanning is performed in the area with sealed pots.

REFERENCES

1. Michener, C. D. 1974. <u>The social behavior of the bees</u>. Harvard University Press, Cambridge, Mass.

2. Zucchi, R. 1977. Aspectos etológico-evolutivos da bionomia dos Meliponinae (Hymenoptera. Apidae). Tese Docência-Livre. USP Riberão Preto. 204 pp.

INTERCASTE OR MOSAIC, THE WORKER MANDIBULAR GLAND SECRETIONS OF CAPE HONEYBEES (*APIS MELLIFERA CAPENSIS*)

R.M. Crewe, T. Wossler and M.H. Allsopp

The mandibular glands(MG) of queens of European honey bees produce large quantities of (*E*)-9-ketodec-2-enoic acid (9ODA), the 'queen substance'; while workers produce (*E*)-10-hydroxydec-2-enoic acid (10HDA), the 'worker substance' (reviewed in Crewe 1988). This distinction between queen and worker MG secretions was challenged when Ruttner et al (1976) and Hemmling et al. (1979) showed that *capensis* workers produce a secretion containing 9ODA. This discovery then lead to the use of *capensis* workers in a variety of behavioural and chemical studies (Velthuis et al., 1989).

In order to investigate the way in which social conditions in groups of bees affect the secretions produced by *capensis* workers, we investigated the head extracts of workers from a queenright colony, workers that had become pseudoqueens and workers that were the offspring of pseudoqueens.

EXPERIMENTAL METHODS

Experimental colonies of *capensis* were used to produce workers that were the offspring of a mated queen. Pseudoqueens were produced by placing individual *capensis* workers in mating nucs containing between 1000-2000 *scutellata* workers.

The sealed brood produced by pseudoqueens and by mated queens was removed from the colony in which it was produced and placed in an incubator until emergence. The emerging bees were placed in groups of various sizes in small cages in an incubator. When they were 8-10 days old they were harvested for chemical analysis. Chemical analysis of the head extracts of the experimental bees was performed using gas chromatography (Crewe and Moritz, 1989).

RESULTS

The chromatograms of the head extracts of the *capensis* workers indicate that they produced secretions that were rich in compounds normally associated with queen mandibular gland secretions (Crewe, 1988). The pattern of component production by the workers is not very different from that of *capensis* mated queens (Table 1). Pseudoqueens produce a signal that is very queenlike in its quantity of 9ODA. Virgin *capensis* queens in contrast have extracts that are relatively poorer in 9ODA. Comparison of pseudoqueen extracts with those of workers reared from queen eggs or laying worker eggs showed that while pseudoqueens have quantitatively greater amounts of the various components, the worker groups produced a secretion that would be unambiguously queenlike in any other honey bee race.

DISCUSSION

Of the compounds used for analysis in this study, most had previously been identified in head extracts by other workers (Free 1987). What has not been fully appreciated previously is that the secretion contains two homologous series of acids - those based on octanoic acid (7OOA, 7HOA & 8HOA) and those on decanoic acid (9ODA, 9HDA & 10HDA).

Apart from the aromatic acids, pseudoqueens produced head extracts that were very similar to those of *capensis* queens. Thus the semiochemical basis for the etinue response (Slessor et al., 1988) may be variable in the various honey bee races.. All of the worker groups produced secretions that were quite different from those of races that had been investigated previously. The most notable difference lies in the samll amounts of 10HDA that are produced (Boch & Shearer, 1967).The analysis of *capensis* worker mandibular gland secretions presented here reinforces the unique nature of their chemical signalling system and indicates why it is possible for these workers to act as pseudoqueens when they are placed in groups of workers of other races. Superficially, capensis workers have the appearance of being intercastes, in that they show some queen-like charcteristics. However, they are clearly at the worker end of the developmental spectrum and show a mosaic of worker and queen characteristics.

REFERENCES

Boch, R & Shearer, D.A. 1967. Z. vergl. Physiol. 54: 1-11.
Crewe, R.M. 1988. In Africanized Honey Bees and Bee Mites. Eds. Needham, G.R., Page, R.E., Delfinado-Baker, M. & Bowman, C.E. Ellis Horwood Ltd., Chichester. pp. 149-158.
Crewe, R.M. & Moritz, R.F.A. 1989. Z. Naturforsch. 44: 590-596.
Free, J.B. 1987. Pheromones of Social bees, Chapman & Hall, London.
Hemmling, C., Koeniger, N. & Ruttner, F. 1979. Apidologie 10: 227-240.
Ruttner, F. Koeniger, N. & Veith, H.J. 1976. Naturwisssenschaften 63: 434.
Slessor, K.H., Kaminski, L.-A., King, .G.S., Borden, J.H. & Winston, M.L. 1988. Nature 332: 354-356.
Velthuis, H.H.W., Ruttner, F. & Crewe, R.M. 1989. In Social Insects: An Evolutionary Approach to Castes and Reproduction. Ed. W. Engels. Springer Verlag, Berlin. (In press).

Table 1: Mean percentage composition of the head extracts of queens, pseudoqueens and caged workers of *A. m. capensis*. The queen data presented in this table is from Crewe, 1988.

Group	n	% composition of major compounds						
		7HOA	8HOA	9ODA	9HDAA	9HDA	10HDAA	10HDA
Queens								
virgin	27	-	2.7	80.1	-	11.1	trace	3.8
laying	6	-	3.0	84.8	-	9.8	0.5	1.0
Pseudoqueens								
Trial 1	8	tr	4.0	85.4	0.2	8.1	0.8	1.4
Trial 2	5	0.1	4.0	90.5	tr	4.9	0.6	1.7
Trial 3	6	0.2	3.0	89.3	0.2	5.3	0.5	1.6
Worker offspring of queen								
Wkr grp1	20	tr	0.2	78.2	0.9	15.3	1.5	2.1
Worker offspring of laying worker								
Wkr grp2	19	0.5	5.0	69.2	0.8	20.3	0.8	3.6
Wkr grp3	10	0.6	5.6	66.3	1.3	22.1	0.9	3.2

THE INHIBITION OF EMERGENCY QUEEN CELL CONSTRUCTION BY CAPE VIRGIN QUEENS (*APIS MELLIFERA CAPENSIS*)

L.A. Whiffler and H.R. Hepburn

Synthetic 9-oxo-decanoic acid (9-ODA) reduces the tendency of worker bees to construct emergency queen cells [1], but not as effectively as a complete extract of or direct physical contact with a mated queen (2, 3). Yet, queens lacking mandibular glands inhibit queen rearing as effectively as intact queens [4]. Since virgin queens also inhibit emergency queen cell construction [5], emergency queen cell construction in colonies each given a mated, virgin (with and without mandibular and/or tergite glands) or dead virgin queen as well as queenless colonies were investigated in the Cape honeybee. 73 queenright, five-framed nuclei were dequeened and given a mated, virgin (with or without mandibular and/or tergite glands) or dead virgin queen 12 hours later or no queen at all. 7 days later, the number of capped queen cells/colony was counted, colony size determined and the queen recaptured for GC-analysis. Cape virgin queens inhibited emergency queen cell construction as well as mated Cape queens ($P < 0.71$), whether the virgins had mandibular and/or tergite glands or not. The dead virgin queen and queenless colonies each constructed more emergency queen cells than the mated or virgin (with and without mandibular and/or tergite glands) queen colonies ($P < 0.001$). Although the dead virgin queen colonies constructed slightly fewer emergency queen cells than the queenless colonies, this difference was not significant ($P < 0.07$). Colony size and number of queen cells in the dead virgin queen and queenless colonies were not correlated , nor was the amount of 9-ODA found on the queen heads and the number of queen cells. Virgin queens without mandibular glands had little 9-ODA but inhibited queen cell construction as well as mated as did intact virgin queens. 9-ODA alone does not inhibit emergency queen cell construction, but may act with other unknown inhibitory substance(s) or factors (cf. [4]).

REFERENCES

1. Butler, C.G. and J. Simpson. 1958. Proceedings of the Royal Entomological Society, London (A), 33: 120-122.
2. Mussbichler, A. 1952. Zeitschrift fur Vergleichende Physiologie, 34: 207-221.
3. Butler, C.G. 1954. Transactions of the Royal Entomological Society of London, 105: 11-29.
4. Free, J.B., Ferguson, A.W. and J. R. Simpkins. 1985. Physiological Entomology, 10: 271-274.
5. Free, J.B., Ferguson, A.W. and J.R. Simpkins. 1985. Physiological Entomology, 10: 271-274.

INSTANT INTRODUCTION OF QUEEN HONEY BEES

J.RAJ
Department of Agricultural Microbiology
University of Agricultural Sciences (GKVK)
Bangalore-560 065, INDIA

Queen honey bees introduced as such are killed by workers. Generally a queen is introduced into the colony through a cage with its opening filled with candy. Presently, a time and labour saving method of queen introduction has been evolved.

Queens removed from 9 colonies of *Apis cerana indica* and two colonies of *A.mellifera* were immersed in tap water for 10 seconds. They were placed in drenched condition amidst workers on combs of other colonies of same species. Within 5 minutes queens were walking freely as water evaporated from their body. The queens of *A.cerana* were seen ovipositing within 15 minutes. Two colonies which had lost their queens for more than 20 days and which had laying workers also accepted virgin or mated queens. All the 5 virgin queens mated normally after introduction as evidenced by the presence of mating sign and subsequent laying.

Queens of both species drenched in honey or otherwise were balled, bitten and stung by foreign workers soon after placement. Eventually most of them died or vanished from the colony within 2 days.

CUTICULAR SURFACE STRUCTURE, A CRITICAL CUE OF THE QUEEN-QUEEN RECOGNITION FOR THE ELIMINATION OF REPRODUCTIVE RIVALS IN HONEYBEES

Takeshi Sonezaki and Masami Sasaki
Laboratory of Entomology, Faculty of Agriculture,
Tamagawa University, Machida, Tokyo, 194 Japan

Monogynic system in honeybee society loosen temporarily in the reproductive season and the selection works among candidate queens for the next generation. The selection process among queens of Apis mellifera was analysed from ethological point of view.

Queen Cell Destruction by Virgin Queen

After careful inspection, the assaultive queen made a hole on the side wall of a target queen cell. If the victim queen inside the cell had already been emerged, the assaulter stung and killed the rival by herself. When the victim had not yet emerged, however, the assaulter did nothing and workers treated later.

Queen-Queen Recognition and The Cue for Stinging

When the rival queen had already been hatched from the queen cell, the assaulter tried to eliminate by 1) rush and ride on the rival with active antennation, 2) grasp the victim by mandible and 3) stinging. Dried specimen or isolate abdomen was enough to release the stinging behavior. Neither 9ODA and 9HDA released the behavior. Instead, the cuticular surface structure was proved to have crucial importance in the queen-queen recognition (Fig. 1). If the queen-specific short, spine-like hairs were removed, no more attack occurred. The cuticular surface dimorphism was found in all three species (A. mellifera, A. cerana and A. florea) so far investigated.

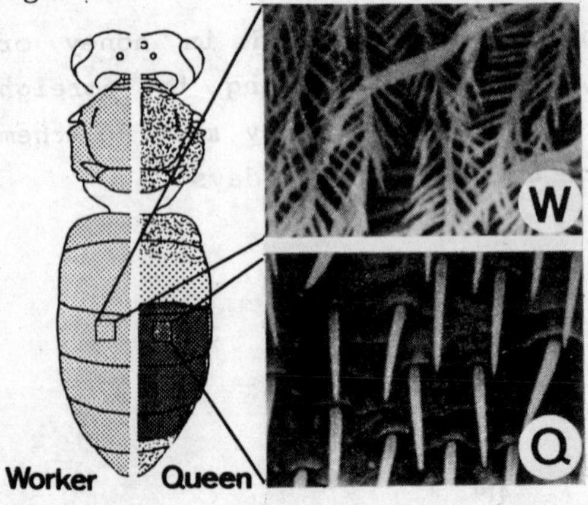

Fig. 1. Cuticular surface dimorphism of the female honeybee, Apis mellifera. Fringed (W) and spine-like(Q) hair which are most remarkable morphological character for discrimination between worker and queen.

COMPARATIVE STUDY OF HYDROCARBON PROFILES ON THE JAPANESE AND THE EUROPEAN HONEYBEE

Hiromi Sasagawa and Yasumasa Kuwahara, Inst. of Appl. Biochem., Univ. of Tsukuba, Tsukuba, Ibaraki 305, Japan

Very little is known about chemical ecology on the Japanese honeybee *Apis cerana japonica* Rad. (ACJ), who can share the same habitat with the European honeybee *Apis mellifera* L. (AM) in Japan. In the present paper, hydrocarbon (HC) components present on the body surface of ACJ were analyzed and compared with those of AM, using GC (CP-Sil 19CB capillary column) and GC/MS, and the double bond position of each alkene was determined by GC/MS analysis of dimethyldisulfide adducts.

Workers and drones of ACJ were collected from three natural nests around the University of Tsukuba. A laboratory colony was used for AM, originated from a beekeeper. Each honeybee was dipped in n-hexane (700μl), and thus collected sample was analyzed.

ACJ contained less variety of HCs than AM did, and consisted of at least 13 clusters (a mixture of saturated and unsaturated HCs of the same carbon) with various chain length from C_{21} to C_{37}, among which C_{23}(16.7%), C_{25}(38.6%) and C_{27}(30.0%) were main components, while AM consisted of 18 clusters of HCs ranging from C_{17} to C_{37}, among which C_{25}(12.3%), C_{27}(25.2%), C_{29}(14.2%), C_{31}(16.4%) and C_{33}(18.3%) were the major. The same was true about saturated HCs; ACJ consisted of 8 kinds from C_{21} to C_{29}, among which C_{23}(20.1%), C_{25}(48.0%) and C_{27}(28.9%) were major constituents, while in AM, 16 saturated HCs were detected and following HCs were the major components; C_{23}(10.8%), C_{25} (21.7%), C_{27}(27.4%), C_{29}(17.8%) and C_{31}(14.8%).

Unsaturated HCs separated by 10% $AgNO_3$-impregnated SiO_2 column chromatography indicated the presence of monoene, diene and triene in both species. Alkanes of ACJ consisted of nearly 72% of the total HCs, and alkenes 28%. Monoene composition of ACJ was also simpler than that of AM. Double bond positions were restricted to 1-, 5-, 7- and 9-en in ACJ, while in AM following 9 positions were found; 1-, 3-, 4-, 7-, 8-, 9-, 10-, 11- and 12-en. It was also noted that lower HCs consisted of less amount of alkanes and higher ones of higher amount of unsaturated HCs in both species.

HC profiles of individual workers and drones of ACJ were not uniform and was able to classify into several groups, depending on each colony. In the case of workers, 5-10 groups were identified in each colony, which may represent the queen's multiple mating, while drones were separated into two groups.

CUTICULAR COMPOUNDS IN *RETICULITERMES* TERMITES : SPECIES, CASTE AND COLONIAL SIGNATURE

A.G. Bagnères*, J.L. Clément*, C. Lange* and M.S. Blum+.**

**CNRS-UPR38.31,Chemin J.Aiguier.13402-Marseille Cedex 9-FR*
***Lab.Ch.Org.St.UA CNRS455-Univ.Paris VI-4,Place Jussieu 75230-Paris Cedex 05-FR*
+Dpt of Entomology-U.G.A-Athens-Georgia 30602-U.S.A

The cuticule of insects constitutes a complex mixture of aliphatic compounds, the majority of which are hydrocarbons, and which play a part in the survival of species as a physical barrier and a recognition factor. Species of *Reticulitermes* are distributed around the world, in the new-world represented principally by *R.flavipes*, and in Western Europe by the complex of species *R.lucifugus*. In France, an other species *R.santonensis* seems to be a colony of *R.flavipes* introduced across the Atlantic. Different variations of the cuticular compounds are studied here, including comparative studies and analysis of intraspecific variations as colonial, caste and seasonal. We present here some of the principal results.

Cuticular extracts were obtained from *R.flavipes*, *R.malletei* (UGA, Georgia, USA), *R.santonensis*, *R.(lucifugus) grassei* and *R.(l.)banyulensis* (France). The cuticular hydrocarbons of Individuals from differents colonies and different castes were extrated, and the cuticular compounds were determined by GC/MS, quantified and the results processed statistically. Each extract consists of dozens of individuals, so, for the principal component analysis (PCA) each vector represents a mean for one caste in one colony, visualized after plotting the different hydrocarbons for every species.

Each species has a particular mixture of 20 to 40 hydrocarbons (containing from 20 to 35 carbon atoms) consisting of n-alkanes, mono and dimethyl-alkanes, monoenes and dienes, and also trienes in the case of *R.malletei*. Only *R.flavipes* and *R.santonensis* show the same pattern. Between species compounds change both qualitatively and quantitatively, but the *lucifugus*-complexe has more common components than other groups of species. Chemical polymorphism between real species in the same genus and in the same genetic group provides supplementary proof that *R.santonensis* and *R.flavipes* are an unique super-species (*R.flavipes sp.*).

Within a species, hydrocarbons vary quantitatively. For example, *R.flavipes* presents two sorts of variations, phenotypic variations following in part exocrines variations, and seasonal variations separated by the third canonical axis of the P.C.A. The caste variations in *R.santonensis are* very interesting as each caste (worker, soldier, nymph, neotenic...) is defined by a family of chemicals (n-alkanes, monenes, etc). *R.malletei* does not present caste variations for these products. *R.(l.)grassei* and *R.(l.)banyulensis*, for which 40 % of hydrocarbons are possessed in commons, are very agressive and present closed societies. The application of an alien mixture of hydrocarbons too a termite-lure induces agression by its congeners against it. Apparently, quantitative variations of mixtures only are the reason for the agression.

Hydrocarbons seems to be, for a lot of species (termites, ants), a real signal of recognition and of communication. Our recent work shows the importance of these compounds combined with exocrine compounds, to form a real "finger-print" for species, castes and colonies.

Symposium 17

POLLINATION ECOLOGY OF SOCIAL INSECTS

Organizers : **LAZARUS W. MACIOR, U.S.A.**
B. MALLIK, India

SOCIAL BEES AND PALM TREES : WHAT DO POLLEN DIETS TELL US?

David W. Roubik and Jorge Enrique Moreno*, Smithsonian Tropical Research Institute, Balboa, Panama (APO Miami 34002-0011, USA)
*current address BIOSS Ltda., A. A. 52514, Bogotá, Colombia

To judge from types of pollen stored in the nests of social bees, it no longer seems reasonable to refer· to honeybees, stingless bees, social halictines, etc. as "generalist pollinators". Various studies of pollen diet in tropical forests show that a small number of flower types are of paramount importance to eusocial bees [1-5]. These are ususally trees producing dense clusters of small flowers that are often unisexual. The social recruitment systems of bees are suited to monopolize such resources [3,4]. Social bees are often not pollinators of these flowers— clarified by examining their relationships with palms [2,3,6,7].

STINGLESS BEES (MELIPONINAE)

Table 1 summarizes the palm pollen contents and total pollen taxa of 18 nests of stingless bees in primary lowland forests in Panama and Amazonian Colombia. Also in the table are results from a study near the Tapajós and Trombetas Rivers of the Amazon [1], which examined pollen subsamples. The similarities are striking, for in each study, although of differing methodology, an average of 15% of the total taxa in the bee nests came from palms. Palm pollen grains tend to be larger than most used by bees, being 38 to142 μ in longest dimension among the 9 spp. in the Panama-Colombia study. Therefore, simple counts of grains do not show their relative representation in the bee diets—the largest grains we saw of any plant were 325μ, and the smallest 9μ. However, in all but 4 of 18 bee nests there were substantial numbers of palm pollen grains, and in the nests of *Melipona fasciata* and *Nogueirapis mirandula* they were almost all of the pollen present, despite there being up to 23 total pollen species. On the other hand, although palms comprised 5-18% of pollen species in nests of *Trigona nigerrima* and *Oxytrigona mellicolor*, the bees' stored pollen was almost all of *Tetrapteris* (Malpighiaceae) and *Gustavia* (Lecythidaceae), respectively. Palm genera seen most often in bee nests in Panama and Colombia were *Iriartea* (11 nests), *Socratea* (10), *Oenocarpus* (5), *Chamaedorea* (5), *Scheelea* (4), *Phytelephus* (2), *Bactris* (2) and *Geonoma* (1). Bees have never been seen visiting the pistillate (seed-producing) flowers of *Socratea, Bactris, Phytelephus,* nor *Elaeis*, [7]. It seems likely that pollination of the other palm genera occurs primarily from bees, with possible exceptions in *Geonoma* and *Oenocarpus* [7]. The stingless bees in these forests are sometimes mutualists, sometimes parasites, and sometimes commensals of the palms [2,3]. In addition, the number of pollen species stored in the bee nests was not high (Table 1), ranging 9 to 45, averaging 21. The total pollen species found at *one season* among the nests of a eusocial halictine *Halictus hesperus* was nearly twice this number in a similar forest, although seen during a time of year when up to 600 species are in bloom [10]. Considering that the forest environment contains nearly 1000 flower species accessible to bees, the highly eusocial bees in their natural habitat do not seem especially generalized.

Table 1. Pollen identified in nests of stingless bees in neotropical forests (see text).

Bee species	locality	spp. palm	Total spp.	notes
Tetragona dorsalis	Panama	9-13%	30-45	3 nests
Nogueirapis mirandula	Panama	23%	13	mostly palm
T.(Cephalo.) capitata	Panama	5-7%	22-28	2 nests
T. nigerrima	Panama	5%	20	mostly malpighs
T. fulviventris	Panama	17%	18	
Oxytrigona mellicolor	Panama	18%	11	mostly lecythids
Scaura latitarsis	Panama	0-25%	9-18	4 nests (2 poor)
Melipona fasciata	Panama	22-23%	30-45	2 nests, mostly palm
M. nebulosa	Colombia	17%	30	
M. fuliginosa	Col, Pan	5-7%	21-30	2 nests
23 spp, 9 supraspecific	Brazil	15%	122	37 nests

HONEYBEES

The honeybee most studied in pollen use is *Apis mellifera*, which in its native Africa, currently can use only 3 palm genera and a few species, although there were likely many more genera and species there before climatic changes that took place after the Tertiary, as there are presently on Madagascar18 genera and >100 spp. In both Africa and the Neotropics, *Apis* intensively utilizes *Elaeis* [4,8,9], which it does not pollinate [7]. In the lowlands of Panama *Apis* often forages at palms, mainly during the rainy season, when it uses *Bactris, Chamaedorea, Elaeis, Socratea* and *Oenocarpus* [2-5]. In terms of the actual fitness of the honeybee population, it was found that in the wetter lowland forests the pollen of *Oenocarpus* was the second most important, and that of *Socratea* the fifth most important pollen source to reproduction of the Africanized honeybee during a year [2]. It has been shown that *A. mellifera* uses approximately 25% of the angiosperm flora in the neotropics, yet the numbers of pollen species that it utilizes in a significant amount are but a few dozen during an entire year [2-5].

REFERENCES

1. Absy, M. L., J. M. F. Camargo, W. E. Kerr and I. P. A. Miranda. 1984. Revista Brasiliera de Biologia, 44:227-237.
2. Roubik, D. W. 1988. In: Africanized Honey Bees and Bee Mites, G. R. Needham, R. E. Page, Jr., M. Delfinado and C. Bowman, Eds, Ellis Horwood, Chichester, pp.45-54.
3. Roubik, D. W. 1989. Ecology and Natural History of Tropical Bees. Cambridge University Press, New York. p. 514.
4. Roubik, D. W., J. E. Moreno, C. Vergara and D. Wittmann. 1986. Journal of Tropical Ecology, 2:97-111.
5. Roubik, D. W., R. Schmalzel and J. E. Moreno. 1984. Technical Bulletin 24, OIRSA (Central America, Mexico, Panamá). p. 73.
6. Búrquez, A., J. Sarukhán and A. L. Pedroza. 1987. Botanical Journal of the Linnean Society, 94:404-419.
7. Henderson, A. 1986. The Botanical Review, 52:221-259.
8. Lobreau-Callen, D., R. Darchen, B. Darchen and A. LeThomas. 1989. Proceedings 4th International Conference on Apiculture in Tropical Climates, Cairo, pp. 410-421.
9. Lobreau-Callen, D., R. Darchen and A. LeThomas. 1986. Apidologie, 17:279-306.
10. Brooks, R. W. and D. W. Roubik. 1983. Sociobiology, 3:263-282.

SOME POLLEN AND NECTAR RESOURCE PLANTS OF *APIS FLOREA* AND *A. CERANA INDICA*

RAJU J.S. ALURI
Department of Biology, University of Akron, Akron, OH 44325, U.S.A.

Regular seasonal observations were made to record forage resource plants for honeybees in their naturally distributed areas at Visakhapatnam, Andhra Pradesh, India. **Apis florea** and **A. cerana indica** usually forage on the same flower species for their nutrition. Bisexual flowers of **Ocimum americanum, O. basilicum, Hyptis suaveolens, Leucas aspera, Tribulus terrestris, Zizyphus mauritiana** and bisexual and male flowers of **Sapindus emarginatus** serve as pollen and nectar source for honeybees. Male flowers of **Croton bonplandianum** and **Phyllanthus pinnatus** are a source for pollen and male and female flowers of **Jatropha gossypiifolia** for nectar only. The female flowers of **C. bonplandianum** and **P. pinnatus** lack nectar and the honeybees ignored them.

Analyses of nectar characteristics of the flower species indicate that the nectar volumes range between 0.4 µl and 11. 47 µl per flower and total sugar concentrations 9-48%. The sugars glucose, fructose and sucrose are the common ones. The honeybees exhibit different foraging behaviours in relation to the floral mechanism and the kind of forage they collected and bring about pollination nototribically and sternotribically in mess and soil fashion. Their foraging activity period was usually observed between 0600-1630 h, the peak being 0700-1030 h. They normally concentrated on the flowers of densely distributed plants, and thus pollen-flow among different populations of each flower species was highly limited. The flower species included here are considered to be the principal forage source seasonally for the honeybees where they are distributed in the same area.

POLLINATION OF A MINT IN RELATION TO HONEYBEE FORAGING BEHAVIOUR

RAJU J.S. ALURI
Department of Biology, University of Akron, Akron, OH 44325, U.S.A.

Introduction

The reproduction of many angiosperms is closely and obligately linked to insect pollinators, especially bees, which forage for nectar and pollen provided by flowers. Reproductive cycles of bees depend on these forage materials, which form their sole nutritional resource; many flowers depend upon bees for pollination. Life cycles of plants and bees are therefore intimately interdependent. This paper examines the interdependency between a mint, **Anisomeles indica** and native honeybees, **Apis florea** and **A. cerana indica** for pollination and for forage source.

Material and Method

Wild populations of **Anisomeles indica** O. Kze. on the banks of irrigation canals at Turimella ($15^{o}10'$N; $18^{o}45'$E) in Andhra Pradesh, India formed the material for study in October-December 1985. Flower blooming process and features were examined in relation to the forager activity and behaviour at the flowers. Nectar of 20 flowers from individual plants was measured and analysed for sugar concentration with a hand-held refractometer and for types of sugars present [1]. Compatibility to self and cross pollen were tested through hand-pollinating flowers.

Observations and Results

Floral biology

Flowers of **Anisomeles indica** are bisexual, gullet-shaped, bilabiate with stamens and stigma roofed over by upper corolla lip, and nectariferous. Nectar volume was 1.6 $\pm$ 0.2 μl per flower, and nectar concentration was 32-43%. Nectar components included glucose, sucrose and fructose. Flowers anthesed during 0530-0730 h and remained open for a day and a night. Hand pollination for in and out breeding systems showed that self pollination was 31% effective compared to 79% cross pollination.

Foraging activity and behaviour of honeybees

The honeybees **Apis florea** and **A. cerana indica** were active foraging on flowers during 0600-1800 h but the peak activity was 0700-1030 h. They foraged for nectar and pollen. The flowers are pollinated sternotribically by bees collecting pollen upside down and are noto-tribically pollinated by bees foraging for nectar. The bees are regular and consistent in their visits to the flowers. The flower visiting rate in a unit time (one minute) and the length of each foraging visit (in seconds) were 30 ± 4; 3.0 ± 0.74 for **A. florea** and 24 ± 5; 3.7 ± 0.90 for **A. c. indica**, respectively.

Discussion

Floral characteristics coupled with breeding systems showed that **Anisomeles indica** is principally dependent on honeybees for reproduction. When bees forage for nectar and pollen, they cause pollen movement within and between flowers of the same and different individual plants thus selfing and crossing flowers. They use nototriby for nectar collection and sternotriby for pollen gathering; both result in the pollination of each flower. Similarly, bumblebees used nototriby and sternotriby for nectar and pollen collection on **Pedicularis grayi** [2]. Such a dual behaviour is used by the same **Apis** sp. only to collecting nectar on **Jatropha gossypiifolia** [3]. Therefore, the behaviour pattern exhibited by honeybees is an attribute of flower behaviour, bee body size, and kind of forage collected [4, 5]. The coadaptations of the floral mechanisms and the behaviour of honeybees insures the reproductive success of **Anisomeles** and the seasonal availability of nectar and pollen forage sources for the honeybees.

References

1. Harborne, J.B. 1973. Phytochemical Methods; Chapman & Hall, London.
2. Macior, L.W. 1973. In: Pollination and Dispersal, N.B.M. Brantjes and H.F. Linskens, Eds., Univ. Nijmegen, Nijmegen, pp. 101-110.
3. Reddi, E.U.B. and Reddi, C.S. 1983. Proc.Indian Acad. Sci.92:215-31.
4. Macior, L.W. 1974. Ann. Missouri Bot. Gard. 61: 760-769.
5. Macior, L.W. 1974. Melanderia 15: 1-59.

DOES FLORAL HELIOTROPISM OF *BRASSICA JUNCEA* INFLUENCE VISITATION PATTERN OF *APIS DORSATA*

B.MALLIK and S.RAMANI
Department of Entomology, University of Agricultural Sciences,
GKVK, Bangalore 560065, INDIA

The infloresence of *Brassica juncea* (L.)Czern. exhibit heliotropism,bending towards east in the morning and west in the evening. We observed that *Apis dorsata, Apis cerana* and *Apis florea*, which visit these inflorescences approach them predominantly from east in the morning and west in the evening.We tested whether such orientation of the bees is influenced by the heliotropic movements of the inflorescence.

MATERIAL and METHODS

We tracked individual *A.dorsata* bees from 600hrs to 1800hrs during March,1989, in a 50 X 50M plot of *B.juncea* at GKVK campus of University of Agricultural Sciences, Bangalore, and recorded the direction of approach, landing and leaving an inflorescence, and movement on an inflorescence. Nectar harvested was determined by comparing nectar content of bagged and open flowers (visited by all species) at two hour intervals. The nectar at the base of the corolla was absorbed by Whatman filter paper discs (5mm dia.) and the sugars were estimated.

RESULTS

a) Sun tracking by the inflorescence: The central axis of the inflorescence bend by 17 ° towards east in the morning and by 19 ° towards west in the evening (table 1).

Table 1: Orientation of Inflorescence of *B.juncea*

Time (hrs)	n	Orientation of axis towards east (E) or west (W)
1600	5	0
0800	97	17 ° E
1000	79	14 ° E
1200	76	5 ° E
1445	120	13 ° W
1515	29	19 ° W

b) Approaching and landing direction of bees: The bees land on flowers in the direction from which they approach the inflorescence. In the morning (805 to 910hrs) 54.5 per cent of the bees land on flowers in east quadrant (table 2), while in the evening (1745hrs) 73.81 per cent land on the west. At mid day (1210 to 1220hrs) when the orientation of the axis of the inflorescence is ~5 ° E 57.15 per cent of the bees land on flowers in the north quadrant.

Table 2 . Landing of *A. dorsata* on flowers in different sectors

Time (hrs)	Percent landing during respective times of day East	West	North	South
805 to 910	54.5	14.6	28.3	2.6
1035 to 1100	9.9	29.6	53.5	7.0
1210 to 1220	21.4	15.0	57.2	6.4
1440	32.1	41.1	12.5	14.3
1745	7.2	73.8	7.1	11.9

c) Movement on the Inflorescence: Most flowers in all the sectors of the inflorescence are visited by the bees, however those in east and west quadrants are more frequently visited (χ^2 - 91.17,p 0.005).Bees landing on north or south quadrants tend to forage more on east quadrant flowers in the morning and equally on east and west quadrant flowers in the evening (table 3).

Table 3. Percentage of flowers visited by A .dorsata in different quadrants

Time (hrs)	East	West	North	South
700 to 900	43.9	22.6	21.9	11.6
1000 to 1100	22.3	27.3	35.2	15.1
1200 to 1300	25.5	24.2	35.1	15.2
1400 to 1500	29.0	29.8	14.5	26.7
1700	18.0	44.0	21.0	16.0

d) Nectar harvesting: Over the entire day, proportion of nectar harvested from either the eastern (79.34%) or western (73.91%) flowers did not differ significantly, though the total nectar harvested from the eastern flowers is high (table 4).

Table 4. Nectar harvested from flowers in east / west quadrants during different times of day

Time (hrs)	east quadrant			west quadrant		
	nectar secreted µ g/flower	nectar harvested µ g/flower	percent harvested	nectar secreted µ g/flower	nectar harvested µ g/flower	percent harvested
0630 to 0830	63.8	48.62	76.2	49.86	41.86	83.95
0800 to 1000	128.22	102.42	79.88	67.4	49.89	74.02
1145 to 1345	34.96	28.93	82.75	36.4	23.73	65.19
1600 to 1800	9.65	7.79	80.73	8.0	4.0	50.00
Total	236.63	187.76		161.66	119.48	

e)Leaving the inflorescence: After visiting the flowers of an inflorescence *A.dorsata* tend to leave from flowers on the same direction on which they intially landed. This rule appears to be stringently followed by bees landing on east and west quadrant flowers and not as much by others. Further the direction of the last flower visited on an inflorescence did not influence the direction of landing on the subsequent inflorescence.

DISCUSSION

The bees followed a well regulated patern of approaching and landing on east quadrant flowers in the morning and west quadrant flowers in the evening. We propose two alternative explanations for such observed pattern:1.Bees are selected to land more often on flowers from which they can maximise the total nectar reward. Our results however indicate that this is true only for visitation to eastern flowers in the morning as these have relatively more nectar than western flowers. On the other hand visiting western flowers in the evening does not fetch them any additional reward compared to eastern flowers, had they visited. Thus differential nectar reward can not explain the approaching and landing pattern observed by us.2.The improved visibility of the flowers in the eastern quadrant in the morning and western quadrant in the evening due to heliotropic movements of the inflorescence could explain alternatively the observed approaching and landing pattern of the bees during morning and evening.

ACKNOWLEDGEMENT

The authors thank Dr.G.K. Veeresh, Professor & Head, Department of Entomology, GKVK, for facilities. The fruitful discussions with Dr.R. Umashaanker, Dr.K.N. Ganeshaiah and other ENRU members have greately helped preparing this paper.

DIFFERENTIAL ATTRACTIVENESS OF TWO CRUCIFEROUS SPECIES TO HONEYBEES

R.C.Sihag
Laboratory of Animal Behaviour and Simulated Ecology
Department of Zoology,
Haryana Agricultural University,
HISAR-125004, INDIA

INTRODUCTION

At Hisar (India), two honeybee species were found to forage more frequently on a lower energy rewarding plant in the presence of a higher energy rewarding one. The reasons for their such a behaviour pattern have been described.

MATERIAL AND METHODS

Population density of two honeybee species viz. Apis dorsata F. and Apis mellifera L. were recorded on two plant species viz. Brassica chinensis L. and Eruca sativa grown in the adjoining fields during Feb. 1989 [2,3], and their floral energy rewards were determined [1,5]. The foraging modes in relation to the floral structures of two plant species were recorded [4] and the foraging rates (number of flowers visited/min.) were also determined.

RESULTS AND DISCUSSION

Notwithstanding the flowers of B. chinensis having higher per cent of dissolved sugars (%) in the nectar (Table 1) the total nectar volume per flower per 24 h (v), total dissolved sugars (wt) and total energy reward (e) per flower were significantly more in

Table 1. Energy reward pattern of two plant species

Plant species	v (µl)	%	wt (µg)	e (cal).
E. sativa	0.56±0.03	28±0.13	0.1567±0.01	0.63±0.04
B. chinensis	0.45±0.02	32±0.15	0.1231±0.01	0.49±0.03
	*	*	*	*

Data based on 40 observations, *P ($\leq$0.05) between plant species: significant (t-test). For abbreviations see text.

E. sativa than in B. chinensis ($P \leqslant 0.05$). Therefore, E. sativa should be more attractive to the visitors than B. chinensis [5].

However, the pattern of attractiveness as reflected in terms of population densities of two honeybee species was reverse of the floral energy reward(s) pattern and, in fact, B. chinensis was more attractive to these two honeybee species (Table 2).

Table 2. Population density of two honeybee species on two plant species

Time (h)	Population density /m^2 (a)			
	A. dorsata		A. mellifera	
	E. sativa	B. chinensis	E. sativa	B. chinensis
10 45	0.2±0.4	0.8±0.4	0.4±0.5	1.4±0.5
11 45	0.6±0.5	1.2±0.8	0.8±0.4	3.6±0.5
12 45	1.4±0.5	3.4±0.5	2.6±0.9	5.8±1.1
13 45	1.0±0.7	2.6±0.5	1.8±0.4	4.6±0.5
14 45	0.6±0.8	1.8±0.4	1.0±0.4	3.8±0.4
15 45	0.2±0.4	0.8±0.8	0.8±0.4	1.6±0.5

(a) Data based on 5 observations for each bee species x crop x time combination; *P ($\leqslant 0.05$) between plant species for a bee species: significant (t-test).

The reason for such a preference could be determined by studying the floral structure and the foraging behaviour of the visitors and their energy harvest rate (energy gained or harvested per min.). Structurally flowers of E. sativa were more tunnelled than those of B. chinensis (Fig.1). The visitors of two honeybee species foraging on

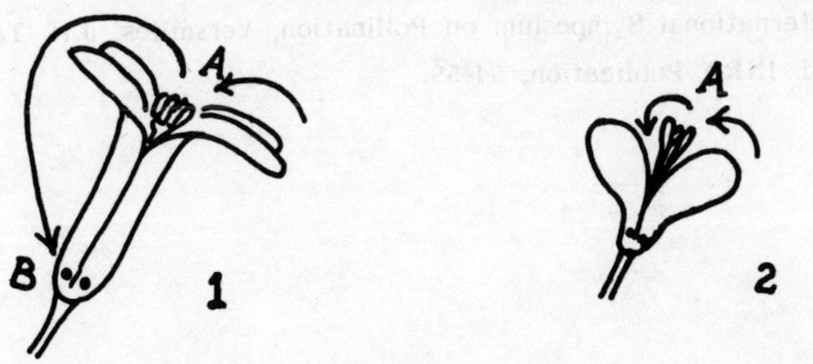

Fig.1. Foraging pattern of two honeybee species on the flowers of two cruciferous species 1.E. sativa 2. B. chinensis.

the flowers of the former plant species first approached the anthers to gather pollen only (Fig.1, 1A). Then the forager moved to the back/posterior end of the flower to gather nectar (Fig. 1, 1B). On B. chinensis, however, as the forager approached the anthers, it rolled over to gather both pollen as well as nectar (Fig.1, 2A). Therefore, foraging rates (r) of the visitors to E. sativa were significantly lower (5.2±0.3 and 4.1±0.2 flowers per min. of A. millifera and A. dorsata respectively, n=40) than the visitors to B. chinensis (10.1±0.9 and 6.4±0.5 flowers per min. respectively) ($P \leq 0.05$ for a bee species visiting two plant species: significant, t-test, n=40). Taking r x e as energy harvest rate, these bees could harvest more energy in unit time on B. chinensis (4.95±0.3 cal./min. and 3.14±0.15 cal./min. by A. mellifera and A. dorsata respectively) than on E. sativa (3.28±0.12 cal./min. and 2.58±0.08 cal./min. respectively), and the differences between plant species for a bee species were significant ($P \leq 0.05$, t-test, n=40).

This should indicate that, in practice, foraging on B. chinensis was more profitable to these bees, notwithstanding higher energy reward in E. sativa. The study also reveals that E. sativa should not be grown in the vicinity of B. chinensis lest the former should suffer reproductively by losing its pollinators to the latter crop plant.

REFERENCES

1. Heinrich, B. and P.H. Raven. 1972. Science, 176: 597-602.
2. Sihag, R.C. 1982. Indian Bee Journal, 44(4) : 89-90.
3. Sihag, R.C. 1986. Journal of Apicultural Research, 25: 121-126.
4. Sihag, R.C. 1988. Bee World, 69: 153-158.
5. Sihag, R.C. and R.P. Kapil. 1983. In Proceedings of Vth International Symposium on Pollination, Versailles, J.N. Tasei, Ed. INRA Publication, 51-59.

ROCKBEE POLLINATION IN TAMARIND (*TAMARINDUS INDICA* L.)

V.Bhaskar and Y.N. Mahadevaiah
Department of farm Forestry, University of Agricultural Sciences, GKVK, Bangalore-560 065, India

Rockbees (Apis dorsata F.) have been observed as the chief pollinators of tamarind during our studies at GKVK forest. The flower structure is also suited to enable rockbee pollination.

The flowers in tamarind (12-32 per inflorescence) are borne on small terminal or axillary drooping racemes. There are 3 unequal petals borne on top of funnel-shaped calyx cup, one of the petals is small, linear and erect and the lateral two are large and bent downwards. The calyx cup bears the nectar for which rockbees visit the flowers. There are 3 fertile stamens alternating with staminodes with connate hood-like filaments. Stamens are exerted either on left or right side of the style on the same tree, creating equal chances of stigma receiving the pollen. The stigmatic part is bent and placed away from anthers to avoid selfing.

Rockbee pushes itself into the flower either through staminal or stylar side. In the former case the hairy thorax of the bee brushes against anthers where pollen gets deposited and when the bee exits from the stylar side the pollen laiden thorax brushes against the stigma thereby effecting self pollination. When a bee enters from the staminal side or from stigmatic side and exits from the same side or through staminal side neither self nor cross pollination is effected. Rockbees visited one flower after other in rapid succession on the same tree or between trees. When the tamaraind trees were growing closeby the the movement of bees between trees was more common resulting in increased cross pollination.

Black ants (Camponotus compressus F.) were common on inflorescences of tamarind trees. When the inflorescences with mature buds were bagged, with and without ants, during the regular fruit setting season, fruits were not set in both the cases. Examination of the body of ants and stigmatic surfaces enclosed in bag revealed no pollen attached to them confirming that ants play little role in pollination (Table I) thus contradicting the earlier report (1). On the contrary, these ants were found foraging on the floral parts, especially the petals. Pollinators are not usually limiting for fruit set in tamarind, but the extent of fruit set was influenced by other factors such as variation in 'early' or 'late' flowering, percentage of self or cross compatibility which may vary from tree to tree and weather conditions (Unpublished data).

Table I. Percentage fruit set due to hand selfing, hand crossing, open pollination and bagging with and without ants.

Tree No.	Hand selfed	Hand crossed	Open pollinated	Bagged with ants	Bagged without ants
1	55.55	44.45	9.12	0	0
2	37.50	62.50	18.99	0	0
3	50.00	50.00	14.72	0	0
4	37.00	63.00	12.87	0	0

Reference:

1. Thimmaraju, K.R., M.R. Narayana Reddy, N.Swami Rao and U.V.Sulladmath. 1975. Studies on the floral biology of Tamarind (Tamarindus indica L.), Mysore J. Agri. Sci., 11: 293-298.

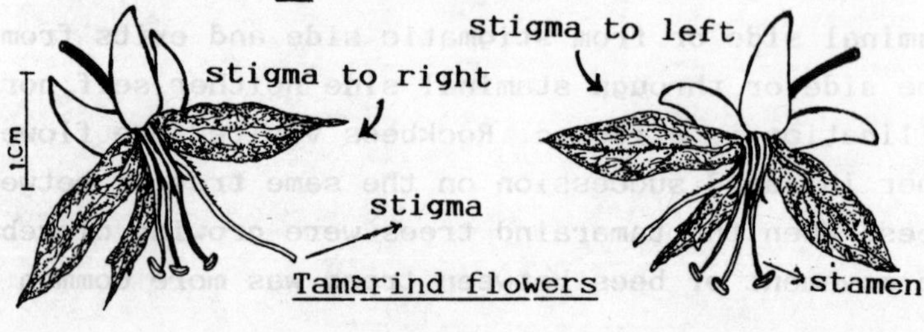

Tamarind flowers

MUTUAL ADAPTATION OF *PEDICULARIS* FLOWERS AND THEIR *BOMBUS* POLLINATORS

Lazarus Walter Macior
Department of Biology
The University of Akron
Akron, Ohio U.S.A. 44325-3908

Great diversity of floral form in over 600 species of **Pedicularis** is attributed by Pennell [7] and Li [3] to selection by pollinators. Macior [5] identified **Bombus** as its primary pollen vector. Four basic forms of the bilabiate flower include (1) nectariferous tube with toothless upper lip (galea), (2) same with two apical galea teeth, (3) short nectarless tube with beaked galea, and (4) very long tube and beak. Pollination of the first three has been studied in North America and Asia. The fourth was first studied last year in the Kashmir Himalaya by Macior [6].

Pollination function in **Pedicularis** varies with floral form. Nectariferous flowers are pollinated by **Bombus** queens in an upright (nototribic) position. As nectar is withdrawn, the stigma, protruding from the galea tip in all species, contacts pollen lodged in the neck crevice of the insect. Pollen is deposited dorsally from anthers within the galea. Pollen is groomed from the body, but residual pollen in the insect's midline contacts the stigma. In the beaked, nectarless form, the insect grasps the basal edge of the upper corolla lip with its mandibles and rotates until the stigma contacts pollen in the midline; pollen is vibrated from the anthers. A variant of this mechanism has a spiral beak that contacts an upright pollinator's face or dorsum covered with pollen.

The seasonal reproductive cycles of **Pedicularis** and **Bombus** are interdependent. Spring-blooming **Pedicularis** species are nectariferous, accommodating nectar-foraging queens. In late spring, **Pedicularis** species provide nectar for nototribically pollinating queens and pollen for small workers scraping it from anthers in a ventral (sternotribic) pollinating position. In summer, when pollen is in greater demand for brood-rearing, all **Pedicularis** species provide pollen only, which workers vibrate nototribically or sternotribically. Variations from this pattern are rare, but **P. lanceolata** is pollinated sternotribically by workers deflecting the lower corolla lip to scrape pollen, and **P. furbishiae** is a nototribically pollinated summer-blooming species containing nectar. Hummingbirds (Trochilidae) pollinate **P. densiflora, P. procera (P. grayi),** and **P. canadensis**, and the solitary bee **Osmia tristella** pollinates **P. semibarbata**. Long-tubed, rostrate flowers were thought to be pollinated by long-tongued, nectar-foraging lepidopterans, for want of empirical evidence, until **P. punctata** was found to be nectarless and pollinated by pollen-vibrating **Bombus** workers. The entire **Pedicularis** genus is fundamentally **Bombus**-related; adaptation to other pollinators is incidental and more recent.

Limpricht [4] described the Holarctic dispersion of **Pedicularis** from a Himalayan center where Hong [1] identified the largest species concentration. **Bombus** is also Holarctic, especially in montane and Arctic-alpine regions. Migration of **Pedicularis** from Asia to North America through Beringia must have been associated with **Bombus** pollinators. Although seasonal adaptations of the reproductive cycles of **Pedicularis** and **Bombus** were probably established early in the plant's history, Li's [2] suggestion that the survival of particular **Pedicularis** species depends upon pollination by particular species of insects is, for the most part, questionable.

References

1. Hong, D. 1983. Annals of Missouri Botanical Garden, 70: 701-713.
2. Li, H.L. 1948-49. Proceedings of the Academy of Natural Sciences of Philadelphia, 100: 205-378; 101: 1-214.
3. Li, H.L. 1951. Evolution,5: 158-164.
4. Limpricht, W. 1924. Repertorum Specierum Novarum, 20: 161-265.
5. Macior, L.W. 1982. In: Pollination and Evolution, J.A. Armstrong, J.M. Powell, A.J. Richards, Eds., Royal Botanic Gardens, Sydney, pp. 29-45.
6. Macior, L.W. 1990. Plant Species Biology (Kyoto), In Press.
7. Pennell, F.W. 1948. Brittonia, 6: 301-308.

INSECTS AS POLLINATOR, OF HORTICULTURAL CROPS

Dr. Kehar Singh, M.Sc. Ph.D.(Hort.), M.A.Eco.(KSB.dh.)
Professor and Head, Department of Horticulture,
Roorkee (Haridwar), U.P., India.

ABSTRACT

Pollination is the transference of the pollen grains from the anther of a flower to the stigma of the same flower.In self-pollination only one parent plant is concerned is producing the off-spring, cross pollination on the other hand is the transference of the pollen grains from one flower to another flower born by two separate plants of the same or allied species invariably. whether the flowers are bisexual or unisexual- The cross pollination is brought about by external agents which carry the pollen grains of one flower and deposit them on the stigma of another flower. They are the agents like insect, some animals, wind, water etc. There are certain types of flowers like, dictiny, self-sterlity, dichogamy, protandry, protoggny, heterostyle, herkogamy. They are pollinated by the different pollinators because they have the various types of the structure of their own.

It is apparent that the insect plays a vital role in the welfare of our general economy through its aid in the production of fruits, vegetables and seed crops. Apart from above, it also aids in the production of cultivation crops. The insect is also a factor in maintaining the native plants that cover water shed, prevent erosion and provide food for wild and domestic animals alike.

Owing to the dessimination of native pollinators they are important for intensive agricultural practices. Honey bee is the only pollinator that can be multiplied and can be multiplied and moved inadequate numbers as needed. The realization of the great need of insect for pollination is rather recent many problems remain to be solved in relationship between grower and the beekeeper. For the most economic production of the many crops, they are benefitted by the service of the insect.

The relation between flower and flower visiting insect is a special kind of phytophagy which is mutually beneficial, the plants offer to the insect food in the shape of nector or pollen and insect serve the plants by bringing about cross pollination of flowers. It has been claimed that the values of bees in pollination of crops is ten to twenty times, the value of honey and wax they produce, and insects act as the social worker for the welfare of human being by helping in environment and balancing the ecosystems. They are the predators like dragon flies feed on aquatic insect, preying-mantids, nymphs and adults of grass-hopper, rubber flies, catch - insects at night and some other insects kill the weed also. They feed also unwanted weed prickly, pear, opuntia dillenii eradicated by cochineal insects and other coccid. Some are useful for soil building also.

POLLINATION POTENTIAL OF THE SOCIAL WASP *ROPALIDIA SPATULATA*

Prof. C. SUBBA REDDI
Department of Environmental Sciences,
Andhra University,
WALTAIR 530 003, INDIA

Studies on pollination ecology of the plant species growing spontaneously at Visakhapatnam (17° 14'N - 82° 18'E), India, revealed that the wasps constitute an important component of the pollinator guild. One such wasp species is the social wasp Ropalidia spatulata vecht. It is a regular visitor to such plant species as Zizyphus oenoplia Mill., Z. mauritiana Lamk., Santalum album L., Tectona grandis L.f., Premna latifolia Roxb. and Vitex negundo L. Its pollination potential in relation to the floral biology of these plant species is assessed on the basis of frequency of visits inter-plant flights, number of flowers visited in a unit time, the time spent at a flower and the number of pollen carried on the body.

In P. latifolia, flowers are small, zygomorphic, (corolla tube 2.5 mm long) erect and arranged in flat-topped corymbose inflorescences. When the wasp forages at the flowers, its legs, abdominal surface, forehead brush against the dehiscent anthers and receptive stigmas. This results in 'mess and soil' pollination. In V. negundo, the flower is almost gullet blossom, and is pollinated nototribically. The wasp forages on nector. It lands on the expanded lower corolla lip and inserts its proboscis slightly pushing its head in. This pollen is deposited on the back of head or thorax. The flowers of S. album are small and shallow with reflexed perianth lobes and with easily accessible nectar. The taxon is obligately xenogamous and the wasp has been observed to fly from plant to plant thereby moving xenogamous pollen. In T. grandis, the flowers are small, relatively shallow with reflexed corolla and well exposed essenticial organs. The taxon is predominantly xenogamous. When the wasp moves on the flower, any part of its ventral side comes into contact with the sexual organs resulting in sternotribic pollination. The flowers of both species of Zizyphus are small, shallow. Nectar is produced in female phase only. Though wasps visit the flowers fornectar only, it was observed that they visit the flowers irrespective of their sex phase.

Of the total flowers visits, these of wasps constituted 19.45% in T. grandis, 35.5% in S. album, 76% in P. latifolia, 43% in V. negundo, 4 - 29% in Z. mauritiana and 18 - 34% in Z. oenoplia. Of the total wasps visits, those of R. spatulata made up 5.35% in T. grandis, 9.27% in S. album, 10.6% in P. latifolia, 11.2% in V. negundo, 16 - 100% in Z. mauritiana, 48 - 100% in Z. oenoplia.

The number of flowers visited in a minute ranged between 4-26 , the time spent at a flower from 1-6 seconds, and the number ofpollen that the wasp carried on its body from 17-488.

The flowers of plant species under consideration are all promiscous and are visited not only by wasps but also by bees, flies and butterflies. However, the wasps have been found to move more of xenogamous pollen, particularly in the obligately xenogamous S. album, and this speaks of the importance of wasp component of the pollinator guild in the biotope of the study area.

POLLINATION OF *PTILOTRICHUM SPINOSUM* BY ANTS : EXPERIMENTAL STUDY OF THE QUALITY COMPONENT

GOMEZ, J.Mª, ZAMORA, R. and TINAUT, A.
Dpto. BIOLOGIA ANIMAL, ECOLOGIA y GENETICA
UNIVERSIDAD DE GRANADA
18001 GRANADA, SPAIN

The intensity of a biological interaction depends on its frequency of occurrence (quantity component) and its fitness consequence (quality component, (1). The quality component refers to the behaviour of a flower visitor as a true polinator. This is measured by the increase in the quality and/or quantity of seeds. It is generally believed that ants have a negative effect on the plant pollinator system, given that their presence can diminish the abundance of other pollinators on the flowers (2), and that pollen carried by ants is damaged by their metapleural gland secretions (3). In this paper, we analyse the quality component of the interaction between ants (mainly **Proformica longiseta**) and **Ptilotrichum spinosum** (Cruciferae). In another paper in these Proceeding, we analyse the quantity component of this interaction.

P. spinosum is a stunted woody plant found in the high mediterranean mountains. In Sierra Nevada (S. of Spain) this species appears along a wide gradient (1600 3340 m a.s.l.). This plant has a typical mass flowering, with small, hermaphroditic flowers, placed in inflorescence of 6-22 flowers. **Proformica longiseta** is a endemic species of the S. Nevada, with a wide altitudinal gradient (4). It is a habitual visitor of some plants of the high mountains.

We quantified the composition and abundance of the assembly of pollinators for labeled plants. **Proformica longiseta** is the most abundant of the 52 species of insects which visit **P. spinosum** flowers. We analysed their movement patterns, and the effect of these insects on the abundance ot the other pollinators. We have also tested the efficiency of ants as pollinators by quantifying their influence on seed production of **P. spinosum** (direct component of fitness). To accomplish this, we designed a selective exclusion experiment of pollinators. Stems of 50 inflorescences received a barrier of glue to exclude ants. Another 50 received a barrier of glue and were covered by a fine screen to exclude all pollinators. And finally, 50 were used as a control.

The ants show a displacement distance between consecutive flowers similar to that of the other winged pollinators (1.5 cm and 1.4 cm, respectively). Ants always visit flowers of the same individuals, making 31.8% of its displacement between flowers of the same inflorescence. The presence of ants does not negatively affect the abundance of the others pollinators ($r = 0.279$, $n = 80$, $p < 0.02$). The results of the experiment show that 100% of the control flowers produced seeds. This quantity is reduced to 90% in the case of flowers excluded of ants only, and 75% for those excluded of all pollinators. Therefore, ants act as a true pollinator, increasing the fruit production of **P. spinosum**. The foraging behaviour of ants causes all the flowers to be fecundated by pollen of the same plant. The self-compatibility of **P. spinosum** favors the development of a mutualistic system where ants play an important role as pollinators.

1. Herrera, C.M. 1987. Oikos 50: 79-90
2. Fritz, R.S. and D.H. Morse. 1981. Oecologia 50: 316-319
3. Beattie, A.J. 1985. The Evolutionary Ecology of Ant-Plant Mutualisms. Cambridge University Press. New York.
4. Tinaut, A. 1979. Bol. Asoc. Esp. Entom., 3: 171-183.

POLLINATION OF *PTILOTRICHUM SPINOSUM* BY ANTS: QUANTITY COMPONENT

GOMEZ, J.Mª, ZAMORA, R. and TINAUT, A.

Dpto. BIOLOGIA ANIMAL, ECOLOGIA y GENETICA

UNIVERSIDAD DE GRANADA

18001 GRANADA, SPAIN

Studies that quantify the abundance of ants as pollinators are scarce, given that traditionally they have been considered as nectar thieving and not true pollinators. In a former paper in these proceeding, we demostrate that ants are true pollinators of **P. spinosum.** In this paper, we will quantify the frecuency of this ant/plant interaction. We selected two populations of **P. spinosum** situated at 2180m and 2550 m a.s.l., labeling 40 and 20 plants, respectively. The relative abundance of the pollinator species was quantified by means of census (durantion of one min) for each labeled individual. The visitation rate of each species was quantified as the number of flowers visited per min. The quantity component of the interaction between **P. longiseta** and **P. spinosum** (NFV sensu 1) refers to the total number of flowers visited by the ants populations, and calculated as the product of the visitation rate by the relative abundance of ants. **P. spinosum** has a pollinator assembly of 52 species, the most abundant being **P. longiseta** (80.1 % of the total). This ant shows a highly variable relative abundance between individual plants (0-90%), related to the amount of flowers per plant ($r = 0.619$, $n = 60$, $p < 0.001$). The average visitation rate of winged pollinators was 15.5 flowers/min, whereas the visitation rate of ants was clearly lower (3.2 flowers/min). However, due to their greater abundance, ants show a interaction frequency with **P. spinosum** (NFV = 3.9) clearly superior to the other pollinator species (NFV $<$ 1 all cases). The spatial variability of ants related to the placement of the ant nests represents an extrinsic factor to the plant which conditions interindividual differences in reproductive success.

1. Herrera, C.M. 1990. Oecologia (Berlin). In press.

EFFECT OF GROWTH REGULATORS ON INSECT FORAGING AND POLLINATION OF SOME ENTOMOPHILOUS CROPS

S.K. SHARMA
DEPARTMENT OF ENTOMOLOGY
HARYANA AGRICULTURAL UNIVERSITY,
HISAR-125004 INDIA

So far no information is available on the effect of growth regulators on nectar secretion and its effect on insect foraging in entomophilous crops. Shuel (1959, 1964) conducted some experiments on the effect of growth regulators on nectar secretion in excised flowers cultured on sugar solution. The secretion of nectar in the flower would greatly determine the foraging activity of insect pollination which would ultimately affect the pollination process in the entomophilous crops. Keeping this fact in view investigations were carried out with four growth regulators viz. GA_3 NAA, IAA, IBA which were used twice at normal doses i.e. 80, 160, 5 and 10 ppm before flowering respectively at an interval of 15 days on three entomophilous crops viz. mustard, cauliflower and onion. The results are summarised in Table 1.

(i) Effect of growth regulators on nectar:- Flowers from plants treated with GA_3 had maximum volume of nectar/flower in mustard (1.25 ul) and cauliflower (1.81 ul) followed by flowers from NAA, IAA and IBA treatments. The untreated plants had the least volume of nectar/flower. Sugar concentration of nectar in mustard (51.75%) and couliflower (31.15%) was maximum in GA_3 treated plants followed by plants treated with NAA IAA, IBA and the control. Among different treatments flowers from GA_3 produced maximum amount of sugar/flower in mustard (4.77 mg) and cauliflower (3.46 mg) while NAA, IAA and IBA produced lesser quantity of sugars/flower. However growth regulators had no effect on volume, concentration and dry nectar sugars in the flowers of onion.

(ii) Effect growth regulators on abundance of insect visitors and their foraging speed:- Relative abundance of flower visiting insects on mustard and cauliflower in GA_3 treated plants was maximum (2.09 and 1.93 insects/m^2/5 min) followed by NAA, IAA and IBA treatments while untreated plants attracted the least number (1.03 and 1.05 in mustard and cauliflower). Time spent/flower irrespective of different visitors was maximum in GA_3 treated plants in mustard (6.54 seconds) and cauliflower (7.09 seconds) while it was lowest in the untreated plants (4.53 seconds in mustard and 6.35 seconds in cauliflower). However, growth regulators did not show any difference in the foraging speed of insect visitors in onion.

(iii) Effect growth regulators on yield parameters:- Besides above effects yield parameters such as percent seed set number of seeds/pod and yield were significantly higher in treated plants than control in mustard and cauliflower.

References

1. Shuel, R.W. 1959. Can. J. Bot 37 1167-1180.
2. Shuel, R.W. 1964. J. Apic. Res. 3 (2) : 99-111.

Table 1. Effect of growth regulators on volume, concentration, dry nectar sugar, insect abundance, and foraging speed in three entomo-philous crops.

Treatments	Nectar Vol./Flower (ul)			Nectar Conc. (%)			Nectar sugar/ flower (mg)			Insect abundance (m^2/5 min)			Time spent/flower* (Seconds)		
	C	M	O	C	M	O	C	M	O	C	M	O	C	M	O
GA_3	1.81	1.25	2.07	35.15	51.75	69.33	3.46	4.77	12.24	1.93	2.09	0.57	7.09	6.54	5.51
NAA	1.71	0.98	2.06	33.47	48.01	68.91	2.37	3.66	12.15	1.64	1.42	0.70	6.82	5.72	5.37
IAA	1.33	0.84	2.04	33.30	46.91	68.61	2.37	3.14	12.17	1.44	1.31	0.58	6.92	5.09	5.40
IBA	1.11	0.99	2.02	33.13	45.93	68.33	2.63	2.52	11.55	1.57	1.38	0.62	6.38	4.69	6.53
Control	0.91	0.56	2.07	31.38	41.68	68.24	1.50	2.26	12.07	1.05	1.03	0.51	6.35	4.53	5.36
CD(0.05)	0.31	0.17	N.S.	0.50	1.60	N.S.	0.40	0.52	N.S.	-	-	-	-	-	-

C (Cauliflower), M - (Mustard) , O - (Onion)

* Foraging speed irrespective of different insect visitors

Symposium 18

PARASITE-HOST RELATIONSHIP OF *VARROA JACOBSONI* AND OTHER ASIAN HONEY BEE MITES

Organizer : **N. KOENIGER, F.R. Germany**

THE PARASITE HOST RELATIONSHIPS OF *VARROA JACOBSONI* AND *APIS* SPECIES

N. Koeniger

Institut für Bienenkunde (Polytechnische Gesellschaft)
Fachbereich Biologie der J.W. Goethe-Universität Frankfurt
Karl-von-Frisch-Weg 2., D 637 Oberursel, F. R. Germany

The parasite host relationship of Varroa jacobsoni (Vj.) is of a three fold nature. Naturally Vj. is associated with Apis cerana (Ac.) and Apis koschevnikovi (Ak.). Recently, Vj. successfully changed over to a new host, the Western Apis mellifera (Am.).

Varroa jacobsoni and Apis koschevnikovi : In a colony of the recently rediscovered red bee Ak. (1) several Vj. were found. Details are not yet reported.

Varroa jacobsoni and Apis cerana : Vj. was discovered 1904 in an Ac. colony in Java and it is generally agreed that this mite is an original parasite of Ac.. No further reports were available until few years ago. The general interest in Vj. increased because of its dramatic damages to beekeeping industry in Europe. This relatively large mite remained unnoticed because it does not cause any apparent damage to the Ac. colony. Three mechanisms are discussed which can contribute to a regulation of Vj. population in an Ac. colony: 1. The reproduction of Vj. is restricted to drone brood. The mites do not reproduce in worker brood cells (2). This general rule seems to have a few exceptions (3). 2. In Java we observed Vj. in a larger apiary with more than 500 Ac. colonies in one location (Sukabumi). We inspected sealed drone brood cells in three colonies and found an infestation rate of 57%. Among the 'normal' cells several 'old' drone cells with dark or black cell cappings were noticed. These cells contained many (up to 28/cell) dead Vj. beneath a dead drone pupa. Hence, Ac. workers did not open sealed drone brood cells after the pupa had died and Vj. got trapped and died. 3. A special grooming behaviour of Ac. seems to cause a elimination of Vj. from adult bees (4).

Apis mellifera, A. Varroa jacobsoni in Europe:

In Am. Vj. reproduces in worker brood cells. But, there is a significant preference for drone brood (5). The reproductive success is higher in drone cells, 2.6 mated daughters per drone cell and 1.3 per worker cell (5). Further, the reproductive success depends on the number of mites which enter into one brood cell. For example, a mite will produce more offspring as single parasite in a worker cell than as fourth mite on a drone pupa (6). The 'choice' of Vj. between worker and drone brood seems to be influenced by selection (7). In Europe the reproduction of Vj. seems to increase the number of mites in an infested colony until finally the colony will collapse (or the beekeeper applies acaricides).

B. Varroa jacobsoni in South America : In South America Vj. does not cause severe damages to bee colonies (8). Several factors are discussed. 1. The africanized bees (Am.scutellata hybrids) had a slower mite reproduction than Am.carnica which were imported from Germany (9). 2. A high percentage of mites found in worker brood cells did not reproduce (10). 3. In a recent experiment in Germany, South American bees and their F1 hybrids eleminated significantly more mites from the winter cluster than pure Am.carnica (11).

LITERATURE: 1.Mathew S.,Mathew K. 1988: Newsl. Beekeepers Tropical and Subtropical Countries 12,10. 2.Koeniger N. Koeniger G. Definado-Baker M. 1983: Apidologie 14,197. 3. De Jong D. 1988: Apidologie 19,241. 4. Peng C. Fang Y. Xu S. Ge. 1987: J.invert. Path. 49,54. 5. Schulz A. 1984: Apidologie 15,401 6. Fuchs S. 1986: Apidologie 17,368. 7. Otten C. Fuchs S. 1988: Proc. EC Epert Meeting (ed.Cavalloro), 69. 8. Ritter W. De Jong D. 1984: Z.angew. Entomol. 98,55. 9. Engels W. 1986: Apidologie 17, 203. 10. Ruttner F. Marx H.&G., 1984 Apidologie 15,43. 11. Koeniger N. 1989: Die Biene 6,321.

ASSOCIATIONS OF MITES (ACARI) WITH BEES (APIDAE)

M. Delfinado-Baker, Beneficial Insects Laboratory-Bldg. 476
ARS, U.S. Department of Agriculture, Beltsville, Maryland 20705

The relation of various mite groups with Apis and meliponine bees are examined, the taxa involved and certain biological aspects related to the development of bee-mite relationship. The nature of associations with bees is treated under the general terms of parasitism and commensalism. Examples of associations are as follows: 1) parasite- the relation of mites to their hosts is one of obligatory parasitism, the mites being dependent on the bees, feeding exclusively on the bee haemolymph, and may cause damage to the bees; 2) phoresy- the mites utilize adult bees solely for transport and dispersion and are morphologically adapted to their hosts mode of life; 3) predator on other mites and/or insect eggs and larvae; none is predaceous on bees, and 4) scavenger, non-parasitic facultative, or obligatory occupant of nests and hives, feeding mostly on organic detritus, pollen and fungi.

Thirty-one genera in 19 families of mites are recorded from Apis hives and nests, or attach to honey bees. For the most part these mites are indirectly involved with the honey bees; few feed directly on bees (haemolymph) and are parasitic, such as mites of the genera *Varroa* and *Euvarroa* (family Varroidae), *Tropilaelaps* (family Laelapidae) and *Acarapis* (family Tarsonemidae). The *Apis* host-mite parasite relationship is genus-specific and not species-specific. *Varroa*, *Euvarroa* and *Tropilaelaps* are specific to bee brood of *Apis cerana-mellifera*, *Apis florea-andreniformis* and *Apis dorsata-laboriosa*, respectively; *Acarapis* is specific to adult bees of *Apis mellifera-cerana*, of which *A. woodi* infests the tracheal system.

Stingless bee nests harbor 29 genera in 10 families of mites; none is parasitic on stingless bees. Mostly the mites are obligatory or specific inhabitants adapted to definite environmental condition of meliponine nests. The mite fauna is unique to the Meliponinae.

There is virtually no exchange of significant mite groups between *Apis* and meliponine bees. A list of the mite taxa and bees is given, and important bee-mite associations are discussed and illustrated.

Table 1. Distribution of mites on Apis and Meliponinae (1, 2).
Number of known genera; parasitic genera in ( ).

Group of Mites	Apinae Apis	Meliponinae Trigona	Meliponinae Melipona
Astigmata			
Acaridae	5		
Aeroglyphidae	1		
Carpoglyphidae	1		
Chaetodactylidae	1		
Gaudiellidae		2	2
Glycyphagidae	1		
Hemisarcoptidae			2
Pyroglyphidae	1		
Suidasiidae	1		
Prostigmata			
Cheyletidae	4		
Ereynetidae			1
Erythraeidae	1		
Pyemotidae	1		
Scutacaridae	2		
Tarsonemidae	3 (1)		
Tydeidae		2	1
Mesostigmata			
Ameroseiidae	2	2	
Ascidae	5		
Diplogyniidae		1	
Laelapidae s. str.	5 (2)		
Laelapidae-Hypoaspidinae		5	5
Macrochelidae	1	2	1
Parasitidae	1		
Triplogyniidae			1
Uropodidae	1		2
Varroidae	(2)		

References

1. Eickwort, G. C. 1988. In: Africanized Honeybees and Bee Mites, G. Needham, R. E. Page, M. Delfinado-Baker and C. E. Bowman, Eds., Ellis Horwood Limited Publishers, Chichester, pp. 327-338.
2. Delfinado-Baker, M. 1989. American Bee Journal, 129 (9): 609-613.

EUVARROA SINHAI AND SOME ASPECTS OF ITS ASSOCIATION WITH ITS HOST *APIS FLOREA*

Kamal Aggarwal
R.B.Mathur

Acarology Laboratory, Department of Zoology,
Haryana Agricultural University, Hisar (Haryana),
India-125004

INTRODUCTION

Apis florea is a valuable pollinator of many crops like alfalfa, beans, cotton, litchi, sunflower, toria and sarson etc.(3); as such its parasitization by Euvarroa sinhai does have some economic repurcussions, which can assume serious dimensions if the mite shifts to another Apis host e.g. A. mellifera. The mite is known from India, Thailand, Sri Lanka and Iran (1,2). Unfortunately not much is known about the biology of the mite and the various aspects of its association with A. florea.

BIOLOGY AND SEASONAL CYCLE

The life cycle passes through the egg, protonymphal and deutonymphal stages and is similar to Varroa jacobsoni except that the Euvarroa mites occur only on drone brood (2). On adult workers, the population levels of female mite showed a cyclic pattern with no or negligible infestation from August-January and high infestation during April-June. Reproduction occurred mainly during March-April and September (5). There is a synchronization of the mite reproduction with the seasonal reproductive cycle of the host bee which rears drone brood mainly during late February-early May and secondarily during September-October.

HOST-PARASITE INTERACTION

Effect of Queen Cells on Mite Infestation

It is believed that the queen cell construction exerts an inhibitory influence on mite infestation. In a March'85 comb, infestation in drone brood away from queen cells was 93.1% and close to queen cells was 40%. A similar trend was observed in another comb (1).

Effect of Mite Infestation on Host Haemolymph Proteins

Preliminary studies show that infestation results in an increase in protein concentration (ug/ul). In pupae with brown eyes, 4-6 mites/cell resulted in an increase of 24.68% and in pupae with tanning on whole body, 1-3 mites/cell resulted in an increase of 56.85% while 4-6 mites/cell resulted in a rise of 62.90%. Our observations show that Tropilaelaps clareae parasitization of A. mellifera drone pupae also results in an increase in protein concentration. Its felt that change occurs primarily due to a new delicate balance reached between JH III and haemolymph protein concentration following mite infestation.

Effect of Mite Infestation on Bee-Behaviour

The bees recognize the infestation to some extent. In one comb, both mites and bee stages were found dead inside some of the brood cells. A few infested combs were found abandoned. Apparently, the bees abandon the heavily infested combs and build a new one at a new site thus dealing with infestation in their own unique way.

HOST SPECIFICITY

Uptil now, the mite has shown a great degree of host-specificity to drone brood of A. florea. Greater biomass of drone larvae, their longer developmental period plus the passive and drifting behaviour of adult drones can be a reason (2).

NEW INFESTATION

In recent years few female mites were collected from A. mellifera hive-debris (4,6). In many parts of the world, A. florea co-exists with A. dorsata, A. cerana and A. mellifera (3). Its feared that just as V. jacobsoni and T. clareae shifted from their native hosts and proved disastrous to A. mellifera; the E. sinhai may also follow suit. Increasing densities of A. mellifera colonies within the range of E. sinhai is likely to encourage this shift in hosts (5). The reason as to why these mites could not be found in brood cells can be that the mite carrying field bees are unlikely to be in direct contact with nurse bees surrounding the brood frame. Thus, the occasional mites that reach the bee hive do not get access to developing bee brood and die. The mite, however, appears capable of parasitizing A. mellifera .

REFERENCES

1. Aggarwal,K. and R.P.Kapil. 1988. In: Africanized honey bees and bee mites, Needham et al. Ed., Ellis Horwood Ltd., England, pp 404-407.
2. Akratanakul,P. and M.Burgett. 1976. Journal of Apicultural Research, 15: 11-13.
3. Annonymous. 1986. FAO Agric.Ser.Bull. 68: 1-283.
4. Kapil,R.P. and K.Aggarwal. 1987. Bee World, 88: 189.
5. Kapil,R.P. and K.Aggarwal. 1988. In: Progress In Acarology, G.P. Channabasavanna and C.A.Viraktamath,Ed., Oxford & IBH Publishing Co.Pvt.Ltd., New Delhi. pp. 277-281.
6. Sihag,R.C. 1988. American Bee Journal, 128: 212-213.

IN VITRO REARING OF HONEY BEE PARASITE *VARROA JACOBSONI*

NILLA DJUWITA ABBAS, Department of Developmental Biology, University of Tübingen, Germany

WOLF ENGELS, Department of Developmental Biology, University of Tübingen, Germany.

Rearing of Varroa jacobsoni in artificial cells is possible on drone and worker brood of Apis mellifera carnica in an incubator under conditions similar to the bee colony. About 2000 female mites with their hosts were transferred into the laboratory 2 hours before, 5 hours and 24 hours after cell capping and kept in the polystyrol cells. On 1108 drone hosts from 1302 eggs laid 326 adult mites were reared, and 22 from 213 eggs laid on worker brood. On drone hosts about 20 % and on worker brood 10 % of the adult mites were males. In the nymphal stages, the rate of in vitro survival was appreciable in deutonymphs, but many mites died before moulting into adult. After feeding with L5 hemolymph through a membrane 2 eggs were laid. The transmembrane feeding technique is promising but still has to be improved.

FEW OBSERVATIONS OF *VARROA JACOBSONI*, MITE PEST OF HONEY BEES IN THE COLONIES OF *APIS CERANA* AND *MELLIFERA*

Dr.Naresh C. Tewarson
Department of Zoology
Ewing Christian College
Allahabad-211003

Varroatosis, brood as well as adult disease of two species of hive bees, Apis cerana and recently introduced, Api mellifera, is still neglected. In Uttar Pradesh (North India) the mite, Varroa jacobsoni is causing much damages to cerana bees. For example the mite associated diseases like 'Sac brood viral disease' and other microrganismal diseases have created real problems for bee keeping industry recently in hills of Uttar Pradesh, Gorakhpur and Allahabad. The vivo transfer experiments in cerana bees, showed some inhibitory factor either in the haemolymph of the worker larve in the capped brood or the topography of the brood cell,

restricts the egg laying or further development of the Varroa eggs, contrary to the normal development inside the drone brood cells. Keeping in mind the previous findings, that most of the hemolymph proteins of host honey bee, reach as macromolecules in the developing oocytes of the mite via it's haemolymph, the inhibitory factor in worker lavae's hamolymph seems to be probable. Studies on bionomics of Varroa revealed the pattern of parasitisation in A. cerana bees and gross effect on colony strength and honey production.

DISTRIBUTION OF THE ACARICIDE PERIZIN IN THE HONEYBEE (*APIS MELLIFERA* L.) AND ITS INFLUENCE ON THE HAEMOLYMPH

Buren, N. W. M. van, Kemps, T. M. A. & J. W. Stegeman
Laboratory for Comparative Physiology, University of Utrecht, P.O.Box 80.086, 3508 TB Utrecht, The Netherlands.

SUMMARY
The systemic agent Perizin (Bayer AG) is used to combat the mite *Varroa jacobsoni* Oud. Its active ingredient coumaphos is believed to spread through the haemolymph of the honeybee. This study reports on the distribution over time of coumaphos over honey stomach, haemolymph, intestine and rectum after bees of *Apis mellifera* have been fed individually. Changes occurring in the haemolymph volume due to the ingestion of Perizin were also studied.
The amount of coumaphos in the haemolymph reaches its maximum after 4 hours, but the amount is still very low (2 % of total amount recovered). After 15 minutes only 30 % of the coumaphos is still in the honey stomach and available for trophallaxis. The coumaphos accumulates in the rectum over time.
The volume of the haemolymph was significantly larger in bees which were fed Perizin than in bees which were fed syrup or in bees which were not fed at all.

INTRODUCTION

Perizin is one of the systemic agents commonly used to combat the parasitic mite *Varroa jacobsoni* Oud. in honeybee colonies. The active ingredient of Perizin is the thiophosphoric ester, coumaphos.
The systemic effect of the agent is dependent on the number of trophallactic interactions within a bee colony. After the bees have been treated with Perizin, they are assumed to be cleaned or to clean themselves by licking. Subsequent trophallaxis probably distributes the active ingredient among the colony members.
The purpose of the present study is to determine how much of the ingested coumaphos reaches the haemolymph, how long it remains there, and how long it is available in the honey stomach for trophallactic transfer.

MATERIAL AND METHODS

Young worker bees were taken from a colony that was maintained in a bee flight room. They were placed in groups of about fifteen in Liebefeld cages in an incubator at 28 oC. They were provided with a pollen/sugar-solution and syrup, and were given the opportunity to fly and to defecate every 3-5 days. When 3, 8 or 14 days old the bees were used for the experiment. At first one hour of flight was allowed. Subsequently they were isolated and deprived of food for another hour. Each bee was fed 10 µl of a syrup (saccharose:water=1:1) to which Perizin containing 14C coumaphos (38 nCi, Bayer AG) had been added (Perizin:syrup=1:49). After 15 or 30 min., or after 1, 2, 4, 6, 12 or 24 hrs. a bee was removed for the recovery of 14C from the three parts of its intestinal tract and from its haemolymph. To determine the volume of the haemolymph the bee was injected between the 2nd and 3rd abdominal segment with 0.5 µl of 120 nCi 3H inulin in Ringer solution. First of all the inulin was allowed to disperse evenly in the body for 15 min., then a haemolymph sample was taken. The inulin allowed us to determine the total haemolymph volume, the 14C revealed the total amount of coumaphos in the haemolymph. After the haemolymph had been sampled the bees were killed in liquid nitrogen and kept at -20 oC. Upon thawing the honey stomach, midgut and rectum were stored separately in 400 µl Lumasolve for 3 days at 50 oC and were vortexed twice a day. Radio-activity was measured using a Packard scintillator (Type 4550).

To evaluate the effect of Perizin on haemolymph volume, control treatments involved feeding the bees with a 10 µl Perizin solution (Perizin:water=1:49) or with 10 µl syrup (saccharose:water =1:1) or not feeding the bees at all. In the same way as in the former experiment the haemolymph volume was analysed after 30 min., or after 1, 2, 4, 8 or 16 hrs.

RESULTS

Distribution of Perizin

No differences were found in the patterns of distribution of Perizin in relation to age. In the following bees are not differentiated according to age.

Perizin was found to have a toxic effect; mortality rates were particularly high in bees that had been exposed for 12 and 24 hrs.

The total amount of coumaphos recovered in the three parts of the intestinal tract and in the haemolymph was on average 68 %.

Fifteen minutes after the bee had ingested Perizin only about 33 % of the amount ingested was recovered from the honey stomach. After 4 hrs. the amount of Perizin in the haemolymph had reached its maximum value (2.0%). After 24 hrs. almost nothing was left in the haemolymph (0.3%). After 12 hrs. the honey stomach was free of radio-activity. In the intestine the highest level of coumaphos was found 6 hrs. (52%) after feeding. The coumaphos moves towards the rectum relatively slowly: after 24 hrs. up to 71% was found in the rectum.

Volume of the Haemolymph

Few of the bees which were fed with Perizin solution survived for longer than 16 hours. Those bees that did survive differed markedly as regards the volume of their haemolymph.

Bees fed on Perizin had a significantly higher haemolymph volume after 1,2,4 and 8 hrs. of exposure than bees which were not fed at all. Compared to bees fed on syrup they also had a higher haemolymph volume after 2,4 and 8 hrs. of exposure. Using regression-analysis we found an increase in the volume of haemolymph in bees which had been fed with Perizin.

DISCUSSION

The fact, that Perizin (coumaphos) is available to the mite for a long time is important for the action of this systemic agent. The acaricide Thiazolin, studied by Eyrich (1986) reaches its maximum value after one hour; Perizin by contrast reaches its maximum after 4 hrs. Like Thiazolin, Perizin has disappeared almost completely from the haemolymph after 24 hrs. Perizin however remained in the honey stomach for a longer time than has been reported for Thiazolin (Eyrich, 1986). In a natural situation, where trophallaxis can take place, this could promote the distribution of the acaricide over the colony members. On the other hand trophallaxis may shorten the time that the active component has to reach the haemolymph of the donorbee.

The presence of Perizin seems to lead to an increase in haemolymph volume. After a long exposure time we recorded fairly high mortality. These results suggest Perizin has a marked effect on the physiology of the bee itself. Further investigations need to be carried out.

ACKNOWLEDGEMENT

This study was was performed in cooperation with Bayer AG and was financed by the Technology Foundation (STW), Utrecht.

Eyrich, U. 1986. Die Verteilung eines systemisch wirkenden Akariziden Thiazolins zur Behandlung der Varroatose im Volk der Honigbiene (*Apis mellifera* L.), thesis, Ludwig-Maximilians-Universität München.

NICOTINIC ACETYLCHOLINE RECEPTORS ASSOCIATED WITH HONEY BEE NEURAL MEMBRANES

Z.-Y. Huang and C.O. Knowles
Department of Entomology, University of Missouri
Columbia, MO 65211, U.S.A.

Although the cholinergic system in insects generally has been extensively studied, there have been few investigations of its components in honey bees, with the exception of acetylcholinesterase which was recently purified. Also, honey bee brain homogenates have been found to contain acetylcholine (ACh). Locust nicotinc ACh receptor antiserum was shown to bind to bee brain components, and it was concluded that the antiserum cross-reacted with honey bee ACh receptors, an interpretation supported by the fact that alpha-bungarotoxin (BGT) binding sites were present in some areas of strong immunoreactivity [1]. To gain additional insight into honey bee ACh receptors, we examined the kinetic and pharmacological properties of [^{125}I]BGT binding to membranes prepared from heads or brains using the filtration method.

Over 97% of the specific binding (defined by nicotine) of [^{125}I]BGT occurred in the 40,000 g pellet and was heat sensitive, linear with tissue concentration, and saturable. B_{max} and K_d value were 69 fmol/head and 743 pM, respectively, from saturation experiments. Binding was reversible with an on rate of 1.38×10^6 $s^{-1}M^{-1}$ and an off-rate of $6.2x10^{-4}$ s^{-1}. The Hill coefficient was 1.09, indicating noncooperativity. Inhibition studies revealed that some (nicotine, ACh+dichlorvos, and tubocurarine) but not all (decamethonium and hexamethonium) nicotinic drugs were potent inhibitors of BGT binding; muscarinic drugs generally were weak inhibitors.

These data suggest that BGT is binding with high affinity to a single class of putative nicotinic ACh receptors. It might be possible to develop acaricides to exploit differences in ACh receptor properties between honey bees and parasitic mites.

1. Kreissl, S. and G. Bicker. 1984. J. Comp. Neurol. 286: 71-84.

Symposium 19

THE BENEFITS OF SOCIAL INSECTS IN AGRICULTURE

Organizer : **PHIL J. WRIGHT, United Kingdom**

THE POLLINATION RESEARCH PROGRAMME IN INDIA

R.C. Mishra
Project Co-ordinator,
All India Co-ordinated Project on Honeybee
Research and Training(ICAR), H.A.U., Hisar

Efforts to increase agricultural production in India are continued and in this venture most energy has been expended on components like crop breeding, fertilizers, pest control and other agronomic practices. The static agricultural production for the last about ten years warrants a need for changed strategies and look for other possible inputs for increasing crop yields. The fact that beekeeping and agricultural productivity are closely interlinked has not been fully realized and appreciated, therefore, sufficient knowledge of pollination ecology and pollination requirements of crops in India is lacking.

Comparative abundance of different insect visitors and their diurnal activity has been observed in most crops. Insects which have been found to visit crop bloom belong to hymenopteran genera; Apis, Bombus, Xylocopa, Helictus, Megachile, Ceratina, Osmia, Braunsapis and syrphid flies. Such observations give valuable indication of insects associated with the crop but most of the insects might not be pollinating them. Keeping this objection in view there is a growing trend in India to work out comparative pollination efficiency of insect-flower visitors. The efficiency is measured by giving weightage to abundance, foraging behaviour, morphometrics of insects and pollen carrying capacity.

ASSESSMENT OF POLLINATION REQUIREMENTS:

Cages of many types of construction material are being used for pollination requirement experiments. These cages can be made to exclude insects and also avoid wind borne pollen grains to pollinate flowers. Under the conditions only selfing is possible. Cages made up of material with appropriate mesh size are used to allow windpollination in addition to selfing. The third set is left open for pollination by selfing, wind and insects. With the help of the three sets of cages the pollination requirements and role played by wind and insects is confirmed.

Another method in use is to enclose small plots in insect proof screen cages, enclose honeybee colonies in second plot and the third of the same size is left exposed.

Results from cage experiments are compared by determining the number or weight of seeds, pods or fruits per cage or percentage of flowers setting fruit or seed. The above studies have been conducted in apple,almond,peach,litchi,onion,cauliflower,Brassica spp., ladies finger, sunflower, cardamom and cotton.

Flowers are enclosed in bud stage and again caged after hand pollination. The cages are removed after the stigma is no more receptive and it is found out whether the inadequate pollination is a factor limiting yield.

Attempts are also afoot to find out the usefulness of variable number of bee visits to flowers. The flowers are enclosed in bud stage with muslin or paper bags. The bags are removed on flower opening and again enclosed after it is visited by bee. Usefulness of multiple visits is also assessed in this process. The information is used to provide bee population so that each flower receives similar number of visits during the period of stigma receptivity.

HONEYBEE COLONIES PER UNIT AREA:

Bee colonies are moved to crops and placed in groups. With the established fact that number of bees decrease with distance from colonies, the corresponding decrease in yield is observed. The area optimally served by the colonies is determined by comparing the yields at different distances from bee colonies. In some cases number of colonies per unit area is increased and number of colonies required per unit area are estimated by comparisons of fields with variable number of colonies.

INSECTICIDES AND POLLINATION:

Estimates of yield gains by controlling pests with insecticides and yield loss by way of reduced pollinators activity are being made.

MEASURING THE FORAGING STRENGTH OF BUMBLE BEE COLONIES

N. Pomeroy and S.R. Stoklosinski

Department of Botany and Zoology, Massey University,
Palmerston North, New Zealand

Commercial decisions to raise colonies of bumble bees for crop pollination must ultimately depend on the value (price) per bee exceeding the the production cost per bee. The value per bee depends on the number "needed" per unit area of crop (in itself a complex question answerable from the non-linear pollinator density - crop yield curve) and on the commercial value of the crop (or more precisely the *increase* in crop value caused by the bees brought to it).

Estimation of how many colonies must be provided depends on knowing the number of bees per colony actually actually working the crop. This is highly variable, depending on total worker population, amount of brood and competing forage sources, and therefore needs to be measured emipirically in each pollination situation.

We have developed a "forager trap" (Figure 1) to collect from each colony the entire foraging force working in the field at a chosen moment. Each hive must have its own trap, which must allow unimpeded foraging when not set. Such a trap must not interfere with the appearance, odour or other features of the entrance or bees will go away again. Hence the trap needs to be on the hive for at least a day before any trapping is done. Once the trap is set no more bees can leave the hive, and as foragers return they are retained in a transparent chamber for counting and examination of pollen loads. We find few foragers arrive later than one hour from setting the trap but this is variable and in the first instance one needs to monitor the rate of return to discover the maximum trip duration.

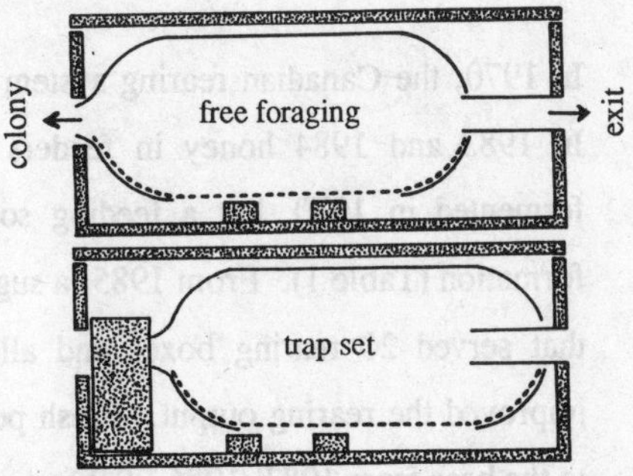

Figure 1. Mid-vertical secion through forager trap.

The trap comprises a wooden box containing a 1 - 2 litre plastic bottle with the neck cut off, a cut-out floor covered with mesh (which extends up to the holes for grip) and a hole (20 mm diameter) in the "bottom" end which fits loosely over a glass tube. To set the trap the bottle is slid towards the exit and a barrier such as a block of soft foam plastic is placed between the hive and the bottle's inner hole. At least 40 mm of tube should project into the bottle allowing bees to enter but not leave. To be effective the outside of the glass tube should be clean and preferably lightly greased. If there is too much light in the box some bees may fly up to the tube and get out. This tendency seems to vary between species and may be reduced by incorporating a light-hood to shade the exit hole. These traps temporarily disrupt a colony's foraging so we recommend that they not be used more than once daily.

The traps should be useful for studies of foraging ecology and pollination, as well as for crop owners who may wish to use them to verify the strength of colonies purchased. Another use is is to procure the field force of a colony if it needs to be relocated in daylight hours.

BUMBLE BEE REARING IN NEW ZEALAND

Griffin, R.P., Macfarlane, R.P. DSIR Plant Protection, Department of Scientific and Industrial Research, Private Bag, Christchurch, New Zealand.

Many variations in the details of bumble bee rearing methods have been described from Europe and North America (1,2,3), but consistent colony formation in quantity has not been achieved especially for pocket-making species. Our paper summarises the outcome of the first New Zealand trials with rearing bumble bees.

Queens of *Bombus terrestris* (L.), *B. ruderatus* (Fabr.) and *B. subterraneus* (L.) were taken from flowers. In 1970, 63% of *B. terrestris* were collected in winter and these queens largely contributed to poor fitness for rearing (Table 1), while the nematode *Sphaerularia bombi* Dufour had sterilised 18% of the queens collected in spring. After 1970, queens were collected in spring from nematode-free areas. In 1987-1989, *B. hortorum* (L.) queens were collected in colonies in autumn, mated and overwintered in a screen cage and 197, 193 and 306 queens respectively emerged for rearing in spring.

In 1970, the Canadian rearing system (1) was used, but without any nest material. In 1983 and 1984 honey in feeders was used, because the honey solution had fermented in 1970, but a feeding solution allowed for better laying and colony formation (Table 1). From 1985, a sugar solution was supplied. Clear plastic tubing that served 20 rearing boxes and allowed for a solution change every six days improved the rearing output. Fresh pellets of pollen from honey bee hives was fed to the bees from 1983- 1986. Subsequently pollen was mixed to a dough with honey. This dough was put in the dishes in the next chamber until eggs were laid and placed directly beside any larval broods, which proved to be critical for rearing pocket-making species. From 1984 onwards, upholsterers wadding, plaster of paris blocks, small pieces of undulating cardboard, and felt were used as nest materials. Heating cables with thermostats set at 25°C were put under nesting materials from 1986, in an attempt to speed up colony formation. Daytime temperatures in the rearing room were 18-25°C. Workers were added to the six queens that laid in 1970 to stimulate colony development, but not subsequently because their use is impractical for commercial rearing. Queens were transferred to hives once 5-10 of their workers had hatched and the hives were put outdoors near forage sources.

Table 1 Rearing performance of *Bombus terrestris* queens

Year	Queens		% Fit Queens*		Mean
	No.	% Fit*	Laid Eggs	Hatched Workers	Rearing Duration#
1970	46	46%	35%	24%	71 days
1983	44	91	13	0	-
1984	33	91	33	3	-
1985	117	88	73	38	70
1988	52	96	56	48	74

*Fecund queens surviving over 15 days in rearing boxes
#Days queens confined until the field release of the colony

In 1985, 3 of 85 *B. ruderatus* queens raised colonies. In 1986, 1987 and 1988, 25, 144 and 100 *B. hortorum* colonies respectively were reared. In 1986 and 1987, 2 of 42 *B. subterraneus* queens raised colonies. In 1970, no colonies of any of these three species were raised from 30 queens confined.

Rearing colonies in confinement fully controls the species obtained and offers the potential to produce many colonies that are timed to match crop flowering unlike the use of field hives, but such rearing needs more capital and labour.

1. Plowright, R.C. and S.C. Jay. 1966. Journal of Apicultural Research 5(3): 155-165.
2. Ptacek, V.B. 1985. Sbornik Vedeckych Praci (9): 56-67.
3. Roseler, P.F. 1979. Erwerbsobstbau 21: 177-178.

BIOLOGICAL LIMITING FACTORS FOR THE BEEKEEPING WITH STINGLESS BEES IN THE CARRIBEAN AND CENTRAL AMERICA

Veen van J.W.[1], M.C. Bootsma[1], H. Arce[2], M.K.I. Hallim[3] and M.J. Sommeijer[1]
[1]Utrecht University, Bee Research Dept., P.O. Box 80.086, 3508 TB Utrecht, The Netherlands. [2]Universidad Nacional, Heredia, Costa Rica. [3]Min. Food Prod., Marine Exploit., Forestry and Environment, Trinidad and Tobago

INTRODUCTION

With the results from our first studies, carried out in Costa Rica, Trinidad and Tobago, we will illustrate the most important biological problems to overcome for beekeeping with stingless bees in Central America.

Beekeeping with stingless bees (Apidae, Meliponinae), called "Meliponi-culture", has been practiced for many centuries in various parts of Latin America. The honey of these bees has always been used as food and medicine by pre-Colombian inhabitants of Central America. This form of beekeeping played an important role in the religious traditions of the ancient Maya culture [1,2].

The Maya indians in Yucatán (Mexico) still discriminate between more than ten species of stingless bees, especially for the medicinal use of their honey [3]. The traditional great value of these bees to the local people, is indicated by the rich nomenclature for these insects in all regions [4].

In contrast to the development of beekeeping with the introduced *Apis mellifera* in the Americas, the traditional beekeeping with the (less productive) indigenous stingless bees, gradually obtained the status of a primitive enterprise [5,6]. Recently, meliponiculture has been given much attention. This is due to the spectacular spread of aggressive African honeybees over South and Central America, after their introduction in Brazil in 1957 [7].

Various indigenous species are still being domesticated in all central american countries. "Meliponiculture" is usually practiced inefficiently, mostly due to the lack of knowledge about their social behaviour [8]. Honey is frequently collected from natural colonies in the forest. This often leads to the destruction of the nest. Domesticated colonies are mostly housed in wooden logs or even cardboard boxes. During the harvest, the food storage pots are removed from the nest and the honey is being squeezed out of it. The difficult accessibility of the nest cavity, usually only a narrow opening at each end of the log, often causes severe destructions of the brood.

In Costa Rica, certain regions are well known for their traditional meliponiculture. Of special importance is the area around Nicoya on the peninsula of Guanacaste. The first Spanish settlers already named it "tierra de la miel" (=land of honey). Wagner (1958) indicated the incidence of this primitive type of beekeeping in this area as: "One of the most characteristic features of the rural household in Nicoya is the bee log which hangs under the eaves of the dwelling. .... Almost every house in the country hamlets and many of the dwellings in the town have such logs".

RESULTS AND DISCUSSION

In Costa Rica we found two types of meliponiculture:

a. The domestication of hives of *Melipona*. The most important species is *Melipona beecheii*, which is commonly domesticated in the peninsula of Guanacaste. Little of this meliponiculture is found at altitudes above 700m or in deforestated areas. In the area of Perez Zeledon colonies of a dark variety of *Melipona fasciata* are being kept. Most owners keep their Meliponas in logs hanging under the roof of the house or under special roofings. Some are being kept for many decades. The amount of honey that is produced depends on the size of the cavity in the bee log, and can be as much as ten liter. Owners are reluctant to transfer a colony to a new housing. They state that the bees do not accept new homes. This probably relates to the severe attack by phorid flies *(Pseudohypocera spec.)* of hives that are not properly sealed. The hive used for *Melipona beecheii* is a short section of hollow log, with an entrance hole midway between the ends; carefully fitted wooden stoppers are used to plug the ends, and any remaining cracks are sealed with mortar made of clay. To harvest the honey, one end of the log is opened, and the storage pots surrounding the broodcombs are squeezed, so that the honey runs out. Normally the beekeepers only extract honey once a year, in the "Semana santa" (Eastern).

b. The domestication of nests of *Trigona*. This type of meliponiculture occurs throughout Costa Rica, also at high altitudes and in urban areas. Especially the nests of the minute *Tetragonisca* are very frequently being kept in small boxes or in bamboo pipes. These nests produce only up to one liter of honey. To this honey great medicinal properties are being assigned [3]. It is sold at a high price to local pharmacies.

In Trinidad and Tobago we held a questionnaire among 47 stingless beekeepers [9]. This showed that meliponiculture in Trinidad is as primitive and traditional as in Costa Rica. Two species are being domesticated: *Melipona trinitatis* and *Melipona favosa*. In spite of the fact that the stingless bees produce far less honey than the honeybee *Apis mellifera*, there is a vivid interest in meliponiculture, also among the younger generation.

In the years 1985, 1986 and 1987, all questioned beekeepers had two or three hives on the average. 51% Of them had their hives for more than ten years, but more interesting, 30% had their hives less than five years. This means that they started beekeeping with stingless bees after the introduction of the Africanised race of the honeybee *(Apis mellifera)* in 1979. The main reasons for beekeeping with stingless bees were: the quality of the honey (89%), and the aggressiveness of the honeybees (60%). After the bees are collected in the forest, they are being transferred in wooden boxes, of all kinds and sizes.

Half of the beekeepers harvest the honey on Good Friday, and the other half in the months April-May. 50% Takes out the food storage pots, and squeezes them. The wax is never given back. The other half squeezes the pots in the nestcavity. Only 11% ever opens the hives to check the colony development. Most beekeepers (78%) extract honey once a year, only 7% more often. The maximum amount of honey extracted from one hive in one year is 5.1 (s.d. 2.5)liter. 90% of the beekeepers use the honey at home and about one quarter of them sell some honey, usually at a high price (US$ 9-15 / bottle [0.75l.]).

It is commonly used as a medicine against coughs, throat diseases, eye illnesses and against heart complaints. It's used purely or mixed with water, juices, rum or even lampoil. 65% Preferes stingless bee honey as a medicine above honeybee honey.
On the question if colonies die out, and why, 57% answered no, 22% told us that the bees leave the hive for unknown reasons, 11% answered to have had troubles with ants and phorid flies. Few attacks of "robber bees" *(Lestrimellita)* were noticed, and in some cases (4%) colonies died out due the use of pesticides near the hive.

REFERENCES

1. WAGNER, H.O., 1960. Haustiere im vorkolumbischen Mexico. *Z. Tierpsych.* 17: 364-375.
2. WEAVER, N. & E.C. WEAVER, 1981. Beekeeping with the stingless bee *Melipona beecheii*, by the Yucatan Maya. *Bee World* 62 (2): 7-19.
3. GONZALEZ ACERETO, J.A., 1984. Historia de la Meliponcultura. *Rev. Univ. Aut. Yucatán* :89-92.
4. GONZALEZ ACERETO, J.A., 1983. Acerca de la regionalización de la nomenclatura maya de las abejas sin aguijon *(Melipona sp)* en Yucatán. *Rev. Geogr. Agra. Anal. Reg. Agri.* 5-6:190-193.
5. NOGUIERA NETO, 1951. Stingless bees and their study. *Bee World* 32: 73-76.
6. NOGUIERA NETO, 1970. A criaçao de abelhas indígenas sem ferrao. *2a Ediçao revista*. São Paulo, Chacaras et Quintas. 365 pp.
7. MICHENER, C.D., 1975. The Brazilean Bee Problem. *Ann. Rev. Entomol.* 20: 399-416.
8. Van Veen J.W., H. Arce & Sommeijer,1990. Tropical beekeeping: The production of males in stingless bees (*Melipona*). *Proc. Exper.& Appl Entomol., N.E.V.*, 1:171-176.
9. SOMMEIJER M.J. & M.C. BOOTSMA, 1988. *Melipona trinitatis* comme productrice de miel a Trinidad, Antilles. *Actes Coll. Insectes Sociaux*, 4: 291-294.

HIGHLY BENEFICIAL FACTOR-HONEY BEE POLLINATION

MRINMOY DAS MODERN APICULTURE
NAGARUKHRA- 741 257, NADIA, W.B. INDIA.

The benefits due to bee pollination is said to be 10 to 25 times more than that from honey and bee wax.

According to survey, in India about 12 M hectares of land including forests, have potential for bee keeping; which is sufficient to sustain at least 50 million bee colonies for honey production alone, and if pollination is also desired several times more than this number would be needed. India today has only about 800 thousand bee colonies.

Many flowering plants and insects are interdependent for their mutual benefits and even for survival. As cross pollinators honey bees are well adopted with hairy body to catch pollen, a specialized tongue to collect nectar and special communication methods by which thousands of woeker bees can be deployed when good food is available. Best results are achieved by bee pollination in ccrps like clover and alfalfa. Now,several farmers take advantageof bees for pollinating their crop by migrating bee colonies. Although, hand pollination is possible, it can cover about 600 flowers during an eight hour working day compared to a quarter of million flowers pollinated by bees in that time. Therefore, there is need to popularise use of bees for crop pollination and higher crop production.

A SIMPLE HIVING METHOD FOR THE LITTLE BEE *APIS FLOREA*

Dr.B.L.Bhamburkar, Retd. Entomologist,
54,Vyankateshnagar, NAGPUR 440 025 (India)

AND

L,N,Peshkar, Assistant Professor in Entomology,
College of Agriculture,NAGPUR 440 010(India).

Apis florea takes part in pollination of a number of agricultural and horticultural crops. Small cultivators who cannot offord to maintain an apiary or hire *Apis cerana hives,* need a cheap and suitable method for handling for desired pollination. Experiments were conducted by the authors at Sindewah in India, to evolve a convenient technique to tame *Apis florea* and shift colonies to places as and when desired for pollination.

A P-shaped wooden frame (Fig.1) was designed for this purpose. The upper bar AB formed the support for the comb. To begin with, the branch of the tree to which the comb of *A. florea* was attached in nature was gently cut alongth with the AB with the help of banana fibre. The frame was then inserted gently in the canopy of a tree preferably *Acacia, Zizyphus,* lemon, mango, sapota, orange (citrus) etc. and placed in such a manner that the entire frame was hidden among leaves and thecomb was at the same angle with respect to sun as in its natural condition on the tree. The pole V was supported from below either by a peg inserted in the soil below or a big stone and then tied to the branch of the tree properly to avoid jerks.

The colony was not disturbed for about a week. The bees accepted the location and within a week attached the comb to the bar AB and increased the comb length. Observations could be made by holding the branches apart from the frame. If needed the frame was brought out in the open and the working of the queen and other bees observed. The colony was examined once in ten days. All the manipulations could be conveniently done as in the case of *A. cerana* hive. The swarming was restricted or completely avoided by clipping of the queen cells.

When the honey was stored by extending the cells in the upper portion of the comb and the cells were sealed by wax, two thin iron strips 12.5" in length and one inch in width were pressed horizontally from two sides of the foundation of the comb at the line demarcating the honey store and brood portion, taking care not to cut comb. The ends of the strips were inserted in the grooves at the same level in pole V and bar BC. The bees attached the cut portion of the comb to the strips within a week. Later, honey portion was separated from the comb by a sharp knife and the brood portionalongwith iron strips was shifted upwards and tied up to the bar AB by placing the ends of strips in the upper most grooves, on both sides. This operation was repeated for every collection of honey. The frame facilitated both maintenance and shifting of *A. florea* hive wherever needed.

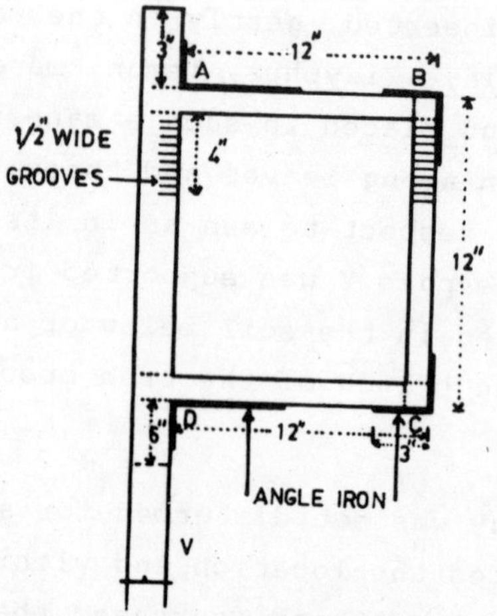

FIG. 1. WOODEN FRAME FOR HIVING LITTLEBEE

INTEGRATION OF APICULTURE IN HORTICULTURE

V.G.Prasad* and Abraham Verghese
Indian Institute of Horticultural Research
Hessaraghatta Lake P.O.Bangalore -560089
INDIA

Apis spp. have helped in the realization of manifold yields in fruits and vegetables. Inspite of several Indian studies in support, (Prasad and Verghese,1983) apiculture has sadly not formed an input in horticultural production. Therefore,this paper is an emphasis to integrate apiculture in horticulture.

An elaborate account of higher yields due to bee pollination in fruits and vegetables has been given by Deodikar and Suryanarayana(1972). They recorded six to 100% increase due to honeybee pollination in vegetable seed production(radish, cabbage, turnip,muskmelon,cucumber,squash,onion and carrot) and fruit yield(litchi,pear,lemon,apple,sweet orange,plum,etc). Introduction of *Apis cerana* at IIHR,Bangalore, significantly enhanced production in *Citrullus lanatus* (Prasad and Verghese,1985). In onion,when *Apis* spp were denied access during flowering the fruit set and number of seeds per capsule decreased from 93.5% to 9.8% and 4.3 to 1.9, respectively (Rao and Lazer,1983). In Karnal,Sinha and Chakrabarti (1983) showed that *Apis cerana* together with *A. indica* and *A. dorsata* helped enhance seed yield by 83.64% and 93.06% in cultivars 328 and Pusa Deepali,respectively. These selected results amply demonstrate the high value of apiculture to horticulture.

In India, there is a need to standardise the foraging pressure (number of bees or bee-visits required to effect desired level of pollination) for different horticultural crops. An example is *Cucumis melo* where 10 bee-visits/flower effected maximum fruit set; >10 made no significant difference (Grewal and Sidhu,1983). Likewise, the number of Apis colony per unit area is to be standardized. In fruits, in general it is one colony/acre and two colonies under high density planting (Johansen,1977). It has been worked out that a hive with full 8-frame brood can effect maximum pollination (Verghese and Prasad,1983). Economic entomologists should schedule insecticidal sprays so as to include safer insecticides(See Johansen,1977) and time the sprays,as done in mango(Verghese,1986) so that insecticidal hazards to pollinators are minimal.

Present Address*: Head,Division of Entomology,IIHR(Retd.)
51/3,Maruthi Nilaya,13th Cross Down,
Temple Street,Malleshwaram,
Bangalore-560055,INDIA.

Horticulturist have been shy of honey bees due to lack of conviction and/or fear of handling them. However, these reasons are unjustified, and this paper calls for a positive outlook to integrate apiculture in horticulture to boost production.

References

1. Deodikar, G.B. and M.C.Suryanarayana. 1972. Indian Bee Journal, 34:53.

2. Grewal, G.S. and A.S.Sidhu. 1983. In: Second International Conference on Apiculture in Tropical Climates. I.A.R.I., New Delhi, pp. 537-544.

3. Johansen, C.A. 1977. Annual Review of Entomology. 22: 177-192.

4. Prasad, V.G. and A.Verghese. 1983. In: Second International Conference on Apiculture in Tropical Climates: IARI, New Delhi, pp. 590-594.

5. Rao, G.M. and Lazar, M. 1983. Ibid. pp. 580-589.

6. Sinha, S.N. and A.K.Chakrabarti, 1983. Ibid. pp. 649-655.

7. Verghese, A. and V.G.Prasad, 1983. Ibid. pp. 659-660.

8. Verghese, A. 1986. Indian Horticulture, 30(4):17-24.

ASSOCIATION OF *APIS CERANA* WITH *APIS FLOREA* AND *ERISTALINUS ARVORUM* DURING MANGO BLOOM

Abraham Verghese* and P.V.Veeraraju*
Department of Entomology,
University of Agricultural Sciences, Bangalore-560065
INDIA

The three important pollinators on mango (cv. Alphonso) during 1990 (January-February) were Apis cerana, A. florea and the syrphids Erastalinus arvorum. The last two were natural pollinators, while A. cerana was introduced; it foraged from a hive kept 50 meters away from the orchard. From a block of 280 10-year old trees, near Bangalore, 15 trees were randomly selected, and ten panicles at random from each tree were observed for 30 seconds per panicle. The presence and absence of these three pollinators were noted and subjected to 2 x 2 contingency table analysis, separately for A. cerana with A. florea and A. cerana with E. arvorum. The test for association between these were based on χ^2 (at 5% for one degree of freedom), as it makes the fewest assumption on the type of distribution. The results are presented in table 1.

Table 1. Association based on 2x2 contingency table**

Association	Components of 2x2 contingency table a	b	c	d	χ^2	Type of association	Remarks
A. cerana Vs. A. florea	20	27	66	37	0.61	ad < bc negative	Non significant
A. cerana Vs. E. arrorum	4	15	83	48	10.52	ad < bc negative	Significant

* * χ^2 at 5%, one degree of freedom = 3.84

* On study leave deputation from Indian Institute of Horticultural Research, Bangalore and Department of Agriculture, Karnataka, respectively.

2

The interpretation of the data is based on Southwood(1978). In both the above cases, the association was negative as ad < bc (See Southwood). But association of A. cerana with A. florea was non significant, implying the result was purely by chance, as A.florea population was too less. However, the association between A. cerana and the syrphid E. arvorum was significantly negative. Behavioural observation showed that A. cerana was a very active forager and while alighting on a panicle would displace the more sedentary E. arvorum. It is pertinent to mention here that the black strain of A. cerana was pollinating the mango panicles, though the yellow strain's hive was also kept near the orchard.

Reference

Southwood, T.R.E. 1978. Ecological Methodsd with Particular Reference to the Study of Insect Populations; ELBS Edition, Chapman and Hall, London.

11th International Congress—IUSSI 1990, India

PREY OF POLISTINE WASPS IN SOUTH INDIA

V.V.Belavadi and G.K.Veeresh
Department of Entomology,
University of Agricultural Sciences,
Bangalore-560 065, India

Polistine wasps invariably take insects as prey to feed their brood. In agro-ecosystems, pest species constitute a large portion of their prey. Several species of **Polistes** have been successfully used to reduce pest populations elsewhere (1,2,3 etc.). Here, we present data on the range of insect pest species preyed upon by three species of polistine wasps in south India.

METERIAL AND METHODS: Observations on the foraging behaviour of Polistes stigma and Ropalidia marginata were recorded for 132 hours during 1980-84 at Bangalore (13° 00'N, 75°32'E) and that of R. montana for 47 hours at Mudigere (13°7'N, 75°37'E). Foraging rates were computed for five nests of each species by recording the number of wasps returning with 'food loads' and the number leaving the nest per hour. All wasps on each nest (excepting R.montana) were marked to identify the nests from which they came to record foraging distances when encountered in the field. Prey species were identified as and when wasps were found handling them in the field.

RESULTS AND DISCUSSION: Foraging rates of the three species were high in afternoon hours. R. montana with its large colony size had the highest foraging rate (341.9), while that of P. stigma and R. marginata was around 18.0. Larve of Lepidoptera formed a major portion of the prey of the latter two species. They preyed heavily (54.18% and 57.14% respectively) on pests of chickpea, soybean and cabbage (Table 1), with mean foraging distances of 5.1 m (3 to 14 m; n=67) and 4.5. m (3 to 10 m; n=82) respectively. R. montana preyed extensively on the 4th and 5th instar nymphs of Heteropsylla cubana (Homoptera: Psyllidae), a serious pest on Leucaena leucocepahala at Mudigere. It foraged for great distances (200 m) compared to the other two species.

The above observations indicate that P. stigma and R. marginata can be utilised in biocontrol programmes, as they take lepidopterous larvae exclusively and can be attracted to nest on artificial nesting structures. R. montana, though capable of reducing H. cubana populations effectively, is difficult to manipulate as it constructs nests in inaccessable places.

Table 1. Prey of Polistinae

Wasp species	Prey	per cent	Host
P. stigma	Heliothis armigera	24.28	Chikpea
X = 69	catapsilia sp.	18.27	Cassia
	Spodoptera litura	12.86	Soybean
FR =18.0*	Crocidolomia sp.	10.00	Cabbage
	Plutella xylostella	7.14	Cabbage
	Unidentified	27.14	-
R. marginata	Heliothis armigera	19.64	Chikpea
X = 72	Catapsilia sp.	23.21	Cassia
	Spodoptera litura	12.50	Soyabean
FR = 17.2*	Crocidolomia sp.	8.93	Cabbage
	Plutella xylostella	5.36	Cabbge
	Unidentified	30.36	-
R. montana	H. cubana		leucaena
X = 4856			
FR = 341.9**			

X refers to mean colony size
* Number of wasps returning with 'food' /hr.
**Number of wasps returning /hr.

References

1. Gould, P. W. and R.L. Jeanne. 1984. Environmental Entomology, 13: 150-156.
2. Gillaspy, J.E. 1979. Southwestern Entomologist, 4:334-350.
3. Nakasuji, F., H.Yamanaka and K.Kiritani. 1976. Kontyu, 44:205-213.

EFFECT OF TERMITE FUNGAL COMBS ON THE GROWTH AND YIELD OF FINGER MILLET (*ELEUSINE CORACANA* (L.) GAERTN.)

D.K. Siddegowda and D. Rajagopal
Department of Entomology
University of Agricultural Sciences
GKVK, Bangalore - 560 065, INDIA

The termite species of Macrotermitinae are known to cultivate fungus combs by depositing their excreta in the nests. The fungal combs are rich in cellulose, lignin, organic carbon, mineral matter and nitrogen apart from major role in plant nutrition, however the information on the utilization of these fungal combs in plant growth is lacking. Hence a field study was conducted to know the effect of termite fungal combs on the growth and yield of finger millet in comparision with recommended does of fertilizers (NPK) (40:20:20/0.4ha) and organic manure (3000 Kg/0.4ha).

An experiment was laid out in randomised block design with six treatments replicated four times imposed in optimum microplots under field conditions at Main Research Station, University of Agricultural Sciences, Hebbal Campus, Bangalore. The fungal combs of Odontotermes obesus (Rambur) were collected and used as alone and in combination with other commonly recommended manures and fertilizers to the finger millet crop. The treatments were a) fungal combs (FC) alone 4000 Kg/0.4ha. b) Farm yard manure (FYM) alone 3000 Kg/0.4ha. c) Recommended NPK (40-20-20/0.4ha) d) 1000 Kg FC+3000 Kg FYM + NPK (30-15-15/0.4ha) e) 2000 Kg FC+2000 Kg FYM + NPK (20-10-10/0.4ha) f) 3000 Kg FC + 1000 Kg FYM+NPK (10-5-5/0.4ha). The method of application was followed as per the package of practices.

The results of this experiment revealed that the incorporation of fungal comb as nutrients had a favourable influence on the growth and yield of fingermillet. Both plant height and number of tillers differed significantly due to application of fungal comb at 30 days after transplanting (DAT) and 90 DAT. The plant height increased from 29.98 to 31.83 cm in case of fungal comb treated plots compared to 24.83 cm (3000 Kg FYM) and 29.5 cm (40:20:20 NPK). Similarly number of tillers also increased in 3000 Kg FYM (6.69) treated at 30 DAT. However at 60 DAT, plant height and number of tillers/hill did not differ significantly between the treatments. Plant height increased from 84.24 to 92.92 cm and number of tillers from 9.07 to 10.99 in case of fungal comb treated plots at 90 DAT.

Per cent productive tillers per hill and ear head weight did not differ between treatments. More number of fingers per ear head (6.77) were recorded in fungus comb treated plots as against 5.75 and 6.40 in 3000 Kg FYM and 40:20:20 NPK respectively.

The grain yield did not differ between treatments indicating same effect of fungal comb (20.16 Q/0.4 ha) with that of FYM (17.56 Q/0.4 ha) and NPK (19.85 Q/0.4 ha). However, higher straw yield was observed in 3000 Kg FYM (19.0 Q/0.4ha) treated plots. The maximum test weight was observed in plots treated with 4000 Kg fungal comb (2.99g) justifying with higher grain yield (20.16 Q/0.4ha) compared to other treatments.

Increased yield and plant growth in fungal comb treated plots may be attributed to the changes in fertility status of soil by adding the fungal comb, as it contains fecal matter of the termites. However the organic manure mineralise nutrients and release them slowly and helps in increased straw yield in finger millet (4). The other factors were fungal comb contains D-Mannitol which helps in maintaining soil temperature (1). In addition to this high amount of major and minor nutrients such as N, Ca, Mg, Cu, Zn, Fe and Mn and high amount of organic carbon (3) and (5) might have influenced the plant growth and higher yield.

Obviously the growth of crops on termite mounds or in areas where mounds have been levelled were showed better growth of fodder sorghum on the mound soil of Odontotermes wallonensis (Wasm.) (2). However, the present investigation has indicated the better growth of finger millet when fungal combs are incorporated alone or incombination with other manures and fertilizers.

1. Abo-Khatwa, N., 1977. Pontificlae Academic Scientarum Scripta Varia. 41(3): 447-479.
2. Rajagopal, D., 1983. Tropical Pest Management, 29 :194-195.
3. Rajgopal, D., 1987. J. Soil. Biol. Ecol. 7(1) : 36-47.
4. Rayachoudhari, S.P., 1969. Indian Fmg. 18 (1): 18-19.
5. Roharmann, G.F., 1978. Pedobiologia, 18 : 89-98.

Symposium 20

EVOLUTION OF SOCIAL BEHAVIOUR IN CARPENTER BEES

Organizer : **HAYO H.W. VELTHUIS, The Netherlands**

SELECTIVE PRESSURES INVOLVED IN THE FACULTATIVE SOCIAL BEHAVIOUR OF *XYLOCOPA PUBESCENS*

K. Hogendoorn.
Department of Comparative Fysiology, University of Utrecht,
P.O.Box 80.086, 3508 TB Utrecht, The Netherlands.

Carpenter bees of the genus Xylocopa are considered to be the intermediates between the highly social Apidae and the solitary Anthophoridae, not only in the taxonomical sense but also with respect to the level of sociality attained. Therefore, they form an interesting group for the study of the evolution of sociality.

All females are long lived, and, as a rule, overlap of generations is found. A general scenario for the evolution of generation overlap in carpenter bees is proposed, in which competition for pollen and for nesting material plays an important role. Once the overlap of generations is complete, a reproductive division of labour can ensue in multivoltine species, resulting in a facultative form of sociality.

The ontogeny of social nests is illustrated using *Xylocopa pubescens* as an example. This species has been studied extensively in the Negev desert in Israël [1, 2, 3]. In the social nests two distinct tasks are to be seen: fouraging and guarding. The fourager is reproductively active.

Two types of guard bees are found: the teneral young, which need pollen for their further development [4], and the old females, which have lost the reproductive dominance in the nest. The teneral bees can usually be found in the nest entrance when the fourager is away from the nest [5]. The old bees, however, only leave the nest entrance in order to collect nectar for their own use; otherwise they will always guard the nest.

Young teneral bees do not seem to guard the nest actively, and thus their guarding seems to be the result of competition for pollen between the young, as was suggested by Velthuis [6].

Old guard bees do protect the nest against possible usurpers as well as against bees attempting to rob pollen. One of the most important reasons for their guarding behaviour seems to be the protection of their own brood and young, which are still present in the nest at the time they lose the reproductive dominance.

REFERENCES

1. Ben Mordechai, Y., R. Cohen, D. Gerling and E. Moscovitz. 1978. Israel Journal of Entomology, 12: 107-121.
2. Gerling, D., P.D. Hurd and A. Hefetz. 1983. Smithsonian contributions to zoology, 369.
3. Velthuis, H.H.W. and D. Gerling. 1983. Behav. Ecol. Sociobiol.12: 209-214.
4. Blom, J. van der and H.H.W. Velthuis, 1988. Ethology, 79: 281-294.
5. Gerling, D., P.D. Hurd and A. Hefetz. 1981. Journal of the Kansas Entomol Soc., 54 (2): 209-218.
6. Velthuis, H.H.W. 1987. In: From Individual to Collective Behaviour in Social Insects, J.M. Pasteels and J.L. Deneubourg, Eds., Brinkhäuser Verlag, Basel, pp. 405-434.

ECOLOGY AND BEHAVIOUR OF A CARPENTER BEE *XYLOCOPA VALGA* GERSTACKER POLLINATING FRUIT BLOSSOMS

D.P. ABROL
Division of Entomology, Sher-e-Kashmir University of Agricultural Sciences and Technology, Shalimar Campus, Srinagar-191 121, Jammu and Kashmir, India.

Xylocopa species are acknowledged pollinators of several field and fruit crops. They nest in thatched houses and hollows of bamboo stems.

Xylocopa valga, recently recorded from the Indian sub-continent, is an important pollinator of several field and fruit crops. The emergence of X. valga coincides with the blooming of almond, apple and cherry flowers during early March when the inclement weather limits the field activities of other flower visiting insects. It works at a critical air temperature of 6-7 °C and on cloudy overcast days while Apis cerana indica F, and Lasioglossum sp. need 8-10°C and 13-14°C respectively for the initiation of the foraging activities. The latter species of bees are most active on clear sunny days. Lassioglossum sp. remains absent on cloudy/overcast days.

X. valga continues its field activities throughout the day with maximum population between 9 and 11 hours. X. valga on an average visited more almond (22.40), apple (21.84) and cherry (23.20) flowers per minute than A.c. indica 7.5, 8.2 and 7.6 flowers respectively. Evidently, X. valga is an important pollinator of fruit crops as it pollinates more flowers in a unit time and works in inclement weather conditions. Foraging population of X. valga visiting almond flowers were recorded for 10 different days, on alternate days, during March-April, 1989. For this purpose, five branches one from each tree were marked and the number of bees visiting each branch counted from 6 to 17 hours. The mean of these five observations constituted the reading for each hour. The environmental factors, viz., air temperature (T), relative humidity (RH), light intensity (LI), solar radiation (SR), soil temperature (ST), and nectar-sugar concentration (NSC) were also recorded simultaneously. The data were analysed for simple correlations and path analysis. Analysis of the data reveal that foraging populations of X. valga correlated significantly and positively with T LI, SR, NSC (r=0.522, 0.490, 0.510, 0.480 respectively) and negatively with RH(-0.421), where as the association with ST was non-significant (0.209). Since all these factors are linearly related, the results were further analysed by path coefficient technique. The estimate of direct and indirect effect path coefficients revealed that SR directly influenced the foraging populations of X. valga. Direct effect of SR was pronounced and positive while the effects of other factors were negative or low in magnitude. The direct effect of SR may be related to the black colour of the body of the insect which absorbs more heat. Abrol (1987) found that black or dull insects correlated negatively with temperature parameters, when T exceeded 35°C and SR 38 mW/cm^2. In the present study I believe that under temperate conditions where very low temperatures (below 22°C) are anticipated during this part of the year, black bodied insects may have a physiological advantage to raise their body temperature and minimize energy expenditure for thermoregulation. This may be the reason that X. valga can work under inclement weather conditions.

1. Abrol, D.P. 1987 Environment and Ecology 5 : 90-93.

SOCIAL BEHAVIOUR IN *EXONEURA*

Michael P. Schwarz
Department of Zoology,
La Trobe University,
Bundoora, Vic. 3083 AUSTRALIA

Recent work on an Australian allodapine bee, Exoneura bicolor, has revealed a more complex level of sociality than had been previously recorded in the Xylocopinae. E. bicolor is a univoltine bee in south-eastern Australia. In montane forests, the majority of both newly-founded and re-used nests are occupied by two or more females during brood-rearing[1,2]. Whilst all cofoundresses appear to be reproductive, some of the females that remain in re-used nests do not become inseminated or lay eggs. Relatedness between adult nestmates is high in both colony types[2] (r = ca. 0.5), indicating that cofounding involves effective kin recognition. However, because adults in brood-rearing nests are from the same generation, rather than matrifilial assemblages, relatedness between altruistic females and the brood they help rear is about half this value[3]. Analyses of genetic population structure indicate that this low value is not enhanced by subdivision or population structuring[3]. Hence, sociality in this species occurs within the context of low altruist-beneficiary relatedness.

Although all cofoundresses lay eggs, reproductive differentiation is well developed among females re-using their natal nest[1]. This differentiation is most apparent during winter, when each colony contains one or two inseminated females with large ovaries and a variable number of uninseminated females with small ovaries. These inseminated and ovarially active females were usually the first pupae to reach adult eclosion in their colony, and this age dependency is evident even in nests where the time separating first and second eclosion events may be a few days or less[4]. This reproductive differentiation occurs even when access to males is precluded, indicating that insemination per se is not responsible for the formation of ovarian hierarchies.

Removal of inseminated reproductives in overwintering colonies leads to ovarian enlargement of uninseminated nestmates within 2 weeks[5]. This appears to be mediated by an ovarian inhibitory pheromone[6]. Agonistic interactions between female nestmates has not been observed during periods of reproductive differentiation. Furthermore, pre-reproductive assemblages display a high degree of task specialization and trophallaxis is frequent and involves all colony members, including males[7].

Hence, it appears that in _E. bicolor_, sophisticated social behaviour occurs without high relatedness between altruists and beneficiaries, without an overlap of generations during brood rearing, and without large colony sizes. Lack of population genetic structure in _E. bicolor_ makes it unlikely that structured-deme and group selection models will explain sociality in this species. On the other hand, because cofounding effectively involves kin association it seems likely that kin selection may play an important role in the maintenance of social behaviour. Further work is required to test whether benefit/cost ratios for altruism in _Exoneura_ are indeed sufficiently large to explain sociality in this species in terms of a kin selection model.

REFERENCES

(1) Schwarz, M.P. 1986. Insectes Sociaux, 33:258-277

(2) Schwarz, M.P. 1987. Behav. Ecol. Sociobiol, 21:387-392

(3) Blows, M.W. and M.P. Schwarz. in review

(4) Schwarz, M.P. and K.J. O'Keefe. in review

(5) Schwarz, M.P., O. Scholz and G. Jensen. 1987. J. Aust. Entom. Soc., 26:355-359

(6) O'Keefe, K.J. and M.P. Schwarz (1989) Naturwiss. in press

(7) Melna, P.A. and M.P. Schwarz in review

VARIATION IN SOCIAL ORGANISATION WITH NESTING SUBSTRATE IN AN *EXONEURA* SPECIES

Letitia Silberbauer

Department of Zoology,
La Trobe University
Bundoora 3083
Victoria, Australia

The life-cycle and social organization of a twig nesting Exoneura species was examined in two nesting substrates occuring in a microgeographic region. The population is primarily univoltine, with brood emerging in late summer/ early autumn. Nest-founding is continual throughout the year, although there are two surges: after brood emergence and again in early spring.

One substrate, the tussock-forming shrub, Xanthorrhoea minor, tends to grow in treeless meadows and as a common, but patchy constituent of the forest understory. The other substrate, the woody shrub Melaleuca sp., grows within and around the Xanthorrhoea meadows scattered throughout the study area.

Although the life-cycle of the species was synchronous in both substrates, the level of social organization in each was different. Approximately 90% of colonies in X. minor flower stalks were solitary, while the majority of nests in Melaleuca branches contained between two and twenty females. Multifemale colonies in both substrates were semisocial. Two hypotheses are proposed to explain this difference. Firstly, Melaleuca wood is more resistant to decay than Xanthorrhoea stalks and provides the opportunity for longer utilization of nest-sites, allowing an annual increase in the number of occupants of the nest as mature brood remain in the natal nest. Secondly, Melaleuca bushes are more protected from disturbance by fire, wind or grazing vertebrates than Xanthorrhoea, and therefore last longer, providing the opportunity for colony growth.

Symposium 21

BEHAVIOURAL GENETICS OF EUSOCIAL HYMENOPTERA

Organizer : **MICHAEL BREED, U.S.A.**

THE BEHAVIORAL GENETICS OF SOCIAL INSECTS : AN OVERVIEW

Michael D. Breed

University of Colorado, Boulder, CO, 80309-0334, USA

In the social insects studies of genetics and behavior are a rich but relatively untapped area of research. Correlations between the expression of behavior and the genetic composition of individuals are usually determined by regressions between parent and offspring behaviors or by comparing within- and between-family variation in behavior (1). In special cases within the social insects it has been possible to study the Mendelian segregation of specific behavioral traits, such as hygienic behavior (2) or defensive behavior (3) in honey bees. Unique circumstances in the social insects create non-random genetic structuring with colony populations; this structuring has interesting implications for creating within-colony behavioral specialization that is influenced by genetic differentiation.

The study of behavioral genetics in social insects presents several interesting and unique problems; in this short paper I focus on the issues of polyandry (multiple mating by the queen) and polygyny (multiple queens in colonies) as a key issue in hymenopteran behavioral genetics.

PREDICTIONS OF EVOLUTIONARY THEORY

In some species the queen mates more than once and stores the sperm for usage through her life. This is most apparent in the honey bee, in which polyandrous mating is the rule. In other species there are multiple queens laying eggs simultaneously in the colony. In these cases of multiple parentage of workers evolutionary theory predicts both cooperation and competition among the workers. In the example of the honey bee, cooperation is expected because all workers share the mother, and hence have a relatively high level of relatedness. On the other hand, relatedness is even higher among individuals who also share a father; consequently we might expect competition among patrilines and cooperation within patrilines.

Similar logic can be applied to cooperation and competition in conditions with multiple queens. An intriguing twist with multiple queens is that assemblages of unrelated queens may cooperate, in anticipation of reduction of queen number (and elimination of competing queens). In this case the worker offspring of the different queens would be no more related than a randomly chosen population.

IDENTIFICATION OF SIB-GROUPS WITHIN COLONIES

Two problems face workers attempting to test these evolutionary hypotheses. First, genetic markers used to differentiate maternal or paternal lines in colonies may have pleiotropic effects on worker behavior (Seeley and Visscher, personal communication). These effects could confound measurement of between line behavioral differentiation or within line preferential behavior. Because of the complexity of social insect activity, and particularly the impact of temporal polyethism, designing appropriate controls for marker effects is very difficult. This is a particular problem in the honey bee, in which markers are artificially imposed by the investigator, rather than being the result of open matings.

Second, the problem of designing control groups extends beyond the issues generated by using markers. For example, five to ten day old honey bees feed larvae, but they also engage in other activities within the hive. If we hypothesize that the larvae-feeders are a non-random genetic sample of the colony, what is the appropriate control group? Is it the entire colony population, the population of bees currently in that age group, the bees that emerged five to ten days previously, or a group defined in some other way? Does matrilineal or patrilineal structure affect the spatial arrangement of individuals in the colony? If so, how should this be taken into account? How the control group is defined could clearly have an effect on the result of statistical analyses and consequently on interpretation of the data. Nevertheless a number of studies using morphological and allozyme markers in honey bees have revealed considerable information concerning honey bee behavioral genetics (4, 5) and in ants protein allozymes provide an effective technique for identifying patrilines and matrilines (6).

In honey bees it is the lack of allozymic variability makes use of allozymic markers problematical; DNA fingerprinting techniques are the best currently available solution to some of these problems for honey bees. These have the advantages of providing useful data in the absence of artificial insemination or other manipulation of markers, of providing information that could be used to assess relatednesses within patrilines or matrilines, and of providing data that would be amenable to analysis for the effects of individual markers (in this case chromosome segments). Even using DNA markers, however, some of the more difficult problems of design of appropriate controls still remain.

PROSPECTS AND ASSESSMENT

Because of the difficulties in determination of maternity and paternity the predictions of cooperation and competition have not yet been tested in a broad array of social insects. The application of new biochemical techniques, particularly DNA analyses, should allow complete testing of the evolutionary hypotheses suggested by colony genetic structure. The papers in this symposium address the issues of testing behavioral genetic hypotheses on social insects.

Literature Cited

1. Falconer, D. S. 1981. Introduction to quantitative genetics. 2nd ed. Longman, New York.
2. Rothenbuhler, W. C. 1964. Anim. Behav. 12:578-583.
3. Stort, A. C. 1975. J. Kansas Entomol. Soc. 48:381-387.
4. Frumhoff, P. C. and S. Schneider. 1987. Anim. Behav. 35:255-262.
5. Robinson, G. E. and R. E. Page. 1988. Nature 333:356-358.
6. Ross, K. G. 1989. In: *The Genetics of Social Evolution.* M. D. Breed and R. E. Page, Jr., eds. Westview Press, Boulder. pp. 149-162.

GENOTYPE INTERACTIONS IN QUANTITATIVE GENETICS OF SOCIAL BEHAVIOR IN HONEYBEES

Robin F.A. Moritz

Bayerische Landesanstalt fur Bienenzucht
Burgbergstr. 70, 8250 Erlangen, FRG

The power of the theories of quantitative genetics, originally developed for plant and animal breeders, has recently also been acknowledged by ecological and evolutionary geneticists. The techniques to analyze artificial selection by man, can also be applied to understand natural selection in the wild. In fact, the estimation of genetic variance components of characters related to fitness reveal many facets of the mode of natural selection that are difficult to analyze by other methodological approaches.

In populations of social insects, selection often operates not only on the individual but also on the colony level [1,3]. The expression of colony or group phenotypes complicates the theoretical framework of quantitative genetics substantially. In the classical individual approach, genetic variance components are estimated from similarities between related individuals. Similarities in groups can be affected by additive and dominance effects but also by nonlinear genotype interactions that are not selectable. In quantitative models we split the variance of group phenotypes, V_P, into additive (V_A), dominance (V_D), and genotype interaction (V_I) components as follows

(1) $V_P = V_A + V_D + V_I$

Selection is impossible if V_I is large, and a phenotype of a group is determined by its genotypic composition rather then its additive genetic value.

We studied genotype interactions on hoarding behavior of honeybees (*Apis mellifera*) in genetically mixed groups [2]. A queen was instrumentally inseminated with four drones each carrying a specific combination of three marker genes (ivory, cordovan, diminutive). The queen was homozygous for the markers which allowed for the discrimination among the four subfamilies in her offspring. Groups (60 workers) consisting of one, two, three or four patrilines were tested for their ability to hoard sugar syrup. With increasing genetic intragroup variance, hoarding behavior was reduced (Fig.1).

The mixed group behaviour did not correspond to the expected mean of the nonmixed groups, and up to 84.8% of the total phenotype expression was due to nonselectable genotype interactions.

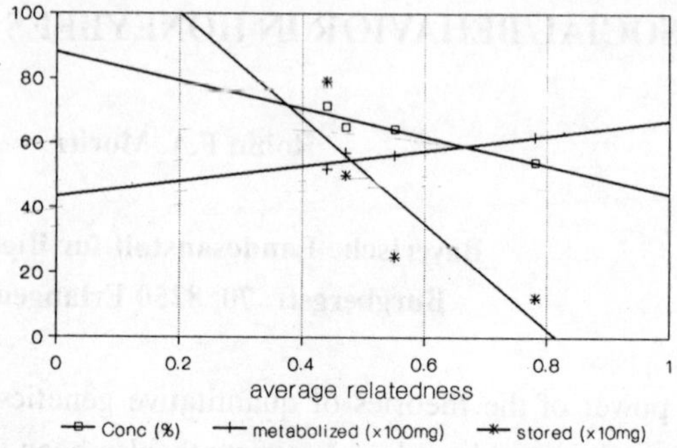

Figure 1. Regression of group traits (Conc. = concentration of hoarded syrup; metabolized sugar and sugar stored in cells) on the average intragroup relatedness.

In a further experiment on hygienic behavior, similar phenomena were found using eight different subfamilies of a colony. Groups of workers were given freeze killed brood for uncapping. In contrast to the hoarding experiment, the genotypically heterogeneous groups proved to be more efficient. There was a positive regression of the number of patrilines per group and the amount of uncapped cells. Both examples show that nonlinear genotype interactions can strongly affect social behaviour of honeybees and interfere with selection of these traits. Even if genotypical variability has a favorable effect on social behaviour closely related to fitness, the trait will not be directly selectable. However, it may indirectly selected via the mating behavior of queens and drones. Polyandry increases genotypic variance in the colony and if this increases colony fitness, it would provide another explanation for the evolution of polyandry in honey bees. But just because of this model, we may not derive that all genotype interactions have to be beneficial. Negative as well as positive effects, depending on the behavioral trait as shown above, seem to be possible.

Acknowledgment: I am grateful to the Deutsche Forschungsgemeinschaft for financial support.

References:

1. Moritz, R.F.A. 1989. Behav. Ecol. Sociobiol. 25:437-444
2. Moritz, R.F.A. and E. Hillesheim 1989. Apidologie, 20:383-390
3. Owen, R.E. 1986. Theor. Popul. Biol. 29:198-234

THE GENETICS OF KIN RECOGNITION IN ANTS

Michael W.J. Crosland

Department of Entomology, University of California, Davis, CA 95616, U.S.A.

In the expanding field of ant kin recognition recent advances have included: the demonstration of larval learning of recognition cues; highlighting the importance of the young callows as a crucial stage for colony integration; and increased research into the importance of cuticular lipids and the queen as sources of nestmate recognition cues.

Genetically-determined odor cues and their relative importance to environmental odors remains an important question. Maintaining the ant Rhytidoponera confusa on markedly different diets and substrates resulted in significantly more aggression against these colony-subsamples by nestmates. However, the increased aggression was slight compared with that towards non-nestmates.

Mixing of genetic odors occurs between nestmates in R. confusa. However, aggression tests indicate that each ant mostly maintains its own individually distinct odor. Research with other ant species (and other social insects) also support the view that, though some odor mixing can occur, genetic cues of nestmates are individually distinct, at least to some extent.

Genetically-related nestmates bear similar genetic cues. Therefore unfamiliar relatives can be recognized (Table 1). Without this mechanism, learning the individually distinct genetic cues of every nestmate individually would clearly present a problem in ant colonies containing hundreds of thousands of workers.

Recent controversy about how polymorphic alleles for kin recognition cues can be maintained against selection might be answered by the suggestion that some individuals within each group or species might show no kin discrimination [1]. A parallel situation may occur

Table 1. Mechanism by which unfamiliar relatives can be discriminated.

Familiar Relatives TOLERATED	Unfamiliar Relatives TOLERATED	Unfamiliar Non-relatives NOT TOLERATED
a	b	c
ABD	ABD	ABZ
ACD	ACD	XYZ
ACF	ABC	
BCE	AEF	
BDF		
BDE		
CEF		
CDE		
DEF		

[a] Nine individuals familiar with each other. Each letter represents an odor produced by a different pair of alleles. (Environmental cues can easily be added to this model by addition of extra letters.)

[b] Four examples of unfamiliar relatives tolerated. (A non-self phenotype matching mechanism.) Individual ABD is tolerated because one of the familiar relatives is also ABD. AEF is tolerated because all the component letters (A, E & F) are present among familiar relatives.

[c] ABZ and XYZ are not tolerated due to unfamiliar odors X, Y & Z. XYZ would be least tolerated because it has no familiar odors.

in R. confusa [2] where preliminary evidence indicates that some colonies consistently show much lower aggressiveness and kin discrimination than others. Furthermore, most conspecific nestmate discrimination is carried out by only a small proportion of the worker force. Many workers in each R. confusa colony show poor or no nestmate discrimination.

References

1. Elgar, M.A. and R.H. Crozier. 1989. Trends Ecol. Evol., 4: 288-289.
2. Crosland, M.W.J. 1990. J. Insect Behavior (in press).

THE GENETICS OF DIVISION OF LABOR AND COLONY PLASTICITY IN *APIS MELLIFERA*

Gene E. Robinson* and Robert E. Page, Jr.+

*Department of Entomology, University of Illinois, Urbana, IL 61801;
+Department of Entomology, University of California, Davis, CA 95616

INTRODUCTION

The processes that integrate worker behavior into coordinated colony patterns are poorly understood. It is possible to gain insight into the regulation of colony behavior by identifying the factors that determine the activities of individual workers. Here we summarize results demonstrating that genetic differences between related colony members affect the division of labor within honey bee colonies (1, 2). We also describe one way in which genetic regulation of individual task performance may influence colonial patterns of behavior.

GENETIC DETERMINATION OF TASK PERFORMANCE

Insect colony members may differ from one another genetically due to recombination, polyandry, and polygyny. Because of polyandry and sperm mixing, honey bee colonies are composed of numerous subfamilies of workers, each subfamily descended from the colony's queen and one of her mates. Using colonies composed of electrophoretically distinguishable subfamilies (derived from instrumentally inseminated queens), we tested the hypothesis that members of some subfamilies are more likely to be sampled at certain tasks than are members of other subfamilies.

Significant genotypic differences were detected for guarding in 9 out of 10 cases (5 out of 5 colonies, two trials each); 4 out of 10 cases for corpse-removal (3 out of 5 colonies); and 3 out of 6 cases for scouting for new nest sites (3 out of 5 colonies). In addition, there were significant differences in subfamily frequencies for samples of guards and undertakers in 5 out of 10 cases (5 out of 5 colonies) and in 12 out of 20 cases for pollen foragers and nectar foragers (9 out of 10 colonies). Additional analyses indicate that these differences occurred too frequently to be explained by sampling error alone; they also cannot be explained on the basis of age polyethism and non-random use of sperm by the queen.

GENETIC CONSTRAINTS ON COLONY PLASTICITY

To shift the focus from the individual worker to the whole colony, consider the colony as a collection of groups of individuals (subfamilies), each group with a different distribution of genetically determined response thresholds for the stimuli that elicit task performance. We hypothesized that intracolonial genetic variation in worker behavior may contribute to a colony's ability to respond to short-term changes in environmental and social conditions. According to this hypothesis, under 'normal' conditions a task is performed by workers with the lowest response thresholds, i.e., those workers most sensitive to the task-associated stimuli. If the need for this task increases due to changes in colony and/or environmental conditions, and there is a concomitant rise in the levels of associated stimuli, then workers with relatively higher response thresholds also perform the task. Through this mechanism, transient increases in the need for a particular task would cause differential recruitment among subfamilies, as more individuals shift from one job to another within the behavioral repertoire of a temporal caste.

The hypothesis was tested by determining whether the genotype frequencies of groups of undertaker bees varied as the need for corpse-removal behavior increased. 'Colony need' was manipulated by adding either 15 corpses every 15 minutes ('low stimulus') or 1000 corpses at one time ('high stimulus'). Using allozymes as paternity markers, workers from some subfamilies were more likely to be sampled as undertakers than workers from other subfamilies in 7 out of 12 cases (6 colonies, 2 trials each) as in the previous study. However, contrary to the predictions of the hypothesis, there were no significant differences in the genotypic composition of corpse-removers responding to the low vs. high stimulus in any colony. Moreover, there was a 30-50% decrease in the rate of corpse removal in each colony during the second trial, which was conducted 7 to 10 days after the first trial. Results of a second experiment support the notion that the observed decrease in colony efficiency in trial 2 occurred because the undertaker population was depleted during trial 1 and there was no measureable, compensatory shift to undertaker activity by other colony members with different genotypes.

DISCUSSION

The discovery that genetic differences between workers influence the division of labor challenges the prevailing model of advanced insect societies in which individual members are equally likely to perform all necessary jobs. The results of the second study suggest that under some conditions, strong genetic influences on the tendency for individual workers to perform a particular task may constrain a colony's response to changing conditions. However, genotypic differences in worker behavior do contribute to colony responses that involve age polyethism plasticity (3). The role of genotype in coordinating a colony's response to changing conditions may vary with task. Individual workers may show relatively less plasticity for behaviors for which there is typically less variance in colony need, such as corpse removal.

REFERENCES

1. Robinson, G.E. and R.E. Page. 1988. Nature 338: 576-578.
2. Robinson, G.E. and R.E. Page. 1989. Behav. Ecol. Sociobiol. 24: 317-323.
3. Robinson, G.E., Page, R.E., Strambi, C., and A. Strambi. 1989. Science 246: 109-112.

We thank M.K. Frondrk for expert technical assistance. Supported by an Ohio State University Postdoctoral Fellowship (G.E.R.), and NSF grants BSR-8800227 (G.E.R.) and BNS-8719283 (R.E.P.).

DIVISION OF LABOR IN A POLYGYNOUS ANT

L.E. Snyder
University of Colorado
Environmental, Population and Organismic Biology
Boulder, Colorado 80309-0334

Division of labor is thought to be responsible for the ecological success of social insects. In the ants, it is well known that a worker's age or size is a determining factor in the particular task she will perform. Recently, the way in which workers divide labor has come under increasing scrutiny because of a great deal of behavioral variability found among workers of the same age or size class. I explored the role of worker size variation in a monomorphic ant, Formica argentea. At the beginning of the active season, over-wintered ants perform all colony tasks. Later in the season pupae begin to eclose into adults. These young workers primarily perform brood related tasks, freeing older workers from this job. I predicted that the importance of worker size to division of labor would shift from one part of the season to the next because of the changes in age structure within the colony.

The behaviors of 300 individually marked workers in 3 F. argentea were observed. This work was done in the laboratory so that the number and kinds of acts amongst individuals of different morphologies could be obtained. The behavioral repertoires for over-wintered adults were compared for two different time periods: before eclosion of adults and after newly-eclosed workers began to tend brood.

A weak, morphological bias in task behavior was apparent for older, overwintered ants following eclosion of young workers. No bias was found among the same over-wintered ants before worker eclosion. A large amount of flexibility in individual behavior probably overcomes a division of labor based on size in order to fulfill certain colony needs such as the rearing of brood. The eclosion of young adults, whose focal activity is brood care, may act as a releasing mechanism, allowing older workers to divide labor by size to some extent.

F. argentea is regularly found with two queens, each contributing her own genetically distinct offspring to a nest. I am currently comparing behaviors between matrilineal lines to determine whether polygyny contributes a genetic component to division of labor.

CHARACTERIZATION OF BIDIERECTIONALLY SELECTED LINES IN HONEYBEES SHOWS THAT INHERITANCE OF LEARNING IS MAINLY DUE TO ADDITIVE GENETIC FACTORS

Ch. Brandes *

Institut fuer Neurobiologie der FU Berlin, Koenigin-Luise Str. 28-30, Berlin and Institut fuer Bienenkunde, Karl-von-Frisch-Weg 2, Oberursel, Fed. Rep. Germany

In queenless colonies of Apis mellifera capensis worker bees produce unfertilized, diploid eggs which develop into worker bees or queens. The diploidism occurs via postmeiotic fusion of pronucleus and polar body resulting in worker bees that are genetically identical to each other and to their mother. Such a genetic system offers many advantages to behavior-genetic studies. A genetic analysis of learning was begun first by selecting for high and low learning and then by comparing estimates of genetic variance among unselected and selected populations.

Learning was characterized by classical conditioning of proboscis extension reflex (PER). In this paradigm bees learn quickly to respond to an odor cue with PER after pairing of odor and sugar water.

A two-step procedure was used for selection of every generation. Worker bees with high or low learning scores were selected and then used to rear queens. Some queens were used to produce drones, which then were used to inseminate queens to produce the next generation of worker bees. Inbreeding during selection was minimized (F<0.3) by crossbreeding several high or low learning lines. Selection was done for 5 generations in the high learning lines and for 4 generations in the low learning lines.

Bidirectional response to selection was asymptotic after 2 generations which suggests only a few (major) genes underlying behavioral variation of learning scores.

Genetic variance in the first selected generation was 4 fold higher than in the source population. In contrast genetic variance in a selected population of parthenogenetic sisters that was produced exclusively with worker bees did not increase. This observation also supports the idea that parthenogenetic populations are genetically homogeneous.

Partitioning of additive and non additive genetic factors was accomplished in the source population using a one way variance analysis and in selected generations using a two way analysis (hierarchical classification). Variation in learning scores was found to be influenced only by additive genetic factors (assuming no epistatic effects or genetic x environmental interactions). This result was confirmed by crossbreeding high and low learning lines from the fourth selected generation. Hybrids showed intermediate scores.

Both genetic observations - a fast response to selection and high additive genetic effects - stand in contrast to genetic characterizationof learning in other insects. The social system in honeybees may contribute to this genetic difference.

* Present address: Brandeis University, Department of Biology, Bassine 235, Waltham, MA 02254-9110, USA

Symposium 22

PHEROMONAL ASPECTS OF NEST AND NESTMATE RECOGNITION IN PRIMITIVELY SOCIAL BEES

Organizers : **JAN TENGO, Sweden**
ABRAHAM HEFETZ, Israel

NESTMATE RECOGNITION IN A GROUP-LIVING NOMIINE BEE (HALICTIDAE, NOMIINAE)

Kukuk, P. F.

Department of Biological Science, University of Montana, Missoula, MT 59812, USA.

INTRODUCTION

The microgeographic structure of populations can be explored using behavioral, genetic and biochemical techniques. At present, no such information is available for the Nomiinae although this halictid subfamily contains many social species (1). Here we report the results of behavioral experiments examining nestmate recognition and population structure for a previously unstudied species of *Nomia (Austronomia).*

MATERIALS AND METHODS

Females were collected by excavating nests from an aggregation in the Cabbboboonee Forest in south-western Victoria, Australia, January I990. Pairs of females were placed in 5 cm clear plastic tubes, plugged at both ends, and were thereby forced to interact. Each pair was scan-sampled (2) at 5 minute intervals (three times) and the following behaviors were recorded for each female. Fighting*, C-posture* (3), Block* (3), Walk, Groom, Sit, Pass (3) and Escape Attempt. If any agonistic behavior (marked with an *) was observed in any of the three scan samples, a pair was scored as positive for agonism, a similar criterion was used to score activity.

RESULTS

There was significantly more agonistic behavior between pairs of females from different nests than between pairs of females from the same nest (Chi^2 = 13.58, df = 1, p = 0.002), but no significant differences were found in agonistic behavior based on the distance between the nest of origin for non-nestmate pairs (see Table 1). Similarly, there was significantly more activity occurring if pairs of females were from different nests as compared to pairs of females from the same nest (Chi^2 = 13.15, df = 1, p = 0.004). Nestmate pairs were less active in several behavioral categories.

Table 1: Presence (+) or absence (-) of agonistic behavior and activity in pairs of *Nomia (Austronomia), sp.* from the same field nest or from different field nests located at three different distances apart in the same nest aggregation.

	Same	Distance Between Nests 50 cm	100 m	200 m
Agonism				
+	2	16	12	17
-	24	19	15	10
Activity				
+	11	25	21	23
–	15	10	6	4

CONCLUSIONS

Individuals of *Nomia (Austronomia), sp.* recognize familiar conspecifics (nestmates) and behave less aggressively toward them than toward non-nestmates. The relationship between these findings and those based on the chemistry of the Dufour's gland secretions along with electrophoretic analysis of allozymes will be discussed at the conference.

REFERENCES

1) Batra, S. W. T. I966. Insectes Sociaux 3:145-154.
2) Altman, J. I974. Behaviour 49:227-217.
3) Smith, B. H. I987. Anim. Behav. 35:211-217.

ODOR BASED INTERINDIVIDUAL AND NEST RECOGNITION IN THE SWEAT BEE *LASIOGLOSSUM MALACHURUM* (HYMENOPTERA: HALICTIDAE)

Manfred Ayasse
Department of Developmental Biology
University of Tübingen, Auf der Morgenstelle 28, D-7400 Tübingen, FRG

In the primitively eusocial bee, *Lasioglossum malachurum* (Hymenoptera: Halictidae), patterns of volatile compounds play an important role in mating biology [1,2] and colonial life [3,4,5].

As part of an outbreeding mating strategy, males discriminate between females of different colonies and populations [2]. They recognize and approach receptive females willing to mate and reject mated females [1]. Young but mated females, old nesting queens as well as foraging workers are not attractive to the males (Ayasse, unpubl.). In field biotests, young virgin females recently killed by freezing, and whole body extracts and cuticular washings of such females elicited mating behavior of males. Dead females devoid of any volatiles, were found to be completely unattractive to the males. However, after having been impregnated with natural extracts or the washings mentioned above, they were pounced by the males. Mixtures of synthetic copies of compounds occuring in extracts were attractive to the males, too. Determination of the active principle in the natural extracts, i.e. the identification of those compounds which actually represent components of the sex pheromone, is presently under investigation.

GC-analyses revealed patterns of volatile compounds specific for young unmated females and old nesting queens, respectively [6]. In young female extracts high concentrations of isopentyl- and isopentenylesters of saturated and unsaturated fatty acids and hydrocarbons were found, whereas macrocyclic lactones were dominant in old queens. In addition, the absolute amounts of the Dufour's gland extracts were found to differ significantly. Old nesting queens showed 100 times higher amounts of volatiles, predominantly lactones, as compared to young and attractive queens.

The presence of macrocyclic lactones in cell lining and impregnating of the nest entrances has been shown several times [3,7]. If not used in odor communication, these compounds should for economic reasons, not be produced before the queen starts nesting.

While the queen has to manifestate her dominant position in the nest she should be recognizable by the foragers. A comparison of both castes showed typical odor compound patterns (Ayasse, unpubl.). The absolute amounts as well as the relative proportions of Dufour's gland secretions were found to be significantly different.

In typical nesting habitats of *L. malachurum* many thousands of nests are spread over a relatively small area. Therefore, a highly developed nesting orientation behavior seems to be necessary. The observations that guarding bees mark the nest entrances with Dufour's gland volatiles, suggest olfactory cues to be involved in nest finding behavior. This presumption was confirmed by behavioral tests in the field [4]. Females returning to their nests were able to discriminate between their own nest entrances and foreign ones even when the visual close range orientation was experimentally made impossible. The main odor compounds of nest entrances, which could be extracted with pentane were found to be alkanes and alkenes (Ayasse, unpubl.). A comparison of the relative proportions of Dufour's gland compounds of nestmates I collected at different nest sites showed some of these compounds to be important as colony specific odor compound patterns.

REFERENCES

1. Ayasse, M. 1987. Apidologie, 18: 371-373.
2. Smith, B.H. and M. Ayasse. 1987. Behavioral Ecology and Sociobiology, 20: 313-318.
3. Hefetz, A., G. Bergström and J. Tengö. (1986). Journal of Chemical Ecology, 12: 197-208.
4. Ayasse, M. 1989. Nestfindeverhalten bei der primitiv eusozialen Furchenbiene *Lasioglossum malachurum* (Hymenoptera: Halictidae). Abstract IUSSI, Bern.
5. Smith, B.H. and C. Weller. 1989. Journal of Insect Behavior, 2: 397-411.
6. Ayasse, M., W. Engels, A. Hefetz, G. Lübke and W. Francke. 1990. Z. Naturforschung, in press.
7. Duffield, R.M., J.W. Wheeler and G.C. Eickwort. (1984). In: Chemical Ecology of Insects, W.J. Bell and R.T. Carde, Eds., Chapman and Hall, London, pp. 387-428.

INDIVIDUAL AND GROUP SPECIFIC ODORS OF BUMBLE BEE FEMALES

J. Tengö (Ecol. Res. Station of Uppsala Univ., Sweden),
A. Hefetz (Dept. Zool., Tel Aviv Univ., Israel) and
W. Francke (Dept. Org. Chem., Hamburg Univ., West Germany)

In a recent review, Michener and Smith (1987) state that nest mate recognition is normally communicated through chemical cues. They also note what little is known of kin recognition in the more advanced primitively social genus Bombus. In this paper we report on the occurrence of individual and group specific odor composition in the Dufour´s gland of Bombus (Pyrobombus) hypnorum (L.).

Nests with colonies of B. hypnorum, collected in SE Sweden, were brought to the laboratory, where each individual was marked and measured for size (thorax width and wing length). Collected nests were transferred to observation boxes and the routes of the exiting and returning foragers were observed.

Twentyfive workers of varying sizes were randomly sampled from four different nests. The Dufour´s gland was dissected from each individual and extracted in pentane. Various notes on the condition of the bees, like worness, presence of sperm in the spermatheca, fatbody, and ovarial development, were taken.

ODOR COMPOSITION

The Dufour´s gland extracts were analysed by combined gas chromatography/ mass spectrometry. The relative intensities of the twenty most abundant volatile components were recorded. The data were subjected to cluster analyses of cases. The volatile secretion is composed by various straight and branched, saturated and unsaturated hydrocarbons, foremost of 23, 25, and 27 carbon atoms chain length. The secretion is species specific (7 species analysed), but also unique for each individual.

ODOR COMPOSITION SUBGROUPS AND BODY SIZE VARIATION

Based on the chemical composition, the workers of a nest colony clustered into subgroups. The medium amalgamation distances between groups were significantly higher than between bees within the groups ($p < 0.001$). The compounds eluting near heptacosane made up the most variable complex.

Workers of the colonies exhibited a size polymorphism ranging from almost queen-sized to small. Using width of thoraces as an indicator of bee size, two main size classes could be distinguished. Comparison of the members of the size subgroups to those of the pheromonal subgroups, revealed that small and large were members of different odor composition groups: one group was composed by large bees only, two by only small ones (see also Table 1).

Table 1. Median amalgamation distances between odor composition subgroups and mean body size within these groups.

Group	A	B	C	D	E	F	G	$\bar{x}$ body size
A	-	60.8	51.8	57.2	133.6	84.5	84.2	2.53
B		-	56.2	90.4	150.9	88.4	57.2	3.41
C			-	98.0	171.4	64.0	70.2	2.56
D				-	82.2	111.3	106.0	2.45
E					-	165.7	163.8	2.50
F						-	113.7	2.85
G							-	3.50

The subgrouping of the workers might be a consequence of multiple mating by the queen and thus representing patrilines. It might also indicate the physiological states of the workers based on age, etc., or reflect the hierachial system within the colony.

ODOR TRAILS

The occurrence of Dufour's gland secretion components in the odor trails between the brood cells and the main nest entrance indicate that this gland might contribute to the guiding of foraging workers at their routes into and out from the nest.

REFERENCE

Michener, C.D. and Smith, B.H., 1987. In: Kin Recognition in Animals, Fletcher, D. and Michener, C.D., John Wiley&Sons, Chichester, pp.209-242.

NEST RECOGNITION BY SCENT IN THE CARPENTER BEE *XYLOCOPA PUBESCENS*

Abraham Hefetz, Daniel Mevoreh and Dan Gerling, Department of Zoology, Tel Aviv University, Ramat Aviv 69978, Israel

INTRODUCTION

The use of scent marks at the nest entrance in bees was inferred from behavioural observation in species of *Osmia* [1] and *Xylocopa* [2]. Behavioural and chemical experiments with *Eucera palestinae* [3] indicated Dufour's gland secretion as a possible source. Dufour's gland compounds could be isolated from nest entrances of *Halictus hesperus* [4] and *Evylaeus malachurum* [5], but their role in nest recognition is still equivocal. During our studies on the behaviour of carpenter bees in Israel [6], we noticed the use of visual and olfactory cues in nest orientation, and having unraveled the chemistry of Dufour's gland secretion [7] we investigated the role of this secretion in nest recognition.

MATERIALS AND METHODS

The bees used for the behavioural experiments were trap-nested in canes, but could forage freely. In this way the positions of the nests could be manipulated and nest entrances could be switched from one nest to the other. In order to separate the visual from the olfactory cues directing the bees to their nests, in the first set of experiments the nests were randomly placed within the site so that the bee could orient to them visually, whereas in the second set all the nests were placed in a row behind a screen so that only the nest entrances were apparent. Each nest was supplied with an artificial nest entrance made of a paper tube that was inserted into the nest. During the experiments the paper tube was either replaced by a clean tube, remained in place, or exchanged with that of another nest entrance. The time elapsed until the bees entered their nest was the criterion for nest recognition.

Chemical analyses of Dufour's glands from individual bees and nest entrances were performed as described before (Hefetz et al.,

1986). In addition, coinjections of nest marking extracts with standard compounds or Dufour's gland exudate were performed.

RESULTS AND DISCUSSION

When visual orientation was made possible the bees did not have any problem in recognizing their own nest despite the fact that the nest entrances were switched, or replaced by a clean paper tube (Table 1). However, when visual orientation was made impossible, nest location by scent was evident. The bee typically scanned all the nest entrances while hovering in front of them. Having their own nest entrance odour, the bees' ability to locate their nest was not impaired, but changing their nest entrances resulted in long hesitation before entering. Hesitation time was longer than the control when their original entrance was replaced by a tube of another nest, and even longer when a new, scentless tube was placed in their nest entrance.

Table 1: Nest recognition by scent in *Xylocopa pubescens*

Treatment	Nest	Time before entry (seconds)		
		Type of nest entrance		
		Own	Strange	Clean
Visual cues present	1	1.98	2.19	-
	2	3.27	-	3.25
Visual cues masked	3	1.06	1.38	>300

Shortly after a clean paper tube was inserted into a nest, it was marked by the bees with fecal material and glandular secretion. Comparative chemical analysis revealed the presence of all Dufour's gland compounds in addition to more volatile ones, of which neither the nature nor the origin were determined.

REFERENCES

1. Steinmann, E. 1981. Bulletin de la Societe Entomologique Suisse 49:635-638.

2. Anzenberger, G. 1986. Ethology, 71:54-62

3. Shimron, O., A. Hefetz & J. Tengo. 1985. Insect Biochem. 15:635-638.

4. Brooks, R.W. & J.H. Cane. 1984. J. Kansas Ent. Soc., 57:161-165

5. Hefetz, A., G. Bergstrom & J. Tengo. 1986. J. Chem. Ecol., 12:197-208.

6. Gerling, D., P.D. Hurd & A . Hefetz. 1983. Smithsonian Contribution to Zoology, 396:1-33

7. Gerling, D., H.H.W. Velthuis & A. Hefetz. 1988. Annual Rev. Ent., 34:163-190

KIN RECOGNITION OF STINGLESS BEE, *MELIPONA FASCIATA*

T. Inoue (Lab. of Ent., Fac. of Agric., Kyoto Univ.) and D. W. Roubik (Smithsonian Trop. Res. Inst.)

INTRODUCTION

Kin recognition of social bees has been intensively studied in this decade but information is confined to some bees, e.g. *Lasioglossum zephyrum* and *Apis mellifera* [1]. Stingless bees are the group that is most diversified in the ritualized oviposition process and in worker's reproductive oviposition at queenright colonies [2]. We studied the kin recognition systems among workers in Panamanian *Melipona fasciata* and found that discrimination of non-kins from kins were quite severe, compared with *A. mellifera*.

MATERIALS AND METHODS

Melipona fasciata is an abundant stingless bee in primary forests in Panama and is a facultative robber [3]. There are 5 sequential age classes in worker's task performance: callows (≤6 days after emergence), young brood carers (5-16), intranidal guards and nest cleaners(15-25), extranidal guards (20-25), and foragers (20-40). Two experiments were done at Curundu, Panama city in 1988-89.

<u>Experiment 1</u> We prepared two colonies (M and A) collected at different locations. We divided M nearly evenly to make the daughter colony, D. The mother queen was left in M. The daughter queen of D mated with male(s) other than from M and A. Then, we made a subcolony from each of M, D and A, for odor conditioning. Callow workers, pupal cells and storage pots were transferred to subcolonies but no immature queen. Emerged workers were marked and kept in subcolonies for up to 3 weeks before introduction to M. Some portion of workers were moved to a sub-colony different from the emerged one. These odor-conditioned bees were carefully introduced into the inside space of M.

<u>Experiment 2</u> In one colony (Mf3), marked callows were introduced into the own colony, daily for 2 months. We daily opened the top glass of the observation hive and removed some parts of the involucrum which covered the brood cells. This treatment simulated continuous attack for the hive by natural enemies.

RESULTS AND DISCUSSION

Experiment 1 A guard bee of colony M first inspected an introduced bee on the body with its antennae and, if it found something strange, it further checked the gut content or glandular secretion at the mouthpart. After these inspection processes, the guard started biting. Other guards joined to this attack, by additional biting or putting of a sticky resin ball on the body. Whether the introduced bee is accepted or rejected (killed) is determined most within 1 day after introduction and, once accepted, even non-kin bees (A) are treated as nestmates. Guard bees discriminated various categories (age, relatedness, etc) of the introduced bees. The percentage of bees killed within 1 day (PK) increased from 7 % for callows (age≤6 days) to 45 % at 20 days, even when the introduced bees were sisters (M). PK of non-kin bees were 46 % for callows and 68 % for young (7-16 days). PK of nieces (D) was just intermediate between the sisters and the non-kins. Thus, the guards could discriminate the degree of relatedness if the other factors were equal. In callows, the conditioning of non-kins in the odor familiar to the guards decreased PK to 9 % but there was no bad effects of the conditioning of sisters in the unfamiliar odor. PK increased to 57 % in elder sisters (7-16 days) when they were kept in the unfamiliar odor. Thus, discrimination became severe for elder bees (switching from habituated label acceptance to foreign label rejection [4]). Accepted non-kins which grew up to the guard class attacked their real sisters introduced later (learning of the reference of discrimination).

Experiment 2 At the beginning of the experiment, PK equaled to that of sister callows (7%) in Experiment 1. We also observed attacks to callows that normally emerged in the colony and, thus, sister killing was not due to bad effects of the experiments. PK increased to 50 % after one-month, daily disturbance to the hive. This shows that the critical value for discrimination [5] changed in response to the invasion risk to the hive; at low risks, guards minimized 'accidental' attacks for sisters while allowing of non-kin's joining, and, at high risks, they minimized the invasion of non-kins (PK=97 %), even killing of sisters.

REFERENCES

1. Fletcher, D. J. C. and C. D. Michener, 1987. *Kin recognition in animals*, John Wileys & Sons, Chichester. p. 465.
2. Sakagami, S. F. 1982. In: *Social Insects III*, H. R. Hermann, Ed., Academic Press, Orlando, pp. 361-423.
3. Roubik, D. W. 1983. *Journal of the Kansas Entomological Society*, 56: 327-355.
4. Getz, W. M. 1982. *Journal of Theoretical Biology*, 99: 585-597.
5. Crozier, R. H. 1987. In: *Kin recognition in animals*, D. J. C. Fletcher and C. D. Michener, Eds., John Wileys & Sons, Chichester, pp. 55-73.

SIZE VARIATION IN MALES OF THE COMMUNAL BEE *ANDRENA FEROX* SM.

Remko Leys,
Vakgroep Populatie- en Evolutiebiologie, Rijksuniversiteit Utrecht, Padualaan 8, 3508 TB Utrecht, The Netherlands.

Male size variation was studied in an isolated population of the communal bee **Andrena ferox** Sm. In this species up to a thousand females may share a perennial colony. A previous study showed that the phenotypic resemblance of females, as measured as the relative concentration of Dufour's gland components, is significantly greater within colonies than between colonies. The same result is now obtained for both males and females using several biometric characters of the forewing and the head. The males, however, showed a much greater variation than the females. The existence of two groups of males is shown by means of regression of the subcosta length (highly correlated with body size) and the length of the distoposterior vein of the second submarginal cell (1Rs) (not correlated with body size). The two groups of males do not only differ in this respect, but they are also significantly different with respect to other pairs of characters which show covariance.

Several explanations of these findings are possible:

- failure of fertilization of eggs on pollen balls intended for female larvae;
- a male egg is laid on a 'female' pollen ball;
- food quantity dependend polymorphic genetic influence on body size characters;
- male diploidy as a consequence of inbreeding.

NEST AND POPULATION SPECIFIC ODOR PATTERNS IN TWO COMMUNAL *ANDRENA* BEES (HYMENOPTERA : ANDRENIDAE)

Manfred Ayasse, Department of Developmental Biology, D-7400 Tübingen, FRG
Remko Leys, Laboratory of Comparative Physiology, 3584 CH Utrecht, The Netherlands
Pekka Pamilo, Department of Genetics and Zoology, Helsinki, Finland
Jan Tengö, Ecological Station of Uppsala University, S-386 00 Färjestaden, Sweden

The question was studied whether in the communally nesting bees Andrena ferox Sm and Andrena jacobi Perkins individuals of different nests, nest patches, and populations differ with respect to Dufour's gland compounds and enzyme patterns. Colonies were different in the relative amounts of Dufour's gland compounds in virgin females of A. ferox, but not between females performing nesting activities. In A. jacobi there was a difference between nest patches and two investigated populations. Striking differences in the quality and quantity of Dufour's gland compounds were found between virgin and mated females of A. ferox. In A. ferox 13 and in A. jacobi 28 recognizable enzyme loci did not show any genetic variation. The reasons for differences in Dufour's gland secretions of both investigated Andrena-species and low levels of electrophoretic variation in hymenopterans is discussed.

CHEMICAL CUES FOR RECOGNITION OF KIN AND NESTMATES IN STINGLESS BEES

T. Suka, T. Inoue (Lab. of Ent., Fac. of Agric., Kyoto Univ.) and R. Yamaoka (Lab. of Biochem., Kyoto Inst. of Technology)

Recognition and discrimination of kin and nestmates were observed in three neotropical, *Melipona fasciata*, *Trigona (Scaptotrigona) barrocoloradensis*, *T. (Tetragonisca) angustula*, and one Southeast Asian, *T. (Tetragonula) minangkabau*, stingless bees. Chemical cues for recognition were analyzed by capillary gas chromatography and GC-MS.

BEHAVIORAL RESPONSE
Newly emerged workers were introduced into conspecific colonies. In all 4 species, aggressive 'inspection' responses by native guard workers were significantly stronger to workers introduced from alien colonies than to nestmates. But eventually, almost all the aliens were accepted, except for *M. fasciata*, in which 50% of aliens were killed by guards.

CHEMICAL ANALYSIS
Among cuticular chemicals, hydrocarbons were main components. Canonical discriminant analysis separated interspecific and inter-colonial differences of cuticular chemicals. For example, the inter-colonial differences in *M. fasciata* were mainly explained by the relative amounts of two hydrocarbon groups (C23:1, C23, C25:1, C25 and C31:1 group, and C27:1, C27 and C31:2 group).

DISCUSSION
Stingless bee species have sufficient variation of cuticular chemicals. And, actually, they can recognize differences between nestmates and non-nestmates. But resulting responses of discrimination are different among species, probably because they have different social traits, e.g. worker reproductive oviposition.

Symposium 23

MODELS AND THEORETICAL APPROACHES TO THE STUDY OF SOCIAL INSECTS

Organizers : **KLAUS JAFFE, Venezuela**
K. CHANDRASHEKHAR, India

EVOLUTION OF SOCIALITY IN A VARYING ENVIRONMENT: NEST SHARING IN PRIMITIVELY SOCIAL WASPS

R.E. Owen[1,2] and D.B. McCorquodale[1]

[1]Department of Biological Sciences, University of Calgary, Calgary, AB, T2N 1N4, Canada, [2]Department of Chemical and Biological Sciences, Mount Royal College, Calgary, AB, T3E 6K6, Canada.

The relative importance of various factors that may promote the evolution of sociality is still unclear (Andersson 1984). Our aim is to elucidate the roles that cost, benefit and relatedness play in the evolution of nest sharing, a probable first step on the road to eusociality. Our model is based loosely on the nest sharing behaviour of sphecid wasps (McCorquodale 1989a,b; Matthews 1990). Our major innovation is that we consider a variable environment. We envisage a system where wasps are either solitary or share nests with others. Sharing is determined by a dominant allele <u>A</u> at a single gene locus. Those sharing nests may or may not share with kin. The environment varies from season to season and can be dry or wet. In a wet season nests are initiated easily and the cost of being solitary is lower than that of sharing a nest. However, in dry seasons nest initiation is delayed and solitary nesting incurs a higher cost. Wasps allowed to share nests in dry seasons do not incur the reproductive cost caused by the delay.

Let genotype i have association-specific fitness $w_{ij}=1+c_i+b_j$, when interacting genotype j, where i directly affects its own fitness by cost c_i and the fitness of its associate by benefit b_i (Michod 1982). The inclusive fitness effect $e_i = c_i + Rb_i$, where R is the relatedness between interactants. Thus we have:

Sharers	<u>AA</u>	$e=c+Rb$
	<u>Aa</u>	$e=c+Rb$
Solitary	<u>aa</u>	$e_2=c_2$

Costs, c, c_2 < 0, vary according to season while benefit b>0 is fixed.

In an outbred population with <u>weak</u> selection gene frequency change is determined by the inclusive fitness effect (Michod 1982). Consider first a <u>single</u> season, then using the methods of Michod (1982) we obtain the condition for the increase of the <u>A</u> allele as $C+Rb>c_2$. Hence even if R=0, <u>A</u> may increase because $c>c_2$ is possible (since costs are negative). Therefore relatedness is not necessary for nest sharing to evolve, but the relative costs of being solitary or sharing are important. (Note in the usual case where <u>aa</u> individuals are non-altruists, $c_2=0$ and $R>-c/b$ is required for <u>A</u> to increase). Now we consider a sequence of T seasons and employ standard results for selection with varying direction (Haldane and Jayakar 1963) with modification for gene frequency change due to the inclusive fitness effect. It follows that for protection of allele $\underline{A}$,

$$\Pi\,(c+Rb) > \Pi c_2$$

and for protection of allele $\underline{a}$

$$\Sigma\, c_2 > \Sigma\,(c+Rb).$$

If both these conditions are met then a polymorphism will result. Although relatedness is not required for the evolution of nest sharing, association of related individuals will favour increase in the $\underline{A}$ allele. With associations of relatives, the conditions for a polymorphism will become more restrictive and fixation of the nest sharing allele will become more likely.

Acknowledgements:

Funding for research and travel by NSERC of Canada.

References:

1. Andersson, M. 1984. Ann. Rev. Ecol. Syst. 15:165-189.
2. Haldane, J.B.S. and S.D. Jayakar. 1963. J. Genet. 58:237-242.
3. Matthews, R.W. 1990. In: The social biology of wasps. K.G. Ross and R.W. Matthews, eds., Cornell Univ. Press, Ithaca, NY.
4. McCorquodale, D.B. 1989a. Ins. Soc. 36:42-50.
5. McCorquodale, D.B. 1989b. Ecol. Entomol. 14:191-196.
6. Michod, R.E. 1982. Ann. Rev. Ecol. Syst. 13:23-55.

PATTERN FORMATION ON THE COMBS OF HONEY BEE COLONIES : SELF-ORGANIZATION BASED ON SIMPLE BEHAVIORAL RULES

Scott Camazine
Cornell University
Section of Neurobiology and Behavior
Seeley G. Mudd Hall
Ithaca, New York, USA 14853-2702

THE PATTERN

The typical feral honey bee colony comprises approximately 25,000 worker bees and a single queen living in a tree cavity. In addition to the adult bees there is immature brood, consisting of developing eggs, larvae and pupae, as well as a variable amount of accumulated food, namely honey and pollen. These are stored within the hive in a series of parallel wax combs subdivided into approximately 100,000 cells. A characteristic well-organized pattern develops on the combs, consisting of three distinct concentric regions - a central brood area, a surrounding rim of pollen, and a large peripheral region of honey. This pattern of cell utilization is the subject of this report.

The well-organized pattern suggests its adaptiveness. A compact brood area may help to ensure a precisely-regulated incubation temperature for the brood and may facilitate efficient egg laying by the queen. Also, a rim of pollen adjacent to the brood area may promote efficient feeding of the larvae by being readily accessible to the nurse bees, the principal consumers of pollen. Assuming a functional significance for this pattern, an important question concerns the mechanism accounting for its development and maintenance. How does a global, colony-level pattern emerge from the activities of thousands of individuals? I propose that the pattern emerges spontaneously through a self-organizing process based on the dynamic interactions of the colony members. Even with each bee acting autonomously, and using only a few, simple behavioral rules based upon limited, local knowledge, an orderly pattern emerges. The key features of such a self-organizing system are that "The patterns arise solely as a result of the dynamics of the system..., with no specific ordering influence from the outside and no homunculus inside" [1]. To explain the process of pattern formation, I first present the behavioral rules of the bees and a set of parameter values which have been estimated from observations of the developing pattern, experimental manipulations, and previously published data. Then, through computer simulation, I combine these rules and parameter values into a model demonstrating the validity of the self-organization hypothesis.

BEHAVIORAL RULES AND PARAMETER VALUES

The following summarizes the key behavioral components of the model: 1) Starting from the center of the frame, the queen moves randomly over the comb and lays eggs in any empty cell that is less than 4 cells to the next nearest brood cell. After the 21 days of honey bee development, the cell is vacated. 2) Honey and pollen are deposited in randomly selected cells, either empty or partially filled with the same substance. 3) Honey and pollen is removed from randomly selected cells. However, the amount removed from each cell is proportional to the number of surrounding cells containing brood. Cells completely surrounded by brood are emptied at 10 times the rate of cells without adjacent brood. Each of these three simple behavioral rules are based entirely on local information, namely the contents of the cell itself and that of its closely neighboring cells. The following parameter values [2,3] specify the rates of honey and pollen input and removal, and the egg laying rate of the queen: 1) The ratio of honey removal to honey input is 0.58. 2) The ratio of pollen removal to pollen input is 0.95. 3) The average ratio of pollen input to honey input is 0.26 (range = 0.06 to 0.83). 4) The maximal egg-laying rate is 1 egg/min.

COMPUTER SIMULATIONS

To appreciate the complex dynamics of brood, pollen and honey deposition and removal, and the effects of various parameter values describing their rates, all the biological details described above have been incorporated into a computer model which simulates the time course of filling one side of a standard Langstroth frame (approximately 3300 cells). The results of the simulations allow us to determine whether the behavioral rules and parameter values described in the preceding section are sufficient to generate the observed pattern, and to determine which components of the model are necessary for the pattern formation. The simulation shows how the characteristic cell utilization pattern develops. Initially pollen and honey are found throughout the comb as bees deposit their loads randomly on the empty frame. At the same time the queen wanders over the frame from her central starting point, and oviposits in suitable empty cells. The result is a random mix of honey and pollen in the periphery, and a central area sparsely occupied with brood. Many of the cells interspersed among the eggs contain honey and pollen. I call this the early "disorganized" stage. Several days later (Figure 1) the characteristic well-organized pattern has formed. The central area is now a compact region of brood. Honey and pollen in the periphery have segregated into a peripheral region, almost entirely honey, and a band of pollen adjacent to the brood area. How has this transformation occurred? Three separate processes contribute to the pattern formation. First, a compact brood area results from the preferential removal of honey and pollen nearby the brood. This continually provides empty cells in the brood area into which the queen oviposits. The second process explains the segregation of honey and pollen in the periphery. Since both are deposited randomly, initially pollen as well as honey appear in the periphery. However, in a typical colony, on average, 95% of the collected pollen is consumed. With the normal fluctuations in pollen availability, much of the time there is even a daily net loss of pollen. Thus, whatever pollen is deposited in the periphery is likely to be consumed at nearly the same rate. However, since honey is brought into the colony at a much greater rate than pollen, cells in the periphery that have been emptied of their pollen are more likely to be replaced with honey. Gradually any pollen deposited in the periphery is removed, leaving this region almost entirely honey. Where, then, is pollen stored ? The only cells available are those with a high turnover rate. These are the cells adjacent to the brood. Once a cell is occupied by an egg, it is "reserved" as a brood cell for the next 21 days of honey bee development. But in the interface zone between the centrally-located brood and the peripheral stores of honey, the preferential removal of honey and pollen continually provides a region where cells are being emptied at a relatively high rate. These cells are available for pollen.

DISCUSSION AND CONCLUSIONS

The model illustrates a pattern formation process whereby brood, pollen and honey segregate into 3 concentric regions. In this self-organizing system, there is no need to specify particular locations for eggs, pollen or honey, nor do the bees need to acquire any global knowledge about the developing pattern to which they are contributing. Following a few simple rules based on local information, the observed pattern in the combs emerges automatically through the dynamic interactions of the participating bees.

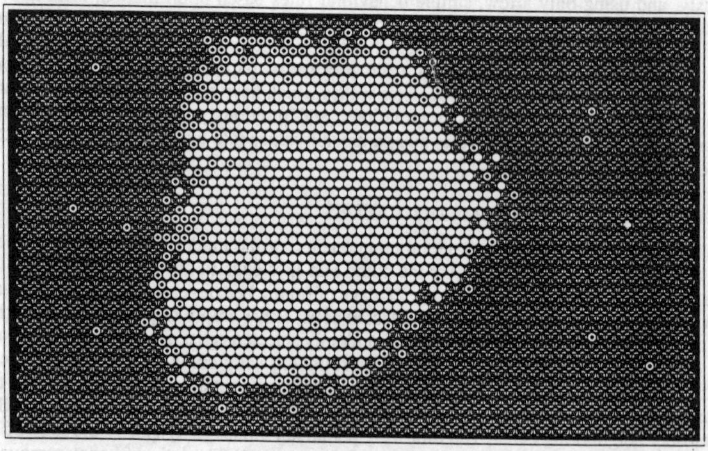

Figure 1. Computer simulation of pattern formation on a frame of comb: Day 7. Open circles represent brood, dotted circles are pollen, and the remaining grey areas are honey.

REFERENCES

1. Schöner, G. and J.A.S. Kelso. 1988. Science, 239: 1513-1520.
2. Seeley, T. 1985. Honeybee Ecology, Princeton Univ. Press, Princeton.
3. Visscher, P.K. and T. D. Seeley. 1982. Ecology, 63(6): 1790-1801.

SOCIAL INTERACTIONS AND EPIDEMOLOGY IN ANT COLONIES

Fowler, H.G. and Costa, A.E. de C., Departmento de Ecologia & Centro Para o Estudo de Insetos Sociais, Instituto de Biociências, UNESP, 13500 Rio Claro, São Paulo, Brazil

In spite of the fact that interest is growing on the use of pathogens in control programs for pest ant species, our present knowledge indicates that we will probably not be successful (1,2). Indeed, in spite of a long history of biological studies, reports of disease epidemics in ant colonies are, at best, non-convincing and few and far between. What are the reasons for this apparently low severity of pathogen epidemics, and the low levels of reported endemic occurrences? Although I do not discount the role of glandular secretions, for example, current knowledge suggests that behavioral interactions among colony members, and behavioral changes in individual ants, are major factors involved in modulating potentially epidemic pathogen outbreaks.

Behavioral Changes

Ethograms of colonies of *Acromyrmex coronatus*, *Monomorium pharaonis*, and *Solenopsis saevissima* of colonies prior to and following application of *Beauveria bassiana* spores to worker subsets demonstrated significantly different task distribution functions over long time periods if colonies were kept enclosed and humidity maintained. However, if colonies were provided with alternate arenas, infected ants left the colony to die at distant sites, and the task distribution functions returned to normal in as short a time period as one-week.

Using plated incidence rates for these groups, it was found that the gamma distribution could be used to characterize worker vulnerability to pathogen-induced mortality. When colonies were open system, worker vulnerability quickly dropped, changing from Types 2 and 3 to Type I. This in itself would indicate stability at the colony level (3).

Behavior and Catastrophe

Once the gamma distribution was obtained, it is a simple step to transform this parameter into one that can be analyzed for stability under castrophe theory. I adapted a simple model (4) relating colony population to colony growth rate under the constraint of a concave exploitation function by pathogens. Interestingly, because of the shift of the gamma distribution, no saddle points emerge, and stable points are obtained only at large colony sizes, which consequently have reduced colony growth rates.

General Insights

Although the experiments used to obtain parameter estimates were simple, leading to singular stable points, this analysis suggests that simple models can lead to interesting hypotheses concerning epidemology in ant colonies. If the results and predictions presented here can be validated in the field, then perhaps we can turn our efforts to other avenues of biological control. This model, by its nature, does not exclude the role of pathogens in the regulation of founding queen numbers, principally because there are no social mechanisms, due to lack of workers, of hygiene, the most varied category of the ethogram analysis. Further studies may perhaps be better directed toward this stage of colony development, although the claustral nature and generally low densities would preclude pathogen epidemics. Nevertheless, the role of modelling should not be discounted, whither the ultimate objectives be applied or basic.

References

1. Jouvenaz, D.P. 1990. In: Applied Myrmecology, R.K. Vandermeer, K. Jaffe and A. Cedeño-Leon, Eds, Westview Press, Boulder, in press.
2. Kermarrec, A. 1990. In: loc cit
3. Bailey, V.A., Nicholson, A.J. & Williams, E.J. 1962. Journal of Theoretical Biology, 3:1-18.
4. Gatto, M. & Rinaldi, S. 1987. Vegetatio, 69:213-222.

TWO-DIMENSIONAL MODELS OF HARVESTER ANT MOVEMENT

James W. Haefner and Thomas O. Crist
Department of Biology and Ecology Center
Utah State University
Logan, UT 84322-5305 U.S.A.

Models of animal movement range from static optimal foraging theory to deterministic partial differential equations to stochastic simulations of individuals. Recently, the last approach has received much emphasis [1], particularly in the theory of insect movement and dispersion. An advantage of these models is their focus on the properties and abilities of individual organisms without making assumptions concerning individual movement competence. Many ordinary differential equation models of finite patches assume that individuals know the location of and can find resource patches. Individual-based models, while also making assumptions, begin at a lower physiological and behavioral level and seek to reconstitute the higher level phenomena (e.g., patch visitation) using individual movement rules. Individual-based models of ant movement have been developed [2]-[4].

To extend these models and to evaluate the movement rules sufficient to explain patch visitation and seed removal, we constructed a general model of individual ant movement in two dimensions for a colony of the western harvester ant, Pogonomyrmex occidentalis. Using 10-sec time steps, the model simulates the dynamic spatial positions of all individuals of three classes: ants without seeds, ants with seeds, and seeds. Each ant foraging above-ground is moved in a non-stationary correlated random walk according to a set of rules that determine the direction and distance of each successive step. We modeled a range of movement rules that hypothesized different degrees of foraging prowess in five general areas: (1) space perception, (2) memory, (3) communication, (4) physiological competence of sensory systems, and (5) stimulation of nestmates. Field experiments that manipulate seed densities were simulated by depositing model seeds in patches at different times and places during the simulation.

Models that best fit a field experiment in the western USA were those having rules in which (a) recruitment stimulation emanates from the locations of the resource, (b) successful ants returning to the nest stimulate additional foragers to leave, (c) individual ants remember the location of seed capture, and (d) not all foraging ants respond to chemical stimuli.

Additional simulations indicated that of the comparisons performed, the greatest rate of seeds returned to the nest from a single, renewing seed patch, occurred when ants used memory, used the locations of captured seeds to stimulate recruitment, stimulated foragers to leave the nest, and where all foragers responded to recruitment stimuli. The lowest rates of seeds returned were obtained by ants that either used no recruitment stimulation or relied solely on memory.

Three additional experiments tested the model's validity in new situations: (1) 150 seeds 3.5 m from nest, 1.0 m from trail: model predicted lower recruitment rates than observed; (2) 2 patches of 75 seeds 3.5 m from nest, 0.25 m each side of trail: model correct on one patch, underestimated recruitment at second patch; (3) 150 seeds 7.0 m from nest, 0.25 from trail: model underestimated recruitment. Other simulations compared 2 different movement rules and 2 different dispersions of 300 seeds: uniformly spaced and clumped in 12 patches. When all ants respond to chemical stimui, more seeds were returned to the nest from clumped dispersions than from regular dispersions. There were no differences if only 0.1 of the ants respond to stimuli. When only a fraction of ants respond, an intermediate number of seeds were returned compared to all ants responding in regular (returning more seeds) or clumped (returning fewer seeds) dispersions. Thus, movement rules that cause some ants to ignore chemical stimuli may be an adaptation to environments in which seeds occur both in clumps and widely spaced.

REFERENCES

[1] Huston, M., D. DeAngelis, and W. Post. 1988. Bioscience 38:682-691.
[2] Jaffe, K. 1980. Journal of Theoretical Biology 84:580-609.
[3] Harkness, R.D. and N.G. Maroudas. 1985. Animal Behavior 33:916-928.
[4] Deneubourg, J.L., S. Goss, N. Franks, and J.M. Pasteels. 1989. Journal of Insect Behavior 2:719-725.

HOW ARGENTINE ANTS ESTABLISH A MINIMAL-SPANNING TREE TO LINK DIFFERENT NESTS

ARON S., DENEUBOURG J.L., GOSS S., and PASTEELS J.M.

Unit of Behavioural Ecology, C.P. 160
Université Libre de Bruxelles
B - 1050 Bruxelles, Belgique

Argentine ant societies are not central structures, but are composed of a number of nests (or sub-societies) connected by a permanent network of chemical trails. Workers, larvae and even queens are regularly exchanged between these "outposts", allowing a flexible allocation of the work-force in the foraging area in response to environmental cues.

Recently, we have shown that much of *Iridomyrmex humilis*' spatial organisation is the result of the workers marking the ground as they move, their direction of movement being influenced by the marks left by preceding workers (Deneubourg et al., 1989; Goss et al., 1989; Aron et al, 1990a). This factor seems also to be central to their inter-nest organisation (Aron et al, 1990b).

The formation of a network of connections between laboratory nests was studied using cardboard bridges in different configurations, in particular branches of equal length arranged in a triangle linking three nests, in a square linking four nests, and two branches of different length linking two nests.

The traffic between the nests, at first evenly distributed over each branch, rapidly becomes asymmetrical. In all the configurations studied so far, the ants use a subset of the available connections that nevertheless links all the nests. For example two branches are used to connect three nests, three branches connect four nests, and one branch (the shortest one) connects two nests.

Different experiments underline the primary role of chemical cues in the establishment of these networks, as opposed to other possible factors such as individual memory or visual cues.

A mathematical model based on the individual workers' simple trail-laying and trail-following behaviour generates colective networks similar to the experimental ones. Certain differences in the case of the square configuration suggest the existence of an additional factor that seem to prevent pairs of nest from becoming isolated from the rest of the network.

Acknowledgements: This work is supported in part by the Belgian program on interuniversity attraction poles, Les Instituts Internationaux de Physique et de Chimie, and a Schlumberger grant from "Les Treilles" Foundation.

References

ARON S., DENEUBOURG J.L., GOSS S. & PASTEELS J.M. (1990 b). Functional self-organisation illustrated by inter-nest traffic in the Argentine ant *Iridomyrmex humilis*. In: *Biological Motion*, Alt W. and Hoffman G. (Eds.). Lecture Notes in Biomathematics, Springer Verlag, (in press).

ARON S., PASTEELS J.M., GOSS S. & DENEUBOURG J.L. (1990 a). Self-organizing spatial patterns in the Argentine ants, *Iridomyrmex humilis* (Mayr). In: *"Applied Myrmecology: A World Perspecive"*. Jaffé K., Cedena A., Vander Meer R.K. (Eds), Westview Press, (in press).

DENEUBOURG J.L., ARON S., GOSS S. & PASTEELS J.M. (1989). The self-organizing exploratory pattern of the Argentine ant, *Iridomyrmex humilis*. *J. Ins. Behav.* 3: 159-168.

GOSS S., ARON S., DENEUBOURG J.L. & PASTEELS J.M. (1989). Self-organised short cuts in the Argentine ant. *Naturwissenchaften*, 76, 579-581.

SELF-ORGANIZATION OF BEHAVIOURAL PROFILES AND TASK ASSIGNMENT BASED ON LOCAL INDIVIDUAL RULES IN THE EUSOCIAL WASP *POLISTES DOMINULUS* CHRIST

Guy THERAULAZ, Simon GOSS†, Jacques GERVET* and Jean-Louis DENEUBOURG†*

* *CNRS - UPR 38, 31 Chemin Joseph Aiguier, 13402 Marseille cedex 09, France*

† *Unit of Theoretical Behavioural Ecology, CP 231, Brussels Free University, 1050 Bruxelles, Belgium*

In eusocial insect societies, individuals have to work together to carry out some tasks, the nature of which depends both on the colony's own needs and on environmental features. Each individual constantly decides, acts and interacts with other individuals and the environment, thus continuously changing the state of the group. The latter is however a place where stable, self-regulated individual behaviours are organized. Studies on the processes which lead to the emergence of a stable collective order in insect societies have focused in recent years on the dynamics of individual interactions (DENEUBOURG et coll. 1987, 1989, 1990). Theses researches showed that by adopting some elementary rules of individual behaviour, a society can generate complex patterns and take decisions when encountering some external constraints.

Our own biological studies deal with the processes underlying **task assignment** in *Polistes dominulus* wasp colonies (THERAULAZ et coll. 1990 a, b, c). In these primitively eusocial species no morphological differences exist between castes at the adult stage, and the social roles are largely determined by social interactions. We have detected two types of interaction which regulate the organization of individual behaviour :

1. Direct interactions of the *hierarchical type*, which determine which individual will have precedence over another individual and organize the society into a linear structure;

2. Indirect interactions of the *trophic type*, which determine individuals' relationships with local environment, consisting of the brood, and which give rise in turn to actions performed by the individual on the brood such as brood-tending and/or foraging outside the nest.

Each of these two types of interaction, when it occurs,has reciprocal effects on the individual which depend on the possible outcome of the interaction (cf. fig.1).

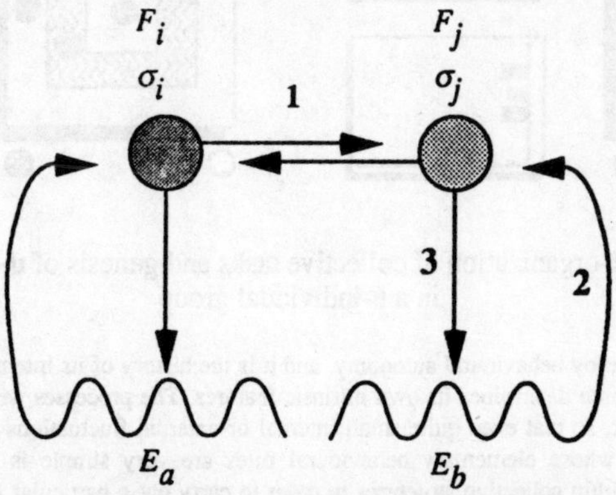

Fig. 1 Individual local action rules in Polistes

When a hierarchical interaction (1) occurs, for instance, the intrinsic probability (F) that the dominant animal will dominate increases by ε at each subsequent meeting, whereas this probability decreases at the same time by ε in the case of the subordinate animal. Likewise, when foraging activity (3) is triggered in an animal which comes into contact with specific brood-related stimuli (2 & E_b), the response threshold of the animal to these stimuli will decrease by η, which increases its probability of responding to them on the next occasion. The model we put forward here, takes the above two local action rules into account. Each component in the system is characterized by 2 variables:

1. a force F, which occurs when random encounters take place in the nest space. Force F determines both the probability of encounter (by increasing the speed of movement on the nest), and the probability of domination at each encounter;

2. a response threshold σ to the stimulation arising from a specific point within the nest space : that containing the larval brood (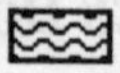).

In addition, the individual random moving component is modulated by a centripetal component which corresponds to a constant individual tendency to invest in egg-laying. The joint action of these two processes generates on one hand a stable hierarchy among the the individuals; one of them, the α individual (here individual No 3, cf. fig. 2), will never leave the nest, where it exerts considerable dominance; at the same time, the α individual's response threshold to brood stimulation increases constantly. And on the other hand, some individuals (here No 1,5 and 6) become specialized in foraging and larva-feeding tasks.

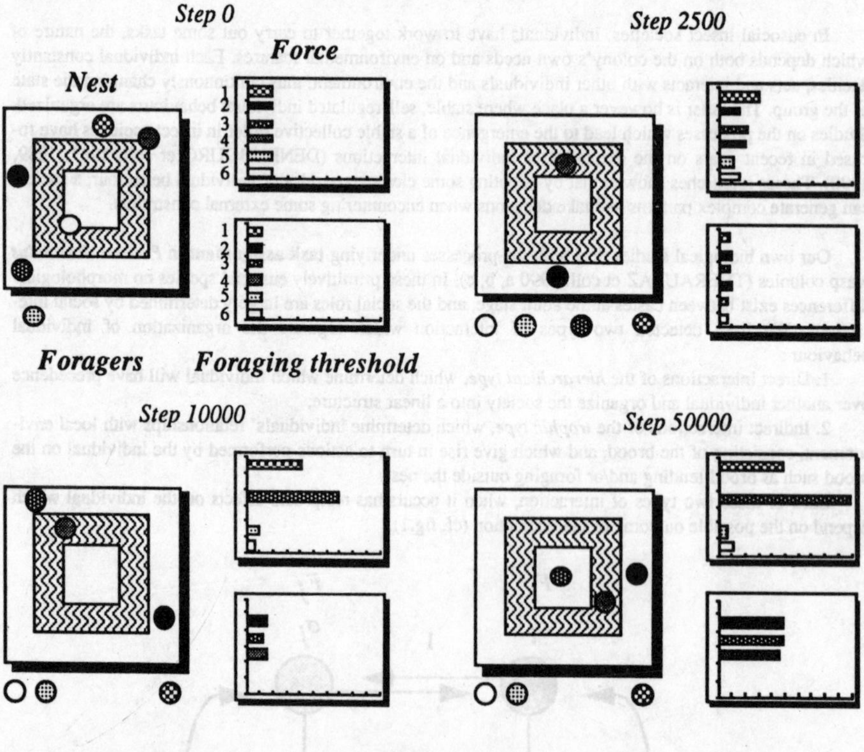

Fig.2 Self-organization of collective tasks and genesis of the hierarchy in a 6-individual group

Each individual enjoy behavioural autonomy, and it is the history of its interactions with other individuals in the colony which determines its own intrinsic features. The processes we have described here act like positive feed-back, so that even quite small internal or external fluctuations are rapidly amplified. A group of individuals whose elementary behavioural rules are very simple is thus able to coordinate individual activities within collective structures in order to carry out a particular general task. The type of task performed will depend on the individual local action rules. Moreover, the interplay between these structures and the environment can generate diverse patterns of collective behaviour starting with the same elementary behavioural algorithms. This type of model, which combines simplicity, reliability and adaptability, can be applied to organizing groups of robots with simple reconfigurable elementary features, so that they perform complex tasks.

References.

DENEUBOURG J.L., GOSS S., PASTEELS J., FRESNEAU D. and J.P. LACHAUD (1987). Self-organisation mecanisms in ants societies (II) : learning in foraging and division of labour, in J.M. PASTEELS & J.L. DENEUBOURG edits. From individual to collective behaviour in social insects. Experientia Supplementum 54. Basel: Birkhäuser Verlag.

DENEUBOURG J.L., ARON S., GOSS S. and J.M. PASTEELS (1989). The self-organising exploratory pattern of the Argentine ant. Jounal of Insect Behaviour (in press).

DENEUBOURG J.L. and S. GOSS (1990). Collective patterns and decision-making. Ethology, Ecology and Evolution (in press).

THERAULAZ G., M. PRATTE and J. GERVET (1990 a). Behavioural profiles in *Polistes dominulus* (Christ) wasp societies : a quantitative study. Behaviour (in press).

THERAULAZ G., J. GERVET, B. THON, M. PRATTE and S. SEMENOFF (1990 b). The dynamics of colony organisation in the primitively eusocial wasp *Polistes dominulus* (Christ). (submitted to Ethology).

THERAULAZ G., J. GERVET and S. SEMENOFF (1990 c). Social regulation of foraging activities in *Polistes dominulus* Christ wasp colonies. A systemic approach of the organization of behaviour. (submitted to Behaviour).

RESPIROMETRY AND THE EVOLUTION OF ORDER : NEGENTROPY CRITERIA APPLIED TO THE EVOLUTION OF ANTS

Klaus Jaffe and Maria Jose Hebling-Beraldo

Univ. Simon Bolivar, Apdo 89000, Caracas 1080, Venezuela
and
Instituto de Biociencias, UNESP, Rio Claro, Brasil

A quantitative measurement of order has always been difficult. With the advent of irreversible thermodynamics, order can be assessed through estimates of entropy. Zotin and Konoplev (1978) proposed a negentropy measurement, based on basal metabolic rates normalized by body weight, as an index of order for living organisms. Zotin and Konoplev showed that this index gives values which reflect the evolutionary gradient among vertebrates and among invertebrates, if comparisons are made at the level of classes. We attempted to use Zotin and Konoplev's index to measure social complexity in ant species. We choose 8 Attini from 4 different genera and measured their basal oxygen consumption. Our results showed the opposite of Zotin and Konoplev's prediction, that is, workers from highly social species are less negentropic that those of primitively social species.

We propose that individuals of highly social ant species are less complex than individuals from simpler ant societies. This finding is in agreement with previous findings of studies on the complexity of the ants nervous system (Jaffe & Perez, 1989). Thus, in order to comply with thermodynamic rules, colonies of highly social species have to have a higher metabolic rate than colonies from less complex ant societies. This can be achieved only if workers of highly complex ant societies are more active than workers from less complex societies, thus, consuming more energy in mean during their lives than workers from socially primitive species. No data to date exist on this aspect.

References

Zotin, A.I. and Konoplev, V.A. 1978. Direction of the evolutionary progress of organisms. In Thermodynamics of Biological Processes, I. Lamprecht and A.I. Zotin eds. Walter de Gruyter, Berlin, pp 341-347.

Jaffe, K. and Perez, E. 1989. A comparative study of brain morphology in ants. Brain Behav Evol, 33, 25-33.

A TEST OF THE ROLE OF HAPLODIPLOIDY IN THE EVOLUTION OF HYMENOPTERAN EUSOCIALITY

RAGHAVENDRA GADAGKAR
Centre for Ecological Sciences and Centre for Theoretical Studies, Indian Institute of Science, Bangalore-560 012, INDIA.

The haplodiploid genetic system found in all Hymenoptera creates an asymmetry in genetic relatedness so that full-sisters are more closely related to each other (coefficient of genetic relatedness, $r = 0.75$) than a female would be to her daughters ($r = 0.5$). The multiple origins of eusociality in the Hymenoptera have therefore been ascribed to haplodiploidy (1). But a Hymenopteran female is related to her brothers by 0.25. A worker who rears equal numbers of brothers and sisters therefore has no advantage over a solitary nest foundress because her average relatedness to brood will be reduced to 0.5. But workers can potentially gain more inclusive fitness than solitary foundresses if they invest more in their sisters than in their brothers (2). When queens mate multiply and simultaneously use sperm from different males, they produce different patrilines of daughters who would only be related to each other by 0.25. This again reduces the inclusive fitness that workers can potentially gain. Whether or not the genetic asymmetry created by haplodiploidy can by itself be sufficient to allow workers to have more inclusive fitness than solitary foundresses thus depends on their relatedness to their sisters and on their ability to skew investment in favour of sisters. A number of estimates of genetic relatedness between sisters in Hymenopteran colonies have now been published. To test the haplodiploidy hypothesis I assume that workers are capable of investing in their brothers and sisters in the ratio that is optimal for them and compute the threshold relatedness to sisters required for them to obtain a weighted mean relatedness to siblings of 0.5 and thus break even with solitary foundresses.

In an outbred Hymenopteran population where workers rear mixtures of sisters and brothers, the optimum number of females that a worker should rear relative to every brother reared is given by $r_f/0.25$ where r_f is her mean relatedness to sisters and 0.25 is that to her brothers. When workers successfully skew investment between sisters and brothers in the ratio $r_f/0.25 : 1$, their weighted mean genetic relatedness to siblings $\bar{r}$ is given by:

$$\bar{r} = [(r_f^2/0.25) + 0.25] \,/\, [(r_f/0.25) + 1] \qquad \text{......(1)}$$

To solve equation (1) for $\bar{r} = 0.5$, I rewrite it as

$$16r_f^2 - 8r_f - 1 = 0 \qquad \text{......(2)}$$

Equation (2) yields a value of 0.604 for r_f. This means that a genetic relatedness between workers and their sisters of 0.604 is required if workers are to gain as much fitness as solitary individuals, inspite of skewing investment between sisters and brothers in the ratio that is optimal for them. I will call this number namely 0.604 the *haplodiploidy threshold.*

Most published estimates of genetic relatedness are accompanied by standard errors and I therefore ask if these estimates are significantly greater than the haplodiploidy threshold. Of 141 such estimates (spread over 35 species) of relatedness between sisters only 16 estimates are significantly higher than the haplodiploidy threshold ($p < 0.05$). Of these, 5 pertain to ants, 4 to primitively eusocial bees and 7 to primitively eusocial wasps. Of 17 species of ants, only three have atleast one estimate which is significantly higher than haplodiploidy threshold. These are *Solenopsis geminata* for which 3 out of 5 estimates are higher, *S.invicta* for which only 1 out of 4 estimates is higher and *S.richteri* for which the only available estimate is higher. *Apis mellifera*, the honey bee does not have relatedness significantly higher than the haplodiploidy threshold. The two species of vespine wasps studied also do not have significantly higer values and of three species of swarm-founding wasps, none have even one estimate that is significantly higher than the threshold. Thus out of 20 species of highly eusocial Hymenopterans studied only three have atleast one estimate significantly higher than the haplodiploidy threshold. I conclude from this that the genetic asymmetry created by haplodiploidy is by itself insufficient to maintain the highly eusocial state.

Of 2 species of primitively eusocial bees only one, namely *Lasioglossum zephyrum* has values significantly higher than the haplodiploidy threshold but even here only 4 out of 14 estimates are significantly higher. Of the 8 estimates for the other primitively eusocial bee namely *Exoneura bicolor*, none are significantly higher. Of 17 species of primitively eusocial wasps, only 2 have estimates significantly higher than the haplodiploidy thrshold. One such species is *Microstigmus comes* in which only 6 out of 18 estimates are higher. The other is *Mischocyttarus immarginatus* in which the only estimate available is significantly higher. Thus of 15 species of primitively eusocial Hymenopterans studied only 3 have at least one estimate significantly higher than the haplodiploidy threshold. I conclude from this that the genetic asymmetry created by haplodiploidy is by itself insufficient to promote the origin of eusociality.

In computing the haplodiploidy threshold I have assumed outbreeding. This appears to be reasonable for most species used in this analysis. I have also assumed that workers are capable of skewing investment in the ratio that is optimal for them. This may or may not hold. If it does not, then I am giving an unfair advantage to the haplodiploidy hypothesis. But that is just as well because it is better to falsify a hypothesis inspite of giving it an unfair advantage.

1. Hamilton, W.D. 1964. J.theor.Biol., 7:1-52.
2. Trivers, R.L. and H. Hare. 1976. Science, 191:249-263.

Symposium 24

FORAGING STRATEGIES OF SOCIAL INSECTS

Organizers : **JAMES TRANIELLO, U.S.A.**
T. VEENA, India

SEARCH BEHAVIOR IN THE ANT *FORMICA SCHAUFUSSI* : SOCIAL REGULATION AND ECOLOGY OF INDIVIDUAL PATTERNS

James F.A. Traniello, Samuel N. Beshers, Vincent Fourcassié, Hui-Shien Loh and Laurie Henneman, Boston University, Department of Biology, 5 Cummington Street, Boston, MA 02215 USA

INTRODUCTION

Many colony-level processes in ants have been shown to arise from social regulation of individual activity. In Formica shaufussi foraging effort is divided between workers which tend homopterans for carbohydrate secretions or search for and retrieve arthropod prey. Is the development of the tendencies to continue to forage for or specialize on either food type based on differences in the distribution in time and space of carbohydrate and protein resources? Do search patterns reflect these differences, and how sensitive are they to social regulation? We studied the organization of individual search patterns, the role of search modification, resource quality, and the nutritional state of the colony to answer these questions.

METHODS

Colonies of Formica schaufussi were collected at Concord Massachusetts and housed in plexiglas nests in the laboratory. Nests were connected to a large foraging arena where food was offered and search patterns recorded. Colony food intake was rigorously controlled, and search behavior was videotaped and analyzed in fine detail with path analysis software. Workers foraging in the arena were fed and allowed to return to the nest to unload and then to re-enter the arena and search at the prior feeding site, or target. Search paths were digitized and analyzed to produce a set of path parameters including velocity, turning rate, average distance to target, initial giving up time, and percentage of time spent in local search.

The effect of sucrose quality was studied using 0.25, 0.5, 1.0, and 2.0 M sucrose, while protein quality was tested with freshly killed 5 and 30 mg crickets. The effect of colony nutritional state was tested using response to 1.0 M sucrose in colonies starved for 0, 3, 9, and 12 days. The search paths were divided into three one-minute segments for analysis. Experimental effects were tested with analysis of variance on each segment separately, and each variable was similarly tested for changes over time.

RESULTS AND DISCUSSION

The size of insect prey had no effect on any search path parameters, and there were no changes over time. For sucrose quality, there were significant differences only between 0.5 and 2.0M in turning rate. Starvation resulted in consistent and highly significant differences in velocity, turning rate, and target distance; in most cases fed and 3 days starved foragers differed from 9 and 12 days starved foragers. The directions of the differences suggested that starvation caused generally prolonged local search with lower spatial deviation while sucrose concentration had little effect. The most consistent differences found were between protein- and sucrose-related search, and were reflected mainly in the greater time spent in local search for sucrose.

Our prior field studies (1,2) on the effect of carbohydrate and protein food showed that workers are less likely to return and search at a location where protein food was found. Recently, we have shown that 1) target distance is significantly lower in response to carbohydrate food, independent of reward number; 2) local search duration increases in response to repeated carbohydrate rewards, whereas protein rewards elicit no such effect; 3) carbohydrate and protein target distance do not vary seasonally; and 4) trail-laying and recruitment in response to the two food types vary seasonally, but are not correlated with search pattern.

These results support the hypothesis that search in Formica schaufussi has evolved principally in response to the distribution and predictability of the two food types. Sucrose elicits fine-scale search where food has been found once and is likely to occur again, while protein elicits ranging search which yields greater coverage of a wide area. Division of foraging effort according to food type may in part be based upon resource dependent rules of organization.

REFERENCES

1. Traniello, J.F.A. 1988. In: Interindividual Variability in Social Insects, R.L. Jeanne, Ed., Westview Press, Boulder, CO.
2. Traniello, J.F.A., Fourcassie, V., and Graham, T. submitted

ENERGY INVESTMENT AND REWARDS IN FORAGING BY *MESSOR CAPITATUS* (LATREILLE)

Mogens Gissel Nielsen
Zoological Laboratory
University of Aarhus, Denmark

Seed harvesting ants are in many ways very good experimental animals for the study of the energetics of foraging, because they carry whole seeds back to the nest. This "benefit" of a foraging trip can easily be identified and the energetic and the nutritive quantity and quality of the food item can be completely analyzed. In the present work the energetics of foraging are analyzed and compared with the energetic reward.

The standard metabolism was measured, using micro-Warburg techniques [1,2] on single individuals of Messor capitatus, and the respiratory rate can be described by:

$$\ln Y = -2.24 + 0.098\ T - 0.258 \ln W$$

where Y is μl O_2 per mg dry weight per hour
T is temperature
W is dry body weight

The Q_{10} value in the temperature interval 20-30 °C was 3.4, a higher value than for many other ants [1].

The "basal" metabolism was measured using the methods of Holm-Jensen et al. [3], where the ants are anaesthetised during the measurements. For Messor capitatus the "basal" metabolism at 30 °C was 1.40 $\pm$ 0.14 μl O_2 per mg d.w. per hour (0.51 $\pm$ 0.05 μl O_2 per mg f.w. per hour), which is 76% of that measured by Warburg techniques. This reduction in respiratory rate is due to lack of motor activity.

By using a modification of the technique used by Nielsen et al.[4], the energetic cost of running (the measured value from the running experiments subtracted from the standard metabolism) was found and could be described by:

$$y = 3.865 + 0.363\ fw$$

where y is Joule per km
fw is fresh body weight

In order to measure the costs of carrying a load, small platinum weights were glued on the head of the ants.

The costs of transport (the extra costs of carrying a load) was obtained by subtracting the values of unloaded running from the value of loaded running. For Messor capitatus we found that the cost of transport was 0.64 times the cost of running, which means that it is energetically more economic to carry as big a load as possible.

The total energy expenditure for a foraging trip could now be calculated from the following parameters: the weight of the ant, its body temperature, the duration of the foraging trip, the distance run in searching for seeds, the weight of the seed, and the distance it was carried back to the nest.

The energetic content of a number of seeds have been analyzed, and the values ranged from 28.1 to 17.8 Joule per mg dry weight. For example, a wheat seed of 32 mg had an energy content of about 600 Joule.

How long can an ant forage on the energy derived from a wheat seed?

According to the data presented here, an ant weighing 20 mg f.w. with a body temperature of 30 °C, searching for an hour over a distance of 10 meters and finding a seed weighing 30 mg which then had to be carried a distance of 5 meters back to the nest in the next hour would consume 1.28 Joule. At this rate a wheat seed contains sufficient energy to keep the ant going for nearly 40 days.

For smaller seeds the "reward" or the benefit of an foraging trip will be less, but the seed has to be extremely small or the time for searching must be extremely long before the cost/benefit will approach one.

References:

[1] Nielsen, M.G. 1986. Journal of Insect Physiology, 32(2):125-131.

[2] Nielsen, M.G. and Baroni-Urbani, C. 1990. Physiological Entomology, (In press).

[3] Holm-Jensen, I., Jensen, T.F. and Nielsen, M.G. 1980. Insectes Sociaux, 22(2) :180-185.

[4] Nielsen, M.G., Jensen, T.F. and Holm-Jensen, I. 1982. Oikos, 39: 137-142.

CONSERVATION REVISITED—RAREFRACTION; LEARNING; AND TRAILS IN LEAFCUTTING ANTS — A RETURN TO OPTIMALITY

H.L. de VASCONCELOS* and H.G. FOWLER**, *Departamento de Ecologia, INPA, 69001 Manaus, Amazonas, Brazil, and ** Departamento de Ecologia, Instituto de Biociências, UNESP, 13500 Rio Claro, São Paulo, Brazil

The hypothesis that leaf-cutting ants conservatively manage the vegetable resources used as fungal substrate within their foraging territories (1,2) has been questioned (3) and reaffirmed (4,5). Given the paucity of published information (6) available on forage selection by leaf-cutter colonies, and the extreme difficulties in obtaining more precise information, we have re-examined the literature in light of current theories of optimal foraging.

Selection as a Function of Relative Abundance

Only three studies have estimates of the relative abundance of vegetation, and none of these demonstrate that the frequency of harvest is dependent upon abundance. As such, this does not falsify the conservative hypothesis.

Rarefraction

We have used rarefraction curves to examine resource usage by colonies in which either time or sampling effort varied in an attempt to standardize our comparative analysis. For species with replicated studies in different sites (see 6), we found that no discernable differences were found between species, or with respect to the uniformity of the habitat and vegetable resources. ANOVA indicated that the within species variation of these standardized data was greater than other identified effects.

Learning

Does learning have a role in the patterning of forage substrate selection? Without trying to operationally define learning, let us consider the foraging strategy of leaf-cutting ants, in which trails lead to an area-restricted search. Firstly, foraging movement will be directional and not random, in accordance with optimal foraging theory although here it is a mass effect and not individual behavior that is important. Secondly, the physical trail structure, forager sub-group specializations, and pheromonal markings determine which trees will be attacked, and these are restricted to a narrow band of foraging trail border. This pattern leads to a Holling Type III functional response, with data from species in homogeneous vegetations showing the same response as those in complex tropical forests. This switching response at the colony level eliminates inexperienced workers cutting vegetation not previously sampled, and further accentuates the Type III response (7).

Given these facts, the conservative foraging strategy is not supported, and our attempts should be directed toward optimal central-place foraging models.

References

1. Cherrett, J.M. 1968. Journal of Animal Ecology, 37: 387-403.
2. Rockwood, L.L. 1976. Ecology, 57:48-61.
3. Fowler, H.G. and Stiles, E.W. 1980. Sociobiology, 5: 25-41.
4. Cherrett, J.M. 1984. In: Tropical Rain Forest: Ecology and Management. S.L. Sutton and T.C. Whitmore, Eds. Blackwell, Oxford. pp. 253-265.
5. Reed, J. and Cherrett, J.M. 1990. In: Applied Myrmecology. R.K. Vander Meer, K. Jaffe and A. Cedeño-Leon, Eds., Westview Press, Boulder, in press.
6. Vasconcelos, H.L. and Fowler, H.G. 1990. In: loc cit.
7. Taylor, R.L. 1984. Predation. Chapman and Hall, New York.

CHANGING FORAGING STRATEGIES INHERENT TO TRAIL RECRUITMENT

R. Beckers, J.L. Deneubourg, S. Goss
Behavioural Ecology Unit, C.P. 231, U.L.B., 1050 Brussels, Belgium

In most ant species, the exploitation of ressources is based on individual foraging. This is also true for species that use a recruitment strategy to exploit large food sources that may not be retrieved by a single forager. In this case a few ants explore independently the foraging area and launch a recruitment after the discovery of a food source.

We will show, how the autocatalytic mechanism of the chemically mediated trail recruitment allows the colony to adjust its foraging behaviour to a changing environment.

Trail recruiting species such as *Lasius niger (L.)* or *Messor rufitarsis,* concentrate on one food source when only a small number are available. Only after this source is exhausted does the colony's activity shift to another one. If the number of sources increases, the colony is no longer able to focus its activity on one food source, but distributes its foragers evenly over different sources. Different foragers, discovering a large number of food sources simultaneously, are not able to attract a sufficient number of recruits to their discoveries. As a consequence, none of trails are reinforced, and all the sources are exploited equally.

What is the sense of this change of strategy ?

In a mathematical model we show that, with an increasing number of food sources, trail recruitment could lead to either a hetero- or a homogenous exploitation of the sources. For a small number of sources the model's solutions predict that the colony concentrates on one of them. As the number of food sources increases, the number of recruited of ants decreases and has a minimum, corresponding to the passage from the hetero- to the homogenous exploitation of the food sources, after which the number increases. The model thus predicts an interesting paradox, whereby increasing the number of food sources can decrease the number of ants recruited (and which reach the food) and thus decrease the colony's foraging efficiency.

In our contribution we present the theoretical assumptions and experimental verifications

PREY FORAGING BY THE ANT *PHEIDOLE PALLIDULA* : DECISION MAKING SYSTEMS IN FOOD RECRUITMENTS

Detrain C.*, Pasteels J.M., Deneubourg J.L. and Goss S.

Unit of Behavioural Ecology CP. 160, Université Libre de Bruxelles, 50 Av.F.D.Roosevelt, 1050 Bruxelles (Belgium)
*Senior research assistant: National Fund for Scientific Research.

Foraging is well known as a sophisticated collective pattern. A generalist ant whose diet varies in size, quality, spatial distribution, etc., exhibits different foraging strategies according to the food characteristics, ranging from individual to collective foraging. In this respect, an important problem is to understand how and when a forager "decides" to recruit or not. Such a decision making system has been studied in the Mediterranean ant *Pheidole pallidula.* Indeed, field studies show that this species is an opportunistic and mainly insectivorous one. Prey range from very small insects such as Collembola to larger ones such as imagos of Coleoptera or Hemiptera. The exploitation of these prey, so diverse in their size, requires the ant colony to use different foraging techniques varying from individual foraging to massive recruitment of nestmates. We have tested in the laboratory the hypothesis that prey retrievability is one of the key decision factors for the induction of recruitment. We have compared, in ant colonies starved for 3 days, food recruitment to large unretrievable cockroaches and to a pile of individually retrievable fruit flies. To suppress any bias due to food preference, we will also compare recruitment to piles of flies covered by a 1 mm mesh wire-netting that allows minors access to the flies but prevents them from taking them away.

Foragers stay 324 ± 98s ($\bar{x} \pm$ S.E., n=17) around cockroaches trying to retrieve them. After several unsuccessful attempts, they return to the nest dragging their abdominal tip over the substrate. Their entrance into the nest induces massive exits of workers, with a mean rate of increase of 14.4 minors /min on the foraging area during the growth phase. On the other hand, foragers stay a much shorter time (132 ± 41s, n=19) near the pile of flies, immediately succeeding in taking one of them between their mandibles. The rate of increase of minors on the foraging area is markedly lower than that observed to cockroaches (2.5 minors /min). Prey retrievability by a single forager thus seems to influence the propensity of foragers to recruit. Indeed, when the pile of flies is covered with a net, the vain efforts of foragers to extract them prolong their stay near the prey (249 ± 62s, n=19) and are followed by a stronger trail-laying behaviour than in the case of uncovered flies. The induced mass recruitment is closer to that to cockroaches, with a rate of increase of 7.1 minors /min. In *Ph. fallax* (1), similar differences in recruitment rates were observed between clumped (a whole hardboiled egg yolk) and dispersed (a finely chopped one) food, although to a lesser extent.

To summarize, for these two *Pheidole* species the prey resistance to transport by ants is a key part of the decision making system. It relies on the length of time the foragers stay at the food source trying to retrieve prey back to the nest. This effort could serve as a "measure" of prey size and weight. Another measurement used to decide whether or not to recruit could be the frequency of contact with prey met successively during a forager's exploration, as suggested in a defensive context for *Oecophylla longinoda* recruitment (2). A third decision rule could be based on an estimation of food persistence more than on size or number, although of course the three measures are correlated. For instance, *Messor rufitarsis* (3) or *Lasius niger* (pers. comm. R. Beckers) begin to form a recruitment trail only after a number of successful foraging trips.

In the experiments described above the foragers are faced with the same kind of food in different circumstances. However, other parameters such as the nature of the food source itself are well known to control recruitment intensity.

We have emphazised the role of individual measurement in the foraging patterns without discussing how the characteristics of a recruitment are able to generate a collective decision. The future challenge is to understand how the colony modulates individual / collective complexity in accordance with its ecological constraints.

1. Itzkowitz, M. and Haley, M. 1983. Insectes Sociaux, 30 (3): 317-322.
2. Hölldobler, B. and Wilson, E.O. 1978. Behav. Ecol. Sociobiol., 3: 19-60.
3. Hahn, M. and Maschwitz, U. 1985. Oecologia, 68: 45-51.

COMPARATIVE LEARNING AND MEMORY OF TWO SYMPATRIC SEED-HARVESTER ANTS

Robert A. Johnson and Steven W. Rissing, Department of Zoology, Arizona State University, Tempe, AZ, 85287-1501,USA.

Messor Pergandei and Pogonomyrmex rugosus are abundant seed-harvester ants throughout much of the Southwest United States. While diets of the two are similar [1], their foraging methods differ: foraging columns of the former change direction almost daily [2], while those of the latter are semi-premanent and may not change directions for months [3]. Therefore, even when co-occurring, colonies of these species experience differing levels of environmental variance as a function of their alternative foraging methods. Because of this, we predicted M. pergandei foragers should learn to recognize a novel seed type faster than P. rugosus foragers. Further, since past information may be less useful to M. pergandei foragers (now foraging in a different part of the environment) and since memory capacity at some point must become limiting for foragers, we also predicted that M. pergandei foragers would lose the ability to recognize a novel seed type faster than P. rugosus foragers. We performed a series of manipulative field experiments to test these predictions.

METHODS and RESULTS

Comparative Learning. We placed a patch of a novel seed type (Kentucky blue grass which does not occur in the environment but is readily harvested by both species in the laboratory and field [4, 5] across the foraging columns of at least 2 M. pergandei and 2 P. rugosus nests at each of 2 sites where the two are sympatric and 1 site where each is allopatric. None of the colonies had experienced this seed previously. We monitored two measures of learning ability: percentage of passing foragers that stopped and handled a seed (% recognition) and percentage of those that handled a seed and subsequently harvested one (% acceptance). Colony-level recognition and acceptance were considered to have reached a maximum on the subsequent "learning curves" generated when percent of the foragers performing a given behavior reached an asymptote; iterative regression procedures were used to determine this asymptote. At both sites, M. pergandei colonies recognized and accepted the novel seed type significantly faster than did P. rugosus colonies.

Comparative Memory. Thirty-two M. pergandei and 22 P. rugosus colonies were provided ad libitum Kentucky blue grass seeds until at least several days after colony-wide recognition and acceptance levels reached an asymptote as described above, after which all unharvested seeds were removed. Each colony was then given a single opportunity to harvest Kentucky blue grass seeds 1 - 100 days after the initial familiarization period. Recognition and acceptance of Kentucky blue grass declined significantly more rapidly against time for M. pergandei than for P. rugosus with the former species reaching the recognition level of naive colonies after 75 days and the latter after >82 days.

DISCUSSION

Learning has classically been studied as "a 'supraspecific' characteristic, (where) the adaptive demands placed on particular species by their environments cannot be relevant to its understanding" [6: p.68]. More recently, however, investigators have begun to view learning as an adaptive response to the environment of a given species with the potential that learning and memory may be more or less developed in species given the environmental variance they experience. Shettleworth [7: p. 1723], for example, has noted that those "studying learning mechanisms have drawn attention to the possibility that these mechanisms are adaptively specialized, and have called for more experimental analysis of naturally occurring examples of learning" while also noting that "the evidence for adaptive specializations of learning is meagre" [7: p. 178].

The ant species we have studied are broadly sympatric and ecologically equivalent, even having similar diets that tend to converge during periods of resource "crunch" [1]. Nonetheless, subtle differences in their foraging behavior, namely the rate at which foraging columns change direction, result in species-specific differences in levels of environmental variance encountered while foraging. These differences appear to have led to related differences in learning and memory abilities of the two species: M. pergandei, the species with more frequent change in foraging direction and subsequent higher levels of environmental variance encountered, learns a novel seed type faster and tends to retain that information for a shorter period of time.

Colony-wide memory of a novel seed type exceeds maximum longevity of foragers captured in the field and held with ad libitum food and water in the laboratory (M. pergandei: 24 days; P. rugosus: 20 days). This suggests that the seed cache in a nest provides an "institutional memory" for the colony transcending succeeding generations of workers as they progress through the normal sequence of age polyethism, first performing tasks inside the nest (especially processing of seeds in the seed cache) and ultimately performing tasks outside the nest, especially foraging.

REFERENCES

1. Rissing, S.W. 1988. Oecologia, 75: 362-366.
2. Rissing, S.W. and J. Wheeler. 1976. Pan-Pacific Entomologist, 52: 63-72.
3. Hölldobler, B. 1976. Behavioral Ecology and Sociobiology, 1: 3-44.
4. Rissing, S.W. 1981. Behavioral Ecology and Sociobiology, 9: 149-152.
5. Rissing, S.W. and G.B. Pollock. 1984. Behavioral Ecology and Sociobiology, 15: 121-126.
6. Johnston, T.D. 1985. In: Issues in the ecological study of learning, T.D. Johnston and A.T. Pietriwicz, Eds., Earlbaum Publishers, Hillsdale, NJ, pp. 1-24.
7. Shettleworth, S. 1984. In: Behavioural ecology: An evolutionary approach, J.R. Krebs and N.B. Davies, Eds., Sinauer Associates, Sunderland, MA, pp. 170-194.

COLLECTIVE EXPLORATION IN THE ANT *PHEIDOLE PALLIDULA*

Detrain Cl.

Senior Research Assistant at the National Fund for Scientific Research. Laboratoire de biologie animale et cellulaire CP.160, Université Libre de Bruxelles, 50 Av.F.D.Roosevelt, 1050 Bruxelles (Belgium)

Collective patterns of exploration to new, unmarked terrains near the nest have been described in *Pheidologeton diversus* (1), *Iridomyrmex humilis* (2) and for the foraging swarms of army ants (3). In the laboratory, we have studied these exploratory collective behaviour in societies of the European polymorphic ant, *Pheidole pallidula*. Their nest is connected by a bridge to a chemically unmarked sand-filled arena (80X80 cm). During 90 min, photographs of this arena are taken every 3 min monitoring the time evolution of exploratory patterns. Experiments are repeated 5 times by changing the sand to re-induce exploration. The first minor workers explore randomly the new area, mainly the part closest to the nest. Recruitment is indicated by the logistical growth of minor explorers on the area. As the exploration increases, a well-pathed trail takes shape linking the exploratory front to the nest. As a clearly defined way to the frontier of the unknown area, the exploratory trail allows a very efficient and fast territorial expansion of the *Ph.pallidula* society. This collective exploration could also be involved in its food searching strategy. During one intense food recruitment to dead mealworms, a short exploratory trail has originated at the food source further away from the nest. In nature, such an exploratory activity in the vicinity of a food source could lead to the discovery of additionnal prey. Chemical similarities/differences between exploratory and food trails remain to be investigated.

1. Moffet,M.W. 1988. Journal of Insect Behavior, 1 (3): 309-331.

2. Deneubourg, J.L., Aron,S., Goss,S. and J.M.Pasteels. In press. Journal of Insect Behavior.

3. Deneubourg, J.L., Goss S., Franks, N. and J.M. Pasteels. 1989. Journal of Insect Behavior, 2: 719-725.

COLONY DISPLACING AS A FORAGING STRATEGY IN FIRE ANTS

Awinash P. Bhatkar, Department of Entomology (TAES), Texas A&M University, College Station, Texas, USA 77843-2475

Many species of the tribe Solenopsidini (Solenopsis invicta, S. richteri, S. geminata, Monomorium minimum, M. pharaonis, etc.) build populous colonies and dominantly exploit their habitat. Their foragers recruit to a proteinaceous food source, eventually carry brood to the resource and develop a satellite colony in proximity. If the resource is persistent the whole colony is displaced, abandoning its original nest site to the satellite site. The extent of this behavior as a foraging strategy in S. invicta was studied by dispersing food baits in 0.1 hectare plots in a rye grass (Lolium perenne) pasture in Central Texas.

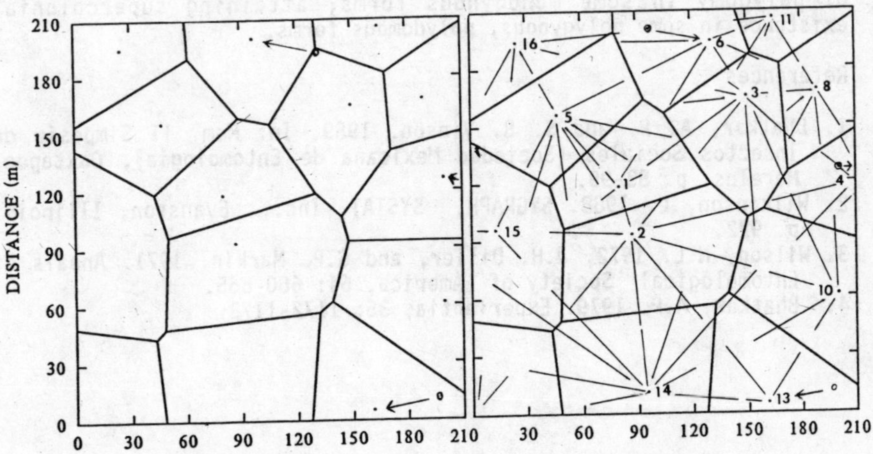

FORAGING IN S. invicta

The original sites of active and inactive mounds were mapped using a surveyor's plain table. About 3,000-5,000 foragers-nest defender ants from each mound (3-5% colony) were marked with distinctive spray paints [1] and recaptured at food palettes (500 mg fish protein, 8.5% lipid) in horizontally placed 20 ml polyethelene scintillation vials placed 25 cm apart. Voronoi tesselation [2] of the spatial nesting pattern was superimposed on the foraging outreach of each colony along the closest internidal distance. The dominantly foraging colonies were ranked (Fig.). Next, the 5 m radius areas around the inactive mounds were provisioned with food for 10 days and the mounds were remapped for displacement.

The colonies in the area were ordinated as 14, 2, 5, 3, 6, 10, 16, 15, 8, 11, 4, 13, according to the foraging distance traversed and total number of foragers at the food sites. Their foragers monopolized the food stations, some adjacent to other mounds and reaching as far as 146 m from the mound. Voronoi tesselation of the spatial nest pattern provides an approximate scaling for ordinating active colonies in time and space. Any ant in the polygon forages closer from its nest than any other during competitive colony packing. The territorial boundaries are not as well defined in monodomous S. invicta colonies as was once suggested [3], and they also may be marred by the occurrence of intercolonial trophallaxis in this species [4]. Rather 'food site monopoly' by the foragers from discrete colonies occurred as a rule. Some colonies stayed resilient while others were dynamic in nesting during the study period. About 18-30% colonies shifted to inactive mounds as a result of the resource richness around them (arrows, Fig.). This behavior suggested foraging-nesting as an economical strategy to exploit resource-rich patches of the habitat. It was further observed that the monodomous, monogynous colonies of S. invicta displace colonies more often than the sympatric polydomous, polygynous form, where colony fissioning and supercolony expansion is common. The adaptive significance of colony displacement may ultimately be the adoption of polydomy in some monogynous forms; attaining supercolonial existence in some polygynous, polydomous forms.

References

1. Bhatkar, A. P. and S. B. Vinson. 1989. In: Mem. II Simposio de Insectos Sociales (Sociedad Mexicana de Entomología), Oaxtepec, Morelos. p. 83-99.
2. Wilkinson, L. 1988. SYGRAPH, SYSTAT, Inc., Evanston, Illinois. p. 922.
3. Wilson, N.L. 1972, J.H. Diller, and G.P. Markin. 1971. Annals of Entomological Society of America, 64: 660-665.
4. Bhatkar, A.P. 1979. Experientia, 35: 1172-1173.

EVOLUTION OF FORAGING DURING PREDATION OF TERMITES BY FOUR AFRICAN PONERINE ANTS

Dejean A. & Corbara B. Laboratoire d'Ethologie, URA CNRS 667, Université Paris XIII, F-93430 Villetaneuse, FRANCE.

Ponerinae are at various levels, predators of termites. The latter have developped antipredatory behaviours which includes constructions, an almost permanent grouping and the action of the soldiers specialized in defence. Termites are quasi-always encountered in groups in such a manner that it is understood that the more an ant species is specialized on termite predation, the more its general hunting organization enables the capture of a large number of prey for a fewer number of displacements.

The hunting workers of *Odontomachus troglodytes* that encounter a group of termites during exploration kill one prey and directly return to the nest. They then undertake a succession of returns between the nest and the prey, the paths being very directional. The arrival of a worker carrying a prey incites a number of congeners to leave the nest.

A *Pachycondyla soror* worker that encounters a group of termites will successively kill a number of prey. There is a change in the parameters of its displacement during exploration and after having killed the first prey, there is an increase in sinuosity and a decrease in speed. The worker then returns with one to three prey between its mandibles and undertakes a series of returns (very direct paths) between the group of prey and the nest. There may follow recruitment of nestmates by the deposit of chemical trail.

The parameters of displacement of workers of *Paltothyreus tarsatus* change when the ground is scattered with pulverised fresh constructions of termites. The presence of a prey is not obligatory. There is an increase in sinuosity of the path, body pivoting and a decrease in speed. A kairomone issued from the termites construction seems to occur. When a worker encounters a group of little termites, after the first capture, the sinuosity of its path increases and the speed decreases. On the encounter of a 2nd, 3rd, etc... prey, the worker seizes them without leaving hold of the former that were captured. It then returns directly to the nest carrying up to 10 termites between its mandibles. It then undertakes a series of returns between the group of prey and the nest. In certain cases, it recruits nestmates by chemical trail.

The *Megaponera foetens* are strict predators of Macrotermitinae termites. The worker caste is polymorphous, the major being foragers. Even though they are about the same size as the *Paltothyreus tarsatus*, the characteristics of their exploratory displacements are very different. The sinuosity is lower, the speed higher. This is in relation with the search of zones where termites forage while *Paltothyreus* look for individuals. Theses zones are protected by constructions that the workers detect thanks to kairomones (Longhurst and Howse 1978). The worker then returns to the nest without prey, leaving a chemical trail. It recruit nestmates which follow the scent trail in columnn. There is attack in group.

Longhurst C. & Howse P. E., 1978. The use of kairomones by *Megaponera foetens* in detection of its termites prey. Animal Behaviour, 26, 1213-1218.

GEOMETRY OF THE FORAGING TRAILS OF *LEPTOGENYS PROCESSIONALIS*-ANALYSIS BASED ON GRAPH THEORY

Ganeshaiah, K.N. Department of Plant Genetics and Breeding, Agril. College, GKVK,
Veena, T. Department of Veterinary Physiology, Veterinary College, Hebbal,
University of Agricultural Sciences
Bangalore 560 065 India

Leptogenys processionalis, a mass foraging predatory ant forages through branched trails. Ganeshaiah and Veena (1990) studied these trails by using Horton's technique originally developed for analysing the branching patterns of streams and rivers. Computing the branching coefficients (which indicate the average number of daughter branches emerging from each parent branch), they showed that the foraging trails are similar to other biological branching systems such as lungs and arteries. In this technique, the order of a branch reflects the latter's functional significance while foraging. However the identity of the branches is not fixed. Further, though the technique describes the growth pattern and provides morphometric analysis of the trails it fails to classify them in to, distinct categories such as 'herringbone' or 'dichotomous' branch types. In this paper we report our analysis of the foraging trails of Leptogenys processionalis using another technique based on graph theory of the trees (Smart, 1978; Werner and Smart, 1973; Fitter, 1986).

A tree can be characterised by three parameters viz., the total number of exterior branches (magnitude, n), the number of links between the base and the farthest branch (diameter, d) and the sum of all the links from all the exterior branches to the base (exterior path length, Pe). It is shown that the slopes, of d and Pe on n, are maximum for herringbone and minimum for dichotomous branching types, others taking the values in between (Werner and Smart, 1973). We traced, ten foraging bouts of 4 nests of Leptogenys processionalis in the open fields in and around the Hebbal Campus of UAS Bangalore. The branches were then labelled following Werner and Smart (1973) and Fitter (1986). From these, we calculated d, n and Pe for each tracing and estimated the regression coefficients of d on n (b1) and that of Pe on n (b2). The b1 was 0.79 (95% confidence interval, 0.50) and did not differ significantly from either random (Ho: b1=0.59) or herringbone (Ho: b1= 1.00) branching patterns. Similarly b2 (1.62; 95% confidence interval, 0.25) also indicated that foraging trails branch randomly or in herringbone pattern.

The branching trees can be further characterised on the basis of the frequency of 4 types of the links viz., EE (the exterior branch that joins another similar one), EI (the exterior branch joining an interior branch), IE (an interior link that joins an exterior branch), and II (an interior link that joins another interior link) (Fitter,1986; Smart,1978). It can be shown that a high frequency of EE and II reflect the dichotomous branching tendency and that of EI and IE reflect the herringbone type. We tested for the significance of the deviations of the observed frequencies from that expected if these four types of links are equally frequent using chi-squared test. One of the foraging trails showed a high frequency of EE and II indicating dichotomous type, the rest showed either equal of more of EI and IE branches. This again shows that the branching pattern of the foraging trails tends to be either random or herringbone type. Pooling the data over all the tracings however showed no significant deviations of the observed frequencies from that expected on the basis of equal occurances.

The dichotomous and herringbone patterns also reflect the space filling and exploratory strategies respectively. Hence we argue that the type of branching pattern followed by the ants is a variable process depending upon the availabilty of food: it pays to resort to the herringbone type in resource poor areas and to dichotomous type in reosource abundant areas.

We further tested whether the branching initaition along the trail is a random process or not by measuring the internode lengths and testing the fit of their frequencies with those expected in a negative exponential distribution. The results indicated that the branching event is generally independent of the length of the trail but is subject to the availability of food along the trail. Since food availabilty perse is random, the branching initiation also occurs randomly.

References

Fitter,A.H. 1986. The topology and geometry of plant root systems: Influence of watering rate on root system topology in Trifolium pratense. Annals of Botany, 58, 91-101.

Ganeshaiah, K.N. and T.Veena. 1990. Topology of the foraging trails of Leptogenys processionalis - Why are they branched ? Behav. Ecol. Soc. Biol.. (Communicated)

Smart, J.S. 1978. The analysis of drainage network composition. Earth Surface Process, 3, 129-170.

Werner,C. and J.S. Smart. 1973. Some new methods of topologic classification of channel networks. Geographical Analysis, 5, 271-295.

FORAGING AND FOOD RECRUITMENT COMMUNICATION IN THE ANT *MYRMICARIA EUMENOIDES* (HYMENOPTERA, MYRMICINAE)

Manfred Kaib and Hubert Dittebrand

Lehrstuhl Tierphysiologie, Universität Bayreuth, Postfach 101251,
D-8580 Bayreuth, Federal Republic of Germany

Myrmicaria eumenoides is one of the most significant ant species of the Ethiopian region. It is reported to feed mainly on social insects with a strong impact on termites. Termite colonies build up high local concentrations of biomass and are rich protein sources. Because of their claustral mode of life termite prey is only temporarily available. In order to exploit successfully termite colonies, predatory ants need fast and effective communication systems recruiting high numbers of nestmates to unprotected gatherings of termites.

In small colonies *M. eumenoides* foragers individually search for food following a visually controlled random foraging scheme. Large colonies (≥30.000 ants), maintain established permanent trunk trails for commuting between the nest and foraging grounds (up to 30m distant). Scout ants use these trails as "highways" and may leave them at any place for individual foraging.

Small and scattered prey insects are captured by scout ants and are individually transported to the nest. However, during combat with larger obstinacy prey the scout ants extrude their blunt and spatulate sting, and deposit topically poison gland secretion onto the antagonist and/or the ground. The chemical signal, which evaporates from that secretion, alerts further ants in the vicinity and attracts them to the sites of combat, where they assemble and take part in overwhelming the prey jointly. The rapid chemical communication during this short-range-recruitment leads in foraging areas from an isotropical ant distribution to a concentration of ants at the site of combat.

Compared with other myrmicine ants the foragers of *M. eumenoides* possess an exceptionable large poison gland reservoir containing up to 0.5 μl of secretion (10% of the total body weight). The poison gland reservoir of newly hatched ants, however, is very small. It enlarges gradually with age and reaches its final size when the ants are ≈6 weeks old. This age-dependent glandular development goes parallel to an age-based division of labour. Newly hatched and young workers perform tasks inside the nest, like care of the queen and the brood. At an age of ≥6 weeks they begin to become foragers, more than 50% of which remain inside the nest as reserve troops and may become active on request.

Poison gland secretion is composed of equal amounts of a high volatile and a low volatile fraction. (+)-limonene, which constitutes >95% of the high volatiles is the only trigger of short-range-recruitment. The ants respond to the limonene signal evaporating from natural poison gland secretion in exactly the same way as to synthetic limonene when the airborne concentrations are identical. This is true for both independent behavioural assays: size of assambly at the scent source as well as running speed of chemically alerted ants. Depending on the doses, both are regulated by the intensity of the limonene signal.

The low volatiles serve as a fixative for limonene and extend the effective period of the chemical signal by modifying its evaporation kinetics. The limonene contents of one ants' secretion (0.2μl) completely evaporates within $\approx$25 min, whereas the corresponding amount of pure limonene would disappear within 10% of this period. Whereas pure limonene would rapidly evaporate with a constant release rate, the evaporation rate of the limonene signal emanating from poison gland secretion declines exponentially to zero. Thus a freshly deposited droplet of secretion will be more active than an older one, which seems to be well adapted for this recruitment communication.

If short-range-recruitment does not attract a sufficient number of ants for fast expoitation of a food source, further ants are recruited from the nest (or trunk-trails) by long-range-recruitment. In this case successful scout ants deposit small droplets of poison gland secretion onto the ground via the sting while returning. In the nest they beat their antennae against those of waiting sister ants, additional to the release of poison gland secretion . Only the combination of the chemical signal with these mechanical signals trigger long-range-recruitment. Scout ants, whose sting apparatus was sealed with wax (no release of a chemical signal), do not recruit nestmates although they perform regular mechanical stimulation by their antennae. Nor is long-range-recruitment initiated, when only odour of poison gland secretion is applied into the nest. However, it can be released, when the antennal beating carried out by gaster-sealed recruiting scout ants is accompanied by an experimental application of vapour of poison gland secretion. By passing on this antennal beating to further nestmates, the few ants initially recruited can carry this mechanical recruitment signal to a large number of individuals. Thus in small laboratory colonies ($\approx$400 ants), for example, one single scout ant recruits within a 2min interval over 60 nestmates, which is 1/3 of the possible foragers.

Ants recruited from the nest (or trunk-trail) reach the site of prey (or combat) following the trail of poison gland secretion, deposited by the returning successful scout ant. This chemical orientation trail remains active for >10 minutes without being reinforced.

In summary we demonstrate: (1) (+)-limonene on its own is the ultimate chemical signal for short-range-recruitment. (2) The low volatile fraction of poison gland secretion elongates the durability of the chemical signal. Due to the exponential decline of the chemical signal, the total number of ants recruited within a short range can be controlled by repeated deposition of poison gland secretion. (3) Long-range-recruitment is only triggered by the simultaneous action of two signals: the mechanical stimulation of nestmates (antennal beating) and the chemical signal (+)-limonene evaporating from the poison gland secretion.

BATCH STRENGTH OF HUNTING GROUPS IN *LEPTOGENYS DIMINUTA* SMITH (FORMICIDAE : PONERINAE)

A.R.V. KUMAR
Depart. of Entomology
University of Agricultural Sciences
G.K.V.K., Bangalore-560 065
India.

Leptogenys diminuta Smith is widely distributed in South India and South-East Asia. This ant generally prefers cool and moist habitats, and is most active during cool hours of morning and evening. It is a group hunting predatory ant.

Data was collected on batch size [BS] of the raidng groups, distance travelled, success of the raid and the number of foragers carrying the prey for 93 recruitments. The BSs were grouped into 20 classes and the data of each class pooled for the analysis.

L. diminuta recruits foragers in batches of 3 to 294 [$\bar{X}$ = 81.98 ± 57; n = 93] of which only a few bring the prey. These foraging batches travel to a distance of 2.22 ± 1.25 mts [n = 91].

The frequency distribution of BS was positively skewed. I tested whether this is optimum with respect to the benefit and cost of harvesting the prey.

BS was positively correlated with the distance travelled [r = 0.4866; n = 20; P< 0.05], per cent raids that were successful [r = 0.588; n = 20; P<0.01] and the mean number of ants that bring back the prey [benefit] [r = 0.627; n = 20; P<0.01].

The cost [product of BS and distance travelled] increases at an exponential rate with BS [$Y = 36.02 e^{0.0156x}$; r = 0.607; n = 20; P<0.01]. However, the benefit to cost ratio decreased monotonically with batch strength [Y = 0.1041 - 0.0172 (ln x); r = -0.473; n = 20; P<0.05] indicating that smaller batches pay more benefit to the unit cost invested in harvesting the prey. Thus the observed positively skewed distribution of BS appears to be an optimum strategy of the species in harvesting the prey.

Further, I also tested whether the observed frequency distribution of BS is a function of the colony size by simulating the BSs based on the size of the colonies studied and the observed percent of workers recruited. The analysis indicated that the BS was independent of the colony size. Thus <u>L</u>, <u>diminuta</u> appears to optimise the BS recruited in its foraging efforts.

DO HONEY BEES NEED TO OPTIMIZE FORAGING FLIGHTS ?

Rainer Krell

ICON, Viale Regina Margherita 239, 00198 Roma, Italy

The foraging behavior of honey bees on natural flower patches has been described by Krell (1990). These observations are used to interpret the energetic efficiencies of the foraging behavior. Selected results are presented in Table 1. Foraging cost is calculated after Beutler's (1936) estimate of 10 mg sugar consumption per hour flight and 0.5 mg sugar consumed per hour during nectar uptake (Weippl, 1928, in Beutler, 1936). Except for the column of estimated efficiency, all calculations are based on the assumption that the foragers extracted all nectar from the flowers. The estimated efficiency is based on extrapolation of experimental evidence that honey bees leave considerable amounts of nectar in flowers depending on their prior foraging and other experiences (Krell, 1990).

It is apparent, that efficiency even at very low reward levels and incomplete extraction of available nectar is fairly high. Small changes in foraging behavior, however, have a large effect on the total energy profit made during a certain period of time. Therefore, the foraging behavior of honey bees, whether unexperienced or not, on flowers is relatively unimportant for the energy efficiency of the individual forager, but of larger consequence to the colony as a whole.

Foraging, particularly in highly variable reward situations, is not always as optimal as possible ih regard to net energy gain. It appears that bees did not change their behavior as much according to individual forager efficiency as that they assumed a foraging pattern, which could be considered scouting or looking for larger floral rewards. High occasional rewards change individual foraging efficiency relatively little, but shorten a foraging trip considerably and thus increase energy gained per time.

In conclusion, beyond a certain efficiency level, inherent to basic behavioral patterns and reward situations, foragers do not have a strong need to optimize for their own energy budget, which might be one of the reasons for the high variability or lack of correlation in measured behavioral responses to reward sizes. Instead, optimization for the benefit of the colony may become more important. Optimization may therefore not be found anymore in flower to flower movements, but in more or less gradual changes of length of foraging trip, recruiting and scouting influenced by the distance to food sources, recruitment for competing food sources, colony food needs, etc. Future foraging studies therefore, need to concentrate more on optimization for energy gain, reproductive capacity and survival of the colony as a whole.

Table 1: Energy efficiency of honey bee foragers on <u>Ilex glabra</u>.

Sugar/fl. (mg)	Foraging cost (mg)	Earnings/ cost ratio	Efficiency (%) *	Efficiency estim. (%)	Time per load (min/30mg sugar)
0.08	4.55	13.17	92.41	78.31	32.52
0.20	4.07	52.54	98.10	96.83	8.58
0.24	4.66	37.34	97.32	96.43	10.63
0.27	7.39	15.53	93.56	91.41	16.77
0.38	6.21	66.09	98.49	97.48	4.45
0.45	4.90	42.59	97.65	96.09	8.84
1.00	4.99	113.75	99.12	98.58	3.20
2.38	5.88	147.49	99.32	98.77	2.09

* net energy gain/total sugar collected.

References

1. Beutler, R. 1936. Uber den Blutzucker der Bienen. Z. vergl. Physiol. 24: 71-115.

2. Krell, R. 1990. Foraging behavior of <u>Apis mellifera ligustica</u> on gallberry (<u>Ilex glabra</u> L.). Behav. Ecol. Sociobiol., submitted.

FROM INDIVIDUAL TO COLONY-LEVEL BEHAVIOR IN HONEY BEES: THE REGULATION OF POLLEN FORAGING

Scott Camazine
Cornell University
Section of Neurobiology and Behavior
Seeley G. Mudd Hall
Ithaca, New York, 14853-2702 USA

INTRODUCTION

Previous studies have clearly demonstrated that pollen is collected in accordance with colony need, and that the colony-level response to a changing need is both precise and rapid [1,2]. This raises the question of what mechanisms individual bees use to make the appropriate behavioral decisions about modulating their foraging rate. This reports examines the activities and behaviors of individual pollen foragers in relation to colony-level patterns of pollen collection. I address the question of what information foragers gather to assess colony pollen requirements and how that information affects their subsequent behavior.

EXPERIMENTAL DESIGN AND RESULTS

Two observation hives were established each with 3 full-depth frames. They were matched in terms of population, food stores and brood. One colony was designated the experimental colony and the other colony served as a control. The upper two frames of each hive contained brood and honey, but essentially no pollen. The contents of the bottom frame was varied according to the experiment. Queen excluder material separated the bottom frame from the upper two frames. In each experiment the colony-level pollen foraging rate was correlated with a number of measures taken of individually identified pollen foragers. Two experiments are reported here.

The first experiment established that colony-level patterns of foraging reflect the magnitude of colony pollen stores and helped distinguish between the following two hypotheses concerning the behavior of individual foragers: 1) Pollen foragers assess colony status directly by monitoring the amount of pollen in the colony. When need is high (colony stores low) foragers collect pollen regardless of available space. 2) Individual pollen foragers assess colony status indirectly by evaluating the amount of available cells in which to deposit their pollen. When there is ample space, bees will forage at a relatively high rate regardless of the amount of pollen already present in the colony. When the colony has few available cells, bees experience difficulty unloading their pollen and thus reduce their rate of foraging regardless of colony need. In this experiment an empty frame was placed in the bottom position of the experimental hive and a frame packed with pollen was placed in the bottom position of the control hive. Over the next 25 days the treatments were periodically reversed as indicated (Figure 1, experimental colony). However as the final treatment, a full frame of honey was placed in the bottom position instead of a full frame of pollen. Thus, this experiment initially co-varied excess pollen with lack of storage space, the situation that typically occurs in nature. However, the final treatment uncouples the two, creating a situation of little pollen but no storage space. The results were as follows: In the first part of the experiment the pollen foraging rate of each colony tracked the changes in colony pollen status. When the bottom frame was empty the pollen foraging rate was high. Conversely when the bottom frame was filled with pollen the foraging rate declined. In the second part of the experiment in which both pollen stores and available space were low the colony continues to track pollen need, foraging intensively for pollen even though there is little space to store the pollen. This result certainly suggests that pollen foragers directly assess the status of colony pollen stores, disregarding storage space availability. However, an analysis of the individual behaviors of the pollen foragers is more revealing. Table 1 (Experiment 1) indicates that when the bottom frame was empty, the bees made fewer inspections to find a cell in which to deposit pollen and more quickly deposited their loads. Under these conditions they were also more likely to perform waggle dances, they performed more dance circuits and they more rapidly returned to the field to forage. Under conditions of ample pollen stores and little storage space, they made more cell inspections and took greater time to deposit. They only slowly returned to the field, if at all, and performed fewer dance circuits. This is the expected result. However, when there were little or no pollen stores but no place to deposit their loads, the individual bees behaved in the same manner as they did when there was excess pollen - foraging slowly, failing to return to forage and dancing poorly. Thus individual foragers respond to space constraints as well as pollen need. Pollen need appears to provide a stimulus for foraging, but in the face of space constraints, foraging is curtailed. This presents a paradox: How is colony-level foraging maintained if individual bees slow down their foraging rate? I return to this question below.

In the second experiment, it remained to determine whether excess pollen stores would depress pollen foraging in the face of ample available storage space. In this experiment, an empty frame was placed in the bottom position of each colony for the first three days to establish the baseline level of pollen foraging. For the next five days half of the bottom frame of the experimental colony was packed with pollen and half of the frame was left empty. The bottom frame of the control colony had no pollen (this half of the frame was covered with a screen) and the other half of the frame was left empty as in the experimental colony. For the following six days the treatments were reversed. In both colonies, the pollen foraging rate was high during the first three days when neither colony had pollen. On day 4, the pollen foraging rate of the experimental colony declined, and continued to do so for the next 4 days. In contrast the foraging rate of the control colony remained high. On day 9 the treatments were reversed, and, as expected, the pollen foraging rates reversed. Excess pollen depresses foraging even when there is ample storage space. A comparison of the individual behaviors of the foragers from the experimental colony during pollen excess and pollen dearth reflected the colony-level pattern (Table 1, Experiment 2) When the colony lacked pollen, made few inspections, unloaded rapidly, returned to forage quickly, frequently performed waggle dances, and danced more circuits than days when there was an excess of pollen, indicating that pollen per se depresses the individual's rate of pollen foraging and recruitment. What was surprising, however, was the finding that the number of cell inspections prior to unloading and the time to unload was different during the two periods despite there being nearly identical amounts of empty space in the colony (half of the bottom frame was empty in both cases). When there was pollen in the bottom frame the bees inspected more cells prior to unloading and took a longer time to deposit. This suggests that the time to deposit not only reflects difficulty unloading but also "motivation" on the part of the bees. During periods of low need, the bees do not forage as enthusiastically as when the need is high. They take their time to unload, and lethargically return to the field to forage.

Both experiments demonstrate the colony-level ability to respond to pollen need. In terms of modulating the foraging behavior of the individual bee, the first experiment showed that the forager takes into account available space and the second experiment showed that foragers assess the colony's pollen stores. Two things remain to be shown: 1) How is the colony-level foraging rate maintained when foragers slow down in the absence of available space?, and 2) How do individual foragers assess pollen stores? With regard to 1), the high pollen foraging rate in the face of pollen need but lack of storage space suggests that new foragers take the place of those that slow down and drop out of the foraging process. With regard to 2), foragers might assess pollen stores by means of odors, their individual "pollen hunger", or by measuring the actual amount of pollen stored in the colony. Experiments are under way to address these questions.

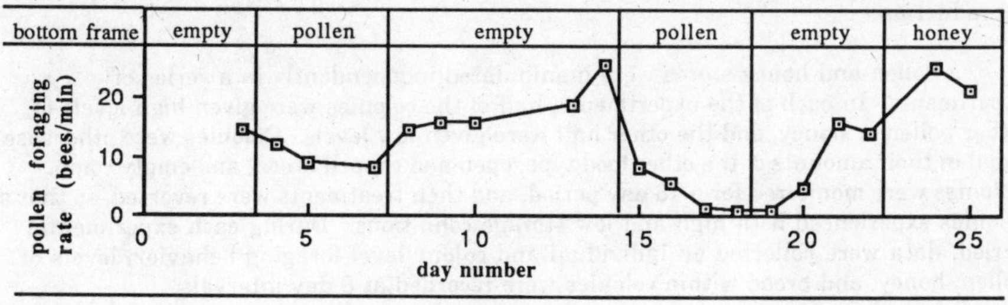

	Experiment 1				Experiment 2	
day of experiment	13-14	16-18	21-22	25	5-6,8	14
bottom frame	empty	pollen	empty	honey	pollen	empty
pollen foraging rate (bees/min)	22.2	2.5	14.8	21.3	7.3	9.3 *
number of inspections	3.6	5.0	4.4	5.1	7.1	2.0
time to deposit (min)	1.8	9.5	3.5	9.9	5.2	2.0
time in nest after deposit (min)	4.8	10.3	7.1	9.0	8.9	5.3
% bees waggle dancing	16.3	6.8	34.2	4.2	7.4	26.3
number of dance circuits/bee	4.4	1.8	11.6	2.4	0.6	7.2
% bees not returning to forage	2.0	35.7	13.2	20.8	7.4	0.0

* afternoon rate (cool weather curtailed morning foraging)

REFERENCES

1. Lindauer, M. 1952. Zeitschrift für Vergleichende Physiologie, 34: 299-345.
2. Free, J. B. 1967. Animal Behaviour, 15: 134-144.

THE RELATIONSHIPS BETWEEN FOOD STORES AND FORAGING EFFORT IN THE HONEY BEE, *APIS MELLIFERA* L.

Jennifer H. Fewell and Mark L. Winston,
Dept Biological Sciences
Simon Fraser University
Burnaby, B.C., Canada

The objective of this study was to determine the effect of variation in food storage levels on individual and colony-level foraging decisions in the honey bee, *Apis mellifera* L. Foraging studies of social insects generally have concentrated on the decisions of individual workers, without examining the relationships between individual foraging strategy and colony-level fitness. However, a detailed understanding of the interactions between individual and colony level behavior are required to place social insect foraging in an evolutionary context, because individuals are the foraging units for the colony, but fitness is measured at the level of the colony as a whole. Interactions between individual foraging strategy and colony state occur at two levels. First, foragers may assess and respond to specific colony attributes. Second, individual foraging responses can affect both colony status and fitness.

Honey bees provide an excellent system for examining relationships between individual foraging strategy and colony state, because both hive contents and individual foragers can be monitored. A number of colony attributes can potentially affect foraging strategy in honey bees [2], including colony demography, colony food stores, and available space within the nest . We examined how levels of pollen and honey storage affect individual foraging behavior, and how individual and colony-level foraging responses, in turn, affect colony attributes important for growth and eventual reproduction.

Pollen and honey stores were manipulated independently in a series of experiments. In each of the experiments, half of the colonies were given high levels of either pollen or honey, and the other half were given low levels. Colonies were otherwise equal in their amounts of the other food type, open and capped brood, and empty comb. Colonies were monitored for a 16 day period, and then treatments were reversed, so that all colonies experienced both high and low storage conditions. During each experimental period, data were collected on individual and colony-level foraging behavior; levels of pollen, honey, and brood within colonies were recorded at 8 day intervals.

The mean amount of pollen stored in colonies before treatments was 2,220 cm^2. High pollen treatment colonies received an average of 4,450 cm^2, while low treatment colonies received approximately 240 cm^2. There was a dramatic foraging response at both the colony and individual levels to the variation in pollen stores. The total foraging rate (bees returning per min) did not differ significantly for the two groups (ANOVA, $F = 0.09$, $P > 0.7$, $N = 11$). However, the proportion of foragers returning with pollen increased significantly, to 36% in colonies with low pollen levels, relative to 26% in colonies with high stores (ANOVA, $F= 12.6$, $P < 0.01$, $N = 11$). This increase in the proportion of the foraging force collecting pollen was due primarily to an increase in recruitment to pollen foraging, rather than to an increase in foraging rates by individuals already collecting pollen. Individual pollen load sizes also increased significantly, from a mean of 14.0 mg for high treatment foragers to 16.8 mg for low treatment foragers (ANOVA $F= 15.17$; $P< 0.001$, $N = 258$). The observed foraging changes occurred both across colonies and within colonies when treatments were reversed. The increased pollen foraging effort by low storage colonies resulted in a 1.6 fold increase in the amount of pollen entering the hives.

The individual and colony-level foraging responses strongly affected colony pollen stores. By day 16, pollen stores in low treatment colonies increased to pretreatment levels. Pollen levels in high treatment colonies also returned to pretreatment levels, so that there was no significant difference between high and low treatment groups at the end of the experimental period (t-test, $T = 1.3$, $P > 0.2$, $N = 11$). Brood stores did not change significantly for low store colonies over the experimental period (t-test, $T = 0.02$, $P > 0.9$, $N = 11$); however, they increased 1.4 fold over this period in high store colonies (t-test, $T= 12.0$, $P < 0.001$).

These results indicate that individual pollen foragers are able to assess stores, and adjust foraging effort accordingly; this adjustment takes place both in the colony and at the flower. Our data also suggest that colonies maintain a homeostatic set point in pollen stores. When pollen storage levels dropped below this level, pollen collection and pollen stores increased. When colonies were given higher pollen levels, the excess stores were converted to increased brood production. Thus, there is a direct interaction between colony pollen levels, foraging effort, and colony reproduction.

The response to differences in honey stores was less dramatic. The number of nectar forager in colonies with low honey stores increased significantly, to a mean of 264 nectar foragers returning in 5 min, relative to 225 nectar foragers per 5 min in colonies with high honey stores (ANOVA, $F= 7.8$, $P < 0.02$, $N = 14$). However, there were no detectable changes in individual foraging rate, or in nectar load size or quality. Changes in colony honey storage were also less dramatic. The amount of honey in high store colonies dropped 16% from a mean of 23,600 cm^2 at the beginning of the experiment, to a mean of 19,700 cm^2 after 16 days (t test, $T = 10.8$, $P < 0.001$). Honey stores in the low treatment group did not change significantly from the initial mean of 5000 cm^2 (t test, $T = 1.3$, $P > 0.2$). These results suggest that individual foragers are less sensitive to changes in honey stores than to changes in pollen stores, when those changes occur independently of variation in other colony attributes. We also failed to establish any direct short-term link between honey stores and brood production; there was no significant difference in brood amounts between the two groups at day 16 (t test, $T = 0.9$, $P > 0.3$, $N = 14$).

Our data suggest that pollen and honey foraging are regulated quite differently from one another. Pollen collection and storage is a tightly regulated system in honeybees; whereas, nectar collection seems to occur relatively independently of actual honey stores until colonies are severely stressed [3]. The difference in regulation may be explained by the difference in use of the two food types. While honey stores are important to colony survival, their significance is more long-term, unless stores are low enough for colonies to approach starvation [4]. In contrast, pollen stores directly and rapidly affect brood production [1,4]. In this study, both colonies and individuals were more responsive to changes in pollen levels than to changes in honey levels, enabling them to correct deficiencies in pollen storage more rapidly.

ACKNOWLEDGEMENTS

This study was supported by NSF Grant No. BSR-8821243 to J.H.F. Jon Harrison made valuable comments on the manuscript.

REFERENCES

1. Allen, M.D., and E.P. Jeffree. 1956. Annals of Applied Biology, 44: 649-656.
2. Houston, A.I., P. Schmid-Hempel and A. Kacelnik. 1988. American Naturalist, 131: 107-114.
3. Seeley, T.D. 1989. Behavioral Ecology and Sociobiology, 22: 229-237.
4. Winston, M.L. 1987. The Biology of the Honey Bee. Harvard University Press, Cambridge, Mass.

EXPLORATION AND SEARCH IN FORAGING HONEY BEES (*APIS MELLIFERA*) AND PAPER WASPS (*VESPULA MACULIFRONS*)

Rudolf Jander
Dept. of Entomology
University of Kansas
Lawrence, KS 66045-2106 USA

Within the foraging behavior of higher Hymenoptera searching and exploring are two readily distinguishable, yet related, activities. By definition, a searching animal moves through, and scans its environment until it finds a pertinent resource - an exploring animal, apparently performs similarly, but is learning constellations of landmarks and other navigation cues, and it terminates this activity after familiarization with such cues. Entomologists customarily refer to exploratory flights as "orientation flights." The search behaviors of *Apis mellifera* and *Vespula maculifrons* similarly comprise two distinct components: Looping for search is a forward flight with highly irregular, laterally alternating loops, and inspecting is a hovering flight with lateral oscillations while being oriented toward some visible object. During prolonged search bees and wasps repeatedly switch between looping and inspecting. Exploration flights are similarly structured in both species: In facing the bee or wasp is oriented toward a visible cue for a resource, they hover with lateral oscillations of increasing amplitude and slowly recede from the focal object; facing typically grades into exploratory looping, which is characterized by forward flying through alternating loops of increasing radius and height above ground (provided the starting point was at ground level). Exploratory looping in *Apis* differs from that in *Vespula* by systematic alternation between smooth, arcuate turns and sharp, counterturning hairpin-loops. Both *Apis m.* and *Vespula m.* exercise searching and exploration near the locations of nests and near various resources. Search typically precedes, and exploration typically follows the use of a relatively permanent resource. For experimental purposes it is convenient to know that the discovery and use of a rich source of food in a novel environment is almost invariably followed by an exploration flight prior to homing. If a conspicuous landmark is presented during such an exploration flight, the bee or wasp will later use it as a navigation cue when returning to the site in question. The mentioned exploration behavior and search behavior, for a series of similarities, can be considered mutually homologous. The similarities in the flight patterns have already been mentioned. Furthermore, both facing and inspecting tend to be oriented away from a source of light, have a negative phototaxis superimposed on them. In a novel environment both searching and exploring can be incited by food, searching when the food is depleted prior to satiation, exploring when satiation precedes depletion. Finally, landmark learning is not restricted to pure exploration; it can be demonstrated experimentally that during searching for a previously experienced rich resource some learning of landmarks takes place, too. Finally, for all the structural and contextual similarities, the exploration flights of *Apis m.* and *Vespula m.* can be considered homologous. This, in turn, implies that the capacity of exploration flights evolved early in the phylogeny of the Aculeata. For insects other than the Aculeata no such exploration flights have ever been noted.

FORAGING PROFILE OF INDIAN HONEYBEE *APIS CERANA* IN RAICHUR, KARNATAKA STATE, INDIA

Shashidhar Viraktamath
Department of Agril. Entomology, University of Agricultural Sciences, Raichur Campus, 584 101, India.

Foraging profile is an index of efficiency of pollination[1]. Hence, the present studies were made under Raichur agro-climatic conditions where beekeeping is being recently introduced for utilising bees for pollination of oil seed crops.

Counts of bees entering three hives in five minutes with and without pollen (assumed as nectar collectors) were made from 0600 to 1800 Hrs at hourly intervals twice in a week during 1987-88. Foraging activity was found throughout the year with wide variations in the levels of activity during different hours of the day and in different months of the year. Major pollen foraging (80 per cent) was observed before noon. However, nectar foraging occurred throughout the day with a major peak during 0600-1100 and a minor peak during 1600-1800 Hrs. The number of pollen and nectar foragers was more during August-February and August-March, respectively. Dearth of pollen and nectar was noticed during May-July.

It is concluded that these honeybees could succesfully be used for pollination of target crops during August-March in this region.

REFERENCE:

1. Erickson, E.H., Whitefoot, L.O., Kissinger, W.A. 1973. Honeybees: a method of delimiting the complete profile of foraging from colonies. Environmental Entomology, 2: 531-535.

OBSERVATIONS ABOUT THE COLLECTING ACTIVITIES AND FLOWER CONSTANCY IN BUMBLE-BEES (HYMENOPTERA)

Peter Schneider

Biologie für Mediziner

In Neuenheimer Feld 504

Fakultät für Biologie

D-6900 HEIDELBERG

The flower-constancy of different bumblebee species was investigated in natural areas with natural flowers.

1. Over five days we observed five species of bumblebees and their visits on 140 flower species in the Botanical garden of the University of Heidelberg. On certain flowers the most frequent species, *B. lapidarius* was never found together with *B. hortorum* or *B. pascuorum.* Nevertheless there are some flowers which are attractive for all investigated bees.

2. Dummies flying or sitting on flowers do not disturb the initiating visit and the flower-constancy.

3. The number of flower-constance bumblebees seems to be limitated and dependent on the number of the flowers in a dense population of the same species.

4. On *Centauria scabiosa* we registrated an individual duration of nectar-collecting bumblebees between 15 minutes and more than 3 hours or from 40 to 743 visits of flowers before returning to the nest.

5. Marked bumblebees show a preference for certain flowers and often they repeat the same flight pathways.

6. Basing on measurements of sucking time, visit duration, number of flight and nectarcontent in different natural flowers the origin and the maintain of flower-constance is discussed.

ENERGY INTAKE AND EXPENDITURE IN CARPENTER BEES *XYLOCOPA FENESTERATA* F. AND *X. PUBESCENS* SPINOLA (HYMENOPTERA: ANTHOPHORIDAE)

D.P.ABROL,
Division of Entomology, Shere-e-Kashmir University of Agricultural Sciences and Technology, Shalimar Campus, Srinagar-191 12-, Jammu and Kashmir, India.

Foraging behaviour of bees is determined largely by their energy requirements and caloric rewards offered bythe flowers. Energetic relationship of Xylocopa fenesterata F. and X.pubescens were studied in relation to Cajanus cajan L., Helianthus annus L., Luffa cylindrica L. and Medicago sativa L. Energy expenditure was calculated by determining the oxygen consumption specific to each temperatures and activity. The oxygen consumption values were converted to energy values considering 1 ml of oxygen consumption is equivalent to 5 cal of energy expenditure. Caloric reward and energy intake were calculated by the formula -

Caloric reward = Nectar volume(ul) x concentration of nectar (%) x 4.0

by considering that 1 mg of sugar irrespective of the type yields 4 cal or 16.74 joules of energy.

Energy intake (G)= u rtf, Where u=standing crop of nectar, r=rate of nectar intake by a bee, Tf total foraging time.

Laboratory measurements on physiologocal energetics compared with field observations on time budget/energy intake to infer daily energy budget balance. The energy balancevaried from one day to another and crop to crop. In general the foraging profitability for both the bee species was in the order C.cajan L.cylindrica H.annus M.sativa. Foraging perferences as determined by the number of bees attracted to each crop also exhibited the similar pattern. C.cajan appears to be competitively superior as forage crop. Evidently polinator preferences is determined largely by energy requirement reward system. The pollinators with high energy requirements may not forage at flowers providing low caloric rewards.

FORAGING ACTIVITY OF THE SUBTERRANEAN TERMITE *MICROTERMES OBESI* HOLMGREN (TERMITIDAE : ISOPTERA)

C.T.Ashok Kumar and G.K.Veeresh
Department of Entomology
UAS, GKVK, Bangalore-560065, India

The foraging activity of Microtermes obesi was studied for 3 years. The number of foraging workers and soldiers varied from 18 to 150 and 1 to 3, respectively in a 3 m row. The ratio was 0.0228. No foragers were seen at the ground level, but the workers made the runways under the soil at a depth of 10-12 cm along the rows of wheat plants.

The workers made their foraging subterranean exit at a 12 cm depth of surrounding surface. The place at which the foraging workers came out of the runways varied from 2 to 15 from the nest according to the studies made on foraging runways of 3 nests. The foraging runways ran in two directions, namely north-south and east-west at different depths (0-30 cm, 30-60 cm, 60-90 cm and 90-120 cm) in 3 nests studied. More number of subterranean runways were recorded in north-south (7.99) than in east-west (4.65) direction. More runways were found in the top soil layer. In both the directions maximum of runways (3.00 to 1.66) were recorded at a depth of 60 cm. The depth at 60 cm had 66.7 per cent of total runways in north-south direction and 64.29 per cent of total runways in east-west direction. About 85-87 per cent of foraging runways were distributed in top 90 cm of soil depth in both directions.

Foraging was brisk during May to August and at a low ebb during December-March and the foraging period varied from season to season. In summer it was during 05.30 to 09.00 h and 18.00 to 21.00 h in the monsoon it was during 06.45 to 10.30 h and 16.00 h to 18.00 h. While in winter, it was during 07.15 to 11.45 h and 15.30 h to 18.00 h.

Symposium 25

BIOGEOGRAPHY, ECOPHYSIOLOGY AND SOCIAL ORGANIZATION OF STINGLESS BEES

Organizers : **DIETER WITTMANN, F.R. Germany**
V.L. IMPERATRIZ FONSECA, Brazil

BIOGEOGRAPHICAL ECOLOGY OF *MELIPONA* (APIDAE: MELIPONINAE)

BIOGEOGRAPHICAL ECOLOGY OF *MELIPONA* (APIDAE: MELIPONINAE)
D. W. Roubik, Smithsonian Tropical Research Institute
Balboa, Panama (mailing address APO Miami 34002-0011, USA)

Melipona consists of more than 40 species and ranges throughout the Neotropics from N. Mexico to N. Argentina. It contains the largest individuals and also the smallest colonies of meliponines. One species, *Melipona variegatipes*, is endemic to the Lesser Antilles, a few (*Melipona yucatanica, Melipona beecheii*, etc.) to Central America and Mexico, and the remainder are primarily S. American. Nonetheless, some occur in both S. and C. America, and the close affinities of other species in these two regions allow interpretations of the biogeography of *Melipona*.

Melipona likely evolved from a member of the *Plebeia*, *Schwarziana* or *Mourella* groups [1,2]. This might have taken place any time in the Tertiary, but geological and biological data [3-5] suggest *Melipona* was present at the end of the Cretaceous and dispersed across a proto-Antillean archipelago. Species that nest in trees in the mangrove forest probably dispersed in floating logs. These are exemplified by *M. compressipes* and *M. favosa*, both S. American. The oldest fossil bees are Cretaceous stingless bees whose change in morphology has been minimal [6]. While there are no known fossil *Melipona*, other fossil stingless bees in Hispañola (which arose in the Miocene) and Mexico demonstrate extinction of genera since an important vicariance event—breakup of the proto-Greater Antilles— which occurred in the early Tertiary. This and other vicariance events since the Cretaceous have isolated previously contiguous populations of *Melipona* to produce some very similar species that do not overlap in distribution in Panama and Costa Rica. They show scant ecological and reproductive divergence, but have formed sister species over the past 70 million years. Distributions of *Melipona* along the Caribbean corroborate the evidence for dispersal between N. and S. America prior to formation of the Panama landbridge 3.2 Ma, amply demonstrated in various organisms [4,7]. *M. yucatanica*, found from Mexico to Costa Rica, is most similar to *M. favosa*, found from N. Argentina to Panama; it recruits very well to resources, unlike *M. favosa*. *Melipona variegatipes* is clearly similar to *M. favosa* and occurs only in the N. Lesser Antilles. The male genitalia of these two species are slightly distinctive, while male genitalia are radically different between *yucatanica* and *favosa*. Their workers show similarities (a wide head, brightly-banded terga) but these might arise from convergence. Diagnostic male genitalia in sibling species is common in *Melipona*, but broad differences are unexpected [5,8,9]. However, *M. beecheii* (Mexico to

Costa Rica) and *M. compressipes* (Amazon to Panama) have similar male genitalia, workers, and nest structure, but are clearly distinct species. The mechanisms preventing the two bees from occupying the same areas are unknown, as are those that maintain *favosa* and *yucatanica* in forests that are structurally similar and barely separated in Costa Rica and Panama. One "subspecies" of *beecheii, fulvipes*, is found in Honduras, Jamaica and the Yucatan peninsula, as well as in Cuba. The former two were joined by the Nicaraguan rise until the Miocene. This suggests natural dispersal via what was probably a "proto-Caribbean arc" [3]. Dispersal to Cuba from Yucatan may have taken place by rafting, but transport by indigenous people cannot be ruled out, since this bee is one of two major honey producing species of Central America. Other C. American *Melipona*, as far N. as Costa Rica and Nicaragua, are *M. fasciata, M. fuliginosa, M. crinita* (*'fuscata'*), and *M. marginata* that came from S. America across the Panama landbridge since the Pliocene.

Why doesn't *Melipona* exist in the subtropics and temperate zone? Highland-adapted species like *Melipona peruviana* are closely related to lowland species and display cold tolerance. Colonies of *Melipona* display ectothermy, as do colonies of *Apis*, but are unable to store food or nest outside of the tropics. Supposedly, the arid southwestern United States is highly favorable for bees because it has several hundred native species, but this is not the case for highly eusocial bees. Food resources for bees are certainly abundant but the absence of suitable nesting sites and nesting materials (such as tree resin) are probably the primary factors that prevent *Melipona* from colonizing the temeprate zone. Failure of highly eusocial species to live outside of the tropics may be the the reason that so many solitary bee species flourish in subtropical America—there are no stingless bees and honeybees to preempt them at food sources.

REFERENCES

1. Moure, J. S. 1951. Ciência e Cultura. 3:40-41.
2. Camargo, J. M. F. 1974. Studia Entomologica 17:433-470.
3. Donnelly, T. W. 1988. In: Zoogeography of Caribbean Insects, J. K. Liebherr, Ed., Cornell University Press, Ithaca, pp.15-37.
4. Briggs, J. C. 1987. Biogeography and Plate Tectonics. Elsevier, Amsterdam.
5. Camargo, J. M. F., J. S. Moure and D. W. Roubik. 1988. Pan-Pacific Entomologist, 64:147-157.
6. Michener, C. D. and D. A. Grimaldi. 1988. Proceedings of the National Academy of Science, USA, 85:6424-6426.
7. Simpson, B. B. and J. L Neff. 1985. In: The Great American Biotic Interchange, F. G. Stelhi and S. D. Webb, Eds., Plenum, New York, pp. 427-452.
8. Moure, J. S. and W. E. Kerr. 1950. Dusenia 1:105-131.
9. Nates, G. and D. W. Roubik. 1990. Journal of the Kansas Entomological Society, 63:124-127.

AN ANALYSIS OF APID BEE RICHNESS IN SUMATRA

Tamiji Inoue (Lab. of Ent., Fac. of Agric., Kyoto Univ.)

INTRODUCTION

Bee fauna in Tropical Southeast Asia is rather poor compared with the Neotropics and Tropical Africa [1] although this region has a rich angiosperm flora [2]. Another important characteristics of bee fauna is that honey bees are diversified at the species level only in this region. For understanding of relationships between the rich angiosperms and the rather poor bees in this region, much information of bees, plants and their interactions should be accumulated. For this purpose, Indonesian collaborators and I studied bee-plant interactions, distribution patterns of bees in various habitat types, and nest site selection of bees in Sumatra for 1980-87.

METHODS

For 300 days, we collected insects on flowers (15 identified species and others) and honey-water baits and recorded bee nests at 107 localities in central Sumatra. This area belongs to the tropical rain forest region with stable rainfalls throughout the year [3]. The distribution pattern of each insect species was analyzed based on two statistics: the degree of commonness as expressed by the percent of localities where at least one individual was found out of the total localities, and the relative abundance as expressed by the average number of individuals collected per one-hour sampling [4,5].

RESULTS AND DISCUSSION

<u>Flower visitors</u> A total of 3385 individuals were collected of 271 species in 57 families of Hymenoptera, Diptera and Coleoptera. Abundant (in number of individuals) families were Apidae (74 %), Anthophoridae (7 %), Halictidae (6 %), Vespidae (4 %) and Megachilidae (1 %). Also in terms of number of species Apidae was abundant, occupying all of the top 10 species with ≥100 individuals. Fifteen flower species were clustered by the frequency distribution of insect visitors. Four types of flowers were distinguished [5]: (1) nectar flowers (typical example = *Stachytarpheta indica*) available only for 'forceful' bee (Anthophoridae), (2) nectar flowers (*Nephelium lappaceum*) available for shorter-tongued and less-forceful bees

(stingless bees and honey bees), (3) pollen flowers (*Mimosa pudica*) available for both large and small bees and (4) pollen flowers (*Cocos nucifera*) available only for small bees.

Distribution of apid bees Twenty-four species of stingless bees, 3 of honey bees and 2 of bumblebees were recorded. Bumblebees *(Bombus rufipes* and *B. senex)* were found only in highlands (≥1400 m above sea level). One stingless bee *(Trigona lieftincki)* was collected only at elevations of 1200-1400 m. The remaining 26 species of stingless bees and honey bees lived at least in lowland primary forests. Both the number of species and the relative abundance decreased in secondary forests and disturbed areas or at higher elevations. Cluster analysis separated bees to 4 groups based on their abundance and distribution patterns [4]. Cluster 1 (12 species) was confined to lowland (elevations ≤ 500 m) primary forests where their density was low. Cluster 2 (5) was also confined to primary forests but was also distributed at elevations of > 500 m. Cluster 3 (6 including *Apis cerana, A. dorsata* and *T. laeviceps*) was most abundant in both low- and highland primary forests and was also found in secondary forests and disturbed areas. Cluster 4 (4) was moderately abundant in lowland primary forests but more abundant in secondary and disturbed areas.

Nest site selection of apid bees We found 252 nests of apid bees. Nest sites were specialized for many bees [4]. *T. moorei* nested exclusively in active *Crematogaster* ant nests. *T. collina* nested only in spaces between tree roots and the ground. *T. canifrons* and *T. thoracica* used only huge cavities in tree trunks. *A. cerana* [6], *T. itama* and 5 stingless bee species nested usually in cavities in tree trunks or branches but occasionally in man-made cavities which resembled natural substrates that they utilized. Average nest volumes of species nesting in tree cavities differed according to colony size; the two extremes were 0.3 *l* for *T. fuscobalteata* and 330 *l* for *T. canifrons*. Abundance and distribution patterns of species found at least in lowland primary forests were explained to some extent by the nest site preference of each taxon [4]. *T. canifrons* and *T. thoracica* were confined to primary forests because huge tree cavities were found only there. *T. fuscobalteata, T. minangkabau* and *T. laeviceps* were abundant even in disturbed areas, because they could use man-made, small cavities.

REFERENCES

[1] Michener, C. D., 1979. *Ann. Missouri Bot. Gard.*, 66: 277-347.

[2] Whitmore, T. C., 1984. *Tropical rain forests of the Far East.* 2nd ed., Clarendon Press, Oxford, p. 282.

[3] Inoue, T. and K. Nakamura, [4] Salmah, S., T. Inoue and S. F. Sakagami, [5] Inoue, T., S. Salmah, S. F. Sakagami, Se. Yamane and M. Kato, [6] Inoue, T., Adri and S. Salmah, all 1990. In: *Natural history of social wasps and bees in equatorial Sumatra,* Sakagami, S. F., R. Ohgushi and D. W. Roubik, Eds., Hokkaido Univ. Press, Sapporo. in press.

PHYLOGENY AND BIOGEOGRAPHY OF THE MELIPONINAE

Charles D. Michener
Snow Entomological Museum, Snow Hall
University of Kansas
Lawrence, Kansas 66045, U.S.A.

The stingless honey bees (Meliponinae) appear to be the lowermost branch of the apid cladogram. Their pantropical distribution suggests antiquity and the oldest known fossil bee is a late Cretaceous meliponine from New Jersey.

Meliponine vagility is low. New nests are established by workers carrying construction materials and food to a new site for weeks before a young queen goes there and a new colony is established. Absconding is virtually unknown. There is thus no possibility of a few individuals being carried to a distant site and establishing themselves there. Results includ (1) species or subspecies separated by Amazonian rivers, (2) absence of stingless bees on Caribbean islands (except for Melipona, probably carried by humans from continents), and (3) limited intercontinental distributions. Across-water dispersal occurred, however, in the Indoaustralian region. Probably nests in wood on floating islands of vegetation account for such ranges.

Figure 1 is a preliminary cladogram of meliponine genera. Below A the workers have setose third valvulae that converge and are not flattened; above A these valvulae have few or no setae but are covered with minute hairs and usually diverge and are flattened. Outgroups show the former characters to be plesiomorphic; the latter, synapomorphic. Thus the cluster of genera below A appears to be ancestral to those above A.

Extreme convergence of external characters is frequent in Meliponinae. Trigona and Dactylurina appear similar; as do Plebeia, Plebeina, and Austroplebeia; etc. Sting rudiments of workers, male genitalia, etc., reveal more reliable relationships.

Some hypotheses indicated by the cladogram are supported by distribution, as shown by large numbers on the figure. There are exceptions; Trigonisca has Old World affinities, and Hypotrigona, in spite of unique features, may have Neotropical affinities.

No genus occurs in both Africa and South America. Thus current faunas must have arisen after the late Cretaceous separation of these continents. Given Trigona's relatives in the Neotropics, it seems likely that Trigona dispersed through what is now the Holarctic region from America to southern Asia. This possibility is supported by the late Cretaceous Trigona from New Jersey. While there is nothing in this fossil to show that it is not a Trigona, one can see its external features only; in Recent meliponines external features do not always indicate relationships accurately.

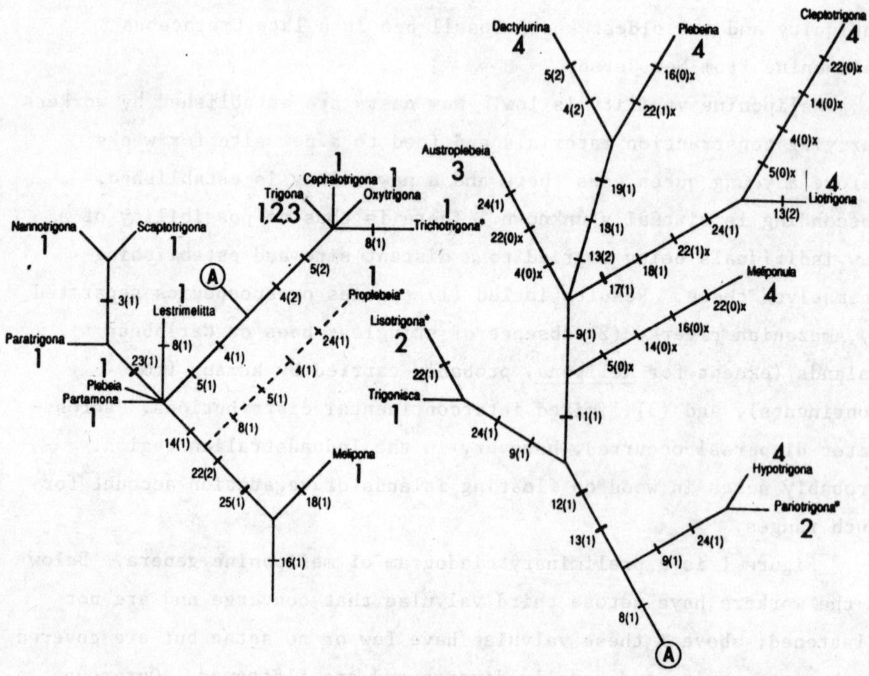

Fig. 1. Preliminary cladogram of meliponine genera. Cross bars indicate synapomorphies; strong bars represent strong characters. Characters are listed by Michener (in prep.). Genera marked by asterisks are unknown in the male. Proplebeia is a fossil (Oligocene, Dominican Republic) added by hand. Present distributions: 1-Neotropics; 2-Southern Asia, Indonesia; 3-Australia, New Guinea; 4-Africa. x indicates a reversal.

DEFENSES AGAINST ANTS IN NINE SPECIES OF STINGLESS BEES

L.K. Johnson, Department of Biology, Princeton University
Princeton, NJ 08544, USA.

S. Appanah, FRIM, Kepong, Selangor, 52109 Kuala Lumpur, Malaysia

Bees, their brood, and their stores constitute potential resources for foraging insects. Ants are a particular threat because of their ability to recruit. We studied the defenses against ants of nine species of stingless bees, genus Trigona, nesting on the grounds of the Forest Research Institute of Malaysia in Kepong.

Weaver ants, Oecophylla sp., hunted for five weeks around the nest entrances of Trigona melanoleuca and Trigona thoracica, but took as prey only the latter bees. We therefore undertook to compare the relative palatability of the nine bee species to weaver ants by pinning ten live individuals from each of eleven colonies in active ant trails. For each pinned bee we counted the number of ants that touched but subsequently rejected the bee before two cumulative ants had attempted predation by pulling and tugging at the test bee. There were significant differences in palatability among the bee species. Trigona terminata was highly acceptable, T. thoracica was taken quickly, and T. melanoleuca was slower to be taken than any other species (Table 1).

To test the hypothesis that T. melanoleuca is protected by a compound on its cuticular surface, we compared the palatability of bee-sized balls of white bread with that of similar balls rubbed on the body surface of fresh T. melanoleuca. Control bread balls pinned onto an Oecophylla trail were seized by two ants after significantly fewer rejections than melanoleuca-rubbed bread balls (3.2 ± 1.0 vs. 16.3 ± 4.3 rejections, $p < 0.02$, two-tailed, Mann-Whitney U test).

All species except T. fuscobalteata and T. collina were observed to attack living ants placed on the nest entrance or held, skewered, in front of the entrance. Ants were attacked with varying degrees of rapidity and ferocity.

TABLE 1. Number of ants rejecting test bee before seizure by a second ant. Nine bee species are compared, including two colonies each of T. terminata, the most acceptable species, and T. melanoleuca the least acceptable.

Species of Trigona test bee	Number of rejections, mean ± S. E.
terminata colony 1	0.4 ± 0.2
terminata colony 2	0.6 ± 0.2
fuscobalteata	4.2 ± 1.0
apicalis	5.6 ± 1.4
thoracica	5.6 ± 1.4
carnifrons	7.6 ± 4.8
fuscibasis	12.0 ± 5.3
laeviceps	21.0 ± 5.7
collina	34.5 ± 12.9
melanoleuca colony 1	70.1 ± 34.2
melanoleuca colony 2	73.0 ± 22.0

Associated behaviors and techniques included buzzing, release of pheromone, biting, dismemberment, bodily removal in the air or on the substrate, and gumming with gobbets of resin.

Features of the species-specific nest entrances were also useful in defense. The entrances of T. apicalis and T. laeviceps had rough, crevice-ridden surfaces that caught ant feet. The entrances of all species were sticky with varying amounts of resin, in every case at least ringing the aperture. The gummiest nest entrances, those of T. thoracica and T. terminata, were clearly effective against ants. Usually, more than one Oecophylla was needed to capture a big thoracica bee, and the patch of resin around the entrance served to slow the rate of arrival of ants assisting a would-be capturer. Heavy resin use is facultative in T. terminata. In response to a raid by arboreal ants, one nest we studied extended the length of its entrance tube 11 cm in 12 hr., studding it distally with ant-entrapping resin balls. As ants applied dirt to the sticky zone, the bees built the tube out further, applying resin frantically, to an eventual distance of 65 cm.

The hypothesis of Seeley and Akratanakul (1982) that resin is a passive defense favored by smaller, weaker bees was not supported for these nine Trigona species.

REFERENCES

1. Seeley, T.D. and P. Akratanakul. 1982. Colony defense strategies of the honeybees in Thailand. Ecol. Monogr. 52:43-63.

OCCURRENCE OF TEGUMENTARY GLANDS IN STINGLESS BEES (HYMENOPTERA, APIDAE, MELIPONINAE)

CRUZ-LANDIM, C. and MOTA, M.H.V.B..Departamento de Biologia and Centro de Estudos de Insetos Sociais, Instituto de Biociências de Rio Claro , 13.500, Rio Claro, SP, Brazil.

The presence of tegumentary glands in bees is known since some last century reports, but extensive comparative studies about variations in their occurrence in different species, in the female castes, or in males, were not done. However these glands must be important to the insect, and particularly to the bees biology since they are of generalized occurrence. Several reports indicate that they produce pheromones responsible, among other taskes, by the queen dominance, and workers orientation during the foraging. Tegumentary glands in bees are present in all tergites and in a few sternites, being more scattered in queens than in workers. In order to have a panel of the variations in the occurrence of these glands in meliponines, workers from all the following species were studied: *Aparatrigona impunctata, Celetrigona longicornis, Duckeola ghilianii, Friesella schrottkyi, Frieseomelitta (silvestrii , varia), Geotrigona mombuca, Lestrimelitta limão, Leurotrigona muelleri, Melipona (bicolor, compressipes, marginata, quadrifasciata, quinquefasciata, rufiventris, seminigra), Mourella caerulea, Nannotrigona(melanocera, minuta, testaceicornis), Oxytrigona tataira, Parapartamona(tungurahuana, zonata), Paratrigona(subnuda, sp), Partamona(cupira, pseudomusarum, testacea), Plebeia(droryana, remota, minima), Ptilotrigona lurida, Scaptotrigona (bipunctata, xanthotricha, depilis, nigrohirta, tubiba), Scaura (latitarsis, tenuis), Schwarziana quadripunctata, Schwarziana flavornata, Tetragona(clavipes, dorsalis, geottei, essequiboensis), Tetragonisca angustula, Trigona* sp, *Trigona crassipes, Trigona (dalattorreana, hypogea, recursa, pallens, branneri, truculenta, fulviventris, spinipes, williana)*, and from some of them were also studied queens and males. Two types of tegumentary glands are found in bees unicellular glands, formed by clusters of round cells, provided indivi -

dually of excretory canaliculi and epithelial glands, formed by secretory epidermal cells. Most of these are tergal glands, i.e., differentiations from the tergite epidermis. Only queens from the species *Trigona crassipes, T. spinipes, Melipona quadrifasciata* and *Scaptotrigona depilis* presented sternal glands from the epithelial type. The tergal unicellular glands appear in workers, as well as, in queens and males with the following distribution: all the studied queens had these glands in three or more tergites; only in the workers from the species *Aparatrigona impunctata, Paratrigona (subnuda,* sp), *Partamona cupira , Plebeia (droryana, minima, remota), Tetragonisca angustula* e *Trigonisca* sp., were the glands absent, the workers of the other species had glands in one, two or more tergites; only the males from the species of *Melipona, Trigona crassipes* and *Lestrimelitta limão* present glands in one or two tergites. The epithelial glands are present only in the females: queens and workers. In the workers they occur from III to VII tergites in determined phases of their life cycle, and are known, as wax glands. In *Scaptotrigona* workers epithelial glands appear in III tergite during all life long therefore, without functional homology with the wax glands. In the queens these glands occur from III to VII tergites, are more developed than in the workers, and their function is unknown.

The studied species were grouped according to the features of the glands distribution, in six groups. The groups maintain some relationship with the species philogenetic position, but the charecteristics of theglands are more related to the species biology. The developmental degree of the glands vary very much along the bees life, even in the same class of individuals. This development may be related to the activities exerced in the precise moment of the life as, for instance, the queen attractiveness to the workers. The developmental degree, also, varies between the castes: the glands from the VIII tergite of queens are more developed than the corresponding glands of workers, while those ones from the II tergite are more developed in workers. The glands size is also different from species to species. For instance, the II tergite glands of *Scaura* and theVIII tergite glands of *Celetrigona longicornis, Friesella shcottkyi, Melipona (rufiventris, seminigra), Mourela caerulea, Partamona (pseudomusarum, testacea), Schwarzula flavornata* appear only as relicts.The present study showed at the same time the generalize occurrence of tegumentary glands in the meliponine abdomen, and a great diversity in the way the glands are presented.

INTRA- AND INTERSPECIFIC CHEMICAL COMMUNICATION DURING PILLAGES OF ROBBER BEES (*LESTRIMELITTA LIMAO*, APIDAE, MELIPONINAE)

Radtke, R., Wittmann, D., Lübke, G.*, Francke, W.*
Zool. Institute, Univ. Tübingen, Auf der Morgenstelle 28, 74 Tübingen, FRG; *Institute of Organic Chemistry and Biochemistry, Martin-Luther-King-Platz 6, 2 Hamburg 13, FRG.

INTRODUCTION

In the Neotropics five species of robber bees (genus Lestrimelitta) are known [1,2,3]. The distribution of four of these cleptobiotic bee species is restricted to Central and northern South America. Only L. limao is widely dispersed and occurs from Mexico to southern Brazil. A broad range of host species with manifold adaptations to the pillaging behavior of the cleptoparasites is assumed. Workers of L. limao do not forage on flowers but rob, in groups of some hundred individuals, food supplies and construction materials from other stingless bees and honeybees. Raids are initiated by a recruiting scout which has localized a host nest. Bees of attacked colonies either exhibit strong defensive behavior [4,5,6] or retreat into the nest interior [7,8]. We studied some of the chemical cues by which raids are initiated and organized and defense or retreat reactions in workers of victim colonies are elicited.

MATERIALS AND METHODS

Volatile compounds from heads of L. limao workers were identified by gas chromatography and mass spectroscopy. Behavioral reactions of robber bees and their victims during recruitment, nest defense or retreat were characterized by frame-by-frame-evaluation of video recordings. In the ensuing bioassay we presented synthetical volatiles on wax discs to the bees in the nest and compared their reactions to those found under natural conditions.

RESULTS

Recruitment: Recruiting scouts rapidly run around in the nest, beat their wings and take up head contacts with nest mates. Within less than a minute, running velocity of worker bees is increased by 60 % (a), the number of wing beating bees around the recruiting scout reaches 100 % (b) and the number of head contacts between bees within a reference area rises from 50 % to 100 % of possible contacts (c). Jostled bees start beating wings and take up body contacts with further nest mates, thus excitation increases within the colony, until the raiding party leaves the nest. Bioassay: The typical reactions (a-c) towards a recruiting scout were released by 6-methyl-5-hepten-2-one and citral. However, the bees did not leave the nest. We assume that the mass recruitment in robber bees is an interplay of chemical and other stimuli.

Guidance To The Victim Nest: Scouts of the robber bees do not land to lay odor trails between the victim and the home nests. Apparently they guide the raiding party in a cloud of volatiles released during flight. Evidence comes from characteristic movements of the group which repeatedly stops in mid-air to subsequently move on again. We interpret these stops as instances when the scout leaves its leading position at the front of the raiding party and flies back to guide those bees back into the group which have lost contact with it.

Defense Of The Nest In *Tetragonisca angustula*: The nest of these bees is defended by groups of guard bees hovering in front of the nest entrance. During attacks of the robber bees the guards recruit hundreds of additional nest mates which fly out, grasp robber bees in mid-air and force them to the ground. Bioassay: Alarming of guard bees and attacks on robber bees are released by 6-methyl-5-hepten-2-one and citral. These reactions are also released when benzaldehyde, a component of the pheromonal bouquet of *T. angustula*, is presented at the nest entrance. Volatiles of the robber bees, therefore, function as kairomone, which, in this interspecific chemical communication, is "translated" into the species specific alarm pheromone.

Alarm And Retreat Into The Nest In *Plebeia emerina*: When an intruder is perceived in the nest, alarm is given until it is caught. Subsequently few bees immobilize the intruder with resin. Alarming bees beat their wings but do not raise their abdomen. Being attacked by robber bees, workers of *Plebeia emerina* do not exhibit strong aggressive reactions towards them. They stay in a secure distance from individual robber bees, raise their abdomens and beat their wings. Soon after, workers retreat and hide in galleries and between brood combs. Bioassay: In the nest of *P. emerina* none of the volatiles of *L. limao* elicited an attack on the wax disc. The bees form rings of about 4 cm diameter around discs with 6-methyl-5-hepten-2-one and citral, raise their abdomen and beat their wings as observed during raids. Total retreat of the bees could not be released by the chemical compounds tested. Here the two volatiles of *L. limao* block aggressive reactions in *P. emerina* workers and, therefore, function as allomones.

DISCUSSION

Wing beating, jostling, raising the abdomen, increasing running velocity and attacking or non-attacking an intruder, are typical reactions of robber bees towards a recruiting scout or of victim bees towards attacking *L. limao*. These behavioral units could be reproducibly elicited by chemical stimuli in the bioassay. Citral and 6-methyl-5-hepten-2-one seem to play a central role in intra- and interspecific communication. Depending on the attacked bee species, these compounds either function as kairomon or allomon. In *Tetragonisca angustula* strong defense of the nest could be elicited by their own alarm pheromone or by volatiles of the robber bees. However, during simulated recruitment in robber bee colonies and faked attacks on nests of *P. emerina*, chemical stimuli alone were not sufficient to trigger bees to leave the nest or to hide in its interior.

REFERENCES

1. Schwarz, H.F. 1948. Bull. Am. Mus. Nat. Hist. 900:1-546.
2. Camargo, J.M.F., J.S. Moure and D.W. Roubik. 1988. Pan-Pac. Entomol. 6:147-157.
3. Sakagami, S.F., D. Roubik and R. Zucchi. 1990. Ecol. Monogr. In press.
4. Wittmann, D. 1985. Behav. Ecol. Sociobiol. 16:111-114.
5. Johnson, L.K. 1987. Biotropica 19:188-189.
6. Wittmann, D., R. Radtke, J. Zeil, G. Lübke and W. Francke. 1990. J. Chem. Ecol. 16:631-641.
7. Moure, J.S., P. Nogueira-Neto and W.E. Kerr. 1956. Proc. Int. Congr. Entomol. 2:481-493.
8. Nogueira-Neto, P. 1970. In: Essays in Memory of T.C. Schneirla:416-434.

ULTRASTRUCTURAL STUDIES OF TERGAL EPITHELIAL GLANDS FROM WORKERS AND QUEENS OF *SCAPTOTRIGONA POSTICA* (HYMENOPTERA, APIDAE, MELIPONINAE)

CRUZ-LANDIM, C.. Departamento de Biologia and Centro de Estudos de Insetos Sociais, Instituto de Biociências de Rio Claro, Universidade Estadual Paulista, 13.500, Rio Claro, SP, Brazil.

All insects epidermal cells have secretory activity linked to the cuticle production and maintenance. However some epidermal cells, with specific location, morphology and characteristc cycle have other secretory functions.

Scaptotrigona postica has epithelial glands in the tergites of adult workers and queens. In workers the glands are present from III to VII tergites and are known as wax glands. In queens the epithelial glands appear only in III tergite. The workers wax glands have a characteristic cycle of functioning, being active between 5 and 15 days of adult life, phase in which the cells are tall. Before and after this time the epithelium of the glandular region are formed of squamous cells as the remainder epidermis. However the epithelial gland of III tergite do not present this cycle. It is already developed in very young workers and remain in this stage even after the withdraw of the glands of the IV to VII tergites. This behavior is very similar to that of queens epithelial glands. In this caste a tall epithelium in III tergite is present since late pupae and remain in this stage during all adult life. The ultrastructural studies of these glands showed that all them have morphological characteristics of lipid synthetising glands . The cells have very developed smooth endoplasmic reticulum and mitochondria oriented along the cell height.However the secretion present in the cells is different. The glands of III tergite of workers and queens store large deposits of secretion within the cells.The deposits are electrondense have irregular form, and present inner regions of vacuolation. The workers glands from IV to VII tergite show enlarged intercellular spaces, bordered by short microvilli. The secretions appear as electrondense granules that are eventually delivered in these spaces.Therefore the III tergite gland of workers is alike to that of the queen.

Symposium 26

HARMFUL EFFECTS OF SOCIAL INSECTS

Organizer : **DAVID E. BIGNELL, United Kingdom**

THE ENVIRONMENTAL IMPACT OF TERMITE DAMAGE TO CROPS, TREES, RANGELAND AND RURAL BUILDINGS IN AFRICA

Dr T G Wood
Head, Entomology Department
Overseas Development Natural Resources Institute
Central Avenue, Chatham Maritime
Chatham, Kent ME4 4TB

ABSTRACT

Termites damage a wide range of crops, forestry trees, wooden rural buildings and, locally rangeland in Africa. The environmental effects of termite damage arises from:

1. The use of persistent organochlorine (cyclodiene) insecticides for termite control. No other insecticides have the required persistence in the field to protect crops and forestry trees. The principal compounds are aldrin, dieldrin, chlordane and heptachlor. The potential human health and environmental hazards of these compounds are well known. These hazards can be minimised by observing recommended handling, application and storage procedures, although the required standards of pesticide management are rarely achieved. There is considerable pressure from National and International donors and from some Governments to ban, reduce or minimise the use of these compounds. It is likely that they will soon be unavailable. Some of the alternative compounds are also hazardous to human health and are more expensive.

2. The wood/straw thatch buildings, characteristic of rural Africa are susceptible to termite attack. Regular rebuilding leads to excessive clearing of native woodlands and forest. Estimates indicate a life expectancy of 5-6 years for thatched roof houses and 8-9 years for corrugated iron roof houses with wooden supports, in some areas of high termite activity.

3. Quick growing trees, particularly _Eucalyptus_, are used throughout Africa for fuelwood plantations and to stabilise already eroded land; they are increasingly used for rural dwellings as natural woodland becomes scarce. Heavy losses of newly transplanted seedlings due to termite damage leads to continued accelerated erosion, with consequent reduced area of farmland and silting and flooding of lowlands. In some areas losses of seedling trees to termites after transplanting

in the field approach 100%.

4. On overgrazed rangeland the foraging of Macrotermes, Odontotermes and Pseudacanthotermes in tropical Africa and of Hodotermes in southern Africa removes the remaining dry grass and grass litter leaving the soil bare and susceptible to accelerated erosion. Termite control is often attempted by mound poisoning or by baiting in spite of the fact that overgrazing is the principal problem. Farmers and Government officials are often reluctant to accept overgrazing as the major cause of denudation: only rarely are termites the primary cause.

5. The interactions of the various facets of termite damage and the human response to damage are complex. The net effects contribute to deforestation and, in combination with overgrazing, denudation and accelerated erosion; misuse of insecticides leads to human health and environmental hazards.

TERMITE DAMAGE AND CONTROL IN CHINA

Dai Zi-rong and Li Gui-xiang
Guangdong Entomological Institute
105 Xingang Road West , Guangzhou,
510260 China

I. Termite damage in China

Recorded history of termite damage in China can be traced back to 2000 years ago. The chinese people started their control against termites since the very early time and have gained a great deal of knowledge and practical experience. According to information gained in South China, termites can damage various products originally made of plant, animal, mineral and synthetic materials. Summarily they are:

a. Timber, bamboo, cotton, linen, rubber and their products;

b. Leather, silk, shell, wool, horn, bone and their products;

c. Some plastics, such as soft PVC, PE, polyester and their products;

d. Synthetic fibres and their products;

e. Sheets of soft metal, such as lead cable sheathing and aluminium foil;

f. Some minerals, such as asphalt, lime, asbestos etc..

Termites cause many problems in China. Perhaps the most important are in buildings and other timber in service, in plastic underground cables, in river banks and reservoir dams, in crops and trees.

1. Damage to buildings

It is difficult to estimate the cost of termite damage to buildings in China. Approximately 40 species of termites do most of the damage to buildings, and among them the most important species are:

Coptotermes formosanus Shiraki ,

Reticulitermes speratus (Kolbe), R. flaviceps (Oshima) ,

R. chinensis Snyder , R. fukienensis Light,

Cryptotermes domesticus (Haviland), C. declivis Tsai et Chen,

Odontotermes formosanus (Shiraki),

Macrotermes barneyi Light.

2. Damage to plastic underground cables

Termites have caused problems with power and telecommunication plastic underground cables in the vast area south to the Yangtze River. The most damaging species is Coptotermes formosanus Shiraki.

3. Damage to river banks and reservoir dams

In river banks and reservoir dams, Odontotermes formosanus (Shiraki) and Macrotermes barneyi Light can build a large number of combs and primary nests. When flood level is high and surpasses the normal level during the rainfall season, the water flow can cause the full of the nest, the seepage, the more serious case can occur, such as the surge of pipes, landslip and the collapse of dam.

4. Damage to agricultural and plantation crops

Sugarcane, peanut, taro, sweet potato and other crops may be damaged by Odontotermes formosanus (Shiraki), Macrotermes barneyi Light and Coptotermes formosanus Shiraki, especially sugarcane in young stage and ripe stage planted in hills is most seriously damaged. In the tree and fruit planletions, young tree of mulberry, tea, eucalyptus, pine, rubber, litchi etc. often are damaged to various extend by termites.

II. Termite control in China

In recent 40 years, a great deal of termite studies and control work has been done and remarkable results have been obtained. Four national conferences were held separately in Shashi of Hubei province, Shaoxing of Zhejiang province, Guangzhou of Guangdong province and Fuzhou of Fujian province in 1960, 1972, 1975 and 1980. The most popular measures currently used to control termites are:

a). Wood preservation.

b). Spraying slow action poisonous powder to the nest and runways of Coptotermes formosanus Shiraki and Reticulitermes spp..

c). Poisonous bait with attractive materials or pheromone to control Coptotermes formosanus Shiraki, Odontotermes formosanus Shiraki, Reticulitermes spp., Macrotermes barneyi Light etc..

d). Fumigation to control Cryptotermes spp..

e). Soil treatment to prevent subterranean termites.

In addition, the use of bio-materials to control termites has been explored. We hope that, in the near future, we can find out a safe method to control termites instead of toxic chemicals.

A TERMITE PEST OF COCOA IN GRENADA

JOHANNA P.E.C. DARLINGTON
ZOOLOGY DEPARTMENT, UNIVERSITY OF THE WEST INDIES,
ST. AUGUSTINE, TRINIDAD.

Grenada is a small Caribbean Island that used to produce a lot of high quality cocoa. About thirty years ago the cocoa estates fell into neglect. No new trees were planted and the older trees were not pruned or weeded. Shade trees and shelter belts were cut down for timber and fuel and not replaced. Cocoa production fell to a fraction of its former level.

In 1983 when attempts began to rehabilitate the cocoa estates, one of the main pests was found to be a termite, Neotermes sp. (Kalotermitidae). In some plantations 80% of trees are infested, and many trees have multiple infestations. Each termite colony is associated with fungal infection which spreads in the heartwood causing a soft, discoloured pocket of rotten wood. This shows on the surface as a dark, moist patch of bark exuding brownish fluid. The combination of termite boring and fungal rot weakens the trunk or limb, which becomes liable to break in the wind. The loss of wind-breaks has aggravated the problem.

A survey in 1989 indicated that nearly all infestations were initiated by primary reproductives settling after the nuptial flight. Secondary infestation from fallen dead or damaged wood did not seem to be important. The infestation levels are so high because the trees are old and neglected. Unpruned dead limbs and untreated limb scars provide easy access to the heartwood, and termites also settle where bark has been damaged by bark beetles, or by cracking in direct sunlight, or by wind shear (again aggravated by the removal of wind-breaks and shade trees).

Old trees will continue to bear if they are carefully pruned and all infected wood removed, but reinfestation readily occurs when the termite alates fly. Preventing exposure to alate settlement does not seem to be a viable option, as the area of neglected cocoa is still far greater than the managed area, and the same termite also infests fruit trees. Shelter and shade should be restored as soon as possible. Ultimately the old plantations will have to be cleared and replanted with new trees. Young trees should be much less susceptible to termite infestation than old trees.

A project is planned to monitor the effectiveness of different pruning regimes in rehabilitating old cocoa trees, and to study the behaviour and success rate of settling termite alates.

TERMITES INJURIOUS TO PLANTATION FORESTRY IN INDIA

M.L.Thakur,
Forest Research Institute,
P.O. New Forest,
Dehra Dun 248006 (U.P.)
INDIA.

INTRODUCTION: Termites constitute the principal silvicultural pests in the tropics, where they cause economic losses in forest nurseries and during the establishment of young out-planted saplings or cuttings in re-forested areas. Though well documented and reliable estimates, in most cases, are lacking, but they can be substantial particularly the exotics, which are susceptible to the native fauna.

TERMITES AS PESTS OF PLANTATION FORESTRY: All forestry crops are prone to termite attack, particularly during the seedling stages in forest nurseries and young plantations, though the incidence and intensity of attack vary from species to species and from one locality to another. Species responsible for damage in plantation forestry, with a few exceptions, belong almost exclusively to family Macrotermitidae and the most wide spread damage is caused by some species of genera Macrotermes (M. estherae) and Odontotermes (O. distans, O.feae, O.indicus, O.microdentatus, O.obesus and O.wallonensis). Localised attack by few other species (Odontotermes horni, O.redemanni, Microtermes incertoides and M. mycophagus) is also common in some areas [1]. The economically important wood-destroying genus Coptotermes (C.heimi) may occasionally attack well grown and mature trees in plantations [2].

Eucalyptus, the most favoured and widely planted species in plantation forestry in India, as elsewhere, suffers severe damage from several species of subterranean and mound building termites in some parts of Andhra Pradesh Maharashtra, Tamil Nadu and Uttar Pradesh [1]. Similarly, clones of some species of poplars (e.g.,P. deltoides, P. eumericana, etc.) have been attacked by Coptotermes heimi, Heterotermes indicola, O.distans and O.obesus [2]. Young seedlings of Acacia albeda, A.tortilis also suffer injuries from Odontotermes microdentatus, O.obesus and Microtermes obesi in forest nurseries in some parts of Haryana and Uttar Pradesh.

INCIDENCE AND NATURE OF DAMAGE: The termite attack in plantation forestry can occur at three stages, viz.,(i) to seedlings or cuttings in forest nurseries, (ii) to young plants (1-3 years) in plantations and (iii) to mature trees in established plantations. The attack in seedlings and young outplanted saplings generally takes place in the upper 20 cm of soil layer. The termites feed on the tap root as well as on rootlets, which results in tapering and complete severance of root system, leading ultimately to the death of the plant. The mortality is greater during the first year (4-6 months in the nursery and 4-6 after planting), when even 100% mortality can occur, especially in harsh areas [3], though the damage during the second year and the subsequent years is not uncommon phenomenon. The ecological characteristics of the land prior to the establishment of a forestry crop and the extent to which it has been subjected to physical disturbances, largely influence the intensity of termite attack. Generally, it is far greater on land, which has been previously under agricultural land use management, rather than on virgin land.

The termite attack in plantation forestry, in the opinion of most of the investigators, may be (i) <u>Primary Causal Factor</u> for mortality of healthy and vigorously growing seedlings in forest nurseries (4-6 months) and in young plantations, (ii) <u>Secondary Causal Factor</u> for mortality in seedlings primarily debilitated and killed by other predisposing factors, notably drought and fungal pathogens and (iii) <u>Complemenntary Causal Factor</u> for mortality in seedlings by various agencies, including termites, where the actual cause is obscure but which may be the result of combined actions of more than one factors.

REFERENCES:

1. Thakur, M.L. and Sen-Sarma, P.K. 1982. Current status of termites as pests of forest nurseries and plantations in India. <u>J. Indian Acad. Wood Sci.</u>, **11** (2)(1980):7-15.

2. Thakur, M.L. 1978. The problem of termite damage in popalrs in the Bhabar-Terai region of Uttar Pradesh in India. <u>Indian J. For.</u>,**1**(3): 217-222.

3. Wardell, B.A. 1987. Control of termites in nurseries and young plantations in Africa; Established practices and alternative course of action. <u>Commonw. For. Rev.</u>, **66** (1): 77-89.

DAMAGES CAUSED BY THE RECENT INFESTATION OF THE SUGAR CANE FIELDS BY THE FUNGUS-GROWING TERMITE *PSEUDACANTHOTERMES MILITARIS*

MORA Ph., ROULAND C., DIBANGOU V., RENOUX J.

Laboratoire de Biologie des populations, Université Paris XII, 94010-CRETEIL cedex. FRANCE

In agroindustrial ecosystems, the reduction of the ecotypes diversity implies for different animal or vegetal species a competition for the energy spare sources. This competition very often leads to the proliferation of one species which adapts better to the new biotop. The sugar cane plantations of the SUCO in Congo are a very good example of this phenomenon.

This plantation is twenty years old and no important modification in the cultural technics have been made during this time. In the last 3 years, we observed a proliferation of fungus-growing termites from the genus *Pseudacanthotermes spiniger*. *P. spiniger* is a termite which has symbiotic relationship with a fungus, *Termitomyces sp*. The fungus grows on structures (fungus comb) built by termites only from pieces of sugar cane leaves. Termites feed at the older fungus comb which is degraded by *Termitomyces sp*. The presence of the nest, principally hypogeous, was revealed by an epigeous mound just over the middle of the undergrounded nest. These mounds were 20 to 120 cm high. The circumference could reach 2 meters. Very often, these mounds were surrounded with several little other ones. It is very important to notice that the mounds are built only one time in a year, after the first rains of the long rainy period (April-May), just before swarmings. These termites mounds cause great damagges to the machines and their number in some fields stopped all mecanic harvest of the sugar cane.

- Quantitative evolution of the nests from 1987 to 1989

The Figure 1 shows that **31%** of the plots presented more than 20 mounds by Ha, rate which pratically prevents the mecanic harvest of the sugar cane. If, in 1987 and 1988, the infestation rates were rarely upper than 35 mounds for one Ha, in 1989 this rate was reached in **12 %** of the plots.

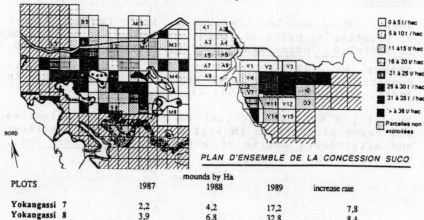

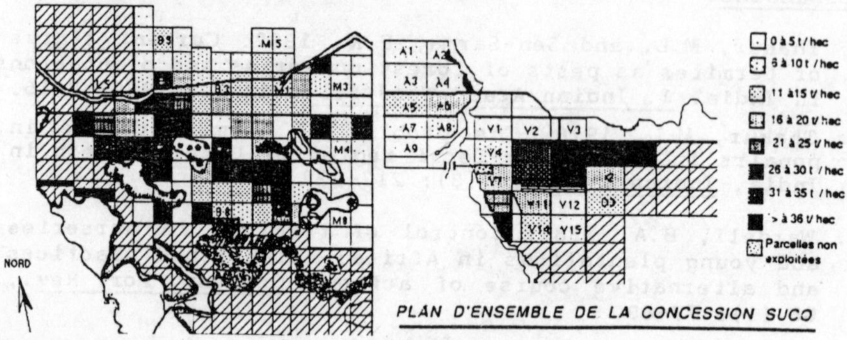

PLOTS	mounds by Ha 1987	1988	1989	increase rate
Yokangassi 7	2,2	4,2	17,2	7,8
Yokangassi 8	3,9	6,8	32,8	8,4
Yokangassi 9	15,3	60,8	93,7	6,1
Yokangassi 11	1,8	3,9	21,5	11,9
Dakar 2	0,6	1,15	8,5	14,2
Dakar 3	1,5	2,3	14,9	9,95
Loudima 4/1	5,1	9,8	38,6	7,6
Loudima 4/2	1,9	3,1	28,6	15
Loudima 4/6	4,2	6,58	39	9,29
Moutela 1/5	0	0,55	1,6	3,2

Table 1 : Evolution of the infestation from 1987 to 1989

The increase rate (1987 density/1989 density) varied in a great extent from a plot to an other but it appeared that the increase was generally more important for the plots less infested (table 1).

- Causes of the pullulation

In some plots, we have noticed a very important infestation when some other places in proximity were not so infested. In the peripheric savanna, some other termites species as *Trinervitermes rhodesiensis*, *Cubitermes sp.*, *Odontotermes sp.* and *Pseudacanthotermes militaris* were present but we never could find *Pseudacanthotermes spiniger*. The sugar cane fields seems to be the favourite biotop for this species. To explain that, we have tried to make some correlations between the rate of infestation of the different plots and the cultural technics, sugar cane harvested alive or burnt (r*=0,02), mecanic or manual harvest (r=0,10); the yield (r = 0) or the place of the plot in the plantation (r=0,62).

It appeared that the yield of the plots have no influence on the infestation rate, but the harvest on green sugar cane and the mecanic harvest seem to slightly increase it. On the other hand, there is a clear correlation between the infestation of the plots and their localization : the more infested plots were the peripheric plots of the plantations just near the deserted plots.

Competition between *P. spiniger* and *P. militaris*

In the fields coexisted two species of *Pseudacanthotermes* (*militaris* and *spiniger*) but actually *P. spiniger* became to pullulate whereas the population of *P. militaris* decreased.

A reason of this pullulation could be a specific diet of the workers of *Pseudacanthotermes spiniger*. Then we have studied the digestive enzymatic set of this species in comparison to that of *P. militaris*. The determination of the osidases have been realized on the termite workers digestive tract and on the *mycotêtes* of the symbiotic fungi according to the method previously described (Rouland *et al.*, 86). In both cases, in the fungi we have only found enzymatic activities present in the digestive tract of the termite workers confirming the importance of "acquirred enzymes" in the fungus-growing termites nutrition (Martin, 83; Rouland *et al.*, 88a, b).

The results obtained (Fig. 2) showed that there are few differences between the enzymatic set of the two species. The very closed diets of these two species showed that they were highly competitives when they are in the same biotop. Then, the capability of *P. spiniger* to degrade plant material was not an efficient reason to explain its pullulation in the sugar cane fields.

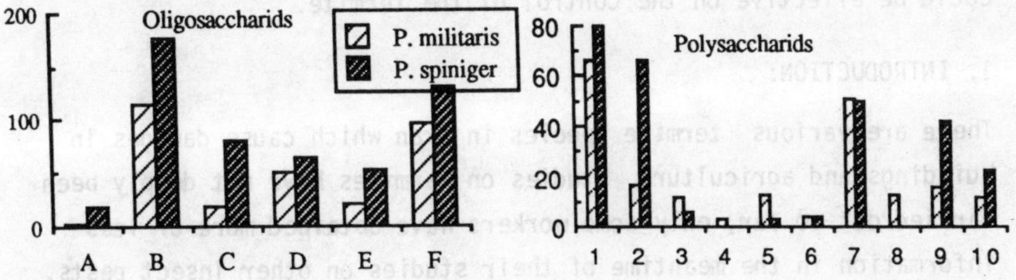

Fig. 2 : Histogrammes giving osidasic activities of *P. militaris* and *P. spiniger* on various oligosaccharids (A-lactose, B-cellobiose, C-maltose, D-saccharose, E-laminaribiose, F-gentiobiose) and polysaccharids (1-starch, 2-laminarin, 3-lichenan, 4-galactomannan, 5-arabinogalactan, 6-xylan, 7-pullulan, 8-CMC, 9-avicellulose).

On the other hand, it appeared that the nest of *P. militaris* was more vulnerable than the *P. spiniger* one because the fungus comb chambers were not so deaper : then, the passage of machins scratched the fungus comb of *militaris* and destroyed a great part of the nest and their only food source.

In conclusion, three principal reasons of this infestation could be proposed :

- The abandonment of several peripheric sugar cane plots have involved a migration of the termites to the cultivated plots;
- The mecanic harvest, harmful to the nests of *P. militaris*, have favoured the installation of the *P. spiniger* because of the disparition of the nutritive competition.
- The presence, after the sugar cane harvest, of a great quantity of sugar cane leaves on the soil give to termites a surabondant food.

References :

MARTIN M.M. 1983.*Comp. Biochem. Physiol.*, **75A**, 313-324.

ROULAND C., CHARARAS C., RENOUX J. 1986. *C. R. Acad. Paris*, **302**, 341-345.

ROULAND C., CIVAS A., RENOUX J., PETEK F. 1988a. *Comp. Biochem. Physiol.*, **31B**, 449-458.

ROULAND C., CIVAS A., RENOUX J., PETEK F. 1988b. *Comp. Biochem. Physiol.*, **31B**, 459-465.

* r = coeficient of linear regression.

ECOLOGICAL STUDIES ON *MICROCEROTERMES* SP. IN TABASS OASIS

HOSSEIN HOOSHMAND & MOHAMMAD BAGHER SHAHROKHI
Plant Pests & Diseases Research Lab. of Mashad, Mashad 91375 IRAN.

ABSTRACT-Microcerotermes sp. has been observedattacking structural commodities and different living plants in agricultural station of Tabass since 1986. Some investigations have been carried out which could be effective on the control of the termite.

1. INTRODUCTION:

There are various termite species in Iran which cause damages in buildings and agriculture. Studies on termites have not deeply been carried out so far, only some workers have obtained more or less information in the meantime of their studies on other insect pests. The outbreak of Microcerotermes sp. in agricultural station of Tabass and severe continuous complaints of local authorities made the plant pests and Diseases Research Lab. of Mashad present a research design to study on it.

2. MATERIAL AND METHODS:

To study the ecology of this termite an experiment in different environmental conditions and different hosts was carried out and the densities of the termite were determined. To accomplish the study the leaf bases of palm trees in sufficiently irrigated and insufficiently irrigated plots were observed. The experiment was most carried out in spring and summer. In each treatment 4 palm trees and from each tree 3 leaf bases were examined and the numbers of termite individuals were counted which is presented in table 1 .

Table 1: The number of termite individuals on palm trees in experimental plots

Replication / Treatment	1	2	3	4	Total	Average
Sufficiently irrigated plots	50	-	40	110	200	50
Insufficiently irrigated plots	2200	2750	2245	5300	13495	3373

CONCLUSION:

Microcerotermes sp. can not have any activities in sufficiently irrigated plots and can say that such conditions are not suitable for its survival, therefore in accordance with this experiment, sufficient and regular irrigation of plots is the best way to control the termite.

References :

1-Chaudhry, M.Ismail ,1972 , Termites of Pakistan Identity, Distribution and Ecological Relationships..

2- Harris ,W.V. 1971 ,Termites their rocognition and control second edition.

3-Krishna ,Kumar and weesner ,F.M.1969 Biology of termites.

EFFECT OF TERMITE FORAGING ON SOIL FERTILITY

Kalidas, P and G.K.Veeresh

1. Scientist, N.R.C. for Citrus , Nagpur-440 006.
2. Department of Entomology,UAS, GKVK, Bangalore-65.

Termites are said to accelerate incorporation of humus in to the soil by carrying sub soil to the surface and organic matter to the underground. In South India much of the farm yard manure applied prior to sowing and crop the residues are eaten away by termites. The present study was an attempt to quantify the loss, if any due to termite foraging under rainfed cultivation.

Material and Methods:

The study was carried out in plots 5 x 5 m,barricaded alround to a depth of 75 cms with stone slab to prevent termite attack in protected plots and with no berricade in plots allowed for foraging. There were four treatments including grass, leaflitter, soft wood and dung in measured quantity, applied twice a year, in two sets, one protected and another allowed to forage by termites,replicated four time, with a check to compare. Soil samples, taken twice a year, were analysed for available carbon, nitrogen, phosphorus and potash and the total biomass measured once a year.

Results and Discussion:

Available nitrogen and organic carbon were found high in plots protected compared to termite foraged, but statistically significant. There was increase in the available potassium in the termite foraged plots, may be deu to termite sheathing which contain more potash compared to surrounding soil (Kalidas, 1966). In case of phosphorus no consistent results were observed between protected and foraged plots.

The protected plots recorded significantly higher biomass production, compared to unprotected plots (Fig.1). Higher organic carbon and available nitrogen in protected plots was mainly responsible for higher production of biomass.

References:

1. Kalidas, P., 1986. Studies on Termites (Odontotermes spp.) with special references to their role in the fertility of soil Ph.D. thesis, UAS, Bangalore. p.123.

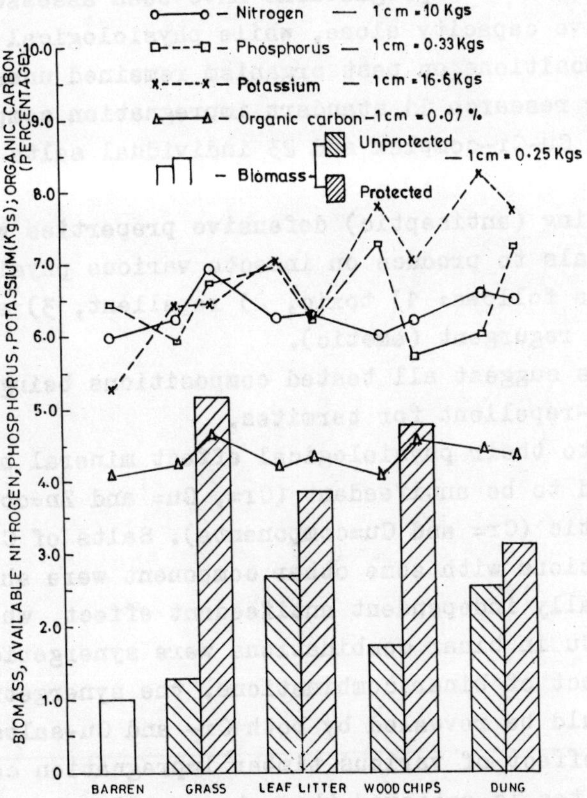

Fig.1. BIOMASS (IN KGS./PLOT) ORGANIC CARBON (PER CENT), AVAILABLE PHOSPHORUS, POTASSIUM AND NITROGEN (IN KGS./HA.), 1983-1985.

THE APPLICATION OF SOME WATER SOLUBLE SALTS AS TIMBER CONSERVANTS AND THEIR PHYSIOLOGICAL EFFECTS ON TERMITES

N.V. Beljaeva
Department of entomology
Biological faculty
Moscow State University

Chemical defanse of timber involves a use of water soluble conservants, containing Cr, Cu, Zn etc. Until now the effectiveness of impregnations have been assessed by their defensive capacity alone, while physiological impact of these compositions on pest organism remeined unexplored

During our research 11 standart impregnation compositions including Cu-Cr-complex and 23 individual salts were tested.

When studying (antiseptic) defensive prepertles we suggested chemicals to produce on insects various physiological effects as follows: 1) toxic, 2) repellent, 3) antifeedant, and 4) regurgent (emetic).

The results suggest all tested compositions being nontoxic and non-repellent for termites.

According to their physiological effect mineral antiseptics proved to be antifeedant (Cr=, Cu= and Zn=components) or emetic (Cr= and Cu=components). Salts of Cr in binar combinations with some other component were shown to produce generally independent antifeedant effect, whereas the salts of Cu in binar combinations were synergetic. As to emetic effect of binar combinations, the synergetic properties could be revealed by both Cr- and Cu-salts.

Thus, the effect of various timber impregnation components on termites is achieved through organoleptic or antibiotic properties. Organoleptic components, even if prevent timber damage, are nontoxic for insect and man, and are not environment pollutants; that is because the use of such a chemicals is generally preferable than of antibiotic ones.

EFFECT OF JUVENILE HORMONE ANALOGUES ON *ODONTOTERMES GUPTAI* (ISOPTERA : TERMITIDAE) UNDER LABORATORY CONDITIONS

R.V. VARMA
Division of Entomology, Kerala Forest Research Institute, Peechi
India 680 653

Introduction

The adverse effects of juvenile hormone analogues (JHAs) on termites include superflous pre-soldier/soldier production, formation of intercastes, defective moulting, feeding inhibition and toxicity. These effects have been demonstrated for a number of termite species (1-6). Information on these aspects on Indian termites is meagre and in the present study the effect of two JHAs was evaluated against the subterranean termite, Odontotermes guptai.

Materials and Methods

The test termite, O. guptai is an economically important species with diverse feeding habits. Termites for the experiment were collected from 6 month-old cultures of O. guptai kept in glass jars (10 cm x 8 cm) and maintained at 25 ± 1°C. Two JHAs-Hydroprene and Methoprene (Gift from Dr. GB Stall, Zoecon Corporation, Palo Alto, California) were tested. The following quantities of the 2 JHAs-3.0 μg, 2.0 μg, 1.5 μg, 1.0 μg and 0.5 μg in 1 ml of actone were applied to Whatman No. 1 filter paper (9 cm x 9 cm). The solvent was removed by evaporation, moistened with distilled water and kept in petri-dishes (9 cm x 9 cm). The control filter papers were treated with 1 ml of acetone alone.

Active termites (2-3 instars) were separated out from the stock cultures and 15 termites each were introduced into the petri dishes. There were 4 replicates. The total duration of the experiment was fixed as 2 weeks because previous experience showed that O. guptai will survive on filter paper only to a maximum of 15 days. All the experimental and control group of insects were maintained in identical conditions and daily observations taken on rate of mortality, morphological changes etc.

Results and Discussion

All the higher concentrations of the 2 JHAs tested, resulted in total mortality of the test termites within a week (Table 1). Most deaths occurred during 2-3 days after the start of the experiment. However, in the lowest dose tested mortality occurred while moulting. Formation of pre-soldiers or intercastes were not observed in any of the concentrations tested. In control, about 65-75% of termites were active even after a week. The feeding activity was very less in all the

Table 1. Effect of Hydroprene and Methoprene on O. guptai

Treatment	No. of termites dead out of 60 and % mortality (in brackets) at the end of 1 week at specified dosage					
	0	0.5 μg	1.0 μg	1.5 μg	2.0 μg	3.0 μg
Hydroprene	14(23.3)	26(43.3)	60(100)	60(100)	60(100)	60(100)
Methoprene	20(33.8)	44(73.3)	60(100)	60(100)	60(100)	60(100)

treated groups as evidenced by less number of nibblings or feeding marks on the filter paper pads compared to the untreated filter paper. The deaths noticed in groups treated with higher doses of both JHAs may be due to the toxic effect of the compounds. Another reason could be lack of feeding activity. It is suggested that in Reticulitermes flavipes, at lower concentrations of methoprene, mortality occurred due to starvation from defaunation (2). Here mortality noticed at the low concentration was due to moulting failures.

It is generally admitted that the feeding and behavioural patterns of the test termite together with the caste composition of the test groups and test conditions may influence the effects of JHAs on termites. It is possible that filter paper pads may not be a suitable feeding substrate for O. guptai. It is also indicated that the biological activity of the JHA treated filter paper may not persist for long under laboratory conditions (3). It seems that further refinement in the testing procedures together with basic information on the biology and behaviour of the test termite would be required to arrive at meaningful conclusions.

References

1. Howard, R.W. and Haverty, M.I. 1978. Sociobiology, 3(2): 73-77.
2. Haverty, M.I. and Howard, R.W. 1979. Annals of Entomological Society of America, 72: 503-508.
3. Hrdý, I. Křečk, J. and Zusková, Z. 1979. Vestnik ceskoslovenske spolecnosti Zoologicke, 43(4): 260-269.
4. Lüscher, M. 1976. In: Phase and caste determination of Insects: Endocrime Aspects, M. Lüscher, Ed. Pergamon Press, Oxford, pp. 91-103.
5. Okot-Kotber, B.M. 1980. Physiological Entomology, 5: 407-416.
6. Su, N-Y..Tamashiro, M. and Haverty, M.I. 1985. Journal of Economic Entomology, 78: 1259-1263.

THE ROLE OF THE BIG-HEADED ANT IN MEALYBUG WILT OF PINEAPPLE

G. C. Jahn
Department of Entomology
University of Hawaii
Honolulu, HI 96826

In most areas of the world where pineapples are grown mealybug wilt disease is a major source of crop loss. Wilt is associated principally with the mealybugs Dysmicoccus neobrevipes Beardsley and Dysmicoccus brevipes (Cockerell) in Hawaii [1].

Mealybugs are associated with BHA (big-headed ants), Pheidole megacephala (Fabr.), in Hawaiian pineapple fields. Carter [2] proposed that BHA may promote mealybug survival by attacking natural enemies, or by consuming Pseudococcid honeydew that mealybugs would otherwise drown in. Mealybug wilt has been controlled mainly by eliminating BHA from pineapple fields. This paper reports upon recent studies to determine the role of BHA in the occurrence of mealybug wilt.

MATERIALS AND METHODS

Using a combination of Mirex and Amdro, BHA were eliminated from half of four 1.4 ha. blocks in an abandoned pineapple field on Maui on 16 June 1989. Thus, there were 4 ant-free and 4 ant-infested plots. Plots were surveyed for BHA, D. neobrevipes, and predators on 9 June, 26 July, 14 Aug., 29 Aug., 26 Oct. 1989, and 24 Jan. 1990.

Mealybug infestation levels in ant-free and ant-infested plots were compared by t tests. Data on the presence of predators in ant-infested and ant-free areas, combined across time, were compared by a one-tailed Fisher's exact test.

RESULTS AND DISCUSSION

Prior to treatment applications, all plots had mealybug infestations on 100% of pineapple fruit surveyed. Ants were present throughout the blocks and very numerous. From 6 to 20 weeks after treatment, ant-free plots had significantly fewer mealybug-infested fruits than plots with ants ($P = 0.001$). From 8 to 20 weeks after treatment mealybugs were not found in ant-free plots indicating that ant presence was necessary for mealybug survival. As the mealybug population declined in ant-free plots there was no evidence that mealybugs were drowning in their own honeydew; nor were any parasitized mealybugs found.

Plots with BHA present had significantly lower predator levels than ant-free plots ($P = 0.0005$), suggesting that BHA suppress predator populations which in turn allows mealybugs to survive. Unpublished related experiments in the laboratory and field support these conclusions.

REFERENCES

1. Beardsley, J.W. 1959. Proceedings of the Hawaiian Entomological Society, 17: 29-37.
2. Carter, W. 1967. Insects and related pests of pineapple in Hawaii. A manual for fieldmen. Pineapple Research Institute (Restricted publication).

THE SPREAD OF THE SOCIAL WASP *VESPULA GERMANICA* IN AUSTRALIA

Michael W.J. Crosland

Department of Entomology, University of California, Davis, CA 95616, U.S.A.

Though partly beneficial due to its predation on other insect pests, the introduced wasp Vespula (=Paravespula) germanica is certainly unwanted by the Australian public. Its painful sting (which is particularly serious for allergic people) and disruption of outdoor activities have resulted in its pest classification. The problem is exacerbated by the high percentage of wasp nests in buildings and the occurrence of large perennial nests.

Vespula germanica first became established in Australia following its accidental introduction into Tasmania in 1959, when two nests were discovered at Hobart (Fig. 1). The wasp most probably came from New Zealand, where it had become abundant following its accidental introduction there in 1944 from Europe. By 1973 the wasp had become established throughout 70% of Tasmania. In 1977 and 1978, V. germanica spread to mainland Australia with nests being found for the first time

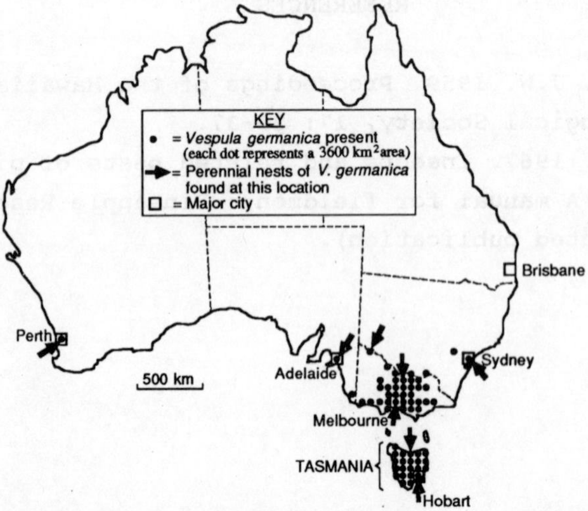

Fig. 1 The distribution of Vespula germanica in Australia in 1989.

in Sydney, Melbourne, Adelaide and Perth. In the 1980's V. germanica had become permanently established in the first three of these cities, with an intensive eradication campaign still continuing in the fourth (Perth). Between 1983 and 1989 the wasp spread (from Melbourne) to most of the major cities in the state of Victoria. Nests are also becoming increasingly widespread and abundant in other states in Australia.

Eradication of the wasp has been attempted in every state in Australia where the wasp has been found. Several examples of success in isolated towns and cities indicate the feasibility of such a strategy. However, frequently wasp nests are already widespread within towns by the time the first nest in that town is located, with consequently little hope of local eradication.

In some localities (e.g. Brighton in Melbourne) over 80% of V. germanica nests occur in buildings (principally in roofs and wall cavities). However, in other localities nesting sites in the ground are more common. Stings from V. germanica have resulted in many hospitalizations in Australia. The wasp presently costs Australian householders approximately $600,000 (Australian dollars) every year to destroy V. germanica nests, with an estimated 23,000 nests destroyed every year (1985-1989 average) in the city of Melbourne alone.

In the future the wasp will almost certainly have an increasing economic impact on beekeepers and soft fruit growers as it continues to spread from town to country areas and become increasingly abundant. This impact will be higher than many other countries due to the presence of large overwintering nests of V. germanica in Australia. To date, large perennial nests have been found in every state in Australia where the wasps are present (Fig. 1).

KEEP DANGEROUS ROCK BEES (*APIS DORSATA*) AWAY FROM HUMAN HABITATION

J.RAJ
Department of Agricultural Microbiology
University of Agricultural Sciences (GKVK)
Bangalore-560 065, INDIA

Enraged rock bees (Apis dorsata) sting humans and animals moving in their neighbourhood, sometimes leading to hospitalisation and death of victims. A preventive method was evolved for discouraging rock bees from frequent colonization of specific sites.

A mixture of Termex (chlordane 20% EC), engine oil and kerosene (1:4:20 v/v) was sprayed to the spots that are vulnerable for frequent colonisation by bees. These included 8 spots under a rock projection and 5 patches on branches of a giant-bee-tree, both had the evidence of rock bee nests. On evaporation of kerosene, a thin film of oil along with the insecticide remained on the surface. A second spray was given after 3 months.

Observation of treated sites for two years at intervals of 4 months revealed no indication of colonisation. However, unavailability of nesting sites due to spraying did not increase the tendency of colonies to occupy the nearby untreated places.

Spray treatment may be widely adopted to save humans from venomous bee hazard and to conserve rock bees which are now being burnt or spray-killed upon being traced amidst human activities.

Symposium 27

BIOLOGICAL CONTROL OF PESTIFEROUS SOCIAL INSECTS

Organizer : **DONALD P. JOUVENAZ, U.S.A.**

BIOLOGICAL CONTROL OF FIRE ANTS : CURRENT RESEARCH

JOUVENAZ, DONALD P.

USDA-ARS, Insects Affecting Man and Animals Research Laboratory, P.O. Box 14565, Gainesville, Florida 32604, USA.

The imported fire ants (IFA), Solenopsis richteri Forel and Solenopsis invicta Buren, were introduced into the United States from South America ca. 1920 and 1940, respectively. These medical and agricultural pests now infest over 10^8 ha in 11 southeastern states and Puerto Rico, and are threatening the west coast. In addition, a polygynous form having denser populations and which is sometimes more difficult to control with chemicals is spreading within the population. The diminished territorial behavior of polygynous colonies, however, may render them more vulnerable to biotic agents.

Fire ants are difficult subjects for biological control due to colony longevity, and the protected environment and high reproductive potential of their queens. In addition, they are ecological generalists, thriving in a wide variety of habitats and environmental conditions. Research has been directed both to the development of microbial formicides and to the study of host-specific natural enemies as candidates for introduction.

Formicides effectively eliminate colonies only if they kill or sterilize the queens. This presents a serious obstacle to the development of a microbial formicide, for fire ant queens are sequestered in subterranean nests, and are surrounded and defended by tens to hundreds of thousands of aggressive workers. They are fed only highly filtered, regurgitated liquids and (probably) glandular products; the gut is usually sterile (Jouvenaz, unpublished). They are groomed meticulously, and their chambers are fumigated with venom, which has antimicrobial properties. IFA also relocate their nests more frequently in response to applications of biotic agents. The straw-itch mite, Pyemotes tritici, and a variety of non-fire ant specific viruses, bacteria, fungi, protozoa, and nematodes have not been effective against IFA. A strain of the fungus Beauveria bassiana under study by Stimac et al (1987) may be an exception. Despite these obstacles, developing a microbial formicide is not a hopeless endeavor. Through biotechnology, the endoparasitic yeasts of fire ants may have potential as microbial formicides, and non-specific nematodes and fungi may yet become efficacious through selection, formulation improvement, or special applications. As an example of the latter, we are currently evaluating the use of commercially available nematodes on nursery stock prior to shipment into uninfested areas.

The major goal of our research is to establish a complex of specific natural enemies (both pathogens and arthropods) of IFA in the United States. The specific natural enemy biota of fire ants includes pathogens, parasites, social parasites, and symbiotic predators. The most promising candidates for introduction appear to be the socially parasitic ant, Solenopsis (Labauchena) daguerri, and the little-known nematode Tetradonema solenopsis. The former reputedly destroys IFA colonies, albeit slowly; the latter is a stressful parasite which may be able to invade adult queens.

The remaining specific natural enemies appear to be debilitating agents well adapted to their hosts; however, the stress they engender may shift the competitive balance in favor of our native ant fauna, or even deliberately introduced exotic species (Buren 1983; also see Jouvenaz, elsewhere in this proceedings).

Biotechnology presents exciting new vistas for biological control research, such as the development of genetically engineered microbial insecticides or avirulent symbionts being made virulent. We have isolated endoparasitic yeasts from fire ants which are prime candidates for genetic engineering. They can be mass-produced, transmitted, and I am optimistic that they will to be prove genus specific. Since they produce no toxins or histopathology, it may be possible to transform them to produce toxins of our choice, insect hormones, or even semiochemicals to disrupt colony organization. We have determined that one species, which appears to be an obligate parasite, is susceptible to the antibiotic hygromycin B, for which a cloned resistance gene is available. We plan to conduct a model transformation test for hygromycin resistance when taxonomic studies in progress are complete. Even without genetic modification, they may prove valuable as stressing agents in concert with other natural enemies or control practices.

Several pesticides are registered for control of IFA (none for use on crops), and will probably always be needed for local suppression of these pests. The Establishment of a complex of specific natural enemies, however, may provide a permanent amelioration of the IFA problem.

Literature Cited

Buren, W. F. 1983. Artificial faunal replacement for imported fire ant control. Fla. Entomol. 66: 93-100.

Stimac, J. L., S. B. Alves, and M. T. Vieira Camargo. 1987. Sucetibilidade de <u>Solenopsis</u> spp. a diferentes especies de fungos entomopatogenicos. An. Soc. Entomol. Brasil 16: 377-387.

A COMPARISON OF FIRE ANT POPULATION DENSITIES IN NORTH AND SOUTH AMERICA

Sanford D. Porter[1], Harold G. Fowler[2] and William P. MacKay[3]

[1]Brackenridge Field Laboratory and the Department of Zoology, The University of Texas at Austin, Austin, Texas, USA 78712; [2]Departamento de Ecologia, Instituto de Biociences, UNESP, 13500 Rio Claro, Sao Paulo, Brazil; [3]Department of Entomology, Texas A&M University, College Station, Texas, USA 77843

Natural biological control offers the ultimate hope for managing the fire ant, *Solenopsis invicta*, in the United States. Such a solution could be permanent, environmentally safe, and economically effective. There are no guarantees of success; nevertheless, parasites and pathogens of fire ants are much less common in the United States than they are in South America [1]. High fire ant populations in the United States may be attributable to this dearth of natural control. Some authors have reported that fire ant densities are indeed much lower in South America [2], while others indicate that densities are about the same [3]. In order to resolve this question, we systematically surveyed fire ant populations in both the United States and Brazil.

MATERIALS AND METHODS

We haphazardly selected 52 roadside sites in the United States along a route from Louisiana through Mississippi, Alabama, Florida, Georgia and then back to Louisiana. In Brazil, we selected 50 roadside sites in the state of Mato Grosso do Sul. Roadside sites were used because they provide appropriate habitat, they were convenient, and they provided structurally equivalent habitat. Sampling was conducted in the spring of 1989 with North American sites being sampled in March and Brazilian sites being sampled in October. Mound densities at each site were estimated from four belt transects (each about 2.7 m by 50 m). All fire ant mounds were checked for activity and tallied into six size categories. At each site, we collected information concerning the soil, the vegetation, and the weather. We also set out 16 bait traps at two-thirds of the sites.

RESULTS AND DISCUSSION

Almost 85% of fire ant colonies at sample sites in Brazil were identified as *S. invicta*. The remaining colonies were either *Solenopsis saevissima* or an unnamed species. All fire ants in the United States were *S. invicta*. For the purposes of this report, all *Solenopsis* species were lumped as ecological equivalents.

Fire ants were much less common in Brazil than they were in the United States (Table 1). Results show that fire ants in Brazil occurred at fewer sites, in lower mound densities, and in smaller mounds. Furthermore, fire ants only accounted for a fraction of the total baits occupied by ants.

Table 1. Abundance of fire ants at haphazardly selected roadside sites in Brazil (n = 50) and the United States (n = 52).

Measure of Abundance	Brazil	United States
Percent of sites with fire ants		
in transects	42%	100%
on baits	47%	100%
found by any method	71%	100%
Mound Density (mounds/ha + S.D.)		
all sites	20+33	170+113*
only sites with fire ants	47+37	170+113*
Percent of Occupied Baits with Fire Ants		
all sites	16%	96%*
only sites with fire ants	19%	96%*
Mound Size (mean diameter cm)	31+13	43+14*

*significantly larger, P < .01.

Overall, fire ants only appear about a tenth as common in the Mato Grosso region of Brazil as they are in the United States. These results are consistent with the hypothesis that biological control agents limit population densities of fire ants in South America. Nevertheless, the differences which we observed could be caused by other factors--influences of climate, season, and ground cover are currently being evaluated. Also, we need to determine if results from Mato Grosso are representative of other parts of *S. invicta's* range in South America. In conclusion, this study does not prove the feasibility of utilizing biological agents to control fire ants, but it does provide encouragement for continued effort in this direction.

1. Jouvenaz, D.P., G.E. Allen, W.A. Banks, and D.P. Wojcik. 1977. A survey for pathogens of fire ants, Solenopsis spp., in the southeastern United States. Florida Entomologist, 60:275-279.

2. Allen, G.E., W.F. Buren, R.N. Williams, M. De Menezes, and W.H. Whitcomb. 1974. The red imported fire ant, Solenopsis invicta; distribution and habitat in Mato Grosso, Brazil. Annals of the Entomological Society of America, 67:43-46.

3. Banks, W.A., D.P. Jouvenaz, D.P. Wojcik, and C.S. Lofgren. 1985. Observations on fire ants, Solenopsis spp., in Mato Grosso, Brazil. Sociobiology, 11:143-152.

THELOHANIA SOLENOPSAE, A MICROSPORIDIAN OF FIRE ANTS : ITS EFFECT ON INDIGENOUS POPULATIONS IN ARGENTINA

R. S. Patterson[1] and J. Briano[2]

1)Insects Affecting Man and Animals Research Laboratory, USDA-ARS, P.O. Box 14565, Gainesville, Florida 32604, USA.

2)Biological Control of Weeds Laboratory, Agricultural Research Service, USDA Buenos Aires, Argentina.

Based on the number of active fire ant mounds in pastures, the density of fire ants in Argentina is similar to that in the United States.. Over 75% of the fire ant nests moved from their original site during a six month period - spring to fall in Argentina. Infection with Thelohania solenopsae did not greatly increase the demise of fire ant colonies. There was a 25% loss of healthy colonies over a six month period. Numerous surveys of pathogens and parasites of fire ants have been made in South America (Jouvenaz 1986), but the fate of individual colonies infected with a single pathogen has not been studied. Beyond a personal communication by Dr. Jerry Stimac, University of Florida, to Carruthers and Hural (1989) that the fungus Beauveria bassiana kills up to 50% of the colonies annually in Brazil, nothing is known about the impact of pathogens in nature. Dr. Stimac is currently evaluating B. bassiana as a microbial formicide.

We began monitoring field populations of fire ants, Solenopsis richteri Forel and Solenopsis quinquecuspis Forel, infected by the microsporidian Thelohania solenopsae Knell et al., in Argentina in October 1988. The data reported herein are complete as of March 1990. Our purpose is to assess the potential of this particular pathogen as a biocontrol agent for importation into the U.S.A. Specifically, we are investigating 1) the pattern of intercolonial transmission of the infection; 2) mortality rates of infected vs. healthy colonies; and 3) stimulation of colony movement with possible elimination of the disease.

Thelohania solenopsae was selected because it is the most common pathogen in fire ant colonies in Argentina. It does not destroy colonies quickly; however, we do not know the fate of individual infected colonies, much less understand fire ant population dynamics in the host county. We established six circular (40 m diameter) plots in unimproved pastures containing cattle and hogs in the vicinity of Saladillo, Buenos Aires Province. Each active fire ant colony was plotted on a map by measuring from a central stake using a compass. The plots were checked monthly for mound numbers and movement. Each colony was examined for the number of major and minor workers present, and for the presence of sexual forms, brood, and disease. The height, width, internal temperature and moisture of the soil (the first 6 inches) of each mound were also recorded. The climate of the test area is similar to that of southeastern United States.

The average number of active colonies/ha was 198 in the six plots. Three of the plots initially had disease rates of 35%; the other three were relatively free of the disease with only 3% of the colonies being infected.

The number of active fire ant colonies/ha in a short grass pasture habitat of the Saladillo area was almost the same as in the southeastern U.S. Adams (1986) reported 60-150 colonies/ha as heavy densities of S. invicta in Florida and Georgia. There was a 25% loss of active fire ant colonies in plots in Argentina from October to March (spring to early fall). This is similar to S. invicta losses in the Southern U.S.A. (Hayes et al. 1982).

Of the active colonies in the six plots, 75% left their original nest and moved to a new location in the plot, usually within a meter of the original site. This also is similar to observations observed in the U.S. for S. invicta and S. richteri (Hayes et al. 1982). There was 10% more movement of colonies infected with Thelohania than non-infected colonies. However, statistically this is not a valid difference.

There was greater loss of colonies with Thelohania, (45%, versus 25%) than of non-infected colonies. In the plots which initially avaraged a 3% infection rate, all colonies were free of the disease after 6 months. However, once a colony was infected and it did not move, it remained infected. We observed no colonies losing their infection. However this is difficult to verify at this time because of colony movement and we were not always sure of the origin of each new colony. We are now using oil-soluble food dyes to follow movement.

T. solenopsae appears to have little potential to suppress fire ants in the field quickly. However, it's long term effect on colony survival and it's interactions in the natural enemy complex of fire ants is currently being examined.

1. Adams, C.T. 1986. Agricultural and Medical Impact of the Imported Fire Ants. p. 48-57. In Fire ants and leaf cutting ants, Biology and Management. Westview Press, Boulder, CO. 434pp.

2. Carruthers, R.I. and K. Hural. 1989. Fungi as naturally occurring entomopathogens. In New Directions in Biological Control. R. Baker and P. Dun (eds.) UCLA Symposia on Molecular and Cellular Biology, New Series, V 112 Allen R. Liss Inc., New York, NY (in press).

3. Hays, S.B., P.M. Morton, J.A. Bass, and D. Stanley. 1982. Colony movement of imported fire ants. J. Ga. Entomol. Soc. 17: 266-274.

4. Jouvenaz, D.P. 1986. Diseases of Fire Ants: Problems and Opportunities. p 327-338. In C.S. Lofgren and R.K. Vander Meer (eds.). Fire ants and Leaf-Cutting ants, Biology and Management. Westview Press, Boulder, CO. 434pp.

5. Knell, J.D., G.E. Allen, and E.I. Hazard. 1977. Light and electron microscope study of Thelohania Solenopsae (Microsporida: Protozoa) in the red imported fire ant, Solenopsis invicta. J. Invertebr. Pathol. 29: 192-200

6. Wojcik, Daniel P. 1989. Behavioral Interactions Between Ants and Their Parasites. Fla. Entomol., Vol. 72, No. 1, 43-51.

LIFE CYCLE OF *SPHECOPHAGA VESPARUM VESPARUM* (HYMENOPTERA : ICHNEUMONIDAE), A PARASITOID OF SOME VESPINAE

B.J. DONOVAN. New Zealand Department of Scientific and Industrial Research, Plant Protection, Lincoln, Private Bag, Christchurch, New Zealand.

Winged female parasitoids must enter wasp nests from the field.[1] They oviposit primarily into cells that have been recently sealed by wasp larvae, and where pupation has recently occurred. Eggs are deposited on the host pupal head or anterior thorax. Eclosion occurs in about a day, and larvae feed as they migrate to the host gaster, and within a few days the host is killed.

Two types of parasitoid cocoon are produced. The first fill the basal third to half of a cell, and are yellowish or reddish brown. They are hard, and are composed of several layers of material. Adults emerge from 10-12 days to 4 years after oviposition, are male or female, and fly strongly.

The second type of cocoon is about one quarter the size of a yellow cocoon, is white, fragile, and composed of only a single thin layer of material. They produce only brachypterous females which emerge 10-11 days after oviposition. These adults oviposit soon after emergence, and progeny are the same as from winged females.

REFERENCE

1. Donovan, B.J., H. Moller, G.M. Plunkett, P.E.C. Read and J.A.V. Tilley. 1989. Release and recovery of the introduced wasp parasitoid, *Sphecophaga vesparum vesparum* (Curtis) (Hymenoptera : Ichneumonidae) in New Zealand. New Zealand Journal of Zoology 15: 355-364.

PROGRESS WITH ATTEMPTED BIOLOGICAL CONTROL OF VESPINAE IN NEW ZEALAND

B.J. DONOVAN, P.E.C. READ, N.C. SCHROEDER. New Zealand Department of Scientific and Industrial Research, Plant Protection, Lincoln, Private Bag, Christchurch, New Zealand.

INTRODUCTION

Vespula germanica (F.) and *V. vulgaris* (L.) are the only Vespinae in New Zealand and both species are adventive.[1] Most vegetated areas are occupied and up to 46 nests/ha have been recorded in native *Nothofagus* forest.[2] High wasp numbers disrupt a wide range of human activities because of the fear of stings, such as tourism, schooling and silviculture. Economic losses result from damage to fruits and attacks on beehives. Severe disruption to native biota is probably being caused by these new predators consuming native insects, honeydew, and even chicks in nests.

PARASITOID PROPAGATION, RELEASE AND RECOVERY

During 1980-81, 8,804 overwintering cocoons of the ichneumonid *Sphecophaga vesparum vesparum* (Curtis) were imported to New Zealand from the CIBC European Station, Delemont, Switzerland. By early 1985, a continuously breeding population of parasitoids had been established under controlled conditions, and by January 1990, 94 consecutive generations had been completed, and 97,580 overwintering cocoons had been distributed over most of the country.

In November 1986, one parasitoid cocoon was recovered from an overwintering nest of *V. germanica*, which was 4.8 km from the nearest parasitoid release site.

On 5 May 1988, 42 *V. vulgaris* nests were examined; 2 contained *S. v. vesparum*. One, 33 m from the nearest release site, contained 471 parasitoid stages, and a second, 625 m from another release site, contained 1034 parasitoid stages. In the two nests a total of at least 1491 wasps had been killed.[3]

At the same site in 1988/89, of 347 nests examined throughout the summer and autumn, 11 contained parasitoid stages, and the minimum number of wasps destroyed ranged from 4 to 153 per nest.[4]

There was no liberation of parasitoids at this site in spring/summer of 1989/90. However, on 15 January 1990, one nest was excavated, and 145 parasitoid stages were present.

CONCLUSIONS

Because the parasitoids recovered in 1990 must have originated from cocoons that overwintered in nests that were attacked during the previous season, *S. v. vesparum* is now considered to have established in New Zealand.

REFERENCES

1 Donovan, B.J. and P.E.C. Read. 1987. Attempted biological control of social wasps, *Vespula* spp., Hymenoptera : Vespidae) with *Sphecophaga vesparum* (Curtis) (Hymenoptera : Ichneumonidae) in New Zealand. New Zealand Journal of Zoology 14: 329-335.

2 Moller, H., G.M. Plunkett, J.A.V. Tilley, P.J. Ward and N.J. Wilson. 1988. The wasp problem on the West Coast: First year research report. Ecology Division report 12: 22 p., DSIR New Zealand.

3 Donovan, B.J., H. Moller, G.M. Plunkett, P.E.C. Read and J.A.V. Tilley. 1989. Release and recovery of the introduced wasp parasitoid, *Sphecophaga vesparum vesparum* (Curtis) (Hymenoptera : Ichneumonidae) in New Zealand. New Zealand Journal of Ecology 16: 355-364.

4 Moller, H., G.M. Plunkett, J.A.V. Tilley, R.J. Toft and N.J. Wilson. 1989. The wasp problem on the West Coast: Second-year research report. Ecology Division report 24: 39 p., DSIR New Zealand.

SELECTED RESEARCH NEEDS IN HONEY BEE PATHOLOGY

J. D. Vandenberg
U. S. Department of Agriculture - Agricultural Research Service
Bee Biology and Systematics Laboratory
Utah State University, Logan, Utah, USA 84322-5310

Some of the speakers at this symposium will address social insects as pests and will consider the use of beneficial microorganisms in controlling them. Honey bees (*Apis* spp.) are, of course, beneficial, and their pathogens are thus pestiferous. Nevertheless, the principles of disease etiology, epizootiology, and pathogenesis remain the same. Honey bees suffer from a variety of diseases of microbial origin. There are also several known parasitic mites attacking honey bees. In this presentation I will address current problems and suggest areas for future research in honey bee pathology.

MITES. Parasitic mites of honey bees will be discussed in another symposium, so I will not address their control per se. However, the mite *Varroa jacobsoni* has recently been shown to transmit pathogenic viruses and bacteria to honey bees (Ball, Strick). Some questions yet to be addressed include determining whether there is any synergism between the mites and the microorganisms. Is colony mortality (or productivity) much worse when both are present? Do the pathogens rely on the mites for their transmission? What pathogens, if any, do other parasitic mites (e.g. *Acarapis woodi*, *Tropilaelaps clareae*) transmit?

VIRUSES. Given the rapid spread of Thai sacbrood virus in *Apis cerana*, can resistant strains of the bee now be found? Are there resistance mechanisms within *A. mellifera* that can be exploited in *A. cerana* through genetic engineering? Although some work has been done (Anderson, Ball), there is a need for rapid detection methods for virus-infected bees. What role, if any, does nutrition play in virus disease etiology? Also, for many viruses, little is known about the effects of infection on colony-level productivity. Are virus diseases worth controlling?

FUNGI. Chalkbrood remains an important disease problem in many areas. Disease resistance has been demonstrated in selected lines (Gilliam and Taber). Can resistant bees be readily established in other parts of the world? Can safe and reliable chemical or antibiotic control methods be found? Although adult nurse bee behavior has been shown to be important in resistance, is there resistance in the larva? There are many aspects of the infection process that are poorly understood. Since many Ascosphaera spp. have been found in other bees, what are the evolutionary relationships among these fungi? Is cross-infectivity (demonstrated in laboratory experiments) important in nature?

BACTERIA. American foulbrood is a well-known and much-studied disease. It is worldwide in distribution. It is controllable, but vigilance is required in many areas in order to keep it in check. Another widespread disease is European foulbrood. Althogh much research has been done on its etiology and epizootiology (Bailey), the role of bee nutrition is not well understood. Why do colonies used for the pollination of certain crops (e.g. blueberry in the U.S.) seem predisposed to European foulbrood? Can nutritional supplements be used to override such predisposition? The role of secondary bacteria in the infection process is not clear. For example, Bacillus alvei is not by itself pathogenic. As a secondary invader it multiplies readily within the cadaver. Do any viable primary pathogens (Melissococcus pluton) remain in such a cadaver? In the case of such a "double infection", does the presence of B. alvei increase or accelerate mortality (synergism)?

PROTOZOA. Another major bee disease problem is caused by Nosema apis. Although the pathogen and the disease have been thoroughly studied, some questions remain. N. apis infection is often associated with other protozoan infections. What is the nature of the relationship between these pathogens? Are they obligately associated through some aspect of their epizootiology? Similarly, N. apis infection can also be associated with certain virus infections (Bailey and Ball). Does infection by one predispose the bee to infection by the other? In temperate climates, N. apis infection is often associated with long periods of winter confinement. However, the disease is truly worldwide in distribution. In the absence of extended confinement in tropical areas, are there certain times within the year when infection rates are higher? If so, when, and why?

CONCLUSIONS. Although I have addressed many different areas of bee pathology, several themes are evident that call for re-emphasis. 1) For research and disease diagnosis, there is a critical need for assay methods: pathogen bioassays for larvae and adults; diagnostic assays for the detection of infection; and colony assays to detect changes in productivity (brood production, honey storage, etc.) following infection. 2) There are fascinating relationships between pathogens (or between acarine parasites and bee pathogens) that are only beginning to be explored. New questions relating to etiology, epizootiology, and pathogenesis will arise from these discoveries. 3) Little is known about diseases of Apis spp. other than A. mellifera. Can knowledge acquired about disease resistance in one of these species be exploited for disease control in another? 4) In the absence of widely-available resistant bee stock and in the absence of chemical disease control methods, are there colony management methods that can be easily used to control disease? Clearly we do not lack important research problems. These should keep us busy for years.

BIOLOGICAL CONTROL OF TERMITES IN AUSTRALIA

J.A.L. Watson

CSIRO, Division of Entomology, GPO Box 1700, Canberra, ACT 2601, Australia

TERMITES AS CANDIDATES FOR BIOLOGICAL CONTROL

Termites are social insects of ancient lineage, dating back at least to the Cretaceous. They derive energy from cellulose, commonly gathered from woody materials. The colonies inhabit soil or timber and the workers usually tunnel through soil to reach feeding sites. Termites have therefore had long contact with the soil-dwelling, insect-pathogenic organisms envisaged as agents for their biological control, and have had opportunity to develop means for coping with them.

Little attention has been paid to two such groups of organisms, bacteria and viruses. Even the literature on fungi and nematodes is meagre. However, both depend on moisture, particularly their infective stages. As termites do not resist desiccation, at least part of the nest is maintained at RH near saturation, thus conducive to infection.

The biological control of termites therefore has to depend on:

(A) The identification of virulent pathogens that can, preferably, induce epizootics in colonies of the target species of termite; and

(B) The development of systems that can deliver an effective dose, rendering the colony inviable either by direct kill or by inducing an epizootic.

TERMITES IN AUSTRALIA

Most termite problems in Australia are caused by subterranean species. The majority are small termites (including the most important genus *Coptotermes*, plus species of *Heterotermes*, *Nasutitermes* and *Schedorhinotermes*). Two genera, *Mastotermes* (which is very destructive) and *Neotermes* (not a subterranean termite), include species with large workers weighing up to 40-50 mg; *Neotermes* colonies are small.

STUDIES WITH *METARHIZIUM*

Various fungi can have deleterious effects on termites [1], but most studies have focussed on *Metarhizium anisopliae*. *Metarhizium* is highly toxic to termites in the laboratory and can control field colonies of *Nasutitermes exitiosus*, but does not always do so [2]. Many, sometimes most, termites taken from colonies that had been treated with conidiospores up to 15 weeks beforehand, but had survived, carried viable spores (indicating limits to the termites' capacity to isolate cadavers and, potentially,sporulation). The spores had failed to germinate, but we do not know if germination was inhibited by the physical environment in the mound, or by materials that the termites produce.

As part of a major new project led by Milner and Watson (CSIRO, Canberra), strains of *Metarhizium*, including some recently isolated from field colonies, are being evaluated for virulence, dependence of germination and sporulation on ambient conditions, and field performance, against species of *Mastotermes*, *Nasutitermes* and *Coptotermes*.

STUDIES WITH NEMATODES

Although some insect-pathogenic nematodes (with their bacterial symbionts) can devastate laboratory groups of termites, their efficacy in large field colonies may be limited by the small size of many termites: the meagre nutrients in each cadaver may prevent the nematode life cycle being completed and stop the development of an epizootic. Our most promising results have come from work with *Heterorhabditis* collected from soil in Darwin. The susceptibility of *Mastotermes* to this *Heterorhabditis* is amongst the highest known in insects, individuals succumbing to less than five nematodes in the laboratory (Bedding, unpubl.). Australian species of *Coptotermes* and *Nasutitermes* are not very susceptible to the *Heterorhabditis* or to *Steinernema carpocapsae* (Bedding, unpubl.). The Darwin *Heterorhabditis* has eliminated small colonies of the relatively small termite *Glyptotermes dilatatus* in tea bushes in Sri Lanka; epizootics were produced [3]. Lenz and Bedding (CSIRO, Canberra) (unpubl.) are experimenting with *Heterorhabditis* and *Steinernema* against *Neotermes* in the south-west Pacific.

Heterorhabditis propagates well in *Mastotermes*, each cadaver producing some 10,000 infective nematodes (Bedding, unpubl.). Miller (CSIRO, Canberra), with Bedding, is experimenting with two delivery systems (release of pre-infested termites, and of large numbers of nematodes, at bait sites) in attempts to eliminate colonies of *Mastotermes* that Miller has delimited by use of fluorescent dyes near Darwin.

SYNOPSIS

The biological control of termites is now a feasible goal. We need reliable pathogens, and reliable delivery systems that can overwhelm the defences that termites appear to have developed against the kinds of organism that we can use. *Metarhizium*, *Heterorhabditis* and *Steinernema*, suitably handled, appear to offer the best chances of success.

REFERENCES

1. Lenz, M. and J.W. Kimbrough. 1982. Botanical Gazette, 143: 546-550.
2. Hänel, H. and J.A.L. Watson. 1983. Bulletin of Entomological Research, 73: 305-313.
3. Danthanarayana, W. and S.I. Vitarana. 1987. Agriculture, Ecosystems,and Environment, 19: 333-342.

ENTOMOGENOUS FUNGI AS CONTROL AGENTS FOR TERMITE *ANACANTHOTERMES AHNGERIANUS* JACOBS (ISOPTERA, HODOTERMITIDAE)

Ljutikova L.I.

Department of Entomology, Faculty of Biology,
Moscow State University, Moscow, 119899, U.S.S.R.

A. ahngerianus may be considered as a classic example of hidden-living insects. The individuals do not go outdoors, apart from short-term period of imaginal flight. Every foraging activity and food transfer are carried out within hypogeal galleries or - on open surface - under termite-built shelters (soil crust). Such a hidden mode of life provides a reliable defense against various unfavourable factors and thereby makes any control (particularly biological one) of this dangerous pest of wood quite difficult task.

Soil as habitat is characterized by some peculiarities being a complex system involving a great diversity of biological, chemical and physical processes. Hence to assure a success of introduced pathogen survival in soil it is important to take for biocontrol only eurybiotic microorganisms.

Among microorganisms, known to be pathogen for A.ahngerianus the fungi are of greatest concern as biocontrol agents. Thus Beauveria bassiana strains and commercial Beauverine induced high termite mortality (≥90%) in laboratory experiments. We furthermore took into consideration that (1) temperature, moisture conditions and pH nest soil are corresponding to ecological requirements of entomogenous fungi, (2) the contamination occurs by contact way, and (3) colony activity requires continuous and diverse interactions between individuals of different castes. The above-mentioned factors would contribute to spread of mycosis in a termite colony. We must however be aware of possible effects of soil fungistasis (ability of soil to inhibit fungal spore germination in spite of suitable soil temperature and moisture conditions). The experiments were conducted to bioassay the fungistatic activity of some nest substrates (mound crust, living chamber lining, soil between living chamber, soil crust from dried plant material in foraging sites). Native soil samples were taken from surface and at depths corresponding to ones of nest substrates for control tests. Target species were entomopathogenic fungi, Beauveria bassiana and Aspergillus flavus, as well as soil saprotrophe Penicillium verrucosum var. cyclopium. Soil substrate effects on germination of conidia were studied by membrane cell method and effects on growth of mycelium - by agar block method. The results confirm current view on soil fungistasis as an universal phenomenon. The predominant soil types in foothill plains of southern Turkmenistan, inhabitat by A. ahngerianus are clay and loam sierozems. Fungistatic activity was shown both in uppermost biogenous layer of native soil and in underground horizon (the depth about 40-50 cm), as evidenced by inhibition of conidial germination and decrease of mycelial growth. Structure and compound of termite nest soil were not identical with corresponding horizons of native soil. Nest soil contains clay brought by workers from deeper horizons into the nest for nest building, as well as considerable amount of organic matter, accumulated on account of life activity of colony. Both conidial germination and mycelial growth were inhibited in test nest substrates as well as in shelter material. The conidia were still viable after at least one year incubation in all test substrates. The nature of termite nest fungistasis is not yet clear, but it is necessary to take this phenomenon into account, when planning the use of fungal pathogenes for the control A. ahngerianus. The spread of mycosis in the colony could be limited by nest fungistasis.

TESTING ENTOMOPATHOGENIC FUNGI AGAINST THE COMMON AGRO-FORESTRY TERMITES

H.Khader Khan, Directorate of Extension, UAS, Bangalore.

S. Jayaraj, Vice-Chancellor TNAU, Coimbatore.

and

M. Gopalan Dept.of Entomology TNAU, Coimbatore.

Entomopathogenic fungi such as *Beauveria bassiana* (Bals.) Vuill., *Metarhizium anisopliae* (Metsch.) Sorokin var. *anisopliae*, *M. flavoviride* Gams Rozsypal var. *minus*, *Verticillium lecanii* Zimm, *Paecilomyces lilacinus* (Thom.) Samson, *P. fumoso-roseus* (Wize) Brown and Smith, *P. farinosus* (Holm ex Gray) Brown and Smith and *Nomuraea rileyi* (Farlow) Samson were tried against three common agro-forestry termites viz. *Odontotermes brunneus* (Hagen) *O. wallonensis* (Wasmann) and *O. obesus* Rambur.

Concentrations of 10^7 conidia ml^{-1} of all mycopathogens at the rate of 3 ml per treatment as direct spray, was used against termites.

Most of the entomopathogenic fungi were pathogenic to termites except *V. lecanii*, *P. farinosus* and *N. rileyi*. The mycopathogens exhibited great variation in pathogenicity against termites. *B. bassiana* was the most virulent followed by *M. anisopliae*, while the others were although pathogenic, but were not virulent. The order of virulence of mycopathogens was *B. bassiana* *M. anisopliae* *M. flavoviride* *P. lilacinus* *P. fumoso-roseus*. Among the three species of termites *O. brunneus* was the most susceptible followed by *O. wallonensis* and *O. obesus*. This difference in the susceptibility of different termite species to mycopathogens might be due to the inherent variation in susceptibility and bio-chemical interactions in the infection process, which might be specific to a particular species-pathogen interaction (Maniania and Fargues, 1984; Keller and Zimmermann, 1989).

REFERENCES

Keller, S. and G.Zimmermann, 1989. In *Insect-Fungus Interactions* (ed.) N.Wilding, N.M.Collins, P.M.Hammond and J.F.Webber. Academic Press, London, New York. pp. 240-321.

Maniania, N.K. and J.Fargues, 1984. *Entomophaga*, **29**: 451-464.

INFLUENCE OF SUCCESSIVE SUB-CULTURING AND MAINTENANCE OF VIRULENCE OF MICOPATHOGENS ON TERMITES

H.Khader Khan, Directorate of Extension, UAS,Bangalore

S.Jayaraj, Vice-Chancellor, TNAU,Coimbatore

and

M.Gopalan, Dept.of Entomology, TNAU,Coimbatore.

The two entomopathogenic fungi of termites viz., Beauveria bassiana (Bals.) Vuill and Metarhizium anisopliae (Metsch.) Sorokin var. anisopliae were sub-cultured on carrot medium seven times successively. At every sub-culture the biometric characteristics of these fungi were studied as per the methods outlined by Hall and Bell 1961, Easwaramoorthy and Jayaraj, 1977 and Schaerffenberg, 1964. The infectivity of these sub-cultured fungi at every sub-culture was tested against Odontotermes brunneus (Hagen) - a mound building termite.

Significantly higher biomass, mycelial growth, conidial count and per cent viable conidia upto the fourth sub-culture was recorded in B.bassiana, while in M.anisopliae it was upto the third sub-culture. After this there was considerable reduction in these growth characteristics in both the fungi.

Mortality data with different sub-cultures showed that there were significantly higher mortalities till the fourth sub-culture in B.bassiana and upto the third sub-culture in M.anisopliae. After this there was significant reduction in Mortalities at every sub-culturing in both the fungi (Table.1)

The same trend was reflected in case of LT50 values in both the fungi. However, when the fungi were passed through termites after the fifth sub-culture, the infectivity was restored to the original level. Further passage of mycopathogens at sixth and seventh sub-cultures did not improve the infectivity (Table.1). Similar loss in virulence due to successive sub-culturing of entomopathogenic fungi has been reported earlier(Schaerffenberg,1964; Ignoffo, 1981; Kmitowa,1982 and Sundarababu et al., 1983).

REFERENCES

Ignoffo, C.M., 1981. In. Microbial control of pests and plant disease, 1970-1980 (ed.) D.Burges,Academic Press,New York. pp. 513-538.

Kmitowa,K.,1982. Polish Ecological Studies, 8 : 433-441.

Schaerffenberg, B., 1964. J. Invertebr Pathol., 8 : 8-20.

Sundarababu, P.C., M.Balasubramanian and S.Jayaraj,1983. Research Bulletin, Tamil Nadu Agric.Univ., Coimbatore. p. 32.

Table 1. Effect of sub-culturing and passing through host on the virulence of termite mycopathogens

Sub-cultures/ passage through host	*B. bassiana*			*M. anisopliae*		
	Per cent corrected mortality +	LT50* (h)	Fiducial limits (95 %)	per cent corrected mortality +	LT50* (h)	Fiducial limits (95 %)
I	87.00^{a}	87.13	84.16- 90.21	80.25^{a}	93.33	89.15- 97.71
II	86.75^{a}	93.12	90.40- 96.96	78.00^{ab}	97.49	92.41-102.84
III	85.25^{a}	101.43	98.13-104.84	77.25^{ab}	108.22	104.49-112.08
IV	84.25^{a}	110.24	104.70-116.07	75.50^{bc}	116.74	112.78-120.83
V	73.75^{b}	122.70	119.10-126.40	66.25^{c}	125.47	120.93-130.18
VI	65.50^{c}	134.48	129.97-139.16	58.00^{d}	134.88	131.01-138.86
VII	57.00^{d}	139.09	134.72-143.60	50.00^{e}	140.49	136.00-145.14
Ist passage after Vth	86.25^{a}	81.15	78.22- 84.20	80.00^{a}	90.45	87.34- 93.68
2nd passage after VIth	84.75^{a}	84.26	80.13- 88.60	78.25^{ab}	93.20	90.05- 96.46
3rd passage after VIIth	85.25^{a}	86.36	81.34- 91.96	77.75^{ab}	96.88	93.44-100.43

* at 4 x 10 conidia ml.

+ Mean separation by DMRT at 5% level

Symposium 28

THE ROLE OF NOURISHMENT IN THE ONTOGENY AND EVOLUTION OF INSECT SOCIETIES

Organizers : **JAMES H. HUNT, U.S.A.**
Christine A. Nalepa, U.S.A.

RESOURCE ACQUISITION AND THE EVOLUTION OF PATERNAL CARE

Douglas W. Tallamy
Dept. Entomology and Applied Ecology
University of Delaware
Newark, DE 19717-1303 USA

In the overwhelming majority of arthropods that nurture offspring, parental care is provided either exclusively by the mother, or less frequently by both parents. Exclusive care by fathers is extremely rare among arthropods and has developed extensively only within a single taxon - the Belastomatidae (1). Parental investment in offspring takes many forms, including the transfer of nutritional resources to females before or during copulation, assisting females with pre- and/or postovipositional parental duties, or assuming all parental responsibilities. The scarcity of exclusive male care has been attributed to the asymmetry in the resources allocated to sperm vs. egg production, to the constraints imposed on females by internal fertilization, to the trade-off between care-giving and opportunities for polygyny, and to the difficulties in guaranteeing paternity (2). When these constraints are relaxed, paternal care is possible.

Here I propose a condition which does not inhibit, but encourage the evolution of male care; in iteroporous arthropods, the likelihood of paternal contributions to offspring care is a positive function of the cost of acquiring nutritional resources. Resources vary in the amount of time and energy required to locate, secure, and consume them. Plant sap and foliage, for example, can be acquired with relatively little cost. Resources exploited by predators, however, usually require considerable time and effort to locate, subdue and consume. Prey are usually dispersed, disjunct resources that individually provide only a small fraction of the caloric and nutritional requirements for continuous oogenesis. Thus, to maximize reproductive output, female predators must hunt repeatedly throughout their lives. The time, energy, and mobility necessary for repeated foraging conflict with the sessile demands of parental care, unless the guardian can forego feeding during periods of care <u>without</u> the loss of reproductive opportunities. Since eggs are physiologically expensive to produce, this is difficult for females, but because sperm are comparatively cheap, males of species with high resource acquisition costs can provide offspring with care without sacrificing gamete production. When parental role reversal occurs, guarding males are sought by females. Thus, opportunities for polygyny are frequent.

It is not by chance that the only arthropods in which exclusive paternal care has evolved suffer high costs of resource acquisition. It is the characteristics of a dispersed and disjunct nutritional resource that have tipped the evolutionary balance from maternal toward paternal care in these species. If paternity can be guaranteed, paternal care is an option for these males because they can improve offspring survival without sacrificing reproductive effort or opportunities for polygyny.

1. Tallamy, D.W. and Wood, T.K., 1986. Convergence patterns in subsocial insects, Annual Review of Entomology, 31 : 369-390.

2. Zeh, D.W. and Smith, R.L., 1985. Paternal investment by terrestrial arthropods, American Zoologist, 25 : 785-805.

LIFE HISTORY CHARACTERISTICS ASSOCIATED WITH A WOOD DIET AND THE EVOLUTION OF TERMITE EUSOCIALITY

Christine A. Nalepa

Entomology Department, North Carolina State University, Raleigh, N.C. 27695-7613

Dependence on a wood diet has led to a suite of life history characteristics in the subsocial woodroach Cryptocercus punctulatus that is unique among insects. If we make the assumption that the prototypical subsocial termite ancestor also shared these traits, a simple behavioral change can result in eusociality.

The most important life history characteristic that would predispose a subsocial termite ancestor to eusociality is a prolonged period of parental care that includes trophic dependence of neonates and leads to semelparity in adults (1), i.e., care givers are subject to a substantial cost in terms of future reproduction. C. punctulatus lays a single brood of eggs, then cares for that brood for at least 3 years, or until the death of the adults (2, 3). This care includes trophallactic feeding of the young, defense of the colony, gallery excavation, and sanitation of the nest. Compare this to an incipient colony (i.e., the subsocial stage) of primitive termites. Adults lay a brood of eggs, then exhibit the same components of parental care as C. punctulatus. As termite neonates age, however, a change in behavior is seen; 4th instars are not only nutritionally independent of adults, but they relieve adults of all brood care duties by assuming reponsibility for younger siblings. At that point, when the cost of brood care shifts from adults to the oldest offspring, the change from subsociality to eusociality occurs. The adult female resumes oviposition, resulting in overlap of worker generations. The assumption of responsibility for younger sibs by 4th instars constitutes brood care, and 4th instars are investing in sibs reserves that could have been channeled into their own reproduction, increasing the duration of their juvenile period and making possible the formation of a non-reproductive caste.

An advantage to thinking of termite eusociality as the transfer ofbrood care duties from parents to offspring is that the conditions surrounding the change in behavior can be analyzed using theories of parent-offspring conflict (4, 5). Adults are in control of dispensing proctodeal food (3) and thus may be in a position to impose subfertility on their offspring. Alternatively, competition among siblings for proctodeal food from adults may result in a high variance in reproductive potential among individuals within families, making it possible for subfertile nymphs to maximize their inclusive fitness by assuming brood care duties (6, 7).

Another advantage to this viewpoint is that Hamilton's (8) equation may lend itself more readily to the analysis of termite eusociality if the costs and benefits of helping behavior are given more emphasis. A delay in reproductive maturity may be a disproportionately small price for a nymph to pay for assuming brood care when compared to the additional siblings such behavior may allow. The benefit to cost ratio may, in fact, be high enough that there is no need to invoke special genetic assymmetries in these diploid insects.

Additional characteristics of C. punctulatus that would be important in a model of the evolution of termite eusociality include: 1) Monogamy. The monogamous mating system of both C. punctulatus and termites was most likely structured by the low quality and scattered nature of their food/nesting material (9) and ensures that all offspring in a family are at least full siblings. 2) A prolonged developmental period of nymphs. Woodfeeders typically have long developmental periods; it takes 5 or 6 years for nymphs of C. punctulatus to reach reproductive maturity (Nalepa, unpublished). It would not be cost effective for a young nymph to desert the family to begin reproducing on its own instead of helping if it takes 3 ot 4 more years before it can reach reproductive maturity. 3) Unlike the Hymenoptera, a hemimetabolous development results in offspring that are capable of work (10). Workers in termite colonies are not adults, but nymphs held in an arrested state of development.

1. Nalepa C.A. 1988. Behavioral Ecology and Sociobiology 23: 135-140.
2. Seelinger, G. and U. Seeliner. 1983. Zeitschrift für Tierpsychologie 61: 315-333.
3. Nalepa C.A. 1984. Behavioral Ecology and Sociobiology 14: 273-279.
4. Michener, C.D. and D,J Brothers. 1974. Proceedings of the National Academy of Sciences 71: 671-674.
5. Trivers, R.L. 1974. American Zoologist 14: 249-264.
6. Craig, R. 1983. Journal of Theoretical Biology 100: 379- 397.
7. Brown J.L. and S.L. Pimm. 1985. Journal of Theoretical Biology 112: 465-477.
8. Hamilton, W.D. 1964. Journal of Theoretical Biology 7: 1-16, 17-52.
9. Nalepa C.A. and S.C. Jones. 1990. In submission.
10. Noirot, C. 1982. Rivista di Biologia 75: 157-196.

FOOD RESOURCES AND THE REGULATION OF COLONY GROWTH AND CASTE DEVELOPMENT IN WOOD-FEEDING TERMITES (ISOPTERA)

Michael Lenz

CSIRO, Division of Entomology, Canberra, Australia

Current understanding of the relationship between the level of nutrition and the regulation of colony growth and caste development in wood-feeding termites is limited.

The importance of the **quantity** of the food supply in shaping termite societies is indicated from a comparison of colony size and other aspects of termite biology between groups of species classified according to their feeding and nesting biology into drywood, dampwood and subterranean termites. Dry- and dampwood termites nest within their food source which is of restricted volume, especially in the case of drywood termites. Subterranean termites establish territories containing the nest and separate feeding sites. Over the three groups the range of population sizes per colony, the number of reproductives a colony can contain, and other features of its biology are scaled to the size of its food base. Drywood termites show the lowest values. The largest colonies, with substantial numbers of reproductives (neotenics) in some species, are found among subterranean termites. Dampwood termites take an intermediate position. Within each of the three groups a similar correlation between colony size and aspects of life history to the quantity of the food supply can be observed in species pairs where each member exploits a food source of different dimensions to the other, e.g. *Porotermes planiceps* living in shrubs as small, short-lived colonies and *P. adamsoni* utilizing mature trees which support much larger colonies over many years.

Experimental data which relate the quantity of available food with events in the termite colony are scarce. Field colonies do not easily lend themselves to relevant experimental manipulation. For most species laboratory studies can only be conducted with non-breeding groups of termites. These may differ significantly in their behaviour from reproducing field colonies, notably in wood consumption rates. Laboratory groups composed largely of foragers remove wood at specific rates. Studies indicate that foragers burdened with dependent castes either increase wood consumption, leaving ample provision for other castes, or do not modify their food intake, leaving insufficient reserves for the dependents. However, in field colonies, reproduction and colony growth could occur to a significant extent only if foragers increase the energy input of the colony to markedly above subsistence level. One of the clues triggering increased food intake appears to be the presence of a large food supply. In laboratory studies on wood consumption, the ratio between the volume (number) of termites and the volume of the wood supply have usually ranged from 1 : 0.4 - 2 (ml/ml). Groups of three species of subterranean termites, composed mainly of foragers, consumed more food if the ratio of termites to wood was 1 : 12 and above. Thus even without the presence of reproductives and other dependent castes they behaved more like field populations.

In the subterranean species *Reticulitermes flavipes* groups produced young (via neotenics) when kept on the largest wooden blocks, wherease many groups failed to breed when provided with timber of smaller dimensions, even though most had produced neotenics. For species of the drywood termites *Cryptotermes*, the size of the wooden blocks influenced the success rate with which groups established themselves as reproducing colonies, the number of their offspring, and the number of last-stage nymphs and alates. In some species brood size peaked below the maximum block size. On larger blocks neotenic production was more intense, in effect reducing the population of helpers that could provision reproductives and the brood. The size of the forager population, hence the quantity of nourishment reaching the colony, also greatly influences colony dynamics in *R. flavipes*.

Termites display marked species-specific preferences for, and prosper best on, certain species of wood, particular parts of woody plants or tissue of a certain condition and wood modified to a degree by certain microorganisms, illustrating the crucial role of the **quality** of nutrition for termite growth and development. Differences in physical and chemical properties of the wood, notably in type and amount of extractives and, most significantly, in the level of nitrogen are assumed to influence these feeding preferences. Maintenance of groups on decayed wood of a specific type has given the best evidence for the impact the quality of the food has on the life history of a termite colony, resulting in significant increases in wood consumption, survival rate, and the number of offspring, accelerating neotenic development and directing the rate of progressive or regressive moults in nymphs.

It is evident that studies of the nutritional ecology of termites can make significant contributions to our understanding of the ontogeny of termite societies.

NOURISHMENT AND SOCIAL DOMINANCE IN COLONIES OF SOME PRIMITIVE ANTS

K. Masuko
Department of Biology, College of Arts and Sciences,
University of Tokyo, Komaba, Meguro-ku, Tokyo 153, Japan

Even among related nestmates of eusocial insects, conflict of interest exists for reproduction [1]. In primitively eusocial species (e.g. Polistes) this conflict often manifests itself as agonistic behavioral interactions, while it is largely concealed in the case of distinct castes (queen and worker) in highly eusocial species. Inequality in nourishment among nestmates, however, prevails in colonies of social insects, and in some cases it may be an indication of the presence of such conflict, because biases in nourishment are connected directly with biases in reproductive output. In this paper I will review this problem focusing chiefly on hemolymph feeding and oophagy in primitive ants.

LARVAL HEMOLYMPH FEEDING

Larval hemolymph feeding (LHF), in which adult ants feed on hemolymph of developed larvae, has been found in Amblyopone, Proceratium, and Leptanilla [2,3]. In A.silvestrii and P.itoi, only queens perform LHF. This monopolization of LHF seems to be related to the queen's dominance, because when colonies are dequeened, dominance hierarchies are formed among workers and only the highest-ranking workers perform LHF and lay male-destined eggs. On the other hand, in the highly migratory ant L.japonica, workers ordinarily perform LHF as well as the colony queen [4]. The occurrence of LHF by workers in L.japonica may be related to the loss of ovaries in this caste and this degeneration can be considered as a result of miniaturization related to its subterranean life [5].

OOPHAGY

In ants, few cases have been reported on differential oophagy among adults as known in social wasps; instead production of trophic eggs is generally known. In more than 20 ant genera workers or foundresses are known to produce such non-reproductive eggs for nutrition of larvae. In some ants, colony queens also entirely depend on trophic eggs as their nutrient. For example, in Myrmecia pilosula (the species with a chromosome number of n=1 [6]), the queen observed in the laboratory fed only on worker-laid trophic eggs [7]. She never obtained eggs without solicitation. The response by workers, however, mostly negative; thus she obtained eggs for only 7.5% of all solicitations. In contrast, the queen of Prionopelta amabilis was always offered trophic eggs without any solicitation [8]. This contrast may reflect some difference in queen-worker relationships in these species. Again in A.silvestrii, workers never develop their ovaries under queenright conditions. This ant is a specialized predator of geophilomorph centipedes and the hunted prey is given as a whole to the larvae. Probably because such prey is too robust for the 1st- and 2nd-instar larvae, they normally use queen-laid reproductive eggs as their food. Since each larva consumes 2 or 3 eggs, about 60-80% of eggs laid by the colony queen are lost due to this oophagy. Here the question arises why this apparently costly oophagy has not been replaced by use of worker trophic eggs. This evolutionary "stagnation" might be related to the queen's suppression of ovary development of workers in connection with her dominance.

REFERENCES

1. Trivers, R.L. and H. Hare. 1976. Science, 191: 249-263.
2. Masuko, K. 1986. Behavioral Ecology and Sociobiology, 19: 249-255.
3. Masuko, K. 1989. ibid., 24: 127-132.
4. Masuko, K. 1990. Insectes Sociaux (in press).
5. Hoelldobler, B., J.M. Palmer, K. Masuko and W.L. Brown, Jr. 1989. Zoomorphology, 108: 255-261.
6. Crosland, M.W.J., R.H. Crozier and H.T. Imai. 1988. Journal of the Australian Entomological Society, 27: 13-14.
7. Masuko, K. 1988. Annual Report of National Institute of Genetics, (38): 84-85.
8. Hoelldobler, B. and E.O. Wilson. 1986. Insectes Sociaux, 33: 45-58.

NOURISHMENT, CASTE THRESHOLDS AND WORKERLESS SOCIAL PARASITIES IN THE FORMICIDAE

John E. Tobin
Museum of Comparative Zoology, Harvard University, Cambridge, Massachusetts 02138, U.S.A.

I hypothesize that two of the distinguishing characteristics of inquiline ant species, their small size relative to their hosts and the partial or total lack of a worker caste, are linked by a common explanation. I argue that selection may have acted on the size of inquilines in order to do away with an unnecessary worker caste, and that being small may simply enable an inquiline to be workerless.

Inquilines, or permanent social parasites, do not found their colonies independently, but invade the colonies of closely related species and take advantage of the host workers' labour. Although both the inquiline and the host queen may produce offspring, in most cases the inquiline brood gives rise solely to sexual offspring. Since inquilines are able to insert their own eggs into the brood production line of the host colony they no longer need workers of their own. Thus selection is expected to suppress the inquiline worker caste, as appears to have occurred, although the mechanisms underlying this evolutionary reversal are not understood. The trend towards miniaturization, on the other hand, does not have any obvious explanation, although several possibilities have been put forth (Sudd and Franks 1987).

In the social Hymenoptera, caste determination is mediated by nutrition. Whether a female larva becomes a worker or a gyne is at least partly a function of the amount of food received during development. A species-specific caste threshold must be crossed if a larva is to become a gyne; if fed less than the threshold amount, the larva will become a worker.

I propose an evolutionary scenario that explains the trend towards miniaturization in the inquilines as an indirect effect of selection for loss of the worker caste. This scenario rests on two important assumptions: caste thresholds are a function of species size, so that a small species requires less food to push it over its caste threshold than a larger one; and, the inquiline queen does not manipulate the host workers in order to obtain preferential treatment for its brood, but the parasite brood are treated and fed as host brood.

If it is in the best interest of an inquiline not to produce any workers of its own, a simple mechanism to accomplish this would be to become smaller over evolutionary time. In doing so, the inquiline would push its nutritional caste threshold towards lower levels, until the amount of food used to produce a host worker suffices to produce a parasite gyne. Apart from any other potential benefits of being small, it ensures that most or all inquiline female larvae become reproductives. Thus for an inquiline being small may be a precondition to becoming workerless, so that if an inquiline were large it would not be workerless.

A number of predictions are derived from this scenario, and data that bear upon them are presented.

REFERENCE

Sudd, J. H. and N. R. Franks. 1987. The Behavioural Ecology of Ants, Chapman and Hall, New York

NOURISHMENT AND CASTE IN SOCIAL WASPS

James H. Hunt
Department of Biology
University of Missouri - St. Louis
St. Louis, Missouri 63121 U.S.A.

The taxonomic and behavioral diversity of insect sociality gives evidence of a variety of pathways by which that sociality has evolved. Levels of social integration vary as greatly as do the routes to it. Eusocial insects meet three criteria (probably first specified as such by Suzanne W. T. Batra [C. D. Michener, personal communication]): cooperative brood care, overlap of generations, and division of reproductive labor. Insect taxa that meet the criteria of eusociality include those that apparently achieve caste differentiation in either (or sometimes both) of two manners: division of imminent reproduction among foundresses of a colony or division of future reproduction among offspring of a colony. For the most part, if not entirely, species that have castes only via division of imminent reporduction among foundresses never acheive large colony size, nor are their colonies long lived. Only in species where division of future reproduction occurs among the progeny of a colony does one find both large colony size and long lived colonies. As the search for the evolutionary origins of eusocial behavior shifts from a focus on inclusive fitness to a focus on the proximate factors of behavior and ecology that affect the evolutionary process, it is important that differences in the two modes of division of reproductive labor be stressed.

Caste differentiation in highly eusocial taxa, exemplified by Apis, Vespinae, and Meliponinae, may involve factors such as larval cell size (and, in Melipona, perhaps genotype). Even so, quantity and sometimes quality of larval nourishment has repeatedly been shown to be a significant contributor to caste differentiation in these taxa. In contrast, the conventional wisdom (at least since Marchal) on caste differentiation in primitively eusocial taxa, exemplified by Polistes, is that division of reproductive labor occurs principally among imagos.

If true, then the achievement of eusociality and the division of future reproductive labor among daughters are separate evolutionary phenomena, with the latter coming later. To include in the definition of eusocial both those species that have caste differentiation only among imagos (be they foundresses or daughters) and those that have caste differentiation via differential larval nourishment of daughters would in that case be wholly appropriate, and to search for the origin of highly eusocial taxa by a focus on caste differentiation among imagos would be similarly appropriate. On the other hand, if caste differentiation among daughter of primitively eusocial taxa such as *Polistes* incorporates differential nourishment in a manner similar to higher eusocial taxa, then this mode of caste differentiation must become the focus of investigations. In that case, caste differentiation among imagos, though interesting in its own right, must be seen as a red herring in the search for the evolutionary origin of highly eusocial insects.

Enough evidence is now at hand to call for serious consideration of the proposition that differential nourishment of larvae in primitively eusocial wasps contributes to caste determination (differentiation of future reproduction) among daughters of a colony in a manner similar to that in highly eusocial taxa. Variables of larval behavior during trophallaxis may be significant contributors to differential larval nourishment in primitively social wasps. The relevant larval behaviors may vary as a function of the general level of colony nourishment. Caste in primitively eusocial wasps is probably incompletely determined by nourishment, with behavior roles upon emergence being at least partially context dependent. Differences in larval nourishment may be, however, the fundamental factor in caste determination.

Full understanding of the evolution of eusocial insects with large, long-lived colonies can only be acheived by a robust explication of the mechanism of caste differentiation (division of future reproduction) among the female offspring of taxa near the "eusociality threshold." That is, to learn *how* daughters in primitively eusocial taxa become workers will be the key to finally, fully understanding the evolution of social insects.

TROPHALLAXIS IN THE COLONY OF STINGLESS BEES

Sommeijer, M.J. and L.L.M. de Bruijn
Utrecht University, Bee Research Dept., P.O.Box 80.086, 3508 TB Utrecht, Netherlands.

INTRODUCTION

Trophallaxis, the exchange of alimentary liquid among members of a colony, has been little studied in stingless bees. Most of the studies related to this behaviour in Meliponinae pertain to the trophallactic transfer of food between the queen and the workers [1, 2, 3].

Trophallaxis of returning and departing foragers, the distribution of nectar and the storage of honey

M. favosa workers returning with nectar expose a drop between their mandibles and perform rapid movements provoking as many interactions as possible with surrounding workers. The returnee continuously offers food to her nestmates. It is only in this context that spontaneous food offering occurs. After this, she proceeds to an open nectar storage pot and discharges her load into this pot by a contraction of her abdomen. Returning pollen foragers do not offer trophallactic food to nestmates. However, both pollen- and nectar foragers do solicit and receive food trophallactically just prior to leaving the hive. As in the honey bee, house bees dehydrate the nectar by taking it up from the pots [4].
In our experimental study of food transfer and division of labour related to the processing of nectar and the storage of honey, we found that 25 % of the food collected from a feeder, was directly stored by the collectors, and the remainder was trophallactically distributed among nestmates [5]. Of the latter part, approximately one third also reached the storage pots during the short feeding period. On average about half of the total amount of collected sugar was stored and the other half remained in workers' stomachs. The bulk of the syrup experimentally fed through this feeder passed through the honeystomachs of workers that did not visit the feeder, before it was deposited into the storage pots. Observations specifically performed on trophallactic transfers within this nest, confirmed this pattern. Uptakers had a higher rate of offering than of receiving food. Workers that only regurgitated in storage pots, received significantly more than they offered. Workers seen both uptaking and regurgitating, had similar scores for both behaviours. Finally, workers not seen near the feeder nor at the honey-pots in the nest, received more food than they gave. Since foragers are relatively old workers, the sugar-flow was directed from older towards younger bees.

Trophallaxis related to the provisioning of the broodcell with larval food

House bees of *M. favosa* never offer food spontaneously to nestmates. All transfers between these non-foragers are initiated by the soliciting behavior of the receiver. Cell-provisioning bees of *M. favosa* obtain most of the larval food that is to be released into the brood cell from other bees. The bees that discharge on a particular day are not involved in the pollen uptake from food pots on that day. The dischargers obtain trophallactically a liquid food with a

high pollen content. Thus, among house bees a group of food-preparing bees, next to the dischargers, can be distinguished.
In colonies of age-marked bees we found that workers regurgitate food more often than once in the provisioning process of a single brood cell [6]. Even during this process of a single cell, discharging *M. favosa* workers perform active food solicitations immediately after they have discharged. They often depart quickly from the cell that is being filled and solicit from bees at some distance from the cell. In this situation these dischargers are strikingly more succesful in soliciting than just before their discharge. Therefore they are often able to perform more than one regurgitation in a single cell. Under favorable conditions the number of discharges per cell is lower even though more individuals participate in ddischarging.

REFERENCES

1. Sakagami, S. F., R. Zucchi and V. de Portugal-Araujo. 1977. Oviposition behavior of an aberrant African stingless bee, Meliponula bocandei, with notes on the mechanism and evolution of oviposition in stingless bees. J. Fac. Sci., Hokkaido Univ. Ser. VI Zool. 20: 647-690.
2. Sommeijer, M. J., F. T. Beuvens and H. J. Verbeek. 1982. Distribution of labour among workers of Melipona favosa F.: Construction and provisioning of brood cells. Insectes Soc. 29: 222-237.
3. Sommeijer, M. J. 1985. The social behavior of Melipona favosa F.: some aspects of the activity of the queen in the nest. J. Kansas Entomol. Soc. 58: 386-396.
4. Sommeijer, M.J., De Rooy, G.A., Punt, W.A. and de Bruijn, L.L.M. (1983). Comparative study of foraging behavior and pollen resources of various stingless bees (Hym., Meliponinae) and honeybees (Hym., Apinae) in Trinidad, West-Indies. Apidologie 14(3): 205-224
5. de Bruijn, L.L.M., Sommeijer, M.J. and Leys, R. (1989) Uptake and distribution of nectar and the storage of honey in Melipona favosa (Apidae, Meliponini). Actes Coll. Insectes Sociaux, 5: 39 - 43.
6. Sommeijer, M.J. , L.L.M. de Bruijn and C. van de Guchte. I985. The social food flow within the colony of the stingless bee, Melipona favosa (F.). Behaviour 92: 39-58.

Symposium 29

NESTS AND NEST BUILDING IN SOCIAL INSECTS

Organizer : **MICHAEL H. HANSELL, United Kingdom**

NEST DESIGN AND SECONDARY FUNCTIONS OF SOCIAL INSECT ARCHITECTURE

John W. Wenzel
Department of Entomology
University of Georgia
Athens, GA, 30602, USA

Animal architecture evolves as an extention of the builder's phenotype. Just as morphological adaptations may acquire secondary functions through time, so can architectural traits. These functions may allow the builder to exploit new opportunities and, when evolving rapidly, may obscure the purpose of the original structure. Perhaps the best understood example is the evolution of the use of silk in spiders, originally to wrap the egg mass, then to line the burrow, then as a device to detect prey near the burrow, and finally as a sophisticated, sticky, hunting tool [1]. The architecture of many social insects includes structures serving functions other than retaining the brood, but how did these structures arise? This overview of social insect nests will stress what appears to be a general trend: embellishment and novelty in design appears most often late in the sequence of construction behaviors or in structures typically built late in the colony cycle.

ANTS. Fossorial habit introduces restrictions that prevent comparison to free-standing nests. Storage chambers are certainly evident. Silk weavers (*Oecophylla, Camponotus, Dendromyrmex, Polyrhachis*) use the silk of the prepupae to build arboreal nests [2,3], a behavior possible only in colonies mature enough to have many last instar larvae.

TERMITES. Storage is less critical in those that nest within the food. The Outer envelope clearly serves as defense in many groups. Late-built, external characteristics of the nest may vary more than the interior [4] and can be used to distinguish close species [5]. Ventilation passages may be derived from building an envelope to isolate the interior from air currents: construction of walls around already moving air reinforces passive flow through these conduits [6].

BEES. The highly social bees demonstrate some of the most sophisticated secondary functions. 1. Euglossines are poorly known, but apparently the nest encloses several offspring that are reared in separate cells, mass provisioned, and there is no storage of provisions. Cells may be reused. Some species build an envelope. 2. *Bombus*: Larvae reared in separate cells, then vacated puparia are reused as pots to provide storage of pollen and nectar (secondary function) [7]. 3. Meliponines: Storage pots must be built prior to pupae evacuating their cells since the nest is established and provisioned before the arrival of the queen. Different designs evolve for each pollen and nectar pots, these are segregated, and the brood cells themselves may be of different types depending upon the caste of the occupant. The envelope includes a specialized and distinctive entrance. An antechamber holds the defense force and at the opposite end of the nest there may be a refuse pit [8].

WASPS. Secondary functions are limited and scattered in wasps, perhaps due to the difficulty of storing meat. Also, social vespid nests are among the most regular and "perfect" structures, being as uniform as honeybee combs, yet with successively larger radii of curvature, etc. Perhaps the controls necessary to effect this uniformity have hindered experimentation that would result in secondary functions. Brood cells may have provided the origin for the umbrellas of certain *Parischnogaster* species [9]. Meat is stored in brood cells of some *Polybia. Protonectarina* and some *Brachygastra* species store honey in the brood cells, and *B. lecheguana* builds special peripheral pockets for this purpose, likely derived from peripheral brood cells. The envelope itself is likely derived from the marginal walls of cells, a transition evident in the sister species *Eustenogaster calyptodoma* and *E. micans* [10] and in some *Protopolybia* [11]. Mechanical reinforcement of the envelope has lead to mimetic forms in *Leipomeles* and defensive, sticky-tipped pillars in *Leipomeles* and *Nectarinella* [12]. What likely originated as a method of expanding the envelope in Vespinae has given rise to a rain-resistant roof, and later to the thermoregulatory paper sheets.

[1] Shear, W.A. (ed.) 1986. Spider Webs, Behavior and Evolution. Stanford Univ. Press, Stanford, California.

[2] Wilson, E. O. 1981. Communal silk-spinning by larvae of Dendromyrmex tree-ants (Hymenoptera: Formicidae). Insectes Sociaux 28: 182-190.

[3] Hölldobler, B. and E. O. Wilson 1983. The evolution of communal nest-weaving in ants. Amer. Sci. 71: 490-499.

[4] Schmidt, R. S. 1964. *Apicotermes* nests. Am. Zool. 4: 221-225.

[5] Thorne, B. L. 1980. Differences in nest architecture between the neotropical arboreal termites *Nasutitermes corniger* and *Nasutitermes ephratae* (Isoptera: Termitidae). Psyche 87: 235-243.

[6] Howse, P. E. 1966. Air movement and termite behavior. Nature 210: 967-968.

[7] Michener, C. D. 1974. The Social Behavior of the Bees. Belknap Press, Cambridge, Massachussetts.

[8] Wille, A. and C. D. Michener 1973. The Nest Architecture of Stingless Bees with Special Reference to Those of Costa Rica. Rev. Biol. Trop. 21: suppl. 1.

[9] Williams, F. X. 1919. Philippine wasp studies. II. Descriptions of new species and life history studies. Bull. Exp. Sta. Hawaii Sugar Planters' Assoc. (Entomol.) 19: 1-179.

[10] Sakagami, S. F. and K. Yoshikawa 1968. A new ethospecies of *Stenogaster* wasps from Sarawak, with a comment on the value of ethological characters in animal taxonomy. Annot. Zool. Japonensis 41: 77-84.

[11] Carpenter, J. M. and J. W. Wenzel, 1989. Synonymy of the genera *Protopolybia* and *Pseudochartergus* (Hymenoptera: Vespidae; Polistinae. Psyche (in press).

[12] Wenzel, J. W. 1990. Evolution of nest architecture in social vespids. In The Social Biology of Wasps, K. G. Ross and R. W. Matthews (eds.) Cornell Univ. Press, Ithaca, New York (in press).

ARBOREAL NESTING IN VARIOUS AFRICAN ANTS

A. Dejean*°, R. Mony°, S. Ngokam° & C. Djieto°

* Laboratoire d'Ethologie et Sociobiologie, URA CNRS 667, Université Paris XIII, F-93430 Villetaneuse, FRANCE.
° Laboratoire de Zoologie, Faculté des Sciences, Yaoundé, CAMEROON

Numerous species of ants, under the tropics, nest in trees where they exploit nectar, honeydew of homopterans and capture different kinds of insects. The plant, in most cases, directly or indirectly provides a dwelling place, but many species of ants built their nests in the tree. Nesting of arboricolous ants may thus be classified in the following manner.

1. Opportunist occupying an existing structure:

1.1. Hollow dead branches : *Tetraponera* (Pseudomyrmicinae), *Cataulacus* (Myrmicinae).

1.2. Dead branches attacked by xylophagous insects : *Camponotus* (Formicinae), *Pheidole* (Myrmicinae), *Platythyrea* (Ponerinae).

1.3 Bark with hollow areas : *Camponotus* (Formicinae), *Discothyrea* (Ponerinae), *Pheidole* (Myrmicinae).

1.4. Sheltered from large fronds of *Platycerium* or the rooots of various other pteridophytes and orchids : *Platythyrea* (Ponerinae), *Pheidole* (Myrmicinae), *Crematogaster* (Myrmicinae), *Camponotus* (Formicinae), *Polyrhachis* (Formicinae).

1.5. Domaties of myrmecophytes : *Barteria* (Passifloraceae) sheltering *Pachysima* (Pseudomyrmicinae) in hollow stem and *Scaphopetalum thonneri* (Sterculiaraceae) sheltering *Technomyrmex* in foliated pockets at the base of the lamina.

2. Semi-independant settling :

The ants built up their nest from elements supplied by the plant.

2.1. Galleries hollowed out by *Melissotarsus* (Myrmicinae) workers in the bark of the trees. They breed Diaspididae (homopterans). The density of ants and diaspines may surpass 20 000 individuals/m^2 of bark on mango trees.

2.2. Galleries hollowed in the wood by *Atopomyrmex* (Myrmicinae).

3. Construction of the nest from leaves of the trees :

3.1. *Oecophylla* (Formicinae) and certain species of *Polyrhachis* (Formicinae) use the silk of larvae, but the structures of their nests differ *Oecophylla* gather various leaves while *Polyrhachis* build their nests under large leaves.

3.2. *Tetramorium* (= *Macromischoides*) (Myrmicinae) build cardboard nests from plant fibres and earth. Their nests are under large leaves or between two leaves.

4. Independance:

Different species of *Crematogaster* (Myrmicinae) build their nests with carboard. In certain cases they can be important defoliations. Most of the species are polydomous. Whereas certain species have very populous nests which have an important impact on the plant (*Camponotus, Crematogaster, Oecophylla, Pachysima, Tetramorium*), others have more reduced populations (*Polyrhachis* and *Technomyrmex*). *Atopomyrmex* and *Melissotarsus* are realy harmfull for the trees.

THERMOREGULATION IN THE FIRE ANT, *SOLENOPSIS INVICTA*

PORTER, SANFORD D.

USDA-ARS, Insects Affecting Man and Animals Research Laboratory, P.O. Box 14565, Gainesville, Florida 32604, USA.

Temperature is a central element in the life of fire ant colonies. Changing temperatures strongly affect all metabolic processes, including rates of activity, development, respiration and even longevity. While temperature is a controlling factor, this does not mean that colonies are completely subject to its vicissitudes. In fact, fire ants are very effective thermoregulators.

There are two keys to fire ant thermoregulation: The first is the mound and the second is behavioral tracking of changing temperatures. The mound itself functions as a solar collecting device. Workers thermoregulate by cycling up and down in the tunnel matrix of their mound as it warms or cools.

Laboratory binary choice tests indicated that well-fed colonies prefer 30-32°C while hungry colonies selected slightly cooler temperatures, apparently in an effort to improve colony growth efficiency.

Field studies of soil temperatures and fire ant thermoregulatory responses indicated that construction of a mound improved a colony's growth potential by about 10% and temperature tracking behavior improved growth potential by another 10%. Higher benefits might have been expected except that thermoregulation is generally only beneficial on sunny days when soil temperatures are within the growth window for fire ant brood (21-32°C).

NEST STRUCTURE, COLONY SIZE AND BEHAVIOR OF THE QUEENLESS PONERINE ANT SPECIES *PACHYCONDYLA SUBLAEVIS*

ITO, F. and HIGASHI, S.

Graduate School of Environmental Science, Hokkaido University, Sapporo 060, JAPAN

In 1987 and 1989 we made an ethological study on a queenless ant species *Pachycondyla sublaevis* in a *Eucalyptus-Acacia* woodland near Mt. Garnest (17° 41'S, 145° 07'E), North Queensland, Australia.

From the nest entrance fringed with 1 to 3 cm high pile of soil, a shaft run 20 to 30 cm horizontally and then downward up to 85 cm deep, with rare occurrence of ramification. The nests had 1 to 6 chambers in which brood was under the care of workers. A few chambers were used for rubbish room in which many pieces of brood, workers and preys were abandoned. Each colony consisted of 7 to 16 workers, one of which had active ovaries. Most of the workers with active ovaries were inseminated, i.e. gamergates; however, some of them were virgin, suggesting that new colonies were founded by budding. Each colony also contained 0-8 eggs, 0-12 larvae, 0-15 cocoons and 0-2 males. The polyethism was recognized among workers. The gamergates also took care of brood, especially of eggs.

NESTING AND DEFENCE IN TROPICAL HONEYBEES AND STINGLESS BEES: YOU ARE WHAT EATS YOU?

D. W. Roubik, Smithsonian Tropical Research Institute
Balboa, Panama (mailing address APO Miami 34002-0011, USA)

A major difference between the two permanently colonial bee groups is that honeybees can rapidly abandon a nest that is under attack, while the stingless bees cannot. Honeybee nesting is probably less regulated by natural selection than is that of meliponines, but strong behavioral similarities exist. The natural products incorporated by the highly eusocial bees, and their nesting behavior, are often correlated with the type of natural enemies with which they evolved. The nearly 500 honeybees and stingless bees are, with the exception of *Apis laboriosa*, tropical species. All have to contend with natural enemies of diverse kinds throughout the year. Both types of colonies combat ants—the major, small colony predators—with resin, although this is partly facultative in honeybees. Stingless bees use sticky resin or make hard resin balls to block the nest entrance. Aggressive behavior in both groups against large vertebrate predators is far more pronounced in species that build exposed or easily visible nests. The major vertebrate predators of bee nests are *Eira* (a tropical mink) and edentates (anteaters, armadillos, tamanduas) in the Neotropics, and primates, *Mellivora*, *Melursus*, *Selenarctos* and probably *Manis* and some civets in the Paleotropics. Honeybees choosing more vulnerable nesting sites seem better prepared to abscond after an attack, particularly some African *A. mellifera*, *A. florea* and *A. andreniformis*. Three of the seven known honeybee species do not build exposed nests, but some varieties are aggressive and some are not [1,2]. In the stingless bees, those nesting in exposed settings are highly aggressive, while many of those inhabiting nests that are virtually impenetrable by the normal suite of predators are not. Perhaps half of stingless bees have no defensive behavior at all against large animals, except the cryptic location of the nest in a living tree, and having a very small opening to the nesting cavity. Those tree-nesting species that will bite if molested seem to select nest cavities with larger openings—through which a vertebrate could extend its limb or muzzle [1,3]. Stingless bees and honeybees shun dead trees for nesting.

Truly exposed nests, those not contained in substrates, are built by very few stingless bees: some *Trigona* (10 spp), some *Paratrigona* (ca. 2 spp), *Dactylurina* (2 spp), some *Partamona* (ca. 3 spp) and one *Plebeia*. Their colony and worker sizes range from the tiny *Plebeia* (100 workers, 3 mm length) to gigantic colonies of > 10^5 (*T. amazonensis*, *T. spinipes*). Their nest materials range from outer coverings of paper (*T. dallatorreana*, *T.*

nigerrima), combined with thick organic "scutellum" [4] inner layers of resin and accumulated bee feces, to resin alone (*P.* aff *minima*), to thick layers of resin. The scutellum is typically thickest under the brood area, where the feces and meconia are deposited. In the Neotropical *Trigona* group, *Cephalotrigona* makes a scutellum although its nests are in tree cavities or narrow caves, and some *Trigona* make an extremely thick scutellum although they nest within substrates—e.g. *T. fulviventris* and *T. crassipes*. The former nests in the ground under large trees and is attacked by armadillos, so that its scutellum is probably needed, but in the latter and a related species, nesting only in trees, the trait is anachronistic [5].

Highly eusocial bees that nest gregariously, or in termite or ant nests, either parasitize the defences of their hosts or neighbors, or contribute to defence of the aggregation. Their presence in the termite or ant nests is permitted by use of resin in nest construction [1,6]. *Trigona moorei*, *Axestotrigona oyani*, *Apotrigona nebulata*, *Scaura latitarsus*, *Nannotrigona mellaria*, *Paratrigona peltata*, *Aparatrigona isopterophila* and *Trigona cilipes* are parasites of the defences of host ants and termites—none of these bees attacks colony predators, in contrast to their host *Crematogaster*, *Nasutitermes*, *Azteca*, etc. [1,7,8]. Aggressive workers of *Ptilotrigona*, *Partamona*, and *Trigona*, however, do attack predators and nest obligately in nests of aggressive hosts. Large nest aggregations, of two or three to several dozens of colonies, are often seen in *Trigona fuscobalteata*, *T. sapiens*, *Scaptotrigona luteipennis*, *Partamona* aff. *cupira* and other *Partamona* [3,9,10]. Only the last two genera are aggressive, but generally make no attack unless their colony is molested. In such aggressive species, the nest entrance is conspicuous and very wide, yet often of brittle, thin material. If broken or molested, many defending bees can exit quickly. Some stingless bees that are totally unaggressive build large nest entrances that look like those of aggressive species.

REFERENCES

1. Roubik, D. W. 1989. Ecology and Natural History of Tropical Bees, Cambridge University Press, New York. p. 514.
2. Seeley, T. D. 1985. Honeybee Ecology, Princeton University Press, Princeton, New Jersey.. p. 201.
3. Roubik, D. W. 1983. Journal of the Kansas Entomological Society, 56:327-355.
4. Nogueira-Neto, P. 1962. Journal of the New York Entomological Society, 70:239-264.
5. Camargo, J. M. F. and D. W. Roubik. in press.
6. Sakagami, S. F., T Inoue, S. Yamane and S. Salmah. 1989. Biotropica, 21:265-274
7. Wille, A. and C. D. Michener. 1973. Revista de Biología Tropical, 21:1-278.
8. Darchen, R. 1970, 1972. Biol. Gab. 5:151-187; Rev. Comp. Anim. 6:201-215.
9. Starr, C. K. and S. F. Sakagami. 1987. Insectes Sociaux, 34:96-107.
10. Camargo, J. M. F. 1980. Acta Amazonica, 10:1-175.

THE RELATIONSHIP BETWEEN NEST ARCHITECTURE AND SOCIALITY IN HALICTINE BEES

George C. Eickwort* and Penelope F. Kukuk**

*Dept. of Entomology, Cornell University, Ithaca N.Y. 14853, U. S. A.
**School of Biological Science, The University of New South Wales, P. O. Box 1, Kensington N. S. W., Australia

The subfamily Halictinae is noteworthy for containing a diversity of social systems, ranging from solitary through communal and semisocial to primitively eusocial, as well as a variety of nest architectures. We investigated how the different types of soil nests are correlated with the phylogeny of the Halictinae, in contrast to their correlation with social systems and with ecological factors.

PRIMITIVE NESTS. The primitive (plesiomorphic) architectural plan, based on the pattern in the Rophitinae, is for cells to be individually placed at the ends of long lateral burrows. This pattern characterizes the Nomioidini and is primitive in the Halictini, characterizing all of the Agapostemon complex, Lasioglossum s. s., and lineages of the Australian complex of Lasioglossum s. l., (e.g., Chilalictus) and the Evylaeus - Dialictus complex of Lasioglossum s. l.. All bees with this nest type are either solitary or communal; none have reproductive castes. The placing of cells in series along a lateral burrow occurs throughout these evolutionary lineages and is not characteristic of social groupings. The energetically expensive digging of lateral burrows may have evolved in soil-nesting bees as a protection against parasitism and cell usurpation by cleptoparasites, including conspecifics. Digging of long laterals has secondarily evolved in Halictus sexcinctus, which also has secondarily evolved solitary nesting from eusocial ancestry (1,3).

SESSILE CELLS. Nests which have shortened lateral burrows, so that the cells are within a cell's length of the main burrow, are primitive in the Augochlorini (2) and in Halictus in the Halictini (1,3), and characterize most Dialictus and many Evylaeus. The shortened laterals lead to greater adult-offspring contact, permitting rudimentary brood care, as well as enhancing control of reproduction by dominant queens (4). Sessile cells are characteristic of all eusocial and semisocial halictine bees, and can be considered a prerequisite to the evolution of castes, although solitary species also can construct nests with sessile cells.

CLUSTERED CELLS. Nests in which sessile cells are constructed or excavated in clusters, typically surrounded by cavities in the soil, are characteristic of most Augochlorini, the "carinate" group of <u>Evylaeus</u>, and secondarily in other lineages of <u>Lasioglossum</u> <u>s. l.</u> and <u>Halictus</u>. Social systems are not correlated with this architectural plan, as solitary, communal, semisocial, and eusocial species all build nests with clustered cells in cavities. The nests are characteristic of poorly drained soils, and may be an energetically expensive means of avoiding cell mortality due to flooding (5). Facultative construction of partial or complete cavities about clustered cells has been observed under laboratory and field conditions in several species of <u>Halictus</u> (1,5).

1. Eickwort, G. C. 1985. Pan-Pacific Entomology 61: 122-137.
2. Eickwort, G. C. and S. F. Sakagami. 1979. Biotropica 11: 28-37.
3. Knerer, G. 1980. Zoologisches Jahrbucher fur Systematik Okologie und Geographie der Tiere 107: 511-536.
4. Kukuk, P. F. and G. C. Eickwort. 1987. In: Chemistry and Biology of Social Insects, J. Eder and H. Rembold, Eds., Verlag J. Peperny, Munich.
5. Packer, L. and G. Knerer. 1986. Insectes Sociaux 33: 190-205.

BIOCHEMICAL SYSTEMATICS AND THE EVOLUTION OF SOCIAL BEHAVIOUR AND NEST ARCHITECTURE IN THE SUBGENUS *EVYLAEUS* (HYMENOPTERA : HALICTIDAE)

Laurence Packer
Department of Biology
York University
4700 Keele Street
North York, Ontario
Canada M3J 1P3

A phylogeny for eight species of sweat bee of the subgenus Evylaeus (genus Lasioglossum) is presented based upon cladistic analysis of 37 allozyme loci. All western palaearctic species for which behavioural data have been published were available for electrophoresis with the exception of one - L. (E.) nigripes. A hypothetical ancestral character suite was generated by comparing electromorphs of the ingroup with those from two species of Dialictus and four of Lasioglossum (s. str.). The latter four species provided no additional outgroup characters not also supported by comparison with the two Dialictus.

Monophyly of the taxon (Dialictus + Evylaeus) was suggested by four uniquely derived electromorphs not present in any of the four Lasioglossum species. Monophyly of Evylaeus was suggested by one allozyme locus.

It is not possible to say whether eusociality arose within Evylaeus because both social and solitary behaviours are present in the outgroup. The cladogram also suggests the following:

1) The bizarre perennial societies of L. (E.) marginatum are quite probably derived from an annual eusocial colony cycle.
2) The ancestral condition for social Evylaeus probably did not involve multiple foundress associations.
3) The habit of excavating a cavity around a cluster of brood cells is ancestral in Evylaeus and may constitute a synapomorphy for the group. This feature has been reversed in L. (E.) marginatum, presumably because of the large colony sizes that this species attains.
4) Some Evylaeus have open brood cells facilitating interaction between adults and immatures. Once thought to be associated with the highest degrees of social evolution in the Halictidae, Plateaux-Quénu has recently demonstrated this behaviour in L. (E.) albipes -- the species with the lowest level of sociality within the subgenus. Mapped upon the cladogram, "open brood cells" shows much homoplasy and it cannot be stated whether it is a primitive character for all species beyond the node leading to pauxillum or whether it has originated several times independently.

DYNAMICS OF BROODCELL CONSTRUCTION IN NESTS OF *M. BEECHEII*, AND ECONOMICALLY IMPORTANT STINGLESS BEE FROM COSTA RICA

Sommeijer, M.J.[1] and H. Arce[2]

[1]Utrecht University, Bee Research Dept., P.O.Box 80.086, 3508 TB Utrecht, Netherlands.
[2]Universidad Nacional, Heredia, Costa Rica.

INTRODUCTION

Melipona beecheii is one of the species of stingless bees that were of great importance for the pre-colombian beekeeping industry in Meso-America. The domestication of this species is still wide-spread in this region [1]. Natural nests are common in most Meso-American countries. This species has been studied by various authors [2,3,4]. The purpose of this presentation is to report the results of our behavioural studies concerning the nest building behaviour of this economic important species.

MATERIAL AND METHODS

Colonies were kept in observation hives [5]. The population of these nests was about 1000 bees. Various aspects of intranidal behavior were recorded by direct observation, by the use of event recorders and video registration. The progress of cell construction and oviposition behaviour was measured continuously during two periods of three days, recording the progress of this process every period of two or three hours. The locomotive pattern of the queen was studied by direct observation. The behaviour of the queen during the 'Provisioning and Oviposition Process (POP)' was studied in various nests. The occurrence of laying worker behaviour was recorded in one series of broodcells that were successively being oviposited.

<u>Oviposition behaviour in *M. beecheii*</u>

M. beecheii queens have a very characteristic position during the fixation period and the subsequent larval food discharging period: they are hanging vertically on the peripherial side of the cell that is being provisioned. The provisioning always starts with drumming behaviour of the queen executed on the bodies of workers who are alternatively inserting their bodies into the brood cell. After a few insertions by workers that are being drummed, one of these starts the discharging of larval food by contracting the abdomen and thus emptying the contents of the honey stomach into the cell. This contraction, even by the first discharging workers, is not so distinct as in other species of the genus, eg. in *M. favosa*. As soon as the food releases have started, the queen withdraws a bit from the rim of the cell, lowering her position on the side of the cell. Her drummings on body-inserting workers now diminish clearly. The total number of food releases per cell is variable. In one representative series we recorded an average number of 24 discharges. It is clear that the duration of the subsequent food releases is diminishing near to the end of the provisioning phase. The duration of the respective releases varies between 6 to 3 seconds. After a number of food discharges the workers, in stead of withdrawing quickly from the cell in which they have discharged food, remain on the cell and take a high standing position on the mouth of the cell. The ovipositions by laying workers occur in this phase of the POP. In *M. beecheii* the occurrence of worker ovipositions is very common. An average number of two worker eggs were released in each brood-cell.

Cell building behaviour

It is very characteristic for the group of the stingless bees that for each egg laid by the queen a new cell has to be built. Old combs are being removed as soon as the bees have emerged. There is no distinct diurnal pattern of building and oviposition activity in *M. beecheii*.. In principle cell-building and oviposition goes on day and night. Cells are being built successively and oviposited facultatively batched [6]. Building material for the construction of new cells is progressively scraped from the older combs. This leads to distinct differences in the thickness of the cell walls. Thus older combs have a light colour, because of the loss of cerumen. New building material is produced by the depositing of wax scales, protruding from the dorsal wax glands, on specific wax deposits. These wax deposits can be found on the comb, but also on storage pots or on other nest structures. Plant resin is brought in by foragers and being dumped on a resin dump, generally very near to the exit of the nest. Small parts of this stored resin are being taken to the wax deposit and being mixed with the wax. This mixture of building material 'cerumen' [7] is used, in addition to the cerumen scraped from existing nest structures, to built new broodcells.

ACKOWLEDGEMENTS

We are grateful for the assistance of Ruud van Leeuwen in carrying out a part of the behavioural observations. Dick Koedam supplied technical assistance and made pictures of the behaviour inside the nests.

REFERENCES

1.Weaver, N., and Weaver E.C. (1981) Beekeeping with the stingless bee Melipona Beecheii, by the Yucatan Maya. Bee World 62 (2): 7-19.

2. Darchen, R. and B. Delage Darchen. 1974. Nouvelles expériences concernant le déterminisme des castes chez les Mélipones (Hyménoptères Apidés). C.R. Acad. Sci. Paris 278 D : 907-910.

3. Darchen, R. and B. Delage Darchen. 1975. Contribution à l'étude d'une abeille du Mexique Melipona beecheii B. (Hymenoptère: Apide). Apidologie 6: 295-339.

4. Darchen, R. and B. Delage Darchen. 1977. Sur le déterminisme des castes chez les Mélipones (Hyménoptères Apidés). Bull. Biol. Fr. et Belg. 111: 91-109.

5. Sommeijer, M. J., J. L. Houtekamer and W. Bos. 1984. Cell construction and egg-laying in Trigona nigra var. paupera Provancher, with a note on the adaptive significance of the typical oviposition behaviour of stingless bees. Insectes Soc. 31: 199-217.

6. Sakagami, S. F. and R. Zucchi. 1974. Oviposition behavior of two dwarf stingless bees, Hypotrigona (Leurotrigona) muelleri and H. (Trigonisca) duckei, with notes on the temporal articulation of oviposition process in stingless bees. J. Fac. Sci. Hokkaido Univ. Ser. VI Zool. 19: 364-421.

7. Michener, C. D. 1974. The social behavior of the bees. Harvard University Press, Cambridge, Mass.

NESTING BEHAVIOUR OF *APIS FLOREA* F. (HYMENOPTERA : APIDAE) IN KHUZESTAN, IRAN

M.S. MOSSADEHG

Department of Entomology, College of Agriculture, Shahid Chamran University, Ahwaz, Iran

Nesting behaviour, swarm and honey production of Apis florea F. was studied for 12 years from sept. 1977 through sept. 1988 in khuzestan, southwest, Iran. Of the 3720 colonies studied, 166 were under constant observation. Nests were built at a height above the ground of ./3 -14m. on trees, bushes, hedges, wells, under the eaves, on windows, entrance to basement of the buildings, cellar's chimney, etc.. Combs were built in a WE direction at the east side of trees and building in spring and summer, and at the south side in winter.

Brood rearing started in the middie of Feb. and reached its peak in June. Then it declined and little or no brood was reared in August. Brood rearing started again about the end of August and continued until the end of Nov.. Drones were reared at the middle of March followed by queen cells production. No brood was reared during December- Junuary.

The largest sesonal comb observed was 30cm. wide and 60cm. deep. An A. florea colony is able to built a wax comb 16cm. to 25cm. wide and 17cm. to 40cm. deep in 30 days from its commencement depending on its population. The upper part (honey storage) of the comb is 6-23cm. Thick when it is built round a twig. The drone brood area occupies about 15% to 40% of the comb in a weil developed colony. The number of queen cells was 4-32/comb. When the queen cells were destroyed, a colony built up 141 cells. Emergency queen cells were observed frequently on both sides of the combs. The number of such cells were 1-14. Swarming occured from the middle of April with two peaks in May and July for winter and spring stablished colonies respeectively. Then it declined but continued until Nov.. UP to 6 swarms were observed. Many of the spring stablished colonies (52%)did not rear drone or produce any queen cells, but yielded a good crop. Copulation of the queen was observed at on elevation of about 4 meters above the colony.

The remained wax or deserted combs atracted honeybee swarms as a lure; 24.1% of the swarms were resettled on the remained wax or deserted combs from previous colonies. From which 18.8% on the remained wax; 4.2% on very old combs which bees than expended it for rearing brood; 1.2% on the clean and rather fresh released combs that bees reused it immediately.

The annual yield of honey per 1-8 months old colony was ./1 to8kg. with an average of 1.33kg. measured for 2350 colonies.This was ./8 to 8kg. for 4-8 months old colonies with a mean of 2.29± 1.39kg. measured for 74 colonies. The maximum migration of colonies occured at the end of Nov. following honey harvesting. When the colonies were not distubed in autumn, 24% migrated in late autumn to a new nearby site and 76% remained at the same place to pass the winter. 14.5% of the colonies remained at the same place for 2-7 years. Honey storage was very important for the survival of colony specially in the cold winter. Some colonies died in Decmber and January due to lack of food. When the colonies without any honey storage were fed with sugar syrup, they survived the winter.

Mismanagement as well as heavy use of insecticides and herbicides drastically declined the number of colonies in the province. On the contrary, the war had a positive effect on population and number of colonies. From Oct. 1980 through Oct. 1984 in some areas, the number of colonies and their population increased rapidely, as there was no disturbance or damage to colonies due to mass evacuation. Up to 11 colonies were found on a christ's thorn (Rhamnus spina-christ) tree. Many giant colonies were found either in buildings or in hedges. The population of some of these colonies was stimated nearly 60000 bees. The constructed combs of two such colonies were 30x70cm. and 30x125cm. respectively.The former was in a cellar's chimney and the latter behind a window inside a room. Colony's miggration, was due to honey harvesting, direct sunshine in summer, hail storm, direct heavy rain on population, smoke and shades in winter.

THE OENOCYTES AND WAX PRODUCTION IN HONEYBEES

H.R. Hepburn , R.T.F. Bernard , B.C. Davidson , P. Lloyd ,

S.P. Kurstjens and S.L. Vincent

INTRODUCTION

Honeybees are unique in the synthesis of beeswax, a material sufficient for constructing a suspended nest. The chemical complexity of this wax indicates an extraordinary metabolic specialization of the cells that secrete it. Cytological (1) and biochemical (2) data suggest a possible oenocyte origin of beeswax hydrocarbons. We now report data using standard techniques, on the ultrastructure and hydrocarbons of wax gland cells over the cycle of wax synthesis. We determined which cells exhibit features that are best correlated with the cycle of synthesis and hydrocarbon content of newly secreted beeswax.

RESULTS AND DISCUSSION

From the time of emergence of the adult worker bee, A. m. capensis, before and during the cycle of wax synthesis and secretion, both the the epidermis and adipocytes of the wax gland complex lacked smooth endoplasmic reticulum (SER), the organelle essential to insect lipogenesis. Only the oenocytes possess SER and only they exhibited ultrastructural and chemical changes (Table 1) correlated with the cycle of wax secretion (3). The oenocyte hydrocarbons (Table 2) correspond with those of scale wax, excepting C_{25} and C_{35}. We believe that the hydrocarbons of the epidermis reflect oenocyte-derived material in transit because the epidermis lacks SER and its age-related changes in hydrocarbons are not synchronized with the cycle of secretion. These results are consistent with those for honeybees (1, 2) and those for hydrocarbon synthesis in other insects (4).

REFERENCES

1. Boehm, B. 1965. Zeitschrift fur Zellforschung, 65, 74-115.
2. Piek, T. 1964. Journal of Insect Physiology, 10, 563-572.
3. Hepburn, H.R. 1986. Honeybees and wax. Springer, Heidelberg.
4. Diehl, P.A. 1973. Nature, 243, 468-470.

Table 1. Oenocyte hydrocarbons (%) as related to worker age.

Hydro-carbon	Bee age, days 3	6	9	12	15	18	21	
25:0	23.8	23.0	21.9	26.9	26.6	30.0	29.9	31.0
27:0	38.8	37.2	35.8	36.1	40.7	44.0	47.8	51.7
29:0	4.5	6.4	7.5	12.0	9.7	8.0	6.3	2.0
31:0	3.0	2.6	2.8	2.8	3.5	2.0	3.0	3.4
33.0	6.0	6.4	4.7	4.6	5.3	4.0	4.51	3.4
35:0	1.5	1.3	0.9	0.9	1.8	2.0	3.0	1.7
Total	77.6	76.9	73.6	83.3	87.6	90.0	92.5	93.2
31:1	4.5	6.4	8.5	6.5	5.3	4.0	4.5	3.4
33:1	17.9	16.7	17.9	10.2	7.1	6.0	3.0	3.4
Total	22.4	23.1	26.4	16.7	12.4	10.0	7.5	6.8

Table 2. Tissue and wax hydrocarbons (%) of Cape workers.

Hydrocarbon	oenocytes	epidermis	scale wax	comb wax
25:0	26.6	22.2	11.4	6.6
27:0	41.5	37.2	39.8	33.3
29:0	7.0	9.0	8.1	13.8
31:0	2.8	3.5	4.1	8.9
33.0	4.6	12.9	14.6	15.4
35:0	1.6	2.8	3.3	3.3
Total	84.3	87.6	81.3	81.3
31:1	5.4	-	-	-
33:1	10.3	4.2	4.9	6.5
35:1	0.0	8.2	13.8	12.2
Total	15.7	12.4	18.7	18.7

BEE GLUE (PROPOLIS) AND ITS ANTIVIRAL PROPERTIES : RECENT ADVANCES AND CURRENT RESEARCH

Koenig ,B. and Dustmann , J.H.
Lower Saxony State Institute for Bee Research
Wehlstr. 4 a , D- 3100 Celle , Federal Republic of Germany (FRG)

Evolution of sociality in insects is influenced - among other factors involved - to a great extent by the choice of nest construction materials , as has been pointed out by HANSELL (1984). Regarding especially the bees , the evolutionary sequence apparently started with nest construction from mud , added plant resins and switched over basically to wax without giving up the use of resins completely - at least in most of the species and races of the highest developed bees of the genus Apis . Thereby the bees make use of the antimicrobial properties (in a broad sense) of those resins , being themselves the results of the chemical evolution of the source plants . Since more than 2000 years ago man has also taken advantage of those properties by using bee glue or propolis for medical purposes - and by the actually worldwide increasing interest in traditional pharmaceuticals from nature this special nest material is gaining more and more economical importance . However , depending on their diverse geographical origin and thereby of course on the diversity of their source plants , propolis samples differ remarkably in their biological properties , among which the antiviral activity is currently the most interesting one . We especially in our lab predominantly are dealing with tests of propolis against animal herpes viruses . Since the appearance of our last paper (KOENIG and DUSTMANN 1989) we tested samples from additional geographic areas and changed and improved our test conditions , enabling us to detect virus titer reductions from four to more than ten orders of magnitude resp. powers of ten by propolis concentrations between 5 and 15 ug/ml cell culture medium for example . Along with the presentation of our own new data important recent advances by others in the field will briefly be summarized , especially relating to inhibition studies on virus-specific enzymes , and thereby insight in current research trends in this field will be given . The significance of these results for honeybee diseases as one of the important topics in their biology will be mentioned .

Hansell,M . 1984 . How to build a social life . New Scientist 102 (1412) 16 - 18

Koenig,B. , Dustmann ,J.H. 1989 . Tree resins , bees and antiviral chemotherapy . Animal Research and Development 29 : 21 - 42

NEST MATERIAL TECHNOLOGY AND COLONY BIOLOGY IN SOCIAL WASPS. IS THERE EVOLUTIONARY CONSTRAINT ?

Michael Hansell, Zoology Department, University of Glasgow, Glasgow, Scotland, U.K.

Artefacts are not only phenotypic extensions of the builder but are also responsible for altering the environment in which the builder lives. The effects of such environmental modification may not simply alter selection pressures acting upon building behaviour but on almost any aspect of the builders biology. This may be seen for example in the loss of body fur and thermoregulatory physiology in the naked mole rat, *Heterocephalus glaber*. Its burrow-dwelling subterranean life now ensures that it lives in a constant temperature environment [1].

Examination of nest building by social insects may therefore reveal major alterations in biology which have arisen as a consequence of the evolution of new nest architectures or new nest materials. A broad inspection of the social Hymenoptera and of the Isoptera suggests that new nest material technologies have contributed to the evolution of greater social complexity in the bees (Apidae) though the evolution of wax as a building material and in the wasps (Vespidae) though the evolution of paper. In the ants (Formicoidea) and termites (Isoptera) new building materials such as carton and larval silk have assisted in the invasion of new habitats [2].

Since evolution by natural selection depends upon the modification of an organisms existing biology, there may be limitations or constraints on possible directions of future evolution [3]. It is therefore to be expected tha nest design or technology may act as such a constraint.

Study of wasps of the subfamily Stenogastrinae (Vespidae) shows that technological constraint is probably the best hypothesis to explain its universally small colony sizes [4]. Setting choice of nest materials against the cladogram of genera for the subfamily it is apparent that the evolution of paper did not coincoide with a sharp increase in social complexity and that the secondary evolution of mud as a nest material has occurred in some cases. Among evidence for the poor quality of paper in the subfamily is the discovery that the nest material of *Anischnogaster laticeps* bears a rich growth of fungal hyphae. This may serve to strengthen the nest although it presumably reduces its life expectancy.

Examination of nest material of Polistinae shows it to be varied but often of a composition typical for the genus. The source of plant material may be woody stem cells, scales of cuticular cells or plant hairs both simple and branched.

Examination of two genera *Polistes* and *Mischocyttarus* was carried out to determine if differences in their nest material was associated with any other clear differences in their biology. They are similar in their nest design and colony size, making them convenient for this comparison; they are, however, typically quite different in the composition of their nest paper. *Polistes* has a highly genus typical paper of woody fibres right across the wide range of body sizes found within it; the paper of *Mischocyttarus* is quite different, being composed of flakes of plant cuticular cells. The most obvious difference in the biology of the two genera is the much smaller body sizes shown by *Mischocyttarus* species compared to *Polistes* species. This raises the question of whether poor quality nest material is responsible for limiting body size in *Mischocyttarus*. Examination of papers across the size range of the two genera shows that one of the larger species of *Mischocyttarus* has in fact achieved fibrous paper though the use of plant hairs, while the mandibular design of larger *Mischocyttarus* species is similar to that of *Polistes*. The technological constraint hypothesis is therefore not supported in this case.

Examination of the relationship between building behaviour and other aspects of biology using a species comparative approach can therefore be used to reveal its influence on social insect colony biology.

REFERENCES

1. McNab, B.K. 1966. The metabolism of fossorial rodents: A study of convergence. Ecology, 47: 712-733.

2. Hansell, M.H. 1989. Les nids des insectes sociaux. La Recherche, 20: 14-22.

3. Maynard Smith, J. *et al*. 1985. Developmental constraints and evolution. Quarterly Review of Biology, 60: 265-287.

4. Hansell, M.H. 1987. Nest building as a facilitating and limiting factor in the evolution of eusociality in the Hymenoptera. Oxford Surveys in Evolutionary Biology. P.H. Harvey & L. Partridge (eds) 4: 155-181.

NEST CONSTRUCTION AND REPAIR AND ASPECTS OF THEIR EVOLUTION IN THE SOCIAL WASP, *POLISTES FUSCATUS*

H. A. Downing
Department of Biology
University of Wisconsin
Whitewater, Wisconsin 53190
U.S.A.

Although the elaborate nest construction of social insects is fascinating, relatively few researchers have investigated their building behavior. Grassé (1) developed the first good theory about the control of construction behavior from his work on termites. His stigmergy theory states that construction is regulated by a linear series of if-then decisions and that the cues regulating those decisions come primarily from construction already completed. Thus in termites and other species that have been investigated, the ability to build a nest comes not from some sort of inherited nest blueprint, but an inherited building program (1,2,3,4,5,6,7). The program consists of the steps of construction, the cues that regulate the transitions between steps, and a responsiveness to those cues at particular thresholds (7).

Stigmergy theory, however, assumes a linear sequence of construction steps, and social insects often build by means of a non-linear sequence of steps. That is, they may have a choice of two or more different building acts at any one time. Little work has been done on how social insects are able to weigh the alternatives and decide where to build next (7).

My work on the construction behavior of Polistes fuscatus indicates that they use multiple cues, both in evaluating where and how to build next, they can modify normal construction to make repairs, and are able to jump flexibly from one type of construction to another (7, 8). In addition, contrary to the normal assumption that construction behavior in social insects is stereotypic, P. fuscatus improve their building repair techniques, reduce the amount of time to make a repair, and decrease the amount of pulp needed for a repair with experience, which indicates learning is occurring (9). Learning in this construction context may have evolved because of the unpredictable nature of nest damage. Programming all of the necessary repair behavior a wasp might need in its lifetime would be cumbersome. Instead, it appears that general repair capabilities are inherited and then modified as needed with experience.

Comparisons between the building programs of social insects and those of solitary bees and wasps indicate that there are distinct differences. Although there is a great deal of variability, the building programs of solitary species appear to be regulated by fewer cues, construction, provisioning, and oviposition are basically linear with dichotomous decision points, and in some, nest construction appears to pass through phases during which the builder will respond to cues of the present phase, but not to those of a previous phase (4, 5, 10, 11, 12, 13). Nest repair ability may be limited, as well (10, 11, 13).

Thus it is apparent that during the evolution of complex building patterns in social insects certain important factors have developed in the building program. First, the non-linearity of social insect programs indicates that they tend to have greater numbers of building subroutines with greater accessibility than do those of solitary wasps and bees. Second, there is a greater separation of different tasks, even to the point that they are performed by different castes. Third, with this separation, the decision of what to do next often involves the analysis of multiple cues, making it necessary for some method of integrating the information coming from different cues. Lastly, because the nests of social insects are long term investments, there must be strong selection for increased building flexibility and repair capabilities (9, 14).

REFERENCES

1. Grassé, P. 1959. Insectes Sociaux, 6: 41-81.
2. Darchen, R. 1962. Insectes Sociaux, 9: 23-38.
3. Crook, J.H. 1964. Proc. zool. Soc. Lond., 142: 217-255.
4. Eickwort, G.C. 1975. Z. Tierpsychol., 37: 237-254.
5. Smith, A.P. 1978. Anim. Behav., 26: 232-240.
6. Downing, H.A. and R.L. Jeanne. 1987. J. Ethol., 5: 53-66.
7. Downing, H.A. and R.L. Jeanne. 1990. Anim. Behav., 39:105-124.
8. Downing, H.A. and R.L. Jeanne. 1988. Anim. Behav., 36:1729-1739.
9. Downing, H.A. in prep. The role of learning in the nest repair of the paper wasp, *Polistes fuscatus* (Hymenoptera, Vespidae).
10. Fabre, J.H. 1914. The Mason Bees. Transl. by A.T. de Matteos. Dodd, Mead, and Co., New York.
11. Raw, A. 1972. Trans. R. ent. Soc. Lond., 124: 213-229.
12. Brockmann, J. 1980. Anim. Behav., 28: 426-445.
13. Downing, H.A. submitted. Nest repair and the building program of the organ-pipe wasp, *Trypoxylon politum* (Hymenoptera: Sphecidae).
14. Richards, O.W. 1971. Biol. Rev., 46: 483-528.

COMB-CUTTING INAN AUSTRALIAN PAPER WASP, *ROPALIDIA PLEBEIANA* :A NEW METHOD OF COLONY FISSION IN SOCIAL WASPS

S. Yamane*, Y. Itô** & J. P. Spradbery***

*Biological Laboratory, Faculty of Education, Ibaraki University, Mito 310, Japan, **Laboratory of Applied Entomology and Nematology, Nagoya University, Nagoya 464, Japan, & ***Division of Entomology, CSIRO, Canberra, ACT 2601, Australia

Nests of Ropalidia plebeiana consist of exposed, single combs suspended by numerous pedicels from the roofs of overhanging cliffs and concrete bridges. In spring, overwintered, inseminated females (foundresses) of this species adopt one of two methods to begin a new colony cycle. Many wasps reutilize old nests within a nest aggregation whereas others establish new nests around the edges of the aggregations, usually with several foundresses per nest. Despite the dense aggregation of nests, each is occupied and defended by a discrete group of females.

Two nest aggregations under a concrete bridge over Sheep Station Creek, near Batemans Bay, N.S.W. were studied by marking all nests with numbered vinyl tags and mapping their shapes and positions in 1987-1989. Many large nests, which had more than ten foundresses, were cut with their mandibles into two or more pieces. Thus, completely independent colonies were produced.

Prior to the division, each of major egg layers possibly with some subordinates tended to occupy different part on a single comb. These females gnawed cells at the intermediate zone between such "territories" and ultimately separated the comb. The independence of separating colonies was established when the groove was made at the border between these "territories". Many other females also built new nests near the nest aggregations, but addition of new nests by the comb-cutting occupied 34.8 % of the increase of nests. This method of colony fission is so far unknown in any eusocial Hymenoptera.

EVOLUTION OF THE AERATION SYSTEMS IN SOME HIGHER TERMITES

Charles NOIROT
Laboratoire de ZOOLOGIE, Université de BOURGOGNE
6 Bd. GABRIEL, 21100 DIJON, FRANCE.

In termites, the nest is faced with contradictory selective forces. For defense, a concentrated and massive structure is favoured, but is opposed by the problems of gas exchange and dissipation of metabolic heat. The tremendous diversity of nest structures points to a multiplicity of solutions, analyzed only in very few cases.

As regards the aeration, the most obvious structural adaptation is the system of large "ventilation canals" fully open to the exterior, often at the tip of turrets or "chimneys", in contrast to the usual closure of the termite nests. These canals anastomose inside the nest in a loose network. Grassé (1944) coined the term exoecie for such a system, in the view of its complete separation from the galleries of the nest proper. Although this separation seems now not so absolute (with differences according to species and colony cycle), the "ventilation system" is not (or rarely) utilized by the nest inhabitants and remains on the whole topologically outside the nest. The role of these systems in gas exchange seems unquestionable, but the air currents in the canals (often suggested by their topography, and postulated as a thermoregulatory device) need a reappraisal.

The exoecie seems species-specific, although this has not always been rigorously demonstrated. It evolved several times independently, as evidenced by its presence in at least a dozen species, dispersed in five genera and three subfamilies. In each genus, nests with and without an exoecie were observed and compared. This comparison is not very informative in some genera for example in Cephalotermes (Termitinae) and Protermes (Macrotermitidae) where the nests devoid of exoecie bear no obvious ventilation system. In other species, some internal cavities of "closed" nests could be, tentatively, considered as "precursors" of an exoecie.

In the genus Macrotermes (Macrotermitinae), several species with a "closed" nest have an inner system of large cavities, involved in aeration and thermoregulation, more or less similar to an exoecie although not open to the exterior. Such a closed ventilation system could easily evolve into an exoecie: the opening of the large canals at the surface could be "compensated for" by an internal closure isolating, at least in part, these canals from the hive. Indeed, in M. bellicosus the nests are always closed, but in some populations (geographical races ? sibling species ?) the old nests acquire several wide openings at their base and the ventilation system becomes then similar with an exoecie. In several subterranean Odontotermes (Macrotermitidae), the "closed" nest is made of large pear -shaped chambers, each with a fungus comb at the bottom. The narrow part of the cavity extends upwards and ends just below the soil surface. Here, it is not too difficult to imagine the development of an exoecie, by opening at the tip and closing just above the fungus comb.

These hypotheses remain speculative, but they can be tested by a comparative study of the development of the nests. From the few detailed observations avalaible, such complex nests undergo, from the establishment of the founding pair till the mature nest, a real ontogeny which may throw some light on the phylogeny of the ventilation system.

1. Grassé, P.-P. 1944. Ann. Sci. Nat., Zool., (11) 6: 97-172.

POLICALIC STRUCTURE OF *ANACANTHOTERMES AHNGERIANUS* JACOBSON (ISOPTERA, HODOTERMITIDAE) SETTLEMENT

Shatov K.S.

Department of Entomology, Faculty of Biology,

Moscow State University, Moscow, 119899, U.S.S.R.

Termite A.ahngerianus inhabiting desert of Middle Asia, is known to build monocalic mounds each with 12-20 thousand of individuals.

Studies of nest structure and population census were conducted in a experimental plot (nest density - 13/ha) situated 160 km eastward Ashkhabad . Among 49 mounds surveyed 8 polycalic settlement with 33 additional foraging nests were detected.

Polycalic colony includes central mould and several (up to 8) foraging nests up to 15 m of distance. Foraging nests are visible as little rising above ground level with the great cell inside containing food storage and small flat cells below them interconnected by vertical galleries. All the foraging nests ave connected with central mound by large underground galleries. Foraging nests population may contain up to 2000 individuals of different castes. Relative caste abundance depends on season: in spring 87% of workers, 4% of soldiers, 1% of nymphs, and 8% of alates; in summer - 88, 7, 5 and 0% respectively. Later instar workers prevailed in the foraging nests (70%), whereas in the central mound this age group was minor (20%). Royal couple, eggs and larvae were found in the central mound only, and were absent in the foraging nests.

Our data suggest that foraging nests in A.ahngerianus serve for aliment storage, termite forager shelter and residence alates under flight period; this phenomenon must be taken in consideration during population structure analysis and development of control measures.

STRUCTURAL VARIATION IN NESTS AND THEIR ADVANTAGES IN CLOSELY RELATED *ODONTOTERMES* SPP. (ISOPTERA: TERMITIDAE)

D.Rajagopal
Department of Entomology
University of Agricultural Sciences
GKVK, Bangalore-560 065, INDIA

A number of fungus growing termite species of Odontotermes build conspicuous nests with a complex design and architecture above the ground level. Eventhough many investigations have been made on the nest construction in termites, there is still a considerable lack in our knowledge and their advantages to the existence of termites. Among several nest building termites of Odontotermes, O.obesus (Rambur), O.redemanni (Wasmann) and O.wallonensis (Wasmann) are dominant and are quite distinct in their nest construction.

Studies on the structural variations and structural mechanism involved in thermoregulation in the three species and the closing and opening mechanism of the nests in relation to season in O.wallonensis were studied. Nest temperature was monitored at the ventilation shafts at 0.60 and 1.20 m depth using a Digitherm 1000C thermometer.

The earthen nest of O.obesus is tall, subcylindrical without any opening on the surface. The basal diameter varied from 1.25 to 2.0 m and attains a height of 3.0 m with a characteristic wall like longitudinal ridges arranged radially with concavity around the nest. Internally, these ridges have hallow spaces extended from a large central chamber with fungus combs lodged in an unilocular fashion.

The nest of O.redemanni is dome shaped at the base, subconical, without any opening on the surface. The basal diameter is irregular with varied size grows to the maximum height of 2.0 m. The ventilation shafts are extended internally from the mound surface and diversified to various parts of the nest. They do not communicate directly to the termites inside. The fungus combs are situated in separate chambers and arranged in multilocular fashion.

O.wallonensis constructs an extremely hard and compact dome shaped earthen structure with diameter of 1.0 to 2.5 m and grows to the height of 1.5 m with open chimney like outgrowths on the surface. The height of the chimneys varies from 15 to 30 cm. The number of chimneys varies with the size of the nest. They do not communicate directly to the termites inside the nest. Internally the ventilation shafts extend from the surface to the interior and diversify to the entire nest. The fungus combs are situated in separate chambers with interconnected galleries for the movement of termites.

The three species are capable of constructing the nests with varied design and structure above the ground level with the mechanism of maintaining the thermoregulation. There are strong indications to believe that the chimney's in O.wallonensis leading to the vertical shaft ramifying internally connecting nests through small perforations play a significant role in regulation of the air movement and maintenance of internal temperature.

During winter months, the termites closed the opened chimneys either partially or completely when the minimum temperature comes down to 13°C. They are capable of maintaining the nest temperature higher than atmospheric temperature from 27°C to 30°C during winter months. However, the chimneys are opened during the remaining period. The maintenance of more or less constant temperature was recorded in the nest of O.wallonensis compared to the fluctuating atmospheric temperature which has given the clue that the regulating mechanism could be partially associated with the increased in the number of chimneys from 1 to 9 in four years of its growth and even the diameter of the chimney also increased from 5 to 7.5 cm (1).

In closed nests, the structure of O.obesus will give a clear indication that they construct the tall nest as the colony grows older with more number of longitudinal ridges with concavity around the nest for gaseous exchange. Similarly O.redemanni constructs many cones on the nest surface in a broader area and the number of cones also increased in bigger nests. The maintenance of thermoregulation is only through the nest wall where simple diffusion of gases appears to be sufficient as thin structure of walls are found to be porous which also promote gaseous exchange through small perforations maintained on the nest wall of these two species.

1. Rajagopal,D., 1979. Ecological studies of the mound building termite, Odontotermes wallonensis (Wasmann) (Isoptera: Termitidae) Ph.D. thesis, UAS, Bangalore, India, P.205.

NESTS DYNAMICS AND NEST BUILDING IN *MACROTERMES BELLICOSUS* (ISOPTERA: MACROTERMITINAE)

Y. Tano (1) & M. Lepage (2)

(1) Laboratoire de Zoologie, Faculté des Sciences et Techniques, 22 BP 582 Abidjan 22, Côte d'Ivoire
(2) Laboratoire d'Ecologie, URA CNRS 258, E.N.S., 46 rue d'Ulm, 75230 Paris Cedex 05, France

INTRODUCTION

Large epigeous mounds built by Macrotermes species are a main characteristic of African landscapes. Several authors have studied the mound architecture [1] [2] and outlined the elaborete structure which was achieved by these insects. It was assumed that such large nests would require several years to be built, being more or less permanent structures. But recent studies revealed dramatic variations in nest density between years: from 14.3 ha^{-1} down to 0.8 nest ha^{-1} on a lateritic plateau of Côte d'Ivoire [3], or from 0.09 up to 0.51 ha^{-1} in Tsavo East National Park in Kenya [4]. These changes have brought up the question of the dynamics of the Macrotermes nests. This problem was studied within a drainage basin followed during 5 years in North-West of Côte d'Ivoire, which environment was described in a previous paper [5].

MATERIAL AND METHODS

Nest dynamics was followed along transects checked at intervals during the study period. 132 Macrotermes nests were checked to determine birth and mortality rates. the same nests have been utilized to calculate growth rates according to the nest parameters (high, basal area). Nest erosion was measured on 37 nests and determined by regular photographs and nest measurements.

NEST ARCHITECTURE AND MOUND MATERIAL

Nests have been excavated down to several meters to described the internal structure of the mound. Characteristics did not differ from what has been found by previous workers. Mound size varied between 0.07 to 3.00 m in high, 0.025 to 7 m^3 in volume and 0.5 to 25.5 in basal area. Young nests exhibited little modification of soil horizons: volume above ground corresponded approximatively to the volume excavated below. Samples of mounds were taken for soil analysis and clay determination and compared with the adjacent soil.

NESTS DYNAMICS

Annual census led to high nest mortality: annual mortality rate varied from 50 to 55%. Measurements showed a great variability of the erosion rate between nests, according to the soil type, the vegetation or the slope. A termitaria could lost from 0 to 0.41 m of its high in one year. The soil lost amounted to 0.14 m^3 per mound in average but up to 0.59 m^3 for some mounds.

Growing rates determined in this study revealed stricking increase of mound size in short intervals: a mound could growth from 0.80 m to 3.20 m in high within one single year, which involved processing of several m^3 of soil.

CONCLUSION

The results obtained drawed a new picture of Macrotermes nest dynamics, which nest populations appeared less stable than expected. Nest density fluctuated greatly within short periods. Growing rate of the nests expressed the great capacity of the colony working force to handle several m^3 of soil per year, being able to built large sized nests within unexpected delays.

REFERENCES

[1] Ruelle, J.E., 1964. In: Etudes sur les Termites Africains, A. Bouillon ed., Leopoldville Univ., 327-362.

[2] Noirot, Ch., 1970. In: Biology of Termites, K. Krishna & F.M. Weesner eds, Academic Press, vol.2: 73-125.

[3] Lepage, M., 1984. J. of Animal Ecology, 53: 107-117.

[4] Pomeroy, D.E., 1983. Kenya J. of Science & Technology, series B, 4: 89-96.

[5] Tano, Y. & Lepage, M., 1987. In: Chemistry and Biology of Social Insects, J. Eder & H. rembold eds, Verlag J. Peperny, München: 613-614.

NEST ARCHITECTURE OF THE SUBTERRANEAN *MICROTERMES OBESI* HOLMGREN (ISOPTERA : TERMITIDAE)

N.G.Kumar and D.Rajagopal
Department of Entomology
University of Agricultural Sciences
GKVK, Bangalore - 560 065, INDIA

The subterranean termite, Microtermes obesi is known to cause heavy damage to agricultural, horticultural crops and forest plantations in oriental and African region (1). However, the detailed information on its nest architecture is very scanty except for brief description on its biology (3).

Investigations were made on the location and the nest architecture of M. obesi by vertical section cutting of the soil at the Regional Research Station, University of Agricultural Sciences, GKVK, Bangalore, India.

The nest architecture of M. obesi consisted of scattered fungus combs found in small chambers interconnected by narrow galleries (1.0 to 3.0 mm) at a depth of 25 cm upto 125 cm placed about 20 cm to 63 cm apart. The fungus combs were small, slightly round or oval in shape, light dark to light yellow in colour and their size varied from 3.6 to 5.0 cm length, 2.8 to 5.0 cm width and 3.5 to 4.0 cm height. The transverse hole size varied from 0.2 to 2.3 cm in length and 0.15 to 0.3 cm height and the fragile fungus layers were arranged one above the other. Further these combs were interconnected to foraging galleries and nursery cells. Majority of the foraging galleries were found between 25 to 60 cm below ground level.

A king and a physogastric queen were found in the nest which consisted of a small, closed semi ellipsoidal cell (3.7 cm x 3.5 cm) at a depth of 120 cm. The chamber was flat at the base with a dome shaped roof, plastered with clay. Cell was constructed in east-west direction and the queen was lying lenghtwise facing its head towards east. Several galleries were found only in the back and left side of the royal cell (1.7 cm l x 0.2 cm w to 0.1 cm d).

Incubation cavities with cluster of eggs were noticed 20 cm away from the royal cell. Egg depository cell was an extended part of nursery cell and irregular in shape with plastered flat base. Its size was 1.6 x 1.3 cm. Further it lead to the irregular shaped main nursery cell (3.2 x0.6 cm) where the activity of immature stages was observed. From this cell two main galleries (0.25 cm diameter) were noticed. Several nymphs including developing alates were noticed in the nursery cell.

Observations on the distribution of multilocular fungus chambers from 25 cm to a depth of 125 cm and distribution of nymphs, minor and major workers and soldiers in these combs were similar to the earlier report (3). The presence of incubation cavities and nursery cell in M. obesi nest confirm the earlier findings (2) and differs with its location, where this species was associated with the mound of Odontotermes wallonensis (Wasmann). This is for the first time the detailed description of a sub-terranean nest architecture of M. obesi is given under Indian conditions.

1. Chatterjee, P.N. and J.L.Thakur. 1964. Indian Forest Rec., 10:219-260.

2. Rajagopal,D. 1979. Ecological studies of the mound building termite, Odontotermes wallonensis (Wasmann) (Isoptera: Termitidae) Ph.D. thesis submitted to the UAS, Bangalore. India, p.205.

3. Roonwal,M.L. 1970. In: Biology of termites, K.Krishna and F.M.Weesner, Ed., Academic press, New York and London 2:pp.315-391.

CHEMICAL COMPOSITION OF THE FUNGUS COMB OF SUBTERRANEAN TERMITE *MICROTERMES OBESI* HOLMGREN (TERMITIDAE : ISOPTERA)

C.T.Ashok Kumar and G.K.Veeresh
Department of Entomology
UAS, GKVK, Bangalore-560065, India

Macrotermitinae termites construct the fungal combs which serve as a source of food.

The moisture content of fresh fungus combs were determined during summer (April-May), rainy season (July-August) and winter (November-December) of 1983. Chemical analysis for cellulose (5), lignin (1), total carbon, total nitrogen, ash (4) and fat (1) content of the fungus combs were also determined.

Fungus combs were present in all the 20 nests observed during this study. The number of fungus combs varied from 1-6 in an area of 2^2 metre.

The moisture content of the fungus comb varied from 45.61 to 55.81%, which favourably compares with 51.30% of the fungus comb of O.obesus (1) and O.wallonensis (6). The cellulose content was higher (22.61-26.35) than the lignin (12.8-15.00) as recorded in Macrotermes sp. and Odontotermes sp. (2) and O.wallonensis (6). The higher cellulose content observed in the combs has been attributed to the lower ability of Macrotermitinae to digest cellulose (2).

The carbon content varied from 29.36 to 33.00% and the nitrogen content from 0.92 to 1.10%. The C:N ratio of the fungal comb varied between 29.91 to 33.00, which agrees with the composition in O.wallonensis (6). The higher C:N ratio may be due to the presence of low amount of nitrogen in the fungal comb (8).

The ash content varied between 20.50 to 22.00% which favourably compares with that (23%) in Macrotermes falciger (3) and 22.87% in O.wallonensis (6). However, ash content was as low as 10.7% in O.redemanni (2). The higher ash content may be due to the presence of higher mineral content in the fungus comb (3).

The crude or total fat content of 2.6 to 3.10% in the fungus comb of M.obesi is comparable to that of 2 to 5% in O.wallonensis (6). The reason for low crude fat content may be due to the greater capacity of this species to degrade the fatty substances or due to the presence of low organic matter. The pH varied from 4.4 to 4.6 which is comparable to the pH reported in Macrotermes spp. (3) and O.wallonensis (6).

On the whole, both moisture content and the chemical composition of the fungal comb of M.obesi does not differ much from fungal combs of other species analysed, except in respect of cellulose:lignin ratio, which in the present study ranged between 1.58 to 2.02 as compared to 0.55 in O.obesus, 0.49 in O.redemanni (2). However, it favourably compared with that in O.wallonensis (1.53)(6).

1. Allen,S.E., Crimshaw,H.M., Parkinson,J.A. and C.Quarnbaj, 1974. Chemical analysis of Ecological materials Ed. S.E.Allen, Blackness Scientific Publications. Oxford, London, pp.565.

2. Becker,G. and K.Seifert, 1962. Insectes Soci., 9:273-281.

3. Hesse,P.R., 1957. E.Afr.agric.J., 23:104-105.

4. Jackson,M.L., 1967. Soil Chemical Analysis. Prentice Hall of India Pvt.Ltd., New Delhi, pp.485.

5. Lossin,R.C., 1971, Compost Sci., 12:12-13.

6. Rajgopal,D., 1979. Ph.D.(Agri) Thesis, UAS, Bangalore, India, p.205.

7. Roonwal,M.L., 1960, Rec.Indian Mus., 58:131-150.

8. Sands,W.A., 1970. In: Biology of Termites, K.Krishna & F.M.Weesner, ed. Academic Press, New York, London, 2:495-504.

Symposium 30

ANT COMMUNITY STRUCTURE

Organizer : **ALAN ANDERSEN, Australia**

THE SPECIAL NATURE OF ANT COMMUNITIES

A.N. ANDERSEN, C.S.I.R.O. Division of Wildlife & Ecology, Tropical Ecosystems Research Centre, PMB 44 Winnellie, NT 0821, Australia.

What is the point of holding a symposium on ant communities - in fact, why bother studying them at all? The answer is that ants have been sadly neglected in contemporary debates about community ecology. This is despite the fact that, on the one hand, ants are model organisms for studies of community ecology, yet on the other they possess special features that can provide unique insights into the general nature of biological communities.

Ants as Model Organisms

Ants are ubiquitous and easy to sample. They can be collected directly by hand, captured in traps, attracted to baits, extracted from litter or counted in quadrats to provide information on species diversity, composition and temporal patterns of foraging activity. Their nests can be mapped to analyse spatial distribution, foragers can be robbed and nest middens sampled to determine diets, and behaviour at baits can be observed to establish competitive hierarchies. In short, comprehensive characterizations of ant communities can be made with comparatively little effort. Removal experiments are readily conducted by eliminating the nests of target species. Ants are therefore ideal organisms for studies of community ecology.

Special Features

Ant communities consist of large numbers of taxonomically related species that are also similar ecologically. Diversity is especially high in lowland tropical rainforests and in hot and open habitats of Australia, where more than 100 species can occur with $100m^2$. All ants belong to a single taxonomic family, which contrasts with the situation in most other invertebrate groups (e.g. communities of spiders, beetles, phytophagous insects) and in vertebrates in general (e.g. mammals, birds, fish). Ants tend to be very similar ecologically -

most nest in or on the ground (except in tropical rainforests) and are general predators and scavengers (except for some specialist predators and seed-harvesters). Nesting and dietary niche separation is therefore often relatively minor. Temporal partitioning, however, is pronounced in most ant communities.

Ant communities are much more than simple collections of species occupying the same habitat. Ants tend to be aggressive creatures and this, combined with the broad ecological overlap of most species, means that direct interactions between species are far more pronounced then in almost all other animal communities. The undeniable importance of interference (and probably also exploitative) competition in structuring ant communities is especially noteworthy in the current climate of questioning the importance of competition in animal communities. Two other features of ants set them apart from most other animals. First, ants are modular organisms, with individual colonies consisting of indeterminate numbers of repeated units. Second, ant colonies occupy fixed positions. These features are more characteristic of plants than of animals, and have extremely important implications for community organization.

General Insights

Ant communities are different from communities of other animals. This, combined with their affinities with plant communities, means that they provide a unique opportunity to gain new insights into the general nature of biological communities. They can make valuable contributions to our understanding of issues such as diversity, competition, niche partitioning, functional groups and ecological strategies. For example, ecological strategies within ant communities in relation to stress and disturbance conform to those in plants, suggesting that these strategies are general properties of ecological communities. Similarly, whereas plant ecologists recognize the importance of events at the regenerative phase in maintaining species diversity, animal ecologists tend to focus on adult behaviour. There appears to be great scope in applying the former view to studies of diversity in ant communities, and probably also to communities of other animals.

Ant communities can also provide insights into the general nature of the ecosystems in which they occur. Ants are ideal bio-indicators in terrestrial ecosystems because they operate at all trophic levels, and integrate a particularly broad range of environmental variables.

CRYPTIC SPECIES ASSEMBLAGES IN TROPICAL AND TEMPERATE LATITUDES

J.H.C. DELABIE* and H.G. FOWLER**, *Secção de Entomologia, CEPEC, CEPLAC, CP 7, 45600 Itabuna, Bahia, Brazil and ** Departamento de Ecologia & Centro Para o Estudo de Insetos Sociais, Instituto de Biociências, UNESP, 13500 Rio Claro, São Paulo,,Brazil

Studies conducted in a 50 yr old 1 ha cocoa plantation in Ilheus, Bahia Brazil, of the ant community of the soil and litter layers, have allowed us to record the highest known species richness of any comparable habitat in the world. Soils in this area are podzols with dry litter calculated to be 6.9 t ha^{-1} (1), which is low for commercial plantations in the region.

Based upon 51 randomly chosen samples of 500 numbered trees through a one year-period, and $1m^2$ samples of leaf litter near the same tree, both of which were extracted by Berlese extractors, we were able to categorize the fauna into three distinct layers: litter, superficial rhizosphere, and subterranean (to the rhizosphere) layers. 69 species were found in litter samples and 93 species were found in soil samples (including the rhizosphere), with a combined total of 112 species present in our 1 ha plot. This value is exactly twice that reported for cocoa plantations in Trinidad (2). The most common subfamilies were the Myrmicinae (49.1%) and Ponerinae (31.2%). In Trinidad these respective values were 60.7% and 26.8% (2).

In our samples, we found species of known epigaeic or arboreal habits, such as *Pheidole*, *Wasmannia auropunctata*, *Azteca* and *Crematogaster*. Through subterranean baiting, we found that some of these ants are able to take advantage of any available food source, often through recruitment while strictly hypogaeic species are generally not able to compete for these food sources.

Species richness varied in the litter with respect to its

depth. We observed that a thickness of 1.0 to 2.5 cm is more favorable to the coexistence of more species than is deeper litter. At litter depths greater than 3 cm, shade is intense or soil depressions exist, which frequently flood. These areas have slower rates of mineralization than areas with direct sunlight or rainfall. These factors may influence food availability through sub-optimal physical conditions (3)

The superficial ground layer, directly under the litter, is of considerable biological importance due to the superficial rhizosphere architecture of cocoa trees. In this strata, some truely cryptic species are present, such as *Acropyga*, which tend symbiotic root mealy-bugs, and others such as *Gnamptogenys*, *Solenopsis (Diplohroptum)*, *Tranopelta*, *Typhlomyrmex*, *Carebarella* and *Anochetus*, similiar to the structure reported for soils of diverse Neotropical crops, such as coffee and cocoa (4,5). For some of these species, climate may be important through modification of food availability of honeydew gathers.

Compared with other studies in temperate areas, we find that species richness may be cued to a greater habitat diversity, and the interfacing of epigaeic faunas with hypogaeic faunas. It should be noted that in this area if we include all resident ant species, we reach a total species richness of more than 225 species, which indicates that at least 50% of the fauna depends entirely or in part upon the soil-litter strata, which distinguishes tropical systems from their temperate counterparts.

References

1. Leite, J. de O. 1987. Revista Brasileira de Ciencias do Solo, 11:45-49.
2. Strickland, A.H. 1945. Journal of Animal Ecology, 14: 1-11.
3. Levings, S.C. and Windsor, D.M. 1984. Biotropica, 16: 125-131.
4. Bunzli, H.G. 1935. Mitteilungen der Schweizerischen Entomologischen Gesellschaft, 16:453-593.
5. Weber, N.A. 1944. Annals of the Entomological Society of America, 37: 89-122.

ANT FORAGING ECOLOGY AND COMMUNITY ORGAINIZATION

H.G. FOWLER, Departmento de Ecologia & Centro Para o Estudo de Insetos Sociais, Instituto de Biociências, UNESP, 13500 Rio Claro, São Paulo, Brazil

One of the principal problems in evaluating the importance of foraging ecology on community organization is that we must define the structure at one level, and look at those processes which contribute to that structure at other levels (1). At the structural level, an analysis of more than 150 published community studies allows me to place species into food guilds, and if we examine guild structure no significant differences are found between deserts, tropical forests, temperate woodlands, or open areas, whether the fauna be hypogaeic or epigaeic. This constant behavior suggests that ecological processes have defined and refined the the trophic relationships in ant assemblages, and therefore permit us to examine at a finer scale what those processes are. Using the same data base, I can also conclude that species assemblages occupying similiar habitats have similiar species richness throughout the world, although this is limited principally to communities of open habitat. This also suggests that mechanisms in organizing communities follow similar rules.

If we move onto an examination of processes and mechanisms, which occur at other levels of ecological organization (1, 2). Spacing patterns are clear evidences of assortative effects in community organization (3), and these are coupled with food exploitation mechanisms (3). It is not only the type of food resource used that is important in ant foraging ecology, but also its size, distribution, predictability, and the presence of other potential competing ant species that determine mechanisms of its exploitation. Principal mechanisms include recruitment processes, scout ant density, and worker size.

Because of the way we examine communities, generally in terms of species-abundance relationships, ant foraging ecology enables us to use a variety of sampling techniques, such as nest counts, pit-fall trapping, hand sorting per known area, etc. to establish null hypotheses of what community structure should be if behavioral processes were not important. By using baits in these same points and examining changes in the abundance-species relations with these other sampling techniques, we can examine the effects of behavior and worker size. By varing bait sizes to simulate naturally occurring prey distributions, we can also examine how resource size determines community structure (3), and by varying bait types, we can further fine tune foraging ecology into community organization.
Highly specalized trophic species, such as leaf-cutting ants can be examined with respect to the foraging strategies used (3). In these ants, foraging strategies involve territories, physical foraging trails and galleries, how plant material is harvested, and interspecific behaviors which lead to assemblage organization. Indeed, if we fine tune our earlier analysis to examine guilds within guilds of ant assemblies, many of which have only weak interactions with other guilds, we can gain an idea of how communities are organized. I suspect that a great many communities are organized in much the same manner, as has been strongly stressed for other types of insect communities (3)

References

1. Allen, T.F.H. & Starr, T.B. 1982. Hierarchy: Perspectives for Ecological Complexity. University of Chicago Press, Chicago.
2. Margalef, R. 1980. La Biosfera. Ediciones Omega, Barcelona.
3. Fowler, H.G., Forti, L.C., Brandão, C.R.F., Delabie, H.C. & Vasconcelos, H.L. 1990. In: Ecologia Nutricional de formigas. A. Panizzi & J.R.P. Parra, Eds, Manole, São Paulo (in press)

THE ANT COMMUNITY OF A TROPICAL LOWLAND RAINFOREST SITE IN PERUVIAN AMAZONIA

Stefan P. Cover, John E. Tobin and Edward O. Wilson
Museum of Comparative Zoology, Harvard University, Cambridge, Massachusetts 02138, U.S.A.

Though ants are known to be diverse and extremely abundant in lowland tropical forests, there have been few detailed studies of restricted areas. This is a preliminary inventory of the ant fauna in lowland rainforest at Cuzco Amazônico, a site in Peruvian Amazonia.

Cuzco Amazónico is a 15,000 hectare private reserve located 15 km. northeast of Puerto Maldonado, Madre de Dios, Peru. The area is approximately 200 m. in elevation and contains both terra firma and seasonally flooded primary forest. During June 1989, ants were collected primarily along four transects of 20 x 500 m., each located about one km. from the station, as well as in the immediate vicinity of the station. This effort was part of a general forest inventory sponsored by the Biotrop program. All major microhabitats for ants were sampled through intensive collecting and litter sifting. Though we collected at fresh treefalls, it is likely that canopy ants are under-represented in our samples. We did not collect all possible series of several of the more common forms, thus they also are under-represented in our material. These include several species of Crematogaster, Dolichoderus, Camponotus, Mycocepurus and Azteca.

The analysis thus far of our 926 collections has yielded 256 species belonging to 64 genera of ants, representing all six subfamilies and nearly 50% of the genera known from the Neotropics (Hölldobler and Wilson 1990). Despite the large number of genera and species at this site, the fauna is clearly dominated by a small minority of genera. By far the most diverse and abundant genus is Pheidole, comprising 49 (19%) of the total number of species collected and 23% of all collections. Of the 49 Pheidole species, as many as 26 are undescribed.

Pheidole is about twice as diverse as its nearest rival, *Camponotus* (23 species). Together the five most species-rich genera (*Pheidole*, *Camponotus*, *Pachycondyla*, *Gnamptogenys* and *Strumigenys*) contain 112 (44%) of the species. Moreover, 21 of the 64 genera were collected only once or are represented by only a single species. Thus, in Amazonian lowland rainforest high generic and species-level richness may not be inconsistent with pronounced ecological dominance.

We have calculated precise species accumulation curves for two important microhabitat types (rotten branches and sticks on the forest floor, and dead hanging branches in understory trees), and a less precise curve for the fauna as a whole. From these we derive estimates of faunal size and compare them to other such data from the literature.

Notable features of the myrmecofauna of Cuzco Amazonico include the nearly complete absence of large leafcutting ants (*Atta* and *Acromyrmex*) and of *Paraponera clavata*. Ant gardens containing *Camponotus femoratus* and *Crematogaster parabiotica* (broadly defined) are very abundant. Some of the more unusual finds include the second known species of *Protalaridris*, and the discovery that land snails are an important prey of *Basiceros conjugans*.

REFERENCE

Hölldobler, B. and E. O. Wilson. 1990. The Ants, The Belknap Press of Harvard University Press, Cambridge, Massachusetts

RAIN FOREST ANT COMMUNITY: SPECIES INTERACTION AT BAITS

PARTHIBA BOSE AND PRIYA DAVIDAR
SALIM ALI SCHOOL OF ECOLOGY, P.O. BOX 154,
PONDICHERRY UNIVERSITY, PONDICHERRY, INDIA.

Interspecific competition for resources has been implicated in organising ant communities (Levins et al. 1973, Davidson 1977).

This study examines the role of interspecific interactions at baits and dominance hierarchy of the ant species in the Western Ghats, India.

METHODS

Two categories of foragers, territorials and non-territorials were observed at the study site. If there were no interspecific dominance hierarchy, then interactions between different species at baits would be haphazard. To test this hypothesis we set baits at different sites. Sites chosen had different combinations of territorials and non-territorials. Honey solution and dried fish were presented as baits at the centre of 1 m^2 frames. All the baits were examined every hour. Soil temperature and light were recorded during the observation session. Behavioral interactions among different species during observation were recorded. Baits were removed at the end of the day.

Three-way contingency tables were used to analyse species

interactions (Feinberg 1970). Contribution of each cell in these tables were examined through Pearson Chi square association test. An aggressiveness index was calculated based on the aggressive behaviors of different species at baits and was used to ascertain the competitive dominance hierarchy in the community. A multiple regression analysis of each species present at baits, soil temperature and light was conducted.

The following null models were considered for analysing each of the three way contingency tables - 1. Mutual independence of the three variables. 2. Variable C is independent of variables A + B 3. Variable A is independent of B + C.

Table 1

Co-occurrence of different species at baits (Pearson X^2 test).

Model	SQ 1	SQ 2	SQ 3	SQ 4
	P C2 T	O C T	C P T	C P C2
I	$p<.001$	$p< .001$	$<.001$	$<.001$
II	$p<.001$	$p< .001$	NS	$<.001$
III	$p<.001$	NS	$<.001$	$<.001$

P = <u>Pheidole sp.</u>, C = <u>Crematogaster sp.</u>, T = <u>Tetramorium sp.</u>, O = <u>Oecophylla smaragdina</u>, C2 = <u>Camponotus sp</u>.

Pearson Chi square test.

The results indicate that most of the species co-occur except <u>Crematogaster</u> and <u>Oecophylla</u> (Table 1). These two are territorial (unpublished data) and appear to exclude each other.

RESULTS

A distinct gradient is present in the index of aggression for different species. Oecophylla smaragdina is the most aggressive and Aphenogaster sp. the least. (Table 2)

Table 2: Index of aggression for different species, calculated as the mean frequencies of aggressive interactions shown by each species.

O.sm	Pheidole sp	Crem.	Tetra2	Camp2	Aph.
0.038	0.0145	0.0144	0.008	0.003	.0016

Activities of Pheidole sp., Tetramorium sp. and Oecophylla smaragdina increased or decreased with temperature from site to site. ($p < 0.05$). When temperature was included as one of the predictor variables in a multiple regression analysis. Oecophylla smaragdina and Crematogaster sp. were negatively associated. No other variables were significant.

DISCUSSIONS

The results show that the more aggressive species do not exclude the submissive species from baits. Only the aggressive territorial ones exclude each other's presence at baits, the exception being the more aggressive Pheidole sp excluding the submissive Tetramorium sp. The present study suggests that aggression alone is not important in determining species coexistence and more factors might determine species co-existence than just interspecific competition.

REFERENCES

1. Levins, R., Pressick, M.L., and Heatwole. 1973, Am. Sci. 61:463-472.
2. Davidson, D.W., 1977, Ecol. 58:711-724.
3. Fienberg, S.E., 1970, Ecol. 51:419-433.

THE COMPOSITION OF ANT COMMUNITIES IN BRASILIAN ATLANTIC RAINFOREST

J.D.Majer* and M.V.B.Queiroz**
*School of Biology, Curtin University of Technology, P.O.Box U1987, Perth, 6001, Australia.
**Department of Animal Biology, Federal University of Vicosa, Vicosa, Minas Gerais, 36570, Brasil.

INTRODUCTION

Although the ant fauna of southern Brasil has been reasonably well studied, few detailed community studies have been performed. Various taxonomists have documented the species occurring within a region [see refs in 2] but the only studies which have incorporated standardised sampling methods are in agroecosystems and in cerrado, or native woodland, formations.

Atlantic rainforest, known locally as 'mata atlantica', is the belt of forest which extends in a crescent around much of the coastal and inland region of south-eastern Brasil. To our knowledge, there are no published studies on the ant communities of this ecosystem, which is regrettable in view of the fact that much of this type of forest has been cleared for agriculture or other types of land-use. The aim of this paper is to document the ant communities in five atlantic rainforest plots and, in the spoken presentation, to make some comparisons with rainforest ant communities which have been surveyed by similar methods in Australia [1].

METHODS

Two areas of rainforest were surveyed at Vicosa, M.G. (20o45'S, 42o, 51'W) and at Ilha do Cardosa, S.P. (25o12'S, 48o, 00'W), while one area was surveyed at Pocos de Caldas, M.G. (21o51'S, 46o, 34'W). The annual rainfalls for these three areas are respectively 1341, 2342 and 1695 mm. The forest on Ilha do Cardosa was primary forest, while that at Pocos de Caldas was a remnant that had been logged at some time in the past ; the Vicosa forest was 30 year old regrowth on the site of an old coffee plantation.

The plots were sampled between August-November 1989 using identical sampling techniques to those used in [1]. Transects of 100m were marked out in each plot and ants surveyed using 10 alcohol pitfall traps, day/night fish/biscuit/honey baits, day/night hand collections and Berlese funnels. Ants were identified to morphospecies and the reference collection is located at the Entomology Museum of the Federal University of Vicosa.

RESULTS

The ants which were sampled in the five plots,plus certain summary parameters, are shown in Table 1.This Table also provides certain summary data from the Australian rainforest plots..

Table 1. Number of species of ants within each genus and various other summary data for the ant communities sampled within five atlantic and eight Australian rainforest plots.

		Vicosa		Pocos de Caldas	Ilha do Cardosa		Overall Values	Australian Values
		1	2		1	2		
Ponerinae	*Acanthostichus*	1		1			1	–
	Amblyopone				1		1	–
	Discothyrea	1		1			1	–
	Ectatomma		1		1	1	1	–
	Gnamptogenys	1	1	1	1	1	2	–
	Heteroponera	1	1	1		1	1	–
	Hypoponera	2	3	3	4	6	6	–
	Odontomachus	1			3	3	3	–
	Pachycondyla			1	1	1	1	–
Myrmicinae	*Acromyrmex*	1		1	1	1	1	–
	Apterostigma				1	1	1	–
	Aspididris					1	1	–
	Atta	1				1	1	–
	Crematogaster	3	4	1	1	2	5	–
	Cyphomyrmex	1	1		2		2	–
	Lachnomyrmex					1	1	–
	Monomorium			1	1	1	1	–
	Octostruma					1	1	–
	Pheidole	11	9	7	9	14	24	–
	Procryptocerus	1		2	1		2	–
	Solenopsis	9	3	4	4	3	11	–
	Strumigenys	1		1	1	2	2	–
	Trachymyrmex	1	1	1		1	1	–
	Wasmannia	2	1	1			2	–
Ecitoninae	*Eciton*		1				1	–
	Labidus	2	1		1	1	2	–
	Neivamyrmex	1	1				1	–
Pseudomyrmecinae	*Pseudomyrmex*	3	2	2			4	–
Formicinae	*Brachymyrmex*	4	3	4	1	1	5	–
	Camponotus	9	7	6	1	3	15	–
	Myrmelachista	1	1	3	1	1	5	–
	Paratrechina	1	2	1	1	1	2	–
Dolichoderinae	*Azteca*	1	1			1	1	–
	Conomyrma	2	2				2	–
	Iridomyrmex	3	2		3	2	5	–
	Tapinoma	1	2	1	1	1	2	–
Ponerinae		7	6	8	11	13	17	19
Myrmicinae		31	19	19	21	29	56	45
Ecitoninae		3	3	0	1	1	4	0
Dorylinae		0	0	0	0	0	0	1
Pseudomyrmecinae		3	2	2	0	0	4	2
Formicinae		15	13	14	4	6	27	25
Dolichoderinae		7	7	1	4	4	10	10
Total species		66	50	44	41	53	118	102
Total genera		27	22	21	22	26	36	35
Species/genera		2.44	2.27	2.09	1.86	2.03	2.13	1.71
Diversity index (H')		0.83	0.98	0.46	1.17	1.19	0.92	0.87
Evenness index (J')		0.58	0.81	0.48	0.87	0.89	0.72	0.73
Abundant surface-active ants		12.1	11.8	9.1	17.1	15.1	–	–
Other surface-active ants		36.4	31.4	27.3	26.8	28.3	–	–
Tree and vegation foragers		24.2	25.5	31.8	9.8	11.3	–	–
Hypogaeic ants		16.7	19.6	25.0	31.7	32.1	–	–
Fungus growers		3.0	0.0	2.3	2.4	3.7	–	–
Large solitary foragers		3.0	3.9	2.3	9.8	7.5	–	–
Soldier ants		4.5	5.9	0.0	2.4	1.9	–	–

REFERENCES

1. Andersen,A.N. and J.D.Majer. in press.In Rainforests of the Kimberley region,Western Australia:ecology and biogeography, N.L.McKenzie, P.J.Kendrick and R.B.Johnston, Eds., W.A. Department of Conservation and Land Management,Perth.
2. Kempf,W.W. 1972.Studia Entomologia,15:1-344.

THE ASSEMBLY OF SAND DUNE ANT COMMUNITIES

L. Gallé

Department of Zoology, JATE University, Szeged, Pf. 659,
H-6701 HUNGARY

The ecological theatre cannot be understood without the knowledge of the process of organization in ecological communities. The studies having been published on ant community succession, e.g.(1),(2),(3),(4),(5), concern mainly particular areas or habitat types only, with restricted possibilities of generalization. This study is a comparative survey on the succession of ant assemblages in sand dune areas in Finland (Tvärminne), Poland (Kampinos) and Hungary (Fülöpháza, Bugac) with the main aim of revealing the species composition, diversity, external correlates and the importance of species interactions. In each area a set of 8-10 successional habitat plots was selected for detailed sampling programs on density of ants, composition of invertebrate fauna, architecture and composition of vegetation and some other habitat properties. Additional ant collections were carried out in other sites in north and middle Finland, Hungary and Yugoslavia.

Taking into account the composition of ant assemblages and the outcome of their multivariate analysis, a possible successional sequence of the different habitat plots can be set up in each studied site. A general trend of increasing total density and species diversity of ant assemblages was found along these successional gradients. In the last phases dominated by high density red wood ant populations, however, the diversity declines.

A pairwise nonparametric correlation analysis between PCA spaces of successional sampling plots ordinated on the basis of their ant assemblage composition and different groups of habitat properties reveals the external factors which ant assemblage compositions are correlated with (Table 1).

Since the importance of interspecific competition is well

Table 1. Significances of Spearman's rank correlation between relative position of habitat plots in PCA space using ants, and various habitat property groups as attributes. Tv = Tvärminne (Finland), Ka = Kampinos (Poland), Fh = Fülöpháza (Hungary)

Habitat property groups	Tv	Ka	Fh
Vegetation architecture	<0.05	<0.02	<0.02
Vegetation composition	n.s.	n.s.	n.s.
Epigeic invertebrates	<0.001	<0.02	<0.001
Invertebrates in the herb layer	n.s.	n.s.	n.s.
Soil properties and microclimate	n.s.	n.s.	<0.001
No and size of stones on the ground	n.s.	n.s.	n.s.
Twigs no, size and condition	<0.01	<0.001	<0.01

documented in advanced stages of ant community development, bait experiments and special sampling were employed in the early succesional stages to study this interaction. The interference competition is weak and insignificant in these ant assemblages, as indicated by the random spatial arrangement of colonies, the absence of postcompetitive niche segregation and the low encounter rates of ant workers on the ground and on baits. From the outcome of the encounters on baits the position of the different species in the aggressive hierarchy was established e.g. from top to bottom Formica sanguinea - Lasius alienus - F. cinerea in Tvärminne and Formica truncorum - F. cinerea and Myrmica rugulosa in Kampinos, respectively.

References

1. Boomsma, J.J. and A.J. Van Loon . 1982. Journal of Animal Ecology, 51: 957-974.

2. Gallé, L. 1990. Holarctic Ecology, in press.

3. Szujecki, A., J. Szyszko, S. Mazur and S. Perlinski 1978. Memorabilia Zoologica, 29: 183-189.

4. Vepsäläinen, K. and B. Pisarski 1982. Annales Zoologici Fennici, 19: 327-335.

5. Zorrilla, J.M., J.M. Serrano, M.A. Casado, F.J. Acosta and F.D. Pineda 1986. Oikos, 47: 346-354.

ANT RECOLONISATION PATTERNS IN REHABILITATED BAUXITE MINES OF SOUTHERN BRASIL

J.D.Majer
School of Biology,Curtin University of Technology,P.O.Box U1987,Perth,6001,Australia.

INTRODUCTION

A number of studies have been performed throughout Australia on the rate of recolonisation of native ants in rehabilitated mines which have been subjected to a range of rehabilitation methods and which are situated in a range of climatic zones.Generally it has been found that succession proceeds more rapidly in rehabilitation which has the greatest floristic and structural diversity and also in climatic zones which experience the most favourable rainfall [1].

Alcoa Aluminio S/A has been mining for bauxite since the mid-1960's at Pocos de Caldas,M.G.,Brasil (21o51'S,46o34'W),which has a mean annual rainfall of 1695 mm..A visit by myself to Brasil in 1989 provided an opportunity to obtain some comparative data on ant succession in Brasilian rehabilitation.

METHODS

Control plots representing the two major vegetation associations in the area were surveyed; atlantic rainforest (mata atlantica) and grassland with small trees (campos sujo).In addition, eleven areas of rehabilitation were selected, ranging from 1-10 years old. The older areas were rehabilitated with simple plantations of <u>Eucalyptus</u> spp. (predominantly <u>E.saligna</u> and <u>E.grandis</u>)or bracatinga(<u>Mimosa scabrella</u>), while the youngest plots were planted with a mixture of predominantly native atlantic rainforest species.Ants were surveyed during October 1989 in an identical way to that used in the Australian mines, namely by pitfall trapping, plus day and night hand collections along 100 m transects.

RESULTS AND DISCUSSION

The number of ant species in the rehabilitated and control plots is shown in Figure 1.The exceptionally high species richness in one of the youngest plots is probably an edge-effect. The main findings of this survey are summarised below.

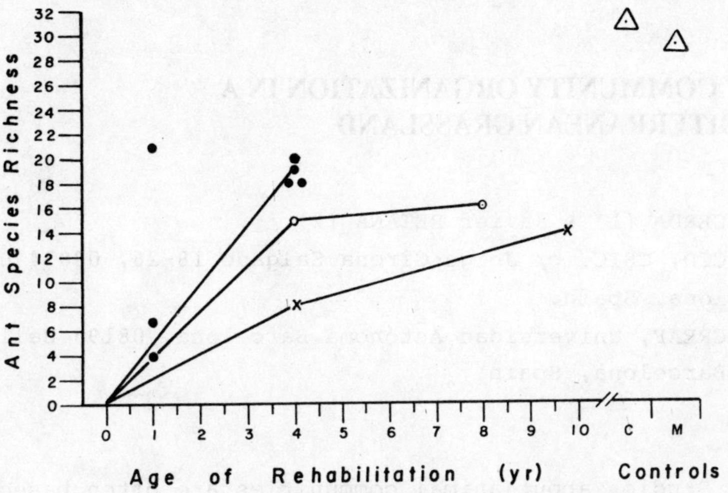

Figure 1. Number of ant species sampled in eleven rehabilitated and two control plots.Key: o = mixed native species; o = bracatinga; x = Eucalyptus; C = campos; M = mata.

o The rate of succession was most rapid in the mixed native species, intermediate in the bracatinga, and slowest in the Eucalyptus rehabilitation options.

o The rate of succession in the native rehabilitaion option was similar to that for matched climatic zones within Australia.

o A greater proportion of the ant species in the rehabilitation was common to the campos vegetation than to the mata.

o Principal co-ordinate analysis of the ant species within control and rehabilitated plots indicated that initial succession was proceeding towards an ant community which was characteristic of the campos type of vegetation.

In the spoken paper, the results of this investigation will be used to make recommendations for enhancing the success of rehabilitation and some generalisations on ant succession in Australia and South America also will be put forward.

REFERENCES

1.Majer, J. D. 1989. In Animals in primary succession: the role of fauna in reclaimed lands, J.D.Majer, Ed., Cambridge University Press, Cambridge,pp143- 174.

ANT COMMUNITY ORGANIZATION IN A MEDITERRANEAN GRASSLAND

Xim CERDA (1) & Javier RETANA (2)
(1) CID, CSIC, c/ Jorge Girona Salgado 18-26, 08034 Barcelona, Spain.
(2) CREAF, Universidad Autónoma Barcelona, 08193 Bellaterra Barcelona, Spain.

Studies about animal communities are often based on the overall analysis of the community, relieving the general trends or factors that contribute to diminish interspecific competition and to structure the whole community. In this study, the starting point has been to analyze the specific methods of resource exploitation of each of the species of an ant community, as a means of elucidating community relationships.

Field work was carried out during the last six years (1984-1989) in a savannah-like grassland in Canet de Mar (Barcelona, Spain), distant 750 m to the Mediterranean coast. The following features of each species have been analyzed: dietary spectrum, prey size, foraging strategies, type of recruitment, speed of finding food, interference at food resources, daily and seasonal activity patterns, etc.

Thirteen ant species have been found in this community. The most abundant are Pheidole pallidula (39.9% of the workers collected at pitfall traps; 40.7% of the nests found at the study area), Tapinoma nigerrimum (16.6%; 17.9%) and Aphaenogaster senilis (13.0%; 10.0%).

For the ten most interesting species, solid diet (items carried by workers to the nest, mainly seeds and arthropod corpses) and liquid diet (sugary liquids collected at the vegetation, mainly nectar of flowers and honeydew of aphids) have been analyzed. By means of multifactorial techniques, ant species have been clustered in four clearly differenciated groups: scavengers, seed-eaters, aphid-tenders and nectar gatherers. Within group similarity of the diets is high (Proportional Similarity Index -PSI- is greater than 0.7), but there are different mechanisms to minimize competition for the same types of food.

One of the mechanisms for prey selection in ants is prey size: ants select prey according to their own size, so then, big species collect big prey, and vice versa. This contributes to reduce competition: e.g., Messor capitatus and Messor bouvieri are two granivorous species that gather almost exclusively for seeds; therefore, they have a great overlap in food type (PSI=0.986). Nevertheless, M.capitatus is quite bigger than M.bouvieri (mean head width of M.capitatus is 1.64 mm, while that of M.bouvieri is 1.31 mm) and this makes it to collect bigger prey. For this reason, both species have a lower overlap in food size (PSI=0.603).

The foraging strategies of the different species are closely related to the exploitation of different-sized resources. From the mean number of workers found at baits, we can distinguish species that forage for food individually, as Cataglyphis cursor, and others that have recruitment mechanisms through which they lead some nestmates to the food. Between the later species, there are those that display group recruitment, that lead some tens of workers to the food, and transport it in cooperation to the nest, and those that display mass recruitment, that lead lots of workers to the food, where they cut it in pieces and carry back individually to the nest.

Foraging activity patterns are another mechanism for reducing competition. The differences in activity rhythms (related with ants' resistence to environmental factors) sometimes reduce interspecific encounters in food resources, as happens in the case of the three Camponotus species found at the study site, that gather for honeydew and nectar, but that have complementary rhythms.

The contribution of all those factors in facilitating pacific coexistence of species and in organizing community structure is analyzed.

ANT COMMUNITIES AT MT. FUJI, WITH SPECIAL REFERENCE TO THE EFFECT OF SCORIA DRIFT INTO THE CONIFEROUS FOREST

Masaki Kondoh
Laboratory of Biology, Shiraume Gakuen College
1-830 Ogawa-cho, Kodaira, Tokyo 187 (Japan)

Mt. Fuji is a dormant conical volcano, covered with thick scoria. The vegetation of this mountain is relatively simple and is a good example of xerarch succession. The climate also relatively typical. These condition seem to be good for analysing the factors which affect the biogeography of ants. The scoria grassland vegetation, such as *Arabido—Polygonetum weirichii alpini* association corresponding with *Myrmica kurokii—Formica lemani* community, deciduous woodland with *Myrmica kurokii—Formica lemani—Leptothorax acervorum* community, coniferous evergreen forest of *Abies* with monospecies community of *Myrmica kurokii* and shrub with *Myrmica kurokii—Formica lemani—Leptothorax acervorum —Formica sanguinea* community were reported (München Congress, 1986).

Two new types of ant community corresponding with coniferous forest and its marginal vegetation where the scoria drift affects the surface of soil, are reported. The study site is located at 2310m—2360m, NNW slope of Mt. Fuji and the survey was done during August 8—9, 1986 and July 27—29, 1987.

METHOD

A set of traps of one-third diluted honey was distributed at intervals of five paces (about 3m) along the route. After 1 hour of exposure, ants in the traps were sorted and counted according to species. The data was then used to determine coverage and constancy. Coverage, defined as the extent of space covered by a species within a given area, was calculated by dividing the number of occupied traps by the total number of traps exposed at a given site. This method is similar to that used by plant sociologists, except that arithmetic, rather than logarithmic, expression was used in my study. This modification enable the researcher to estimate probable utility within a given area. The calculated ratios of coverage at each site were converted into a table of indexes using single characteres (+, 0 to 9), to express ten levels of utilization at interval of 10%. Constancy, or roman numerals (I through V) representing percentage of occurence in five intervals of 20% each. This scale was also adopted from the method used by plant sociologists.

RESULTS

The tabular analysis of 39 sets of trap data revealed 5 clusters of ant community, which corresponded with 5 types of vegetation. Three types of vegetation, such as scoria grassland, deciduous woodland and coniferous evergreen forest, corresponded with similar ant communities mentioned above. But, the coniferous evergreen forest which disturbed by scoria drift and its marginal vegetation corresponded with two types of new ant communities, one of which consist of *Myrmica kurokii*, *Formica lemani*, *Leptothorax acervorum*, *Formica sanguinea*, *Camponotus sachalinensis* and *Lasius niger*, and other of which consist of *Formica lemani* *Leptothorax acervorum*, *Formica sanguinea* and *Camponotus sachalinensis*.

Table 1. Ant fauna and vegetations at 2310m—2360m, NNW slope, Mt. Fuji

Sampling site	Coniferous forest					Deciduous woodland				
	C_{26}	C_{24}	C_{27}	C_{61}	Constancy	C_{37}	C_{53}	C_{48}	C_{47}	C_{50}
Whole ant coverage (ratio)	0/40	1/32	4/50	5/84		1/20	2/12	14/29	9/29	4/11
(index)	.	+	+	+		+	1	4	3	3
Coverage of each species (index)										
Myrmica kurokii	.	+	+	+	(IV)	+	1	4	2	3
Formica lemani	.	.	+	.	(I)	.	.	.	.	.
Leptothorax acervorum	.	.	.	.		.	.	.	.	.
Formica sanguinea	.	.	.	.		.	.	.	.	.
Camponotus sachalinensis	.	.	.	.		.	.	+	1	.
Lasius niger	.	.	.	.		.	.	.	.	.

	Deciduous woodland									Coniferous forest disturbed							
	C_{52}	C_{58}	C_{60}	C_{45}	C_{43}	C_{49}	C_{59}	C_{55}	Constancy	C_{28}	C_{32}	C_{23}	C_{30}	C_{31}	C_{33}	C_{34}	Constancy
Whole ant coverage (ratio)	7/14	10/18	10/30	1/10	5/14	3/8	10/20	3/5		7/11	24/50	24/30	20/46	21/40	3/15	8/49	
(index)	4	5	3	+	3	3	4	5		6	4	7	4	5	1	1	
Myrmica kurokii	4	5	3	+	2	3	4	3	(V)	2	3	5	4	1	1	+	(V)
Formica lemani	.	.	.	.	+	.	+	.	(I)	+	+	1	+	+	.	+	(V)
Leptothorax acervorum	.	.	.	.	.	.	.	.		4	1	+	1	2	+	+	(V)
Formica sanguinea	.	.	.	.	.	.	.	.		.	.	1	+	.	.	.	(II)
Camponotus sachalinensis	.	.	.	.	.	.	+	.	(II)	.	.	.	.	+	.	+	(II)
Lasius niger	.	.	.	.	1	.	.	1	(I)	.	.	.	.	1	.	.	(I)

	Margin of shrub									Scoria grassland							
	C_{29}	C_{38}	C_{36}	C_{46}	C_{41}	C_{39}	C_{40}	C_{42}	Constancy	C_{25}	C_{51}	C_{56}	C_{54}	C_{57}	C_{35}	C_{44}	Constancy
Whole ant coverage (ratio)	4/17	5/18	16/31	3/4	9/30	3/53	3/10	12/39		4/10	2/21	10/13	28/70	3/30	0/15	0/1	
(index)	2	2	5	7	2	+	2	3		3	+	7	3	+	.	.	
Myrmica kurokii	.	.	.	.	.	.	.	.		+	+	.	.	.	.	.	(II)
Formica lemani	.	.	4	4	2	+	.	2	(IV)	2	+	7	3	+	.	.	(IV)
Leptothorax acervorum	1	.	+	.	.	.	.	.	(II)	.	.	.	.	.	.	.	
Formica sanguinea	.	.	+	.	+	+	2	+	(IV)	.	.	.	.	.	.	.	
Camponotus sachalinensis	+	2	+	2	+	.	.	.	(IV)	.	.	.	.	.	.	.	
Lasius niger	.	.	.	.	.	.	.	.		.	.	.	.	.	.	.	

A YEAR IN THE LIFE OF AN OLD WORLD ARMY ANT COLONY: SPATIAL PATTENS IN FORAGING AND EMIGRATION

William H. Gotwald, Jr.

Department of Biology, Utica College of Syracuse University, New York 13502, USA

G.R. Cunningham-van Someren, Ornithologist Emeritus

The National Museums of Kenya, P.O. Box 24947, Karen, Nairobi, Kenya

INTRODUCTION

The so-called "true army ants" are thought to be diphyletic, consisting of the New World subfamily Ecitoninae and the Old World subfamily Dorylinae [1, 2]. Army ants are characterized by two inextricably linked behavioral features that enhance their effectiveness as predators: group predation and nomadism [1]. Spatial patterns in foraging and emigration, at least in some surface-active species of the New World genus *Eciton*, permit colonies to search previously unexploited segments of specific trophophoric fields or of the larger trophic arena [3]. However, the ubiquity of these spatial phenomena in the army ants is debatable [1].

During a study of the behavioral ecology of the East African army ant, *Dorylus molesta*, one colony was observed for 432 consecutive days. The extensiveness of the data gathered invites comparison of this colony's spatial foraging and emigration patterns to those so well-documented in *Eciton burchelli* [3]. This report represents the first intensive field study of *D. molesta* and the longest continuous observation of an Old World army ant colony.

METHODS

Colony KC-506, the subject of this paper, was observed from 16 August 1972 to 16 February 1974. The study area of approximately 5 hectares was essentially limited by property boundaries and is located near Nairobi, Kenya, at an elevation of 1828 meters. The tract is best described as an internally diverse but greatly modified habitat. Colonies were observed and censused at least once a day, usually more often.

RESULTS

Colony KC-506 emigrated 38 times. The time between emigrations varied greatly. The shortest stay at a nest site was three days, the longest 45 days, with an average stay of 13.4 days. Rainfall was considered as a possible environmental cue for colony movement, but the null hypothesis that emigration is independent of rainfall, using the Fisher-Irwin test, could not be rejected. Colony KC-506 occupied 20 different nest sites. It used one nest site, 075B, on 10 separate occasions for a total of 170 days. Prior to colony KC-506's arrival at the research area, nest site 075B had been occupied periodically by three other colonies. Trails were also used repeatedly by the same and different colonies. Emigrations have the appearance of random events, both in occurrence and direction. In three instances, for example, the colony returned to its most recent nest site.

A total of ten additional colonies of *D. molesta* nested and foraged within the research area, at various times, contemporaneously with KC-506. On two occasions, there were 3 colonies present simultaneously, in addition to KC-506. Straight-line distances between KC-506 nesting sites and other colony nesting sites ranged from 61 m to 254 m (average = 147 m).

Colony KC-506 conducted at least one foraging raid on each of 389 days, and emigrations ensured access to foraging areas not recently exploited. Still, considerable overlap with segments in the previous trophophoric field occurred. For example, KC-506 conducted 31 foraging raids from nest 075A before emigrating to nest 075B. Of the 15 raids operated from 075B, 26.7% overlapped with raids conducted from 075A. Two subsequent emigrations resulted in foraging overlap of 29.6% and 25.0%. The calculated angles between consecutive raids from three nest sites revealed no pattern that guaranteed separation of neighboring raids. From nest site 075A, the angles ranged from 0°-107° (average=37.6°); from nest 075B, the range was 11°-160° (average=67.2°); and for nest 075E, the range was 7°-122° (average=52°). Colony KC-506 commonly conducted multiple raids, during both day and night, and consecutive raids sometimes overlapped with only small temporal separation. When coexisting with other colonies within the research area, KC-506 commonly overlapped the foraging areas of these colonies. KC-506 made physical contact with foraging workers from two other colonies on 5 occasions. In two instances agonistic behaviour resulted.

DISCUSSION-CONCLUSIONS

This study provides confirming evidence that *D. molesta*, like other surface-active species of *Dorylus*, emigrates irregularly and lacks the predictable, brood stimulated, statary and nomadic phases of *E. burchelli* and other New World army ants [1]. Factors other than brood development apparently serve as the proximate cues for colony emigration. As in the ecitonine *Neivamyrmex nigrescens*, the frequency and direction of emigrations in *D. molesta* may be influenced by the amount and location of food [4]. Although emigrations occur over foraging trails, the direction emigrations take is not predictable, unlike the emigrations of *E. burchelli* that take the same compass bearing as each day's raid [3]. It is clear, from the data, that nest sites and trails of *D. molesta* become permanent features utilized and re-utilized in common by numerous colonies.

The foraging arena was shared by KC-506 with as many as three other colonies. The occurrence of 5 "collisions" between this colony and two others happened at a much higher frequency than the predicted rate for *E. burchelli* of one collision per colony per 250 days [3].

Eciton burchelli was found to scatter consecutive raids by, on the average, 123°, resulting in the separation of neighboring raids by about 19° [3]. No such spatial patterning could be detected in *D. molesta*, although temporal separation of overlapping raids does occur. Overlap of raiding areas from two consecutive nest sites also happens, so emigrations do not assure access to exclusively unexploited trophophoric fields.

ACKNOWLEDGMENTS

The research was supported, in part, by National Science Foundation grant GB-22856. We are grateful to Mwaniki, who became an intrepid and talented stalker of army ants, and we thank Dr. Mark Deyrup, Archbold Biological Station, Lake Placid, Florida, for critically reading the manuscript, and Ms. Marcia Moretto, for typing the camera-ready copy.

REFERENCES

1. Gotwald, W.H. Jr. 1982. Army ants. Chapt. 3. In: Hermann, H.R., editor. Social Insects Vol. IV. New York: Academic Press, Inc. pp. 157-254.
2. Snelling, R.R. 1981. Systematics of the social Hymenoptera. Chapt. 1. In: Hermann, H.R., editor. Social Insects Vol. II. New York: Academic Press, Inc. pp. 369-453.
3. Franks, N.R.,and C.R. Fletcher. 1983. Spatial patterns in army ant foraging and emigration: *Eciton burchelli* on Barro Colorado Island, Panama. Behavioral Ecology and Sociobiology. 12:261-270.
4. Topoff, H., and J. Miranda. 1980. Army ants on the move: relation between food supply and emigration frequency. Science. 207:1099-1100.

ADDITIONAL PAPERS

THE TERMITE POPULATION OF THE LAC DE GUIERS REGION (SENEGAL)

AGBOGBA C.

Département de Biologie Animale - Faculté des Sciences

Université Cheikh Anta Diop - Dakar, SENEGAL

INTRODUCTION : In this paper we study the Termites, a easily observable biological indicator which enables us to locate the healthy population in space. The inventory made in that area led us to show evidence of the species of termites living on the ligneous and subligneous vegetation, their behaviour and distribution, the trees they attack and the extent the damage done.

METHODS : The method used in this study is based on the straight line technique. With the help of a car the vegetation perpendicular to the track is examined on 10 meters on each side. Seven transects have been covered, each measuring 17 kilometers of length and totalling 34 hectares. The trees showing evidence of an attack from termites (thick layers, tunnel-galleries) have been classified. Samplings were made an species determined at the laboratory.

RESULTS : Our finding shows clearly the existance of two types of termites on the vegetation. *Psammotermes hybostoma* (Desn) and *Microcerotermes sp.*

1. The way these termites attack trees varies according to species. Under the thick layers of earth built on trees, *Psammotermes hybostoma* not only attacks the bark but also the wood itself. All or part of this latter is consumed and the eaten up parts are then filled with earth pellets by the termites. As for *Microcerotermes sp.* it moves inside the upward tunnel-galleries it has built on trees, bores more or less deep galleries in order to penetrate the wood and thus makes a real havoc. The inventory of *Psammotermes*'s thick layers and *Microceroterme*'s tunnel-galleries enabled us to see that the former is present in all transects and the latter in 5 out of 7. The number of thick layers built varies from one transect to another. We have found that the areas sparsely covered with trees are more attacked by termites, on the other

hand the area on the West bank which has more *Acacia senegal* and whose forest species is thicker and more diversified is less attacked by termites.

2. Out of the 31 species of trees that have been taken stock of, 11 were attacked. These are, in descending order and for 1000 trees : *Balanites aegyptiaca* (71 attacks for 1000), *Cadaba farinosa* (34), *Acacia albida* (22), *Commiphora africana* and *Acacia seyal* (18), *Zizyphus mauritiana* (16), *Combretum glutinosum* (11), *Acacia raddiana* (7), *Salvadora persica* (3), *Boscia senegalensis* (2) and *Acacia senegal* (1). Main species are primarily attacked ; quantitywise they are best represented in the stock taking, with the exception of *Jatropha chevalieri* a non-attacked, constant and main species, of *Acacia albida* a secondary and variable species, of *Cadaba farinosa* a secondary and constant species and of *Zizyphus mauritiana* a secondary and sporadic species. On the basis of our observations we can note that only 14 % of secondary species are attacked while the percentage for main species is 89 %. As regards the damage done, *Balanites aegyptiaca* is wrecked by the attack of *Psammotermes* (93 % of cases) which in some cases lead to the death of the tree. On *Acacia* trees the damage is done either by *Psammotermes* or by *Microcerotermes*. As for *Commiphora africana* it is destroyed by *Psammotermes*.

DISCUSSION : *Psammotermes*, which is a very rare species on green wood in the Cap-Vert peninsula (AGBOGBA and ROY-NOËL, 1986) has appeared from the study as a prevailing and constant species. Its impact is significant in 4 transects out of 7, and is more particularly confined on *Balanites aegyptiaca*. The extent of the damage is inversely proportional to the increase in the variety of vegetale species. *Psammotermes hybostoma*, which here attacks the green wood was known as a species feeding on dead vegetable matter (LE PAGE, 1974). Are we in front of a change in food habits due to environmental alterations ?

REFERENCES

1. AGBOGBA, C. et ROY-NOËL, J. 1986. L'attaque des arbres par les termites dans la presqu'île du Cap-Vert (Sénégal). Bull. I.F.A.N., 44, sér. A, pp. 341-364.

2. LE PAGE, M., 1972. Recherches écologiques sur une savane sahélienne du Ferlo septentrional. Sénégal. Terre et Vie, 26, pp. 383-409.

PHORETIC ASSOCIATION OF THE PSEUDOSCORPION *ELLINGSENIUS INDICUS* WITH *APIS CERANA INDICA*

D. SUDARSANAM & V.A. MURTHY

DEPARTMENT OF ZOOLOGY, LOYOLA COLLEGE,
MADRAS 600 034, SOUTH INDIA.

The inter-specific relationship between honey bees and pseudoscorpions is quite fascinating and it draws the attention of bee keepers. This association is termed **'PHORESY'**, a non-parasitic association meant for transportation under extreme condition.

A number of pests, parasites and enemies are associated with the hill variety of honey bee, **Apis cerana indica** Fab. in Nilgris (TN). Chief among them are: **Varroa jacobsoni**, **Euvarroa** sp., **Neocypholaelaps indica**, **Termes** sp., **Liposcelis** sp., and the larvae of wax moth, **Galleria melonella.** The pseudoscorpion, **Ellingsenius indicus** J.C.Chamberlin [3] controls these enemies by feeding on them.

The role of the pseudoscorpions as predators on insects and other arthropods of economic importance has been well recognised. The pseudoscorpions occupy restricted niches depending on the microclimatic conditions. **Ellingsenius indicus** inhabits bee colonies/hives in which a variety of micro-arthropods including the enemies of the bees coexist. Alfonsus [1] considers them as the 'guests' of the bee colonies.

Among the micro-arthropods **Varroa jacobsoni** is a parasitic mite affecting 90% of the colonies of Asian bees. These mites penetrate the thoracic membrane and feed on the haemolymph of the bees leading to the extreme condition of varroatosis as well. **Ellingsenius indicus** is a voracious feeder on this mite. Since the nymphal stages of this mite are limited to the closed cells of the hives, the predatory activity of the pseudoscorpions is restricted to the adults only which are found freely moving on the frames and floor boards as well.

Neocypholaelaps indica is a pollen feeding and phoretic mite found abundantly in the thoracic-abdominal junction of the bee where the latter temporarily accumulates the pollen grains. The source of infestation is from the flowers of eucalyptus by the foraging bees. The nymphal stages of **E.indicus** have greater feeding preference to these mites as they are delicate. Similarly different stages of the wax moth larvae are also predated on by the pseudoscorpions in the bee colonies.

It is evident that **E.indicus** is exposed to a large number of prey species of the bee colony. Nevertheless, there is a perfect synchronisation between the feeding habits of **E.indicus** and bees as they do not interfere with each other for feeding. This may be one of the reasons that the pseudoscorpions are not disturbed by the bees. Interestingly, hives harbouring these pseudoscorpions show minimum infestation of these mites and even a total absence of the wax moth larvae. Hence, these pseudoscorpions are considered to be the enemies of mites [2].

A comphrehensive observation on the feeding strategy of the pseudoscorpions both in the natural hives and laboratory condition of test-feeding have proved beyond doubt that they do not harm either the adult or the larval stages of the bees.

The food chain of **E.indicus**, behavioural complexes and the prey-predator interactions are quite significant. The predatory role of the pseudoscorpions on many species of mites, psocids, wax moth larvae etc., suggest that this arachnid may prove to be of practical value in the general control of these enemies of the bees & the acarine diseases in particular.

References

1. Alfonsus, A. 1891: Der Feind der Bienenlaus. **Dtsch. Ill. 8**: 503 - 506.
2. ,, 1922: An enemy of the mites in the bee hive., **Bee World, 4**: 2- 3.
3. Chamberlin, J.C. 1932: A synoptic classification of the generic classification of the chelonethid family Cheliferidae Simon (Arachnida), **Canad. Entomol., 64**: 35 - 39.
4. Murthy, V.A. & Anantha-kirshnan, T.N., 1977: Indian Chelonethi., **Oriental Ins. Monogr., 4**: 1 - 210.

* * *

PHEROMONES OF LEAF-CUTTING ANTS: USE IN BAITS

P.E. HOWSE & J.J. KNAPP
Chemical Entomology Unit, University of Southampton, U.K.

Synthetic baits for leaf-cutting ants are finding greater acceptance today. Exploiting the ants' plant-harvesting behaviour to get toxins into the nest has potentially considerable advantages over any other method, particularly if the bait can be made specific. The main alternative technique of thermal fogging with insecticides is labour-intensive and hazardous to the operator, especially in humid tropical conditions where protective clothing is often neglected.

The problems that must be overcome in formulating an effective bait are primarily those of facilitating discovery, pickup, transport and acceptance within the nest. One weak link within this chain of events renders the method useless. Importantly also, the bait needs to be harvested before it loses its properties or suffers from biodegradation in hot humid environments.

The probability of discovery of a randomly-distributed bait in a given foraging area that is searched efficiently by ants will vary with the square of the radius of attraction, assuming the radii of baits do not overlap. ($P = N. \ r^2/A$, where N is the number of baits, r the radius of attraction, and A the total area in which they are distributed). Thus if the radius of attraction of the baits is doubled from a to 2a, the probability of discovery in a given time will be increased 4-fold. Clearly, it is of advantage to include a volatile attractant in the bait.

Until now, the most effective volatile attractants have proved to be food attractants, for example citrus volatiles (e.g. Cherrett & Seaforth, 1968). Trials with trail pheromone components have been disappointing. For example, Robinson et al. (1982) found 4-methyl-pyrrole-2-carboxylate only slightly improved the attractance of food baits, and was repellent at high concentrations. Brood pheromones were also ineffective as extracts added to test baits of filter-paper discs (Robinson & Cherrett, 1974).

Part of the difficulty encountered is that attraction does not always lead to pick-up. In attempts to understand this, Bradshaw et al. (1986) discovered a leaf-marking pheromone. This derives from the Dufour's gland and leaf particles marked with it are picked sooner and transported to the nest more rapidly than control particles. However, in field tests with Atta sexdens rubropilosa, Vilela & Howse (1988) found no enhancement of pick-up of citrus-pulp baits to which (Z)-9-nonadecene (the main component of the pick-up pheromone) or various trail pheromone components had been added.

Other types of pheromone bait have recently been tested by the authors in Brazil and Guadeloupe, and have given promising results, in which the discovery and transport times appear to be at least an order of magnitude faster than with simple citrus-pulp baits.

1. Bradshaw, J.W.S., Howse, P.E. and Baker, R. 1986. Animal Behaviour, 34, 234-240.
2. Cherrett, J.M. and Seaforth, C.E. 1968. Bull. ent. Res. 59, 615-625.
3. Robinson, S.W. and Cherrett, J.M. 1974. Bull. ent. Res. 63, 519-529.
4. Robinson, S.W., Jutsum, A.R., Cherrett, J.M. and Quinlan, R.J. 1982. Bull. ent. Res. 72, 345-356.
5. Vilela, E.F. and Howse, P.E. 1988. An. Soc. ent. Brasil, 17(supl.)

11th International Congress — IUSSI 1990, India

Symposium on ***Ecology and Evolution of Honey Bee Behaviour***

SOME ECOLOGICAL AND PHYSIOLOGICAL FACTORS AFFECTING THE THERMOREGULATION OF THE ASIATIC GIANT HONEY BEE (*A.. DORSATA* F.)

Makhdzir Mardan and Abdul Halim Ashaari
Plant Protection Department, Faculty of Agriculture
Universiti Pertanian Malaysia, Serdang 43400, Selangor D. E. Malaysia

Abstract

Some aspects on the nesting biology of honeybees, like nesting site and nest architecture, directly affect thermoregulation of the colony. A study on the thermoregulation of the Asiatic Giant honey bee (*Apis dorsata* F.) reveals that the open nesting species employs a subset of strategies for thermoregulation of the colony. Some of these strategies, which are inconspicuous or minimally employed by the cavity-nesting species, are greatly relied upon for thermoregulation by *A. dorsata* .

Being single-combed and open-nesting, *A. dorsata* is constantly subjected to the direct effects of solar radiation, breezes or winds and rain, hence, coupled with the conditions hot and humid of the tropics pose problems to *A. dorsata* in maintaining the colony temperature below the threshold of upper lethal temperature. *A. dorsata* relies on an efficient mechanism of dispensing excess heat by employing a set of behavioral repertoires, such as: rhythmic extrusion of honeycrop fluid (fluid gobbetting); periodic defecations *en masse*; clustering; wing comportment; fanning; alignment of light colored young bees on the surface of the curtain; and regulating the number of layers of curtain bees covering the brood comb of the colony. To prevent the colony temperature from exceeding 38 °C, *A. dorsata* resorts to either one or a set of the following behaviours for cooling: fluid gobbetting, limited fanning (*breathing holes*), thinning of the layers of bee curtain, forming the outer most layer of the bee curtain mainly of callows and en masse flight and defecations. At cooler temperatures, during the the nights (18-20 °C), and rainy days (23-24 °C), heat in the colony is retained by compaction of the bee cluster and rain is kept away by spanning and underlapping of the fore wings behind the fore wings.

The defensive strategy of nesting low on tree branches surrounded by vines in shady or concealed places cuts out the heat radiation from the sun. Otherwise, when nesting high on tall trees, cliff faces, eaves of lofty buildings in the open by *A. dorsata* colony is exposed to the sun, however, the colony is constantly exposed to intermittent breezes which could be exploited for fortuitous cooling by fluid gobbetting, concomitant with loose clustering (regulating the layers of bee curtain) and wing spanning during hot and windy conditions.

At the individual level, examination on the architecture of the aorta of the circulatory system (in the petiole), which is the internal mechanism of the honeybee thermoregulation, showed the possibility that the curley-coil of the aorta is capable to function as a dynamic facility of regulating heat between the body parts of the honeybees. The dual and facultative function of the abdomen in thermoregulation is afforded by the flexible and dynamic architecture of the curley-coil (can be stretched and retracted as a result of the regulation of the size of the honeycrop - abdomen), which in turn directly regulates the rate of heat transfer in the hemolymph between the abdomen and the thorax, respectively. When there is excess heat generated in the thorax, the abdomen (pumping of honeycrop fluid) serves as a temporary storage of heat cumulatively transfered from the thorax by rhythmic action of fluid gobbetting. Under the condition, the abdomen is distended together with the juxtaposed aorta (stretched) and thus the rate of heat transfer into the abdomen is more as compared to the situation when the aorta is coiled when the abdomen is retracted during cooler temperatures. Quantitatively, heat transfer into the abdomen diminishes by the transience of stretched to coiled architecture of the aorta (change from hot to cool temperatures), thus heat into the abdomen is lesser. This finding on the dynamic architecture of the curley-coil of the aorta is a deviation from the previously held view that the it is fixed in shape, thus its static role in the thermoregulation of the individual honeybee. Later the excess heat is dumped to the external of the body by defecation during flight near the colony. That will only occur when the excess heat load in the body is insurmountable for the evaporative cooling system of rythmic fluid gobbetting to dispense via the head. The defecation by the colony restores the effectiveness of the cooling cooling system by the sudden loss of almost two thousand calories from the bodies of the majority of the bees in the colony via the feces. Continued evaporative cooling (fluid gobbetting) thereafter, enhances the capacity of the evaporative cooling of fluid gobbetting.

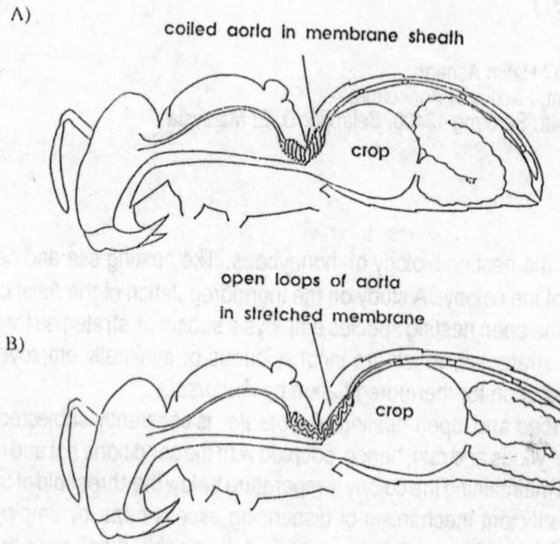

Diagram The dynamic architecture of the curley-coil of the aorta is accounted by the retraction (cool)(A) and distension (hot)(B) of the abdominal segments which stretches or enhances the coil of the aorta covered by a membrane sheath in the petiole of *A. dorsata*. The change in the architecture of the aorta from tightly coiled (cool) to loose or open loops (hot) at the juncture between the thorax and body regions results to increase in the rate of heat exchange between the thorax and the abdomen, thus undertones the pivotal role of the curley-coil in the thermoregulation of the individual honey bee.

References

1. Dyer, F.C. and Seeley, T.D. 1987. Interspecific comparisons of endothermy in honey bees (*Apis* spp.): Deviations from the expected sizee-related patterns. *J. Exp. Biol.* 127:1-26.
2. Heinrich, B. 1979. Keeping a cool head: honey bee thermoregulation. *Science* 205:1269-1271
3. Mardan, M. 1989. Thermoregulation in the Asiatic giant honeybee, Apis dorsata (Hymenoptera:Apidae) *Ph. D.* thesis, *University of Guelph.*
4. Mardan, M. and Kevan, P.G. 1989. Bees and Yellow Rain. *Nature* 341:191
5. Seeley, T.D., Nowicke, J.W., Meselson, M., Guillemin, J. and Akratanakul, P. 1985. Yellow Rain. *Scientific American* 253 (3): 128-137.

AGE COHORTS IN CARPENTER ANTS : IS INDIVIDUAL AGE OR COLONY AGE STRUCTURE A BETTER PREDICTOR OF DIVISION OF LABOR ?

PRASSEDE CALABI* and NORMAN F. CARLIN+

* Harvard University, Graduate Program in Extension, Cambridge, MA 02138 U.S.A.
\+ Harvard University, Museum of Comparative Zoology, Cambridge, MA 02138 U.S.A.

Young ant workers of a given size class typically perform brood care and nest maintenance, while older workers of the same size forage. This age polyethism appears to be the consequence of an ontogenetic behavioral program, which causes workers to perform specific behaviors disproportionately at each age. Recent studies demonstrate that the associations between age and behavior are labile (1). Under experimental manipulation, workers can shift from behaviors characteristic of their own age class to those of other classes. Evidence suggests the existence of age-typical thresholds of sensitivity to the stimuli associated with their usual tasks. If that age class is experimentally removed, the accumulating stimuli trigger responses by other age classes (2).

Do age-typical stimulus thresholds impose a limit on behavioral flexibility, or are they themselves flexible? The ontogenetic program might set thresholds in a strict progression, so that when individuals reach a given age, they will exhibit the same behaviors. Alternatively, each individual might develop low thresholds to stimuli associated with colony needs when it enters the labor force, overriding the "typical" ontogenetic sequence of sensitivities. In the latter case, membership in an age cohort -- those workers which eclosed within a brief time interval and experienced the same initial stimuli -- would be a better predictor of an individual's behavior than its calendar age.

To test this hypothesis, we observed the behavioral ontogenies of individually-marked minor workers in 2 queenright colonies of Camponotus floridanus. More than 700 hours of concurrent focal sampling data were collected over 10 weeks. Labor needs were kept constant by controlling the amount of brood and food present. The amount of time invested in different behaviors is compared across individual workers by cluster analysis. If thresholds follow a rigid age progression, we expect unvarying, sequential replacement of workers performing a given behavior by subsequent groups of younger workers. In a given week, the youngest workers should always appear at the bottom of the dendrogram, moving up in subsequent weeks (Fig. A). However, if experience alters the thresholds, we predict that each new age cohort will take up tasks receiving less attention at the time, and will continue to perform those tasks in subsequent weeks. Successive cohorts, taking up different tasks in turn, will not always insert at the bottom of the dendrogram (Fig. B).

[1] Calabi, P. 1988. In: Advances in Myrmecology, J.C. Trager, Ed., E.J.Brill Press, Leiden, pp. 237-258.

[2] Calabi, P. and J.F. Traniello. 1989. Journal of Insect Behavior, 2(5) : 663-667.

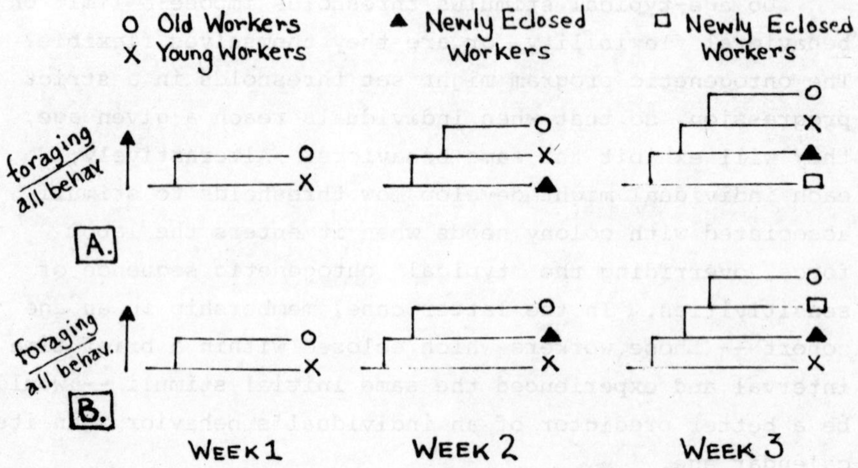

STUDY OF THE PROBABLE OCCURRENCE OF *ACARAI WOODI* IN IRANIAN HONEY BEES

Mohammad Javad GHASSEMI and Siavosh TIRGARI. Irar search Organization for Science and Technology. I Forsat Street, Enghelab Avenue, Tehran, IRAN.

A sudden outbreak of Varroa jacobsoni in Iranian apiaries in 1984, caused a heavy economic loss to bee-keeping industry. That was favoured by migratory bee-keeping which is a common practice for more than two decates in this country.

The presence of Acarapis woodi in India, Pakistan and USSR, and the fear of the above experience made us to investigate the probability of its presence in IRAN.

FAO in its former report in 1980 named IRAN among the infested areas by Acarapis woodi and since then that was quoted by several Iranian investigators.

We studied more than 2000 samples, from 43 apiaries, obtained from different parts of IRAN and they were carefully inspected. We followed the procedure of Baker and Delfinado (1984) in preparation of thoracic tracheae for microscopic examinations. However, we have not found any sign of Acarapis woodi in Iranian honeybees up to September 1989. But we did find several external mites and one case of Apimyiasis, which are being identified.

It is also probable that applications of different fumigants against Varroa has kept the presence of A.woodi unsighted in IRAN.

SENTRIES AT ALL GATES OF LIFE

W.D. HAMILTON
DEPARTMENT OF ZOOLOGY
SOUTH PARKS ROAD
OXFORD OX1 3PS
UNITED KINGDOM

Origins and limitations of the recognition processes of life will be discussed. The field is uncertain but holds hope of general principles.

Once eucaryotic cells are born, and especially after their organisation into multicellular bodies, processes of cell recognition serve (1) to reject fusion with obligate parasites and/or with peer cells which may prove facultative parasites, and (2) to select mates that optimise prospects of genetic recombination. Purposes of the first category directly defend the cell or organism and those of the second further the defence indirectly by allowing best combination offspring to be made from the extant gene polymorphisms.

In Phylum Chordata, molecules of the immunoglobulin superfamily serve to mediate all of the above functions, besides others. The occurrence of homologue molecules in insects and nematodes indicate even more ancient origins. But as functions are diverse, so are the molecular forms; it may prove only coincidental that all three functions are mediated in part through a restricted group within the superfamily in the case of the MHC system of higher vertebrates. In the consanguinity avoidance of mating rats and mice it is now known that the odour cues come from the gut flora, not from MHC directly: MHC affects the flora but does not provide odiferous molecules directly. It is in fact unclear whether the earliest functions for the superfamily were defence or communication in the organisation of tissue building. Nevertheless, needs for defence would undoubtedly have been present from earliest times especially among clones of cells separating temporarily, as happens in Acrasiales (cellular slime molds) in which the expected parasitisms are now known.

Social insects repeat scenarios of the first metazoa at a higher level but with new handicaps. Especially they repeat of the Acrasiales since separate individuals and separate foraging are intrinsic to their organisation. Again it has been found in one instance (Evylaeus) that the same system of recognition that mediates rejection of strangers sometimes determines male enthusiasm to mate. However, in so far as innate cues are used, the substances are completely different from those of the superfamily referred to above: convergence may exist in the system characteristics but there is no homology. Moreover the insects seem to avoid heavy reliance on cues from genetic polymorphisms; instead much use is made of externally acquired chemical blends. Theoretical reasons for expecting such adaptive caution with respect to genetic cues that are to be used in colonies that are not clonal will be discussed. Again it can be argued that the need to defend against microbial parasites plays a fundamental role in restricting the options.

Social insects sometimes communicate colony identity at short distances by means of odour but seem hardly to have begun use of the acoustic and visual cues that have become so important for social (though not eusocial) higher vertebrates. Many colour polymorphisms, however, are known in wasps and a few in bees : their possible significance will be discussed.

THE OPTIMIZATION OF MATURATION COSTS IN THE GYNES OF THE ANTS *LASIUS FLAVUS*

ANT LASIUS FLAVUS

G J Peakin and M G Nielsen
Thames Polytechnic, England and Aarhus University, Denmark

The cost of production of sexuals is high and only mature colonies possess suffcient worker resources to absorb it. The individual gynes are particularly demanding, since not only are they often larger than the males but they have also to possess sufficient resources of their own to enable them to found a new colony.

There are two readily identified stages in the development of sexuals when substantial resources have to be channelled to them by the workers. The larval stage requires resources for both sexes, but often more for each gyne larva if sexual dimorphism is marked.

The beginning of the adult stage, when the sexuals mature in preparation for the mating flight, is marked by the very different requirements of each sex. Males require little, if any, food. They utilise reserves acquired during larval development while waiting for the mating flight. All they have to do is harden the cuticle, mature flight muscles and a batch of sperm for fertilizing the gynes. Gynes, on the other hand, use the period to build up the reserves necessary for colony foundation.

After maturation the sexuals have to wait in the nest until there are suitable environmental conditions for the mating flight. Males will continue to bide their time at minimum cost. Gynes, having acquired reserves, will now need to conserve them, so that, having undergone a period of considerable metabolic activity to process nutrients for their reserves, they should now reduce it to a minimum to conserve them.

This paper will explore the options open to gynes and their adaptations for an optimum solution to the problems faced during the period of their adult life spent in the parental nest.

DISTRIBUTION PATTERN OF THE POLYGYNOUS ANT *FORMICA MONTANA*

Gregg Henderson
University of Wisconsin-Madison
Department of Entomology
237 Russell Laboratories
1630 Linden Drive
Madison-Wisconsin 53706, **USA.**

Distribution and life history data on the prairie ant, Formica montana, from nine open field and prairie sites in southern Wisconsin indicate that nests are clumped. However, although polydomy exists in this species, and the close packing of nests suggests unicolonialism, territoriality through nestmate recognition makes this species multicolonial. Nest spacing patterns appear to be regulated by three factors. (1) Food resources, especially homopteran honeydew, show a significant relationship with mound distance. Edaphic conditions around the mound favoring food plants and protection of Homoptera from predators apper responsible for thid positive correlation. (2) Tolerance of nearby colonies of territorial conspecifics results in the subsequent monopoly of a limited habitat by Formica montana. This monopolization increases the protective boundry for Homoptera and increases their persistence and reliability for the ants. (3) Prairie ant dispersal is severely limited because of the alate females' inability to fly. Mating takes place on the mound top or on nearby vegetation and the new queens either move to a new nest location on the ground or are incorporated into the natal nest. Worker ants prepare nest sites prior to the mating season close to the natal nest.

ROLE OF CUTICULAR HYDROCARBONS IN NEST MATE RECOGNITION IN *FORMICA MONTANA*

Gregg Henderson
University of Wisconsin-Madison
Department of Entomology
237 Russell Laboratories
1630 Linden Drive
Madison-Wisconsin 53706, **USA.**

Polygynous ant species often monopolize patchily distributed habitats and tolerate neighboring conspecifics while aggressively attacking other ant species. We determined that inter-nest aggression occurs in the polygynous ant, Formica montana. We report for the first time the identities of cuticular hydrocarbons of F. montana and present results of their possible role in nestmate recognition. Cuticular hydrocarbons contribute differentially to class discrimination, certain hydrocarbons being more class distinct.

STINGLESS BEES OF THE AMAZON

STINGLESS BEES OF THE AMAZON

João M.F. Camargo
Dep. Biologia
Fac. Fil., Ciências e Letras de Ribeirão Preto, USP
14.049 - Ribeirão Preto, S.P., Brasil.

In the Amazon region at least 26 supra-specific groups (genera following some authors) of Meliponinae exist. These groups amount more than 140 spp. and belong to several large phyletic lines: Tetragona-Tetragonisca (9 genera); Trigonisca (4 genera); Lestrimelitta; Plebeia (s. str.) and groups probably related to the Plebeia line, such as, Scaura, Schwarzula, Partamona, Nogueirapis, Paratrigona, Aparatrigona, Nannotrigona, Scaptotrigona and Melipona (see table). Parapartamona is exclusively Andean. From these only 16 genera with ca. 28 spp. reach South Brazil and 16 genera with ca. 27 spp. reach South Mexico. Schwarzula, Celetrigona, Duckeola, Ptilotrigona, Trichotrigona and Camargoia are exclusively North Brazilian and Guianean.

On the other side, the monotypical genera belonging to the Plebeia line (Schwarziana, Mourella and Friesella remarkable by several autapomorphies) are restricted to South Brazil, North of Argentina and East of Paraguay. Mourella has its "core" area at the cristalline shields of the Serra do Sudeste in Rio Grande do Sul and to the North it never exceeds the Capricorn Tropic. Concerning stingless bees this is the shortest distributional range known.

Within such group, Plebeia (s. str.) is the only politypic genus, with an ample geographic distribution reaching South of Mexico, but with several species restricted to South Brazil, North of Argentina and East of Paraguay. This is the only group of Meliponinae in which such a distributional pattern occurs.

A striking point lies in the contrast between those two main centers of dispersion. While at the northern cratons a large diversity not only at the level of the large phyletic lines but in relation to species numbers also is evidenced (suggesting a relatively dynamic situation), at the southern side, a remarkable specialization mainly in relation to Mourella and Schwarziana occurs (suggesting a long isolation period in a restricted area and slow changing environmental situation).

On the origin of the stingless bees there are only a few hypotesis. A few evidence suggest that the North Brazil and Guianean cratons constituted Pre and Post-Gondwanian dispersion centers. During the oligo-miocene period the possibility of migration through the Antillean bridge (Camargo, Roubik and Moure,1988, Pan-Pacific Entomol., 64:145-157) do not support that the Meliponinae had reached the northern South America coming from the Neartic region as suggested by Michener and Grimaldi (1988, Proc.Natl.Acad.Sci.USA,85:6424-6426). The representatives of the most ancient phyletic lines, being absent in C. America, do not warrant the acceptance of such hypothesis.

The faunal make-up of the Amazon basin is relatively recent and probably originated from stocks coming from the nearby cratons.The wide drainage-basin where the Amazon forest thrives had its early outlines defined (after the final separation between the South American and African plates at the end of Campanian ca. 75 m.y. B.P.) in connection with the progressive uplifting of the continental plataform and formation of the Andes, in the period from late Cretaceous to Pliocene. During all Tertiary the Amazon valley received a strong mass of continental sediments. Large forest masses on such period are not supported in the presence of intense erosion process.After those facts, it is suggested that the forest thrives widely at the sedimentar Amazon valley, after the Pleistocene. So, most of the Amazonic Meliponinae differentiation phenomena (mainly at the specific level) can be explained through recent causal events.

For some of the speciation or sub - speciation events, simple allopatric mechanisms are evoked, that is, some of the actual observed polytypic patterns are correlated with the scenaria determined by the last glaciation (Würm - Wisconsin, 13.000 - 18.000 y.p.), dry period, with retrocession of the sea level and with wide interferences on the Amazonic geographical space (cf. Camargo, 1980, Acta Amazonica, 10 (4): Suplemento, 175 pp.).

TABLE 1. Distribution of Neotropical genera of Meliponinae.MEX.=South of Mexico;A.CENT.=Central America and Panama;R.AND.=Andean region below 1600 m alt.approx.in the Bolívia,Perú,Ecuador,Colombia and Venezuela;ANDES=Andean superhumid woods between 1600 and 2700 m alt.approx.in the Ecuador and Colombia;GU.AM=Guiana and Amazonian basin; NORD.BR.=northeastern Brazil; BR.CENT.=Central Brazil; BR.SUD.=Southeastern Brazil;BR.SUL=Southern Brazil;Northeastern Argentina and east of Paraguay.

GÊNEROS	Nº SPP	MEX.	A.CENT.	R.AND.	ANDES	GU-AM.	NORD.BR.	BR.CENT.	BR.SUD.	BR.SUL
1. *Lestrimelitta*	5	1	1	1?	0	4	1	2	2	2
2. *Leurotrigona*	2	0	0	0	0	1	1?	1	1	1
3. *Celetrigona*	1	0	0	0	0	1	0	1	0	0
4. *Dolichotrigona*	1	1	1	?	0	1	0	1	0	0
5. *Trigonisca*	ca.13	1	4?	4-5	0	8	2?	6	0	0
6. *Trigona*	ca.28	5	9	ca.15	0	ca.18	5	5	5	2
7. *Tetragona*	ca.13	1?	2	ca.5	0	ca.8	1?	5	1	1
8. *Frieseomelitta*	ca.17	1	1?	3?	0	ca.14	4	5-6	1	1
9. *Geotrigona*	ca.14	3	4	3	0	ca.9	3	5	3	1
10. *Duckeola*	2	0	0	0	0	2	0	0	0	0
11. *Tetragonisca*	3	1	1	3?	0	1	1	1	1	1
12. *Ptilotrigona*	1	0	1	1	0	1	0	0	0	0
13. *Trichotrigona*	1	0	0	0	0	1	0	0	0	0
14. *Camargoia*	2	0	0	0	0	2	1	0	0	0
15. *Scaura*	3	1	1	1	0	3	1	1	1	1?
16. *Schwarziana*	1	0	0	0	0	1	0	1	0	0
17. *Plebeia* (*s.str.*)	ca.27	1	ca.5	ca. 6	0	ca.10	2-3	ca.7	ca.4	ca.7
18. *Friesella*	1	0	0	0	0	0	0	0	1	1
19. *Mourella*	1	0	0	0	0	0	0	0	0	1
20. *Schwarziana*	1	0	0	0	0	0	0	0	1	1
21. *Partamona*	ca.16	1	ca.4	ca.5	0	ca.12	2	3	1	1
22. *Parapartamona*	1	0	0	0	1	0	0	0	0	0
23. *Nogueirapis*	3	0	1	1	0	2	0	0	0	0
24. *Paratrigona*	ca.15	?	3	ca.5	0	ca.6	1	2	2	1
25. *Aparatrigona*	2	0	1	1	0	1	0	0	0	0
26. *Nannotrigona*	ca.10	1	2?	ca.4	0	8	2	ca.4	1	1
27. *Scaptotrigona*	ca.29	ca.5	ca.6	ca.7	0	ca.6	ca.4	ca.5	ca.5	2
28. *Cephalotrigona*	1	1	1	1	0	1	1	1	1	1
29. *Oxytrigona*	ca.5	1	2	ca.4	0	ca.4	ca.3	2	2	1
30. *Melipona*	ca.40	3	ca.7	ca.8	0	ca.20	ca.6	ca.6	ca.5	4
Total spp	259	28	57	79	1	145	42	65	38	31
Total genera	-	16	20	19-20	1	26	18	20	18	19

THE 'CLOSED' TYPE MOUNDS OF THE TERMITES GENUS *MACROTERMES* IN THE COAST PROVINCE OF KENYA

RICHARD K. N. BAGINE

NATIONAL MUSEUMS OF KENYA, P.O. BOX 40658,

NAIROBI, KENYA.

Macrotermes species are dominant in savanna grasslands. Of six species recorded in Kenya, four are known to construct dome or conical shaped mounds. In the Rift Valley, Kajiado - Bissel area (1650 m asl), M. michaelseni (Sjöstedt) builds bare conical mounds of 70 - 100 cm high and 100 - 150 m basal diameter.

In Kilifi District around Malindi ($<$100 m asl) species of Macrotermes build tall and bare conical mounds, typically 100 - 500 cm high and 100 - 200 cm in diameter. These mounds have been referred to as 'Malindi' type and are common in the coastal forests and in cultivated areas. Mounds in thickets or forested areas tend to grow taller than those in open, disturbed and settled areas, which often have subsidiary mounds built on their pediments. The interior structure of the nest is characterised by a central air passage which is connected by subsidiary passages and leads down into the hive. The hive is hemispherical and the royal cell mass is usually centrally placed in the bottom of the nursery zone. The fungus combs are usually above and beside the nursery either in separate chambers or within the air passages.

Some taxonomic work carried out on the termites including enzyme characterization of major soldiers have shown some morphological and genetic differences between the 'Malindi' mound builders and the Kajiado - Bissel, M. michaelseni. However, when isoenzymes of other Macrotermes species are compared the 'Malindi' type Macrotermes is closely related to M. michaelseni.

The nest structures and morphological variations including genetic divergence noted in this study could be due to geographical, or other environmental factors. Further investigations using a multidisplinary approach need to be carried out to explain the variations observed.

IMPACT OF QUEEN PHEROMONE ON THE PHYSIOLOGICAL STATUS OF WORKER HONEY BEES (*APIS MELLIFERA* L.)

Herbert H. Hildebrandt and Hans-Hinrich Kaatz
Dept. of Developmental Biology, University of Tübingen,
Auf der Morgenstelle 28, D-7400 Tübingen, FR Germany

Honey bee caste differentiation takes place during larval and pupal development. Of the resulting adult morphs only the queen is fully fertile. She maintains her reproductive dominance by inhibiting the oogenesis of workers pheromonally. Major components of this queen pheromone, mainly 9-oxo-2-decenoic acid (9-ODA), are secreted by the mandibular glands. In the absence of the pheromone, ovarian development starts in most of the workers and some of them lay unfertilized eggs. For this reason, queenless workers achieve an intermediate position between the two distinct castes - not on a morphological, but on a physiological scale. They differ from queenright workers not only in their ovarian status but also in related physiological parameters such as vitellogenin synthesis of the fat body (Engels et al., 1990). It is not yet understood how the queen pheromone influences the reproductive and physiological status of the workers. We therefore investigated whether the neuroendocrine system, especially juvenile hormone (JH) and neurosecretory cells, are involved in the transmission of the pheromonal signal.

MATERIALS AND METHODS

JH-biosynthesis was measured by a radiochemical in vitro assay (PRATT and TOBE, 1974) modified for adult honey bee corpora allata (Kaatz et al., 1985). The pheromonal influence on JH-synthesis was analyzed by exposing small groups of newly emerged worker bees for eight days to queen pheromone, mandibular glands extracts or pheromonal components. Total protein synthesis of the median neurosecretory cells was determined in vivo by injection of a mixture of ^{3}H-labelled amino acids. After one hour of incubation, the cells were explanted and homogenized. Their proteins were precipitated with TCA and counted for tracer incorporation in a LSC.

RESULTS AND DISCUSSION

During early imaginal development, queenless and queenright workers differ distinctly in the two physiological parameters investigated, JH synthesis by the corpora allata and protein synthesis in the neurosecretory cells. Thus, JH synthesis in young workers in the presence of the queen amounts to only half the value for queenless workers of the same age group (Tab. 1). This queen primer effect also can be evoked by exposing queenless workers to mandibular gland extracts or even to single 9-ODA. In both cases JH synthesis of queenless workers is reduced in the same dose-dependent manner to the level of queenright ones. Besides that, the queen effects the activity of the neurosecretory cells. Eight day old queenright workers synthesize twice as much protein in these cells than queenless workers (Tab. 2). This difference is probably caused at least by two factors. On the one hand the queen pheromone impact on the CA might be mediated by polypeptides from the median neurosecretory cells, since neurosecretory factors are supposed to regulate the activity of the CA (Tobe and Stay, 1985). On the other hand neuropeptides may be involved directly in the control of the physiological status of workers as supposed for stimulation of vitellogenesis in queens (Kaatz, 1988).

Tab. 1 Juvenile hormone synthesis in the corpora allata of workers

Age [days after emergence]	JH-synthesis [pmol h^{-1} per pair CA] queenright $\bar{x}$	s.d.	queenless $\bar{x}$	s.d.	stat. diff.*
4	0,59	0,16	1,19	0,17	s.
6	0,76	0,20	2,27	0,43	s.
8	1,46	0,26	2,75	0,50	s.

* statistical difference see Tab. 2, n=5

Tab. 2 Total protein synthesis in the median neurosecretory cells of workers

Age [days after emergence]	Incorporation rate [dpm h^{-1}] queenright $\bar{x}$	s.d.	queenless $\bar{x}$	s.d.	stat. diff.*
4	477	215	377	150	n.s.
8	550	187	295	126	s.

* statistical difference: U-test, =0,01. s.= significant, n.s.= not significant, n=8

The evolutionary tendency to maintain the dominance hierarchy by pheromones instead of aggressive behaviour seems to be closely connected in many social insects to the central physiological function of juvenile hormone as a morphogenetic, gonadotropic or even behaviour modifying effector. As shown here in honey bees, the queen controls the physiological and reproductive status of the subordinate caste by pheromones which are transmitted by the the neuroendocrine system. Thus, juvenile hormone brings along not only age-dependent changes in queenright worker polyethism (Robinson et al., 1989) but also differences in queenright and queenless workers. It may play the central role in causing the intermediate position of queenless workers in this transitional caste development in the adult. However, a functional correlation between this hormone and oogenesis has not been proven for honey bees yet. Nevertheless, this pheromonal impact is not only limited to the Apinae but is also evident in the related Bombinae. In Bombus terrestris, the activity of the corpora allata in queenless workers is also higher than in queenright ones and can be influenced by extracts of the queens mandibular glands (Röseler et al., 1981). Similarily it is possibly true for some wasp species of the Vespinae subfamily (Strambi, 1990). Obviously, this mode of action to control the dominance hierarchy pheromonally via the neuroendocrine system and hence maintain colony homeostasis, is a general phenomenon in many social insects.

REFERENCES

Engels W. et al. 1990. Adv. Invertebrate Reprod. 5, in press.
Kaatz, H.H. 1988. Verh. Dt. Zool. Ges. 81, 272-273.
Kaatz, H.H., Hagedorn, H.H. and W. Engels. 1985. In Vitro 21, 347-352.
Pratt, P.E. and S.S. Tobe. 1974. Life Sci. 14, 575-586.
Robinson, G., Page, R.E., Strambi C. and A. Strambi. 1989. Science 246, 109-112.
Röseler, P.-F., Röseler, I. and C.G.J. van Honk. 1981. Experientia 37, 348-351.
Strambi, A. 1990. In: Social Insects, W. Engels,Ed., Springer, pp. 59-75.
Tobe, S.S. and B. Stay. 1985. Adv. Insect Physiol. 18, 305-432.

INTEREST CONFLICT IN COLONIES OF *MELIPONA MARGINATA* LEPELETIER (APIDAE, MELIPONINAE)

Astrid Kleinert-Giovannini
Depto de Ecologia Geral, Instituto de Biociências,
Universidade de São Paulo, 05508 São Paulo, Brazil.

In stingless bees, where worker oviposition in queenright colonies is almost a rule (Sakagami et al. 1963), distinctive strategies were adopted for reproductive control of the colonies, thus minimizing the expression of interest conflict between queen and workers.

Through an analysis of the supersedure process in colonies where queens mate only once, as those of *Melipona marginata*, it is possible to identify some important aspects involved in this conflict. With this aim, natural and induced (by queens' removal) supersedure processes were observed in colonies of this species, classified according to colony size and daily rate of queen oviposition.

Two natural supersedure processes were observed: in the first one, in a weak colony, supersedure occured after hatching of males and gynes, being the new queen already fertilized the day after her mother disappearence. The first oviposition occured 4 days after, with removal of larval food in the first attempt. In the other, a medium-weak colony, where the queen was accidentaly injured, the interval between the last queen observation and the first new queen oviposition was longer (18 days), similar to what was verified in medium and medium-strong colonies, after queen removal.

From 9 removal experiments, supersedure was verified in 6, its occurence and speed being related to colony conditions, independently of queen age. In weak colonies supersedure was not observed. Only in strong colonies workers conspicuously altered their behaviour after queen removal.

Worker oviposition was only observed in one of the strong colonies, with males and gynes. In one of the weak colonies, worker oviposition was verified after 68 days of orphanhood. In other, it occured after one month.

In medium and medium-strong colonies, the first oviposition of the newly fertilized queen was observed between the 11th and 13th day after removal. In strong colonies, between the 28th and 30th day.

On supersedure day, trophallaxis between the gyne with distended abdomen and workers was common. The interval between observation of the newly fertilized queen and her first oviposition was characterized by frequent trophallatic and antennae contacts with workers. The hatching of the new queen first brood brought about a decrease in her oviposition rate and even its whole interruption.

Soon after supersedure, queen control is not yet efficient. Interactions between queen and workers would directly reflect the degree of queen control: manifestation of agonistic behaviour would indicate control decrease, and a more effective control would be evident by more frequent exhibition of submissive behaviour by workers.

It is hypothesized the existence of pheromonal thresholds above or below which attempts of workers to be at control could be made, followed or not by queen supersedure, depending upon colony conditions. In strong colonies, the vicinity of the upper threshold could be verified by the presence of gynes in the brood area and by male production. In weak colonies, the queen would be next to the lower threshold when, due to low pheromonal production, workers could take over colony control.

In strong colonies with a still efficient queen control, when the queen is removed her pheromones probably remain for a relatively long period, inhibiting worker oviposition. In strong colonies, without such an efficient queen control, evident by the presence of many gynes walking freely throughout the colony and of males, after queen removal worker oviposition begins almost immediately. In medium colonies, the queen "substances" remain for a shorter time in the colony, but still hinder worker oviposition. Here, the acceptance and fertilization of a new queen is faster than in strong colonies. In strong and medium colonies the relatively high no. of workers probably restricts reproductive competition to workers with grandular and ovarian development in agreement with the performance of tasks in the brood area. The acceptance of a new queen would be done by workers not involved in reproduction. Afterwards, supersedure would depend on pheromonal action of the new queen. In weak colonies, reproductive competition occurs among its few workers which, after queen removal, remain almost exclusively in the brood combs; opening of cells already closed and intake of its contents are frequent, and also attack to any hatching gynes.

References

Sakagami, S.F.; Beig, D; Zucchi, R. & Akahira, Y. 1963 Occurence of ovary-developed workers in queenright colonies of stingless bees. *Revta brasil. Biol.*, Rio de Janeiro, **23**(2): 115-129.

SWARMING ACTIVITY IN *SCHWARZIANA QUADRIPUNCTATA* (APIDAE, MELPONINAE)

V. L.Imperatriz-Fonseca
Departamento de Ecologia Geral, Instituto de Biociências
Universidade de São Paulo, 05508 São Paulo, Brazil

Swarming process in stingless bees has been first studied by Nogueira-Neto (1954), who emphasized the gradual foundation of the nest: the workers begin to swarm looking for a suitable place to establish the new nest, behaviour concerned in the mother colony with an unusual external and internal activity. Once decided where to go, cerumen transportation begins, in order to construct the entrance of the new nest and to limit the cavity size.

Swarming activity was observed in *Schwarziana quadripunctata,* in a nest moved two weeks before to our Bee Lab, in S.Paulo University campus. This species, native in the studied area, has subterranean nests, probably established in cavities abandoned by ants. Cerumen transportation was observed from August 14th to October 28th, with greater intensity from August 14th to 27th. As it decreased, the number of bees going out with their stomachs full of a mixture of pollen and nectar increased, maintaining a regular rate until the end of observations. These relationships between both colonies occupied the greater majority of workers that went out of the colony until september 10th, when the quantity of bees going out of the colony without any material was greater than with food, cerumen, detritus or resin.

Inside the mother nest, the observation of one or more workers starting to take out pieces of cerumen from involucrum, pots full of honey, combs, and filling their corbicullae with it, is the first signal confirming the agitation due to a preparation for swarming. Some pots were marked with small cerumen pellets. Many contacts between workers were common. Active gynes were observed: a big imprisoned gyne was very active on August 14th, and 4 days later another, medium sized (as Camargo (1974) pointed out there are several sizes of gynes in this species). Both prisons were destroyed on August 24th, probably when swarm occured.

The swarm may depart in the first or second week as soon as the new nest has some food stored and is ready to receive workers, some males and gynes. Inside the mother nest, males become very excited with the presence of very attractive gynes, and sometimes are observed trying to copulate with the physogastric queen, when she is fixing a brood cell.Nevertheless, queen fertilization should occur in the new nest, and clouds of males are not observed at the entrance of the mother colony. Males are tolerated in the colony, and dehydrate nectar almost all the time. They are found in groups at the hottest places of the colony and their size can be very different (from very small to giant).

During all swarming period many gynes hatch. Their behaviour may be different, according to colony conditions and queen attractiveness. Generally big gynes are

the most attractive, but medium size gynes are more common. After the swarming departure, 5 gynes more hatched in September 14th. From the same comb three gynes more hatched on Sept. 15th, two gynes on Sept. 18th, one on sept. 21th. A royal cell was constructed in the mother nest on September 18th. On October 5th, six gynes of medium size were found imprisoned, one next to the other, near the hive glass cover, and another tried to construct a prison around her. Most of them were killed by workers some hours later, in an attack which lasted for 8 hours. Only one remained undisturbed, inside the prison. When these conflicts are observed in the nest, workers grooming themselves or attacking each other can be observed. The oviposition process is completely disturbed by the presence of these gynes. The surplus of gynes produced by the mother colony assures the swarming process and provides an special condition for supersedure process.
Normal rate of oviposition was established in the mother nest after this period. Pollen collection by foragers began again, and most of them were working for the mother nest. At the end of October, cerumen transportation, as well as food, was almost rare.

References

Camargo, J.M.F. de 1974. Notas sobre a morfologia e a biologia de *Plebeia (Schwarziana) quadripunctata quadripunctata* (Hym., Apidae). *Studia Entomologica*, **17**(1-4):433-470

Nogueira-Neto, P. 1954. Notas bionômicas sobre os meliponíneos. III - Sobre a enxameagem. *Arquivos Museu Nacional*, **42**: 419-451.

FORAGING ACTIVITY OF *APIS CERANA* F. AND *APIS MELLIFERA* L. ON *PLECTRANTHUS* BLOOM IN RELATION TO THE SUGAR CONCENTRATION IN ITS NECTAR

V. K. Mattu, Neelam Mattu and L. R. Verma
Department of Bio-Sciences, Himachal Pradesh University
Shimla-171005, INDIA.

Plectranthus rugosus is an excellent source of nectar for honeybees in northern India. Foraging studies on this perennial shrub showed that peak hours of foraging activity were between 1000-1200 for *A. mellifera* and 0900-1100 hours for *A. cerana*. Both the species showed diurnal fluctuations in the percentage of pollen, nectar and pollen plus nectar collectors. Percentage of nectar collectors was significantly greater than those collecting pollen and pollen plus nectar. Mean ratio of pollen and nectar collectors was 1:2.41 for *A. cerana* and such ratio was 1:1.51 for *A. mellifera*. European bee, *A. mellifera* carried significantly heavier pollen loads and remained longer on individual flower than native bee *A. cerana*. There was no significant difference between these two species of honeybees with regards to distance covered from flower to flower and number of flowering branches visited. Sugar content in *Plectranthus rugosus* varied from 0.132 to 0.397 mg per flower and this content was more during the evening hours than in morning time.

FORAGING ACTIVITY OF *APIS CERANA* F. AND *APIS MELLIFERA* L. ON APPLE AND PLUM BLOOM

L.R. Verma, V.K. Mattu, R.S. Rana and P.C. Dulta
Department of Bio-Sciences, Himachal Pradesh University, Shimla-171005, INDIA

Foraging studies revealed that A. cerana began to forage earlier in the morning and stopped later in the evening as compared to A. mellifera. Maximum foraging activity was recorded for A. cerana between 0900-1130 and 1000-1300, where ,A. mellifera showed peak foraging activity between 1100-1330 and 1100-1200 hours on apple and plum flowers respectively. This is remarkable because by placing both the species in the same orchard the duration of the peak periods of foraging activity might be prolonged and better pollination obtained. In both species, nectar collectors outnumbered pollen collectors. A. mellifera carried heavier pollen loads and duration of foraging trip was significantly longer for this species than A. cerana. However, there were no significant differences between A. cerana and A. mellifera with regard to other parameters like number of flowers visited per minute, time spent per flower, their visits of apple trees in same or different rows and in the ratio of side and top worker bees.

AUTHOR INDEX

GENERIC AND SPECIES INDEX TO ARTHROPODS

INDEX OF ASSOCIATED TAXA